Lecture Notes in Computer Science 16084

The series Lecture Notes in Computer Science (LNCS), including its subseries Lecture Notes in Artificial Intelligence (LNAI) and Lecture Notes in Bioinformatics (LNBI), has established itself as a medium for the publication of new developments in computer science and information technology research, teaching, and education.

LNCS enjoys close cooperation with the computer science R & D community, the series counts many renowned academics among its volume editors and paper authors, and collaborates with prestigious societies. Its mission is to serve this international community by providing an invaluable service, mainly focused on the publication of conference and workshop proceedings and postproceedings. LNCS commenced publication in 1973.

Min Li · Mingji Xia · Peng Zhang
Editors

Theory and Applications of Models of Computation

19th Annual Conference, TAMC 2025
Jinan, China, September 19–21, 2025
Proceedings

Springer

Editors
Min Li
Shandong Normal University
Jinan, China

Mingji Xia
Institute of Software Chinese Academy of Sciences
Beijing, China

Peng Zhang
Shandong University
Jinan, China

ISSN 0302-9743 ISSN 1611-3349 (electronic)
Lecture Notes in Computer Science
ISBN 978-981-95-4838-5 ISBN 978-981-95-4839-2 (eBook)
https://doi.org/10.1007/978-981-95-4839-2

This Springer imprint is published by the registered company Springer Nature Singapore Pte Ltd.
The registered company address is: 152 Beach Road, #21-01/04 Gateway East, Singapore 189721, Singapore

Preface

It is with great pleasure that we present the proceedings of the 2025 Annual Conference on Theory and Applications of Models of Computation (TAMC 2025), held in Jinan, China, from September 19–21, 2025. This volume contains all contributed papers presented at TAMC 2025. Over the last 19 years, the TAMC conference series has served as a valuable platform for researchers specializing in computational theory, information theory, and their applications. This year's conference centered around key themes including computability, complexity, algorithms, information theory, as well as their integration with machine learning theory and the foundations of artificial intelligence.

The Program Committee, composed of 41 active researchers from the field, reviewed 71 submissions and selected 32 papers for presentation at TAMC 2025. Every submission underwent a rigorous double-blind peer review process. Each paper had at least three reviews, with additional reviews solicited as needed.

We would like to express our heartfelt gratitude to everyone who dedicated time and effort to make TAMC 2025 a resounding success: the authors, whose innovative research enriches our collective knowledge, the Program Committee members, whose judgments are invaluable, and the reviewers, whose insights help to ensure scientific quality.

Our gratitude is also directed to Springer for accepting TAMC 2025 for publication in this Lecture Notes in Computer Science (LNCS) proceedings volume. Our special appreciation goes to the conference chairs Donglei Du and Xiaodi Li, and the local organization team for their outstanding dedication and exceptional work.

September 2025

Min Li
Mingji Xia
Peng Zhang

Organization

Steering Committee

Manindra Agrawal	Indian Institute of Technology, Kanpur, India
Jin-Yi Cai	University of Wisconsin — Madison, USA
John Hopcroft	Cornell University, USA
Angsheng Li	Beihang University, China
Zhiyong Liu	Institute of Computing Technology, Chinese Academy of Sciences, China

Conference Chairs

Donglei Du	University of New Brunswick, Canada
Xiaodi Li	Shandong Normal University, China

Program Committee Chairs

Min Li	Shandong Normal University, China
Mingji Xia	Institute of Software, Chinese Academy of Sciences, Beijing, China
Peng Zhang	Shandong University, China

Publication Chairs

Xiang Liu	Southeast University, China
Yang Zhou	Shandong Normal University, China

Local Organization Chairs

Qian Liu	Shandong Normal University, China
Xia Zhang	Shandong Normal University, China
Qiang Zhao	Shandong Normal University, China
Zhaojie Zhou	Shandong Normal University, China

Program Committe

Yixin Cao	Hong Kong Polytechnic University, China
Vincent Chau	Southeast University, China
Jianer Chen	Texas A & M University, USA
Ke Chen	Pennsylvania State University, USA
Xujin Chen	Academy of Mathematics and Systems Science, Chinese Academy of Sciences, China
Yongxi Cheng	Xi'an Jiaotong University, China
Yukun Cheng	Jiangnan University, China
Yi Deng	Institute of Information Engineering, Chinese Academy of Sciences, China
Hu Ding	University of Science and Technology of China, China
Dingzhu Du	University of Texas at Dallas, USA
Donglei Du	University of New Brunswick, Canada
Thomas Erlebach	University of Leicester, UK
Qilong Feng	Central South University, China
Lu Han	Beijing University of Posts and Telecommunications, China
Jianfeng Hou	Fuzhou University, China
Jianping Li	Yunnan University, China
Min Li	Shandong Normal University, China
Minming Li	City University of Hong Kong, China
Xianyue Li	Lanzhou University, China
Qian Liu	Shandong Normal University, China
Wanwei Liu	National University of Defense Technology, China
Xiang Liu	Southeast University, China
Anil Seth	Indian Institute of Technology, Kanpur, India
Zhongzheng Tang	Beijing University of Posts and Telecommunications, China
Chenhao Wang	Beijing Normal University at Zhuhai, China
Guanghui Wang	Shandong University, China
Chenchen Wu	Tianjin University of Technology, China
Zijun Wu	Hefei University of Technology, China
Ge Xia	Lafayette College, USA
Mingji Xia	Institute of Software, Chinese Academy of Sciences, China
Dachuan Xu	Beijing University of Technology, China
Jinhui Xu	University at Buffalo, State University of New York, USA

Yicheng Xu	Shenzhen Institutes of Advanced Technology, Chinese Academy of Sciences, China
Gexin Yu	College of William & Mary, USA
Huaming Zhang	University of Alabama in Huntsville, USA
Peng Zhang	Shandong University, China
Xin Zhang	Xidian University, China
Zhao Zhang	Zhejiang Normal University, China
Xiangqian Zhou	Wright State University, USA
Yang Zhou	Shandong Normal University, China
Martin Ziegler	KAIST, South Korea

Additional Reviewers

Longkun Guo
Ziyun Huang
Jiachen Ju
Bi Li
Bo Li
Heng Li
Weian Li
Yuefang Lian
Wei Liang
Chensheng Ma
Lili Mei
Cuixia Miao
Runjie Miao
Jason Milionis
Chunying Ren
Hao Xiao
Fan Yuan
Peng Zhang
Ruilong Zhang
Lei Zhao
Juan Zou

Contents

Computational Complexity and Algorithms

Complexity and Approximation Algorithms for Fixed Charge Transportation Problems

Yong Chen[1], Shi Li[2(✉)], and Zihao Liang[2]

[1] Department of Mathematics, Hangzhou Dianzi University, Hangzhou, China
chenyong@hdu.edu.cn

[2] School of Computer Science, Nanjing University, Nanjing, China
shili@nju.edu.cn, zhliang@smail.nju.edu.cn

Abstract. The Fixed Charge Transportation (FCT) problem models transportation scenarios where we need to send a commodity from n sources to m sinks, and the cost of sending a commodity from a source to a sink consists of a linear component and a fixed component. Despite extensive research on exponential time exact algorithms and heuristic algorithms for FCT and its variants, their approximability and computational complexity are not well understood.

In this work, we initiate a systematic study of the approximability and complexity of these problems. When there are no linear costs, we call the problem the Pure Fixed Charge Transportation (PFCT) problem. We also distinguish between cases with general, sink-independent, and uniform fixed costs; we use the suffixes "-S" and "-U" to denote the latter two cases, respectively. This gives us six variants of the FCT problem.

We give a complete characterization of the existence of $O(1)$-approximation algorithms for these variants. In particular, we give 2-approximation algorithms for FCT-U and PFCT-S, and a $(6/5 + \epsilon)$-approximation for PFCT-U. On the negative side, we prove that FCT and PFCT are NP-hard to approximate within a factor of $O(\log^{2-\epsilon}(\max\{n, m\}))$ for any constant $\epsilon > 0$, FCT-S is NP-hard to approximate within a factor of $c\log(\max\{n, m\})$ for some constant $c > 0$, and PFCT-U is APX-hard. Additionally, we design an Efficient Parameterized Approximation Scheme (EPAS) for PFCT when parameterized by the number n of sources, and an $O(1/\epsilon)$-bicriteria approximation for the FCT problem, when we are allowed to violate the demand constraints for sinks by a factor of $1 \pm \epsilon$.

Keywords: Fixed Charge Transportation problem · Approximation Algorithm · Complexity

1 Introduction

In 1954, Hirsch and Dantzig [1,2] introduced the Fixed Charge Transportation (FCT) problem, where the cost of transportation from a source to a sink contains

M. Li et al. (Eds.): TAMC 2025, LNCS 16084, pp. 3–14, 2026.
https://doi.org/10.1007/978-981-95-4839-2_1

a linear component, which is proportional to the amount transported, and a fixed cost, which is incurred whenever the transportation occurs. The objective is to allocate all supplies of the commodity from sources to sinks in a way that minimizes the total cost, which is the sum of linear and fixed costs.

The FCT problem is a special case of the single-commodity uncapacitated fixed-charge network problem [3,4], which itself belongs to a broader class of network design problems [5–7]. The single-commodity uncapacitated fixed-charge network problem is defined on an arbitrary directed graph (V, E), where each edge $e \in E$ has a linear cost and a fixed cost as in the FCT problem. The commodity can be transported only along edges in E.

The goal is to match the given supplies and demands so as to minimize the sum of variable and fixed costs. The FCT problem can be seen as a special case of this problem on complete bipartite directed graphs from sources to sinks. As a byproduct in this paper, we show that the two problems are indeed equivalent.

When there are only fixed costs (i.e., linear costs are zero), the FCT problem is called the Pure Fixed Charge Transportation (PFCT) problem [8,9]. The PFCT problem arises in scenarios where transportation costs are negligible compared to the costs of establishing transportation routes, such as pipelines and electric power systems. Fisk and McKeown [8] proposed a search algorithm to solve the PFCT problem exactly, and Göthe-Lundgren and Larsson [9] reformulated it as a set-cover problem, and designed heuristic algorithms by relating it to the maximum flow problem.

Despite considerable research on exact algorithms [10–14], branch-and-bound algorithms [15–18], and heuristic algorithms [19–22] for the FCT and PFCT problems, their computational complexity and approximation algorithms were not well studied. To the best of our knowledge, the only known results in this direction are that both FCT [23] and PFCT [24,25] are NP-hard.

In this paper, we initiate a systematic study of the approximability and complexity of FCT problems. Before we state our results, we give formal definitions of the problems.

1.1 Formal Definitions of Problems

In the Fixed Charge Transportation (FCT) problem, we are given a set S of n sources and a set T of m sinks. Each source $i \in S$ has a supply of $a_i \in \mathbb{Z}_{>0}$ units for a commodity, and each sink $j \in T$ has a demand of $b_j \in \mathbb{Z}_{>0}$ units. It is guaranteed that $\sum_{i \in S} a_i = \sum_{j \in T} b_j$. For every pair $i \in S$ and $j \in T$, there is an edge ij with a fixed cost $f_{ij} \in \mathbb{R}_{\geq 0}$ and a linear cost ratio $c_{ij} \in \mathbb{R}_{\geq 0}$. The objective is to satisfy all demands by sending the commodity from sources to sinks while minimizing the total cost. The formal mathematical formulation is as follows:

$$\min \quad \sum_{i \in S, j \in T} \left(1_{x_{ij} > 0} \cdot f_{ij} + c_{ij} \cdot x_{ij}\right) \qquad \text{s.t.} \qquad x \in \mathcal{X},$$

where throughout the paper we use

$$\mathcal{X} := \left\{x \in \mathbb{R}_{\geq 0}^{S \times T} \quad : \quad \sum_{j \in T} x_{ij} = a_i, \forall i \in S; \quad \sum_{i \in S} x_{ij} = b_j, \forall j \in T\right\}$$

to denote the set of all feasible solutions to the given FCT instance.

The Pure Fixed Charge Transportation (PFCT) problem is the special case where there are no linear costs, i.e., $c_{ij} = 0$ for every $i \in S$ and $j \in T$. We shall show that FCT and PFCT problems capture the Directed Steiner Tree (DST) and Set Cover problems respectively, and thus do not admit constant approximation unless P = NP. To gain a more refined understanding of its complexity, we study two special cases of the fixed cost vector f:

- *Sink-independent fixed costs.* The fixed costs $(f_{ij})_{i\in S, j\in T}$ are said to be sink-independent if there exists a cost vector $f \in \mathbb{R}^S_{\geq 0}$ such that $f_{ij} = f_i$ for all $i \in S$ and $j \in T$. We denote this special case using the suffix "-S".
- *Uniform fixed costs.* The fixed costs are uniform if $f_{ij} = 1$ for all $i \in S$ and $j \in T$. This case is denoted by the suffix "-U".

So, combining these variations whether linear costs are present and the properties of the fixed costs results in six variants of the FCT problem, denoted as FCT, FCT-S, FCT-U, PFCT, PFCT-S and PFCT-U.

1.2 Our Results: $O(1)$-Approximability

We provide a complete characterization of the existence of $O(1)$-approximation algorithms for the six variants of FCT. The results are listed in Table 1 for convenience.

Table 1. Approximability and Hardness of Variants of FCT.

	FCT		PFCT	
	Approx. Ratio	Hardness	Approx. Ratio	Hardness
general		$\Omega(\log^{2-\epsilon}(mn))$		$\Omega(\log^{2-\epsilon}(mn))$
-S		$\Omega(\log(mn))$	2	APX-hard
-U	2	APX-hard	$\frac{6}{5}+\epsilon$	APX-hard

We first discuss the approximation results. For both PFCT-S and FCT-U, we give 2-approximation algorithms:

Theorem 1. *There is a 2-approximation algorithm for the Pure Fixed Charge Transportation with Sink-Independent Fixed Costs (PFCT-S) problem.*

Theorem 2. *There is a 2-approximation algorithm for the Fixed Charge Transportation with Uniform Fixed Costs (FCT-U) problem.*

For the most restricted problem PFCT-U, we present a better approximation algorithm:

Theorem 3. *For any constant $\epsilon > 0$, there is a $(\frac{6}{5}+\epsilon)$ approximation algorithm for the Pure Fixed Charge Transportation with Uniform Fixed Costs (PFCT-U) problem.*

Then we complement the results by showing the other variants are NP-hard to approximate within a constant factor. First, we give a reduction from the Directed Steiner Tree (DST) problem to the PFCT problem. As the polynomial-time approximability of DST remains open, we include the reduction in the theorem statement; any future improvements in the hardness of DST automatically carry over to PFCT.

Theorem 4. *Let $\alpha : \mathbb{Z}_{>0} \to \mathbb{R}_{>0}$ be a monotone function. If there is a polynomial-time $\alpha(\max\{n,m\})$-approximation algorithm for the Pure Fixed Charge Transportation (PFCT) problem in polynomial time, then there is a polynomial-time $\alpha(n)$-approximation for the Directed Steiner Tree (DST) problem.*

Using the $\Omega(\log^{2-\epsilon}(n))$-hardness of DST [31], this gives us

Corollary 5. *It is NP-hard to approximate the Pure Fixed Charge Transportation (PFCT) problem within a factor of $O(\log^{2-\epsilon}(\max\{n,m\}))$, for any constant $\epsilon > 0$.*

Since PFCT is a special case of FCT, the results extend to FCT as well.

We provide a reduction from the Set Cover problem to FCT-S, proving the following result:

Theorem 6. *There is a constant $c > 0$ such that the following holds. It is NP-hard to approximate the Fixed Charge Transportation with Sink-Independent Fixed Costs (FCT-S) problem within a factor of $c\ln(\max\{n,m\})$.*

Finally, we rule out the possibility of a PTAS for the PFCT-U problem, the most restricted variant we study. This is achieved via a reduction from the 3-Dimensional Matching with Bounded Frequency (3DM-B) problem.

Theorem 7. *There is an absolute constant $c > 1$ such that Pure Fixed Charge Transportation with Uniform Fixed Costs (PFCT-U) problem does not have a c-approximation, unless NP = BPP.*

The APX-hardness also extends to the more general problems FCT-U and PFCT-S.

1.3 Our Results: Efficient Polynomial Time Approximation Scheme (EPAS) and Bicriteria Approximation Algorithm

We then consider approximation algorithms parameterized by the number n of sources. We show that there is an Efficient Parameterized Approximation Scheme (EPAS) for the PFCT problem:

Theorem 8. *Given any constant $\epsilon > 0$, there is a $g(n, \epsilon) \cdot \text{poly}(m)$-time $(1 + \epsilon)$-approximation algorithm for the Pure Fixed Charge Transportation (PFCT) problem, for some efficiently computable function g.*

For the FCT problem, we show a simple $(O(1/\epsilon), 1 \pm \epsilon)$-bicriteria approximation algorithm in polynomial time. Such algorithms output solutions which may violate the sink demand constraints by a factor of $1 \pm \epsilon$.

Theorem 9. *Given a Fixed Charge Transportation (FCT) instance defined by $S, T, (a_i)_i, (b_j)_j, (f_{ij})_{i,j}, (c_{ij})_{i,j}$, and a constant $\epsilon > 0$, we can efficiently output a vector $x \in \mathbb{R}_{\geq 0}^{S \times T}$ subject to $\sum_{j \in T} x_{ij} = a_i$ for every $i \in S$, $\sum_{i \in S} x_{ij} \in (1 \pm \epsilon) b_j$ for every $j \in T$, and $\sum_{i \in S, j \in T} \left(1_{x_{ij} > 0} \cdot f_{ij} + c_{ij} x_{ij}\right)$ is at most $O(1/\epsilon)$ times the optimal cost of the instance.*

Organization The rest of the paper is organized as follows. In Sect. 2, we introduce the problems used in our hardness reductions. We provide an overview of our techniques in Sect. 3. In Sect. 4, we describe the 2-approximation for PFCT-S, proving Theorem 1. Due to page limitations, all other results are placed in the appendix, including all proofs of Theorem 2 through Theorem 9.

2 Preliminaries

We define several problems that we use in our reductions for proving the hardness results.

Directed Steiner Tree. In the Directed Steiner Tree (DST) problem, we are given a directed graph $G = (V, E)$ of $n = |V|$ vertices, with a root $r \in V$ and a set $T \subseteq V \setminus \{r\}$ of k terminals. Every edge $e \in E$ has a cost $c_e \in \mathbb{R}_{\geq 0}$. The goal of the problem is to find a subgraph H of G with the minimum cost, that contains a path from r to t for every $t \in T$. A minimal subgraph satisfying this property is a directed tree rooted at r (i.e., a out arborescence), hence the name Directed Steiner Tree (DST).

A simple reduction from Set Cover to DST gives a $(1 - \epsilon) \ln k$-hardness for the latter problem, which translates to a hardness of $\Omega(\log n)$ in terms of n. Stronger inapproximability results are known: it is NP-hard to approximate DST within a factor of $O(\log^{2-\epsilon} n)$ for any constant $\epsilon > 0$ [31]. Moreover, under the stronger assumption that NP $\not\subseteq$ BPPTime$(2^{\log^{O(1)} n})$, the problem does not admit an $o(\log^2 k / \log\log k)$-approximation algorithm. On the positive side, DST admits a quasi-polynomial time $O(\log^2 n / \log\log n)$-approximation [30,32,33], and a $n^{O(1/\epsilon)}$-time $O(n^{\epsilon})$-approximation algorithm for any constant $\epsilon > 0$ [34]. Whether there is a polynomial-time poly-logarithmic approximation for DST is a notorious open problem.

Maximum 3-Dimensional Matching. In the 3-Dimensional Matching (3DM) problem, we are given a set $M \subseteq X \times Y \times Z$ of triples, where X, Y and Z are disjoint sets. The goal of the problem is to find the largest subset $M' \subseteq M$ such that no two tuples in M' have the same coordinate. This is one of the Karp's 21 NP-hard problems.

In our reduction, it is important to ensure that the number of times an element in $X \cup Y \cup Z$ appears in M is bounded by an absolute constant B. We call such the problem the 3-Dimensional Matching with bounded frequency (3DM-B) problem. The 3DM-B problem was proven to be APX-hard by Kann [26], and later was shown to be NP-hard to approximate within a factor of $\frac{98}{97} - \epsilon$ for any constant $\epsilon > 0$ [27]:

Theorem 10 ([27]). *There is a large enough integer B such that the following is true for any small constant $\epsilon > 0$. Given a 3DM instance (X, Y, Z, E) where $|X| = |Y| = |Z| = n$ and every $v \in X \cup Y \cup Z$ appears in between 2 and B triples in E, it is NP-hard to distinguish between the following two cases:*

- *(X, Y, Z, E) is a yes-instance : there is a matching $M \subseteq E$ of size n,*
- *(X, Y, Z, E) is a no-instance : every matching $M \subseteq E$ has size at most $\left(\frac{97}{98} + \epsilon\right)n$.*

Maximum k-Set Packing. In the Maximum k-Set Packing problem, we are given a collection of n sets, each containing at most k elements from a given ground set. The objective is to find the largest collection of pairwise disjoint sets. This problem generalizes the k-Dimensional Matching problem, which involves finding a maximum matching in a k-partite k-uniform hypergraph. In particular, Maximum 3-Set Packing generalizes 3-Dimensional Matching. The best-known approximation algorithm for the Maximum k-Set Packing problem achieves a $\frac{k+1+\epsilon}{3}$-approximation ratio, given by Cygan [28], with subsequent improvements by Fürer and Yu [29] on the algorithm's running time dependence on ϵ.

Other Notations. Given a real vector $x \in \mathbb{R}^{\mathbb{D}}$ for some domain $\mathbb{D}$, we use $\text{supp}(x) := \{i \in \mathbb{D} : x_i \neq 0\}$ to denote the set of its coordinates with non-zero values. For any subset $U \subseteq \mathbb{D}$, we define $x(U) := \sum_{i \in U} x_i$, unless specified otherwise.

3 Overview of Techniques

In this section, we provide a brief overview of the techniques used in our results.

2-Approximation Algorithm for PFCT. We formulate an LP relaxation by considering a linear objective $\sum_{i \in S,\ j \in T} \frac{x_{ij}}{b_j} \cdot f_i$. This is upper bounded by its actual cost $\sum_{i \in S, j \in T} \left\lceil \frac{x_{ij}}{b_j} \right\rceil \cdot f_i$ of x. The LP can be solved using a greedy algorithm that processes sources in descending orders of f_i values, and sinks in descending order of b_j values. The solution obtained will not contain a "crossing" pair of edges. Using this property, we show that the gap between the actual cost and its LP cost is small.

$(\frac{6}{5}+\epsilon)$-Approximation Algorithm for PFCT. Our goal is to partition $S \cup T$ into the maximum number of balanced sets, where a set is balanced if its total supply equals its total demand. The cost of a solution is $n+m$ minus the number of balanced sets. The maximization problem is a special case of the set-packing problem. Intuitively, the optimum solution in a bad instance of the PFCT-U problem should contain many small balanced sets. Fortunately, for the k-set-packing problem, the case where each given set has size at most k, a local search algorithm can give a $(k + 1 + \epsilon)/3$-approximation [29]. Then we restrict our attention to balanced sets of sizes $3, 4, 5, 6$. For each size k, we run the local search $(k+1+\epsilon)/3$-approximation algorithm and output the best solution. The final approximation ratio $\frac{6}{5}$ is then obtained by a small linear program.

Directed-Steiner-Tree-Hardness of PFCT. To prove the hardness of PFCT, we introduce an intermediate problem called the PFCT-Digraph problem, where we are given a directed graph and the flows can only be sent along the edges. The reduction from Directed Steiner Tree to PFCT-Digraph is immediate: we just treat the root as a source with supply k, and every terminal as a sink with demand 1, and assign a fixed cost of 1 to each edge. We then show that PFCT-Digraph is equivalent to the original PFCT problem, by splitting each vertex v into an edge $(v_{\text{out}}, v_{\text{in}})$.

Set-Cover-Hardness of FCT-S. To prove the hardness of FCT-S, we only need to use the special case where each c_{ij} is either 0 or ∞. In this setting, the linear cost ratios determine which edges can be used. The reduction from Set Cover follows the similar vertex-splitting approach as in the previous reduction. Since the fixed costs depend only on the sources, we can only use Set Cover (rather than Directed Steiner Tree) as the basis for the reduction, resulting in a weaker hardness result.

Reduction from 3DM-B to PFCT-U. For a given 3DM-B instance (X, Y, Z, E), we randomly assign each element $v \in \mathrm{X} \cup Y \cup Z$ a demand b_v, from the integer interval $(\Delta, 2\Delta]$. For every triple $ijk \in E$, we create a source ijk with supply $a_{ijk} := b_i + b_j + b_k$. The balanced set $\{i, j, k, ijk\}$ will correspond to the triple $ijk \in E$. A dummy sink is added to ensure total supply equals total demand. Ideally, a solution to the 3DM-B instance should choose the maximum number of balanced sets of the form $\{i, j, k, ijk\}$, and put the remaining sources and sinks into a big balanced set. To prevent "cheating", we choose a sufficiently large Δ so that all the b_v values for $v \in X \cup Y \cup Z$ are sufficiently "independent": no two small disjoint multi-sets of $X \cup Y \cup Z$ have the same total b value. This will make cheating expensive. Note that it is important to have a bounded frequency B, so that the total number of sources and sinks is not too big compared to the total number of balanced sets in the target solution, which is 1 plus the size of the maximum 3-dimensional matching.

2-Approximation for FCT-S. The 2-approximation algorithm for FCT-S is straightforward. We first ignore the fixed costs to obtain a solution x. As the fixed costs are uniformly 1, the total fixed cost is simply the number of edges we used. As $\text{supp}(x)$ is a forest, this yields a 2-approximation for the fixed-cost component, leading to an overall 2-approximation.

EPAS for PFCT Parameterized by n. For the PFCT problem, a PTAS is easy to obtain when $n = O(1)$: we can guess the set P of n/ϵ most expensive edges in the solution, and use the 2-approximation greedy algorithm. The difference between the actual cost and the LP cost of the greedy solution can be bounded by n times the cost of the most expensive undecided edge. This is small compared to the total cost of P. To improve the running time to $g(n, \epsilon) \cdot \text{poly}(m)$, we reduce the number of candidates for the n/ϵ most expensive edges. This is achieved by discretizing the fixed costs, and partitioning the sinks into $\left(O\left(\frac{\log m}{\epsilon}\right)\right)^n$ classes such that all sinks in a class have the same incident fixed cost vector. Each class can be treated as a PFCT-S instance. In this case, we show there is an approximately optimum solution where the n/ϵ most expensive edges are incident to a few sinks with the largest demands. This reduces the size of the candidate edges for the set P.

4 2-Approximation for Pure Fixed Charge Transportation with Sink-Independent Fixed Costs (PFCT-S)

In this section, we give our 2-approximation for the Pure Fixed Charge Transportation with Sink-Independent Costs (PFCT-S) problem. Recall that we are given sources S with $|S| = n$, sinks T with $|T| = m$, supply vector $a \in \mathbb{Z}^S_{>0}$, demand vector $b \in \mathbb{Z}^T_{>0}$ with $a(S) = b(T)$ and fixed cost vector $(f_i \geq 0)_{i \in S}$.

We rename the sources as $[n]$ and sort them in descending order of f_i values. Also, we rename the sinks as $[m]$ and sort them in descending order of b_j values. So we have $f_1 \geq f_2 \geq \cdots \geq f_n$ and $b_1 \geq b_2 \geq \cdots \geq b_m$.

We consider the following linear program for the problem.

$$\min \sum_{i \in S, j \in T} \frac{x_{ij}}{b_j} \cdot f_i \qquad \text{s.t.} \quad x \in \mathcal{X} \tag{1}$$

Recall that $\mathcal{X}$ is defined in Sect. 1.1. The actual cost of the solution x is $\sum_{i \in S, j \in T} \left\lceil \frac{x_{ij}}{b_j} \right\rceil \cdot f_i$, which is non-linear in x. In the LP, we use the linear function $\sum_{i \in S, j \in T} \frac{x_{ij}}{b_j} \cdot f_i$ as the objective. Clearly, this is upper bounded by the actual cost.

We prove the following claim, which will lead to the greedy algorithm for solving the LP.

Claim Focus on a solution x to (1). Assume $x_{ij'} > 0, x_{i'j} > 0$ for some $i < i'$ and $j < j'$. Let $\epsilon = \min\{x_{ij'}, x_{i'j}\}$. Then the following operation will maintain the validity of x without increasing its cost to (1): decrease $x_{ij'}$ and $x_{i'j}$ by ϵ, and increase x_{ij} and $x_{i'j'}$ by ϵ.

Proof Clearly, the operation will maintain the feasibility of x. It increases the cost by

$$\epsilon\left(\frac{f_i}{b_j}+\frac{f_{i'}}{b_{j'}}-\frac{f_i}{b_{j'}}-\frac{f_{i'}}{b_j}\right)=\epsilon(f_i-f_{i'})\left(\frac{1}{b_j}-\frac{1}{b_{j'}}\right)\leq 0.$$

The last inequality holds as $f_i \geq f_{i'}$ and $b_j \geq b_{j'}$.

Therefore, there is an optimum solution x to (1), whose support does not contain a "crossing" pair of edges as described in the Claim. Such a solution is unique, and can be obtained by the following greedy algorithm:

Algorithm 1 Greedy Algorithm for PFCT-S

$x_{ij} \leftarrow 0$ for all $i \in S, j \in T$
$i \leftarrow 1, j \leftarrow 1$
while $i \leq n$ and $j \leq m$ **do**
  $x_{ij} \leftarrow \min\{a_i, b_j\}, a_i \leftarrow a_i - x_{ij}, b_j \leftarrow b_j - x_{ij}$
  if $a_i = 0$ **then** $i \leftarrow i+1$
  if $b_j = 0$ **then** $j \leftarrow j+1$

Till the end of this section, we call the x obtained by the greedy algorithm the greedy solution, and we fix this x. We output x as the solution to the PFCT-S instance.

For real $t \in (0, b([m])]$, let $\pi(t)$ be the smallest j such that $b([j]) \geq t$, i.e., the smallest j such that the first j sinks have total demand at least t. Let $f_{n+1} = 0$ for convenience.

Lemma 1 *The optimum solution of the PFCT instance has cost at least* $\sum_{i=1}^{n}(f_i - f_{i+1})\pi(a([i]))$.

Proof We analyze the cost of the optimum solution as follows. When some sink is connected to a source i, we split the cost f_i into segments $(f_i - f_{i+1}) + (f_{i+1} - f_{i+2}) + \cdots + (f_n - f_{n+1})$. The segment $f_i - f_{i+1}$ will be paid if we connect a sink to a source in $[i]$. The number of sinks connected to $[i]$ is at least $\pi(a[i])$ by the definition of π and that $b_1 \geq b_2 \geq \cdots \geq b_m$. The lemma then follows.

Lemma 2 *The cost of x to the PFCT-S instance is at most* $\sum_{i=1}^{n}(f_i - f_{i+1})\pi(a([i])) + \sum_{i=2}^{n} f_i$.

Proof Similar to the proof of Lemma 1, when x incurs a cost of f_i, we split it into segments $(f_i - f_{i+1}) + (f_{i+1} - f_{i+2}) + \cdots + (f_n - f_{n+1})$. So, the segment $f_i - f_{i+1}$ is paid whenever we make a connection to sources in $[i]$. Notice that the greedy algorithm will only connect $[i]$ to sinks in $[\pi(a[i])]$. As supp(x) is a forest, the number of edges incident to $[i]$ in $\mathrm{supp}(x)$ is at most $\pi(a[i]) + i - 1$. So, the cost of x to the PFCT-S instance is at most

$$\sum_{i=1}^{n}(f_i - f_{i+1})\big(\pi(a[i]) + i - 1\big) = \sum_{i=1}^{n}(f_i - f_{i+1})\pi(a[i]) + \sum_{i,i' \in [n]: i' \le i-1}(f_i - f_{i+1})$$

$$= \sum_{i=1}^{n}(f_i - f_{i+1})\pi(a[i]) + \sum_{i'=1}^{n}\sum_{i=i'+1}^{n}(f_i - f_{i+1}) = \sum_{i=1}^{n}(f_i - f_{i+1})\pi(a[i]) + \sum_{i'=1}^{n} f_{i'+1}.$$

Combining Lemmas 1 and 2 and that the cost of the optimum solution is at least $\sum_{i=2}^{n} f_i$, we conclude that the cost of x is at most 2 times optimum cost. This finishes the proof of Theorem 1.

The following corollary is implied by Lemmas 1 and 2. It will be used in our EPAS for the PFCT problem. For convenience, we prove it here.

Corollary 11 *Suppose we have two PFCT-S instances with the same sources $[n]$, same supplies $a_1, a_2, \cdots, a_n$ and same costs $f_1 \ge f_2 \ge \cdots \ge f_n$. One instance has m sinks with demands $b_1, b_2, \cdots, b_m$, and the other has m' sinks with demands $b'_1, b'_2, \cdots, b'_{m'}$. For every $t \in (0, a([n])]$, let $\pi(t)$ be the smallest j such that $b([j]) \ge t$ let $\pi'(t)$ be the smallest j such that $b'([j]) \ge t$. Assume $\pi'(t) \le \pi(t) + \Delta$ holds for some integer $\Delta \ge 0$ and every $t \in (0, a([n])]$.*

Then, the cost of the greedy solution to the second instance is at most the cost of the optimum solution to the first instance plus $\Delta f_1 + \sum_{i=2}^{n} f_i$.

Proof By Lemma 2, the cost of the greedy solution to the second instance is at most $\sum_{i=1}^{n}(f_i - f_{i+1})\pi'(a[i]) + \sum_{i=2}^{n} f_i$. By Lemma 1, the optimum cost to the first instance is at least $\sum_{i=1}^{n}(f_i - f_{i+1})\pi(a[i])$. The difference of the two quantities is

$$\sum_{i=1}^{n}(f_i - f_{i+1})(\pi'(a[i]) - \pi(a[i])) + \sum_{i=2}^{n} f_i \le \sum_{i=1}^{n}(f_i - f_{i+1}) \cdot \Delta + \sum_{i=2}^{n} f_i = \Delta f_1 + \sum_{i=2}^{n} f_i.$$

The inequality used that $\pi'(t) \le \pi(t) + \Delta$ for every t.

5 Discussion

In this paper, we initiated a systematic study of variants of the Fixed Charge Transportation problem, and provided a complete characterization of the existence of $O(1)$-approximation algorithms for them. Although our techniques are simple, we believe this work opens a new research direction on the problem.

There are several interesting open problems. For the PFCT-S and FCT-U problems, can we achieve an approximation ratio better than 2? For the FCT-S problem, is it possible to obtain a polylogarithmic approximation? For the PFCT problem, can we show that it is not significantly harder than the Directed Steiner Tree problem? For example, by allowing quasi-polynomial time algorithms, can we obtain a polylogarithmic approximation? Finally, can we establish much stronger hardness of approximation results for the two problems, or the most general FCT problem?

Acknowledgments. The research is supported by the National Natural Science Foundation of China (No.12471301, 12371316), the State Key Laboratory for Novel Software Technology and the New Cornerstone Science Laboratory.

References

1. Hirsch, W.M., Dantzig, G.B.: Notes on linear programming: part XIX—the fixed charge problem. RAND Research Memorandum 1383, RAND Corporation, Santa Monica, CA (1954)
2. Hirsch, W.M., Dantzig, G.B.: The fixed charge problem. Naval Res. Logist. Q. **15**(3), 413–424 (1968)
3. Rardin, R.L., Wolsey, L.A.: Valid inequalities and projecting the multicommodity extended formulation for uncapacitated fixed charge network flow problems. Eur. J. Oper. Res. **71**(1), 95–109 (1993)
4. Ortega, F., Wolsey, L.A.: A branch-and-cut algorithm for the single commodity, uncapacitated, fixed-charge network flow problem. Networks **41**(3), 143–158 (2003)
5. Agarwal, Y.: K-partition-based facets of the network design problem. Networks **47**(3), 123–139 (2006)
6. Frangioni, A., Gendron, B.: Reformulations of the multicommodity capacitated network design problem. Discret. Appl. Math. **157**(6), 1229–1241 (2009)
7. Raack, C., Koster, A.M.C.A., Orlowski, S., Wessaly, R.: On cut-based inequalities for capacitated network design polyhedra. Networks **57**(2), 141–156 (2011)
8. Fisk, J., McKeown, P.G.: The pure fixed charge transportation problem. Naval Res. Logist. Q. **26**(4), 631–641 (1979)
9. Göthe-Lundgren, M., Larsson, T.: A set covering reformulation of the pure fixed charge transportation problem. Discret. Appl. Math. **48**(3), 245–259 (1994)
10. Murty, K.G.: Solving the fixed charge problem by ranking the extreme points. Oper. Res. **16**(2), 268–279 (1968)
11. Gray, P.: Exact solution of the fixed-charge transportation problem. Oper. Res. **19**(6), 1529–1538 (1971)
12. Agarwal, Y., Aneja, Y.: Fixed-charge transportation problem: Facets of the projection polyhedron. Oper. Res. **60**(3), 638–654 (2012)
13. Zhao, Y., Larsson, T., Ronnberg, E., Pardalos, P.M.: The fixed charge transportation problem: a strong formulation based on Lagrangian decomposition and column generation. J. Global Optim. **72**(3), 517–538 (2018). https://doi.org/10.1007/s10898-018-0661-y
14. Mingozzi, A., Roberti, R.: An exact algorithm for the fixed charge transportation problem based on matching source and sink patterns. Transp. Sci. **52**(2), 229–496 (2018)
15. Kennigton, J., Unger, E.: A new branch-and-bound algorithm for the fixed-charge transportation problem. Manage. Sci. **22**(10), 1116–1126 (1976)
16. Barr, R.S., Glover, F., Klingman, D.: A new optimization method for large scale fixed charge transportation problems. Oper. Res. **29**(3), 448–463 (1981)
17. Cabot, A.V., Erenguc, S.S.: Some branch and bound procedures for fixed cost transportation problems. Naval Res. Logist. Q. **31**(1), 145–154 (1984)
18. Roberti, R., Bartolini, E., Mingozzi, A.: The fixed charge transportation problem: an exact algorithm based on a new integer programming formulation. Manage. Sci. **61**(6), 1197–1471 (2015)

19. Wright, D.D., von Lanzenauer, C.H.: Solving the fixed charge problem with Lagrangian relaxation and cost allocation heuristics. Eur. J. Oper. Res. **42**(3), 305–312 (1989)
20. Sun, M., Aronson, J.E., McKeown, P.G., Drinka, D.: A tabu search heuristic procedure for the fixed charge transportation problem. Eur. J. Oper. Res. **106**(2–3), 441–456 (1998)
21. Aguado, J.S.: Fixed charge transportation problems: A new heuristic approach based on Lagrangean relaxation and the solving of core problems. Ann. Oper. Res. **172**(1), 45–69 (2009)
22. Buson, E., Roberti, R., Toth, P.: A reduced-cost iterated local search heuristic for the fixed-charge transportation problem. Oper. Res. **62**(5), 973–1201 (2014)
23. Guisewite, G., Pardalos, P.: Minimum concave-cost network flow problems: applications, complexity, and algorithms. Ann. Oper. Res. **25**(1), 75–99 (1990)
24. Kowalski, K.: On the structure of the fixed charge transportation problem. Int. J. Math. Educ. Sci. Technol. **36**(8), 879–888 (2005)
25. Schrenk, S., Finke, G., Cung, V.D.: Two classical transportation problems revisited: pure constant fixed charges and the paradox. Math. Comput. Model. **54**, 2306–2315 (2011)
26. Kann, K.: Maximum bounded 3-dimensional matching is MAX SNP-complete. Inf. Process. Lett. **37**, 27–35 (1991)
27. Berman, P., Karpinski, M.: Improved approximation lower bounds on small occurrence optimization. Electron. Colloq. Comput. Complexity (ECCC) **10**(008) (2003)
28. Cygan, M.: Improved approximation for 3-dimensional matching via bounded pathwidth local search. In: ACM-SIAM Symposium on Discrete Algorithms (SODA), pp. 509–518 (2013)
29. Furer, M., Yu, H.: Approximating the k-set packing problem by local improvements. In: International Symposium on Combinatorial Optimization (ISCO), pp. 408–420 (2014)
30. Grandoni, F., Laekhanukit, B., Li, S.: $O(log^2k/loglogk)$-approximation algorithm for directed Steiner tree: a tight quasi-polynomial time algorithm. SIAM J. Comput. **52**(2), 298–322 (2022)
31. Halperin, E., Krauthgamer, R.: Polylogarithmic inapproximability. In: ACM Symposium on Theory of Computing (STOC), pp. 585–594 (2003)
32. Ghuge, R., Nagarajan, V.: Quasi-polynomial algorithms for submodular tree orienteering and directed network design problems. Math. Oper. Res. **47**(2), 1612–1630 (2022)
33. Charikar, M., Chekuri, C., Cheung, T., Dai, Z., Goel, A., Guha, S., Li, M.: Approximation algorithms for directed steiner problems. J. Algorithms **33**(1), 73–91 (1999)
34. Zelikovsky, A.: A series of approximation algorithms for the acyclic directed steiner tree problem. Algorithmica **18**(1), 99–110 (1997)

Closure Properties and Characterizations of TotP

Yaroslav Ivanashev(✉)

Faculty of Computer Science, HSE University, Moscow 101000, Russia
yivanashev@hse.ru

Abstract. The class TotP consists of functions that count the number of all paths of a nondeterministic polynomial-time Turing machine. In this paper, we give a predicate based definition of TotP, analogous to a standard definition of #P. From a new characterization of TotP it follows that many well known #P problems belong to TotP, and TotP = #P if and only if P = NP. We show that TotP has several closure properties of #P and GapP, and also properties that are not known to hold for #P and GapP. We also prove that the closure of TotP under left composition with FP_+ is equivalent to TotP = FP_+ and P = PP, and give examples of FP_+-functions such that if TotP is closed under composition with them, then it is closed under composition with FP_+.

Keywords: computational complexity · counting problems · closure properties · #P

1 Introduction

The class TotP, which was introduced in [17], is a subclass of #P [23] containing functions that count the number of all paths of a nondeterministic polynomial-time Turing machine (NPTM). In Sect. 3, we give a predicate based definition of TotP, in which it is described as a class of functions equal to the number of witnesses for NP-problems of a certain type. From a new characterization of TotP it follows that many #P-problems belong to TotP, including NonNegative Permanent, #Perfect Matchings, #DNF-Sat, and #NonCliques, which are #P-complete under 1-Cook reductions. It also follows that TotP = #P if and only if P = NP. TotP has complete problems under Karp reductions [3], and it coincides with the closure under Karp reductions of the class of self-reducible functions from the class #PE [20]. TotP is characterized using interval-size functions in [7]. In [6], the relationship of TotP with classes of approximable functions is studied. Logical characterizations of TotP and some of its subclasses are given in [1,4,6].

TotP has natural closure properties, for example, it is closed under addition, multiplication, and decrement by one [5]. In Sect. 4, we give several closure properties of TotP that #P and GapP have [8,10], and also properties that are

M. Li et al. (Eds.): TAMC 2025, LNCS 16084, pp. 15–24, 2026.
https://doi.org/10.1007/978-981-95-4839-2_2

unknown to hold for $\#\mathsf{P}$ and GapP. In Sect. 5, we give a characterization of GapP with TotP-functions that is more precise than $\mathsf{GapP} = \mathsf{TotP} - \mathsf{FP}$ [5].

In Sect. 6, we study the closure of TotP under composition with FP_+ (i.e., whether $\mathsf{FP}_+ \circ \mathsf{TotP} \subseteq \mathsf{TotP}$), and prove that this statement is equivalent to $\mathsf{TotP} = \mathsf{FP}_+$ and $\mathsf{P} = \mathsf{PP}$. For the class $\#\mathsf{P}$, the closure under left composition with FP_+ is equivalent to $\mathsf{PP} = \mathsf{UP}$ and $\#\mathsf{P} = \mathsf{UPSV_t}$ [15,19]; for GapP is equivalent[1] to $\mathsf{PP} = \mathsf{SPP}$ and $\mathsf{GapP} = \mathsf{FP}^{\mathsf{SPP}}$ [10,13,15,22].

We also give examples of FP_+-functions of several arguments such that if TotP is closed under composition with them, then it is closed under composition with FP_+. Our proof is similar to that given in [19], where analogous result for the class $\#\mathsf{P}$ is obtained.

2 Preliminaries

In this paper, we use standard definitions of a deterministic and a nondeterministic Turing machine. Below we give definitions of function and language classes that we will consider:

Definition 2.1. *1.* [23] $\#\mathsf{P} = \{acc_M \mid M \text{ is a NPTM}\}$, *where* $acc_M(x)$ *is the number of accepting paths of* M *on input* $x \in \Sigma^*$.
2. [10] $\mathsf{GapP} = \{gap_M \mid M \text{ is a NPTM}\}$, *where* $gap_M(x) = acc_M(x) - rej_M(x)$ *and* $rej_M(x)$ *is the number of rejecting paths of* M *on input* $x \in \Sigma^*$.
3. [17] $\mathsf{TotP} = \{tot_M \mid M \text{ is a NPTM}\}$, *where* $tot_M(x) = Total_M(x) - 1$ *and* $Total_M(x)$ *is the number of all paths of* M *on input* $x \in \Sigma^*$.[2]
4. [20] $\#\mathsf{PE} = \{f \mid f \in \#\mathsf{P} \text{ and } \{x \mid f(x) > 0\} \in \mathsf{P}\}$.
5. $\mathsf{FP} = \{f : \Sigma^* \to \mathbb{Z} \mid f \text{ is computable in polynomial time}\}$.
6. $\mathsf{FP}_+ = \{f \mid f \in \mathsf{FP} \text{ and } f \geqslant 0\}$.

Definition 2.2. *1.* [9,18] *The class* NP *consists of all languages* L *for which there exists a function* $f \in \#\mathsf{P}$ *such that for any* $x \in \Sigma^*$:

$$x \in L \Leftrightarrow f(x) > 0.$$

2. [21,25] *The class* $\mathsf{C_{=}P}$ *consists of all languages* L *for which there exist functions* $f \in \#\mathsf{P}$ *and* $g \in \mathsf{FP}$ *such that for any* $x \in \Sigma^*$:

$$x \in L \Leftrightarrow f(x) = g(x).$$

3. [12,21] *The class* PP *consists of all languages* L *for which there exist functions* $f \in \#\mathsf{P}$ *and* $g \in \mathsf{FP}$ *such that for any* $x \in \Sigma^*$:

$$x \in L \Leftrightarrow f(x) > g(x).$$

[1] For GapP, the closure under decrement by one is also equivalent to $\mathsf{PP} = \mathsf{SPP}$ [22]. For the class $\#\mathsf{P}$, it is only known that the closure under decrement by one implies $\mathsf{NP} \subseteq \mathsf{SPP}$ [19].

[2] The "-1" in the definition of TotP allows these functions to take a zero value.

The relations between classes of functions are given in Proposition 2.3:

Proposition 2.3. *1.* [20] $\mathsf{FP}_+ \subseteq \mathsf{TotP} \subseteq \#\mathsf{PE} \subseteq \#\mathsf{P}$. *All inclusions are proper unless* $\mathsf{P} = \mathsf{NP}$.
2. [5,10] $\mathsf{GapP} = \#\mathsf{P} - \#\mathsf{P} = \#\mathsf{P} - \mathsf{FP} = \mathsf{TotP} - \mathsf{TotP} = \mathsf{TotP} - \mathsf{FP}$.

3 Characterizations of TotP

In this section, we give a predicate based definition of TotP, analogous to the following definition of #P:

Proposition 3.1. *The class* #P *consists of all functions* f *for which there exist a polynomial* p *and a language* $A \in \mathsf{P}$ *such that for any* $x \in \Sigma^*$*:*

$$f(x) = |\{y \mid |y| = p(|x|) \wedge (x, y) \in A\}|.$$

To obtain a characterization of TotP, we add an additional condition to the above characterization of #P: there exists an algorithm that on any input $(x, c) \in \Sigma^*$ determines in polynomial time the existence of a witness for x with a given prefix c.

Theorem 3.2. *The class* TotP *consists of all functions* f *for which there exist a polynomial* p *and a language* $A \in \mathsf{P}$ *such that:*

1. *For any* $x \in \Sigma^*$*:* $f(x) = |\{y \mid |y| = p(|x|) \wedge (x, y) \in A\}|$.
2. *The language* $\{(x, c) \mid \exists z(|cz| = p(|x|) \wedge (x, cz) \in A)\} \in \mathsf{P}$*, where* cz *denotes a concatenation of strings* c *and* z.

Proof. ($\subseteq$): Let $f(x) = tot_M(x)$ for a NPTM M. There a exists a polynomial-time algorithm that accepts (x, y) if and only if y is a path of $M(x)$ that is not the lexicographically largest path of $M(x)$, where paths of $M(x)$ are encoded so that their length is always equal to a fixed polynomial $p(|x|)$. A polynomial-time algorithm can determine whether a given string is a prefix of a path of $M(x)$ that is not the lexicographically largest path of $M(x)$.

($\supseteq$): Let f be a function such that the above conditions hold. Then $f(x) = tot_M(x)$, where M is the NPTM that behaves as follows on input $x \in \Sigma^*$:

(a) If the empty string is not a prefix of any witness for x, M halts.
(b) M takes one nondeterministic step and halts on the first path.
(c) On the second path, M nondeterministically guesses a witness for x. Initially, y is the empty string. If y0 and y1 are both prefixes of a witness for x, M takes one nondeterministic step. On one path, it replaces y with $y1$, and on the other path, it replaces y with $y0$. If y can be extended only in one way, M doesn't take a nondeterministic step.
(d) If the length of y becomes equal to $p(|x|)$, M halts. □

If $\mathsf{P} = \mathsf{NP}$, then the second condition in Theorem 3.2 holds for all #P-functions. This gives the following corollary:

Corollary 3.3. $\mathsf{TotP} = \#\mathsf{P}$ *if and only if* $\mathsf{P} = \mathsf{NP}$.

From Theorem 3.2 it follows[3] that many well known #P-problems belong to TotP, including #PERFECT MATCHINGS, NONNEGATIVE PERMANENT, #DNF-SAT, #MONOTONESAT, #NONCLIQUES, #NONINDEPENDENT SETS, #NFA, which are #P-complete under 1-Cook reductions [2,20,24]. These problems were initially shown to be in TotP by Theorem 3.4 [20], which gives another way to prove that a function belongs to TotP. It suffices to prove that a function belongs to #PE and it is self-reducible. In [20], it is shown that TotP contains all self-reducible #PE-functions.

Theorem 3.4 ([20]). TotP *is the closure under Karp reductions of the class of self-reducible* #PE *functions.*

The second condition in Theorem 3.2, which states that the existence of a witness with a given prefix can be determined in polynomial time, can be replaced by equivalent conditions:

- There exists a polynomial-time algorithm that on input $(x, c) \in \Sigma^*$ outputs a witness for x that is lexicographically closest to a given string c, or outputs that there is no witness for x.
- There exists a polynomial-time algorithm that on input $(x, a, b) \in \Sigma^*$ determines the existence of a witness for x in a given interval (a, b).

The above characterizations are similar to the characterization of TotP in terms of interval-size functions defined on total p-orders with efficiently computable lexicographically nearest function (see [7] for precise definitions).

Theorem 3.5 ([7]). $\mathsf{TotP} = \mathsf{IF}_{\mathsf{t}}^{\mathsf{LN}}$.

4 Closure Properties of TotP

In this section, we present closure properties of TotP that complement the results from [5] (Proposition 4.2).

Definition 4.1. *For integers a and b, the decrement operation is defined as follows:*

$$a \ominus b = \begin{cases} a - b, & \text{if } a \geqslant b; \\ 0, & \text{otherwise.} \end{cases}$$

[3] In [24, p.413], Valiant observes that many #P-complete problems have the second property from Theorem 3.2. It follows from Fact 7 on p.413 that for all TotP-problems there exists an algorithm that for any $x \in \Sigma^*$ outputs all witnesses in time $p(|x|) \cdot N$, where N is the number of witnesses and p is some polynomial (a term "algorithm with a polynomial delay" is used in [16]). For all #P-problems the same is not true unless $\mathsf{P} = \mathsf{UP}$ [23].

Proposition 4.2 ([5]). *1. For any functions $f, g \in \mathsf{TotP}$, the functions $f + g$, $f \cdot g$, and $f \ominus 1$ are in* TotP.

2. Let $f \in \mathsf{TotP}$. If there exists a polynomial p such that for any $x \in \Sigma^*$: $f(x) \leqslant p(|x|)$, then $f \in \mathsf{FP}$.

The following properties are not known to hold for #P and GapP:

Proposition 4.3. *For any functions $f \in \mathsf{TotP}$ and $g \in \mathsf{FP}_+$, such that $g(x) \leqslant$ poly$(|x|)$:*

1. *The languages $\{x \mid f(x) = g(x)\}$ and $\{x \mid f(x) > g(x)\}$ are in* P.
2. *The function $f(x) \ominus g(x)$ is in* TotP.

The properties from Proposition 4.4 also hold for the classes #P [8] and GapP [10].

Proposition 4.4. *For any functions $f \in \mathsf{TotP}$ and $g \in \mathsf{FP}_+$, such that $g(x) \leqslant$ poly$(|x|)$:*

1. *The function $f(x)^{g(x)}$ is in* TotP.
2. *The function $\binom{f(x)}{g(x)}$ is in* TotP.

Proof. Let $f(x) = tot_M(x)$ for a NPTM M.

1. Let N be the NPTM that behaves as follows on input $x \in \Sigma^*$:
   (a) If $f(x) = 0$, N halts.
   (b) N takes one nondeterministic step and halts on the first path. On the second path, N guesses $g(x)$ paths of $M(x)$ (with duplicates) that are lexicographically larger than the first path of $M(x)$.
   $tot_N(x) = f(x)^{g(x)}$.
2. Let N be the NPTM that behaves as follows on input $x \in \Sigma^*$:
   (a) If $f(x) = 0$, N halts.
   (b) If $M(x)$ has at least $g(x) + 1$ paths, N takes one nondeterministic step and halts on the first path. On the second path, N guesses a set of $g(x)$ distinct paths of $M(x)$ that are lexicographically larger than the first path of $M(x)$.
   $tot_N(x) = \binom{f(x)}{g(x)}$. □

The classes #P and GapP are closed under polynomial summation and polynomial product [8,10]. The same is true for TotP (Proposition 4.5).

Proposition 4.5. *For any functions $f \in \mathsf{TotP}$ and $g \in \mathsf{FP}_+$, such that $g(x) \leqslant$ poly$(|x|)$:*

1. *The function $s(x) = \sum_{0 \leqslant y \leqslant g(x)} f(x, y)$ is in* TotP.
2. *The function $m(x) = \prod_{0 \leqslant y \leqslant g(x)} f(x, y)$ is in* TotP.

Proof. Let $f(x, y) = tot_M(x, y)$ for a NPTM M.

1. Let N be the NPTM that behaves as follows on input $x \in \Sigma^*$:

For y in range $\{0,\ldots,g(x)\}$:
(a) If $f(x,y)=0$, N proceeds to the next iteration of the loop.
(b) If $f(x,y)>0$, N takes one nondeterministic step. On the first path, N guesses one path of $M(x,y)$ that is lexicographically larger than the first path of $M(x,y)$.
(c) If $y=g(x)$, N halts on the second path. Otherwise, N performs the next iteration of the loop on the second path.

$tot_N(x)=\sum_{0\leqslant y\leqslant g(x)} f(x,y)$.

2. Let N be the NPTM that behaves as follows on input $x\in\Sigma^*$:
   (a) For y in range $\{0,\ldots,g(x)\}$:
   If $f(x,y)=0$, N halts.
   (b) N takes one nondeterministic step, halts on the first path, and proceeds on the second path.
   (c) For y in range $\{0,\ldots,g(x)\}$:
   N guesses one path of $M(x,y)$ that is lexicographically larger than the first path of $M(x,y)$.

$tot_N(x)=\prod_{0\leqslant y\leqslant g(x)} f(x,y)$. □

The classes #P and GapP are also closed under exponential summation [8,10]. In Proposition 4.6 we prove that TotP doesn't have this property unless P = NP.

Proposition 4.6. *The following statements are equivalent:*

1. *For any functions* $f\in\mathsf{TotP}$ *and* $g\in\mathsf{FP}_+$, *such that* $g(x)\leqslant poly(|x|)$, *the function* $s(x)=\sum_{|y|\leqslant g(x)} f(x,y)$ *is in* TotP.
2. P = NP.

Proof. We will use the equivalent TotP = #P instead of P = NP. If TotP = #P, then the first statement holds, because #P has this property [8]. Let h be a function in #P. There exist an FP-function $V\colon \Sigma^*\times\Sigma^*\to\{0,1\}$ and a polynomial p such that for any $x\in\Sigma^*$: $h(x)=|\{y \mid |y|\leqslant p(|x|)\wedge V(x,y)=1\}|$. By assumption, the function $h(x)=\sum_{|y|\leqslant p(|x|)} V(x,y)$ is in TotP. □

5 Characterization of GapP with TotP Functions

In this section, we give a more precise characterization of GapP with TotP-functions, than GapP = TotP − FP [5], given in Proposition 2.3. In the second part of Lemma 5.1 the class TotP can be replaced by the class SpanL [2] (the proof is given in [14]). SpanL is a subclass of TotP, so the proof of the second part of Lemma 5.1 also follows from [14]. We give a simpler proof of this statement.

Lemma 5.1. *For any NPTM* M *and any polynomial* p *such that for any* $x\in\Sigma^*$ $p(|x|)$ *is greater than the running time of* $M(x)$*:*

1. [17] *There exists a function* $g_1\in\mathsf{TotP}$ *such that for any* $x\in\Sigma^*$: $acc_M(x)=g_1(x)-2^{p(|x|)}$.

2. *There exists a function* $g_2 \in \mathsf{TotP}$ *such that for any* $x \in \Sigma^*$: $acc_M(x) = 2^{p(|x|)} - g_2(x)$.

Proof. 1. We repeat the proof from [17]. There exists a NPTM M' such that $acc_M(x) = acc_{M'}(x)$ and $Total_{M'}(x) = 2^{p(|x|)}$. Let N be the NPTM that simulates M', but instead of going into an accepting state, generates two accepting paths.
$acc_M(x) = Total_N(x) - Total_{M'}(x) = (tot_N(x) + 1) - 2^{p(|x|)}$.
2. There exists a NPTM M' such that $acc_M(x) = acc_{M'}(x)$ and $Total_{M'}(x) = 2^{p(|x|)-1}$. Let N be the NPTM that simulates M', but instead of going into a rejecting state, generates two rejecting paths.
$acc_M(x) = 2 \cdot Total_{M'}(x) - Total_N(x) = 2^{p(|x|)} - (tot_N(x) + 1)$. □

Proposition 5.2. *For any function* $f \in \mathsf{GapP}$ *there exists a function* $g \in \mathsf{TotP}$ *and a polynomial* p *such that for any* $x \in \Sigma^*$: $f(x) = g(x) - 2^{p(|x|)}$.

Proof. Let $f(x) = acc_M(x) - acc_N(x)$ for NPTMs M and N. By Lemma 5.1, there exist functions $g_1, g_2 \in \mathsf{TotP}$ such that $acc_M(x) = g_1(x) - 2^{p(|x|)}$ and $acc_N(x) = 2^{p(|x|)} - g_2(x)$, where p is a polynomial that is greater than the running time of M and N.
$f(x) = (g_1(x) - 2^{p(|x|)}) - (2^{p(|x|)} - g_2(x)) = (g_1(x) + g_2(x)) - 2^{p(|x|)+1}$. □

In [11], it is shown that for every language $L \in \mathsf{C_=P}$ there exists a function $f \in \mathsf{GapP}$ such that for any $x \in \Sigma^*$:

$$x \in L \Leftrightarrow f(x) = 0,$$
$$x \notin L \Leftrightarrow f(x) < 0.$$

The following characterization of $\mathsf{C_=P}$ we use in the next section:

Lemma 5.3. *For any language* $L \in \mathsf{C_=P}$ *there exist a function* $g \in \mathsf{TotP}$ *and a polynomial* p *such that for any* $x \in \Sigma^*$:

$$x \in L \Leftrightarrow g(x) = 2^{p(|x|)},$$
$$x \notin L \Leftrightarrow g(x) < 2^{p(|x|)}.$$

6 Closure Under Composition with FP

In this section, we prove that the closure of TotP under left composition with FP_+ is equivalent to $\mathsf{TotP} = \mathsf{FP}_+$ and $\mathsf{PP} = \mathsf{P}$. In the next theorem, we will also consider FP_+-functions of several arguments. A function h is in the class $\mathsf{FP}_+ \circ (\mathsf{TotP} \times ... \times \mathsf{TotP})$ if there exist functions $f \in \mathsf{FP}_+$ and $g_1, \dots, g_k \in \mathsf{TotP}$ such that for any $x \in \Sigma^*$: $h(x) = f(g_1(x), \dots, g_k(x))$.

Theorem 6.1. *The following statements are equivalent:*

1. $\mathsf{FP}_+ \circ \mathsf{TotP} \subseteq \mathsf{TotP}$.
2. $\mathsf{FP}_+ \circ (\mathsf{TotP} \times \ldots \times \mathsf{TotP}) \subseteq \mathsf{TotP}$.

3. $\mathsf{TotP} = \mathsf{FP}_+$.
4. $\mathsf{PP} = \mathsf{P}$.

Proof. (2 $\Rightarrow$ 1): (1) is equivalent to (2), because the classes $\mathsf{FP}_+ \circ \mathsf{TotP}$ and $\mathsf{FP}_+ \circ (\mathsf{TotP} \times \ldots \times \mathsf{TotP})$ are equal [15].

(1 $\Rightarrow$ 4): Let $L = \{x \mid f(x) > g(x)\}$, where $f \in \mathsf{TotP}$, $g \in \mathsf{FP}$. The function

$$h(x) = \begin{cases} 1, & f(x) > g(x); \\ 0, & \text{otherwise.} \end{cases}$$

is in $\mathsf{FP}_+{}^{\mathsf{TotP}[1]}$, which is equal to the class $\mathsf{FP}_+ \circ \mathsf{TotP}$ [15]. By assumption, $h \in \mathsf{TotP}$. By Proposition 4.2, $h \in \mathsf{FP}$ and $L \in \mathsf{P}$.

(4 $\Rightarrow$ 3): $\mathsf{TotP} = \mathsf{FP}_+$ is equivalent to $\#\mathsf{P} = \mathsf{FP}_+$, which is well known to be equivalent to $\mathsf{PP} = \mathsf{P}$.

(3 $\Rightarrow$ 2): Follows from $\mathsf{FP}_+ \circ (\mathsf{FP}_+ \times \ldots \times \mathsf{FP}_+) \subseteq \mathsf{FP}_+$. □

For the class $\#\mathsf{P}$, there exists FP_+-functions of several arguments such that if $\#\mathsf{P}$ is closed under composition with them, then it is closed under composition with FP_+ [19]. In the next theorem, we show that these functions have the same property when applied to the class TotP. We repeat the proofs from [19] that are slightly modified for the case of TotP.

Definition 6.2. *1. $span(a_1, \ldots, a_k)$ is the number of distinct elements in a set $\{a_1, \ldots, a_k\}$.*
2. $plu(a_1, \ldots, a_k)$ is the set of most frequently occurring elements in a set $\{a_1, \ldots, a_k\}$.

Theorem 6.3. *The following statements are equivalent:*

1. $\mathsf{FP}_+ \circ (\mathsf{TotP} \times \ldots \times \mathsf{TotP}) \subseteq \mathsf{TotP}$.
2. $\mathsf{C_=P} = \mathsf{P}$.
3. *For any functions $f \in \mathsf{TotP}$, $g \in \mathsf{FP}$, the function $f(x) \ominus g(x)$ is in* TotP.
4. *For any functions $f \in \mathsf{TotP}$, $g \in \mathsf{FP}_+$, such that $g(x) \neq 0$, the function $\lfloor f(x)/g(x) \rfloor$ is in* TotP.
5. *For any functions $f_1, \ldots, f_k \in \mathsf{TotP}$ the function $g(x) = span(f_1(x), \ldots, f_k(x))$ is in* TotP.
6. *For any functions $f_1, \ldots, f_k \in \mathsf{TotP}$ there exists a function $g \in \mathsf{TotP}$ such that for any $x \in \Sigma^*$: $g(x) \in plu(f_1(x), \ldots, f_k(x))$.*

Proof. (1) and (2) are equivalent by Theorem 6.1 (instead of $\mathsf{PP} = \mathsf{P}$ we use the equivalent $\mathsf{C_=P} = \mathsf{P}$ [19]). (1) implies (3), (4), (5), and (6) because these functions are polynomial time computable (for (6) it suffices to consider the function $min(plu(a_1, \ldots, a_k))$).

Let $L \in \mathsf{C_=P}$. By Lemma 5.3, there exist a function $f \in \mathsf{TotP}$ and a polynomial p such that for any $x \in \Sigma^*$:

$$x \in L \Leftrightarrow f(x) = 2^{p(|x|)},$$
$$x \notin L \Leftrightarrow f(x) < 2^{p(|x|)}.$$

(3 ⇒ 2): Let $g(x) = f(x) \ominus (2^{p(|x|)} - 1)$. By assumption, $g \in \mathsf{TotP}$, and for any $x \in \Sigma^*$:

$$x \in L \Leftrightarrow g(x) = 1,$$
$$x \notin L \Leftrightarrow g(x) = 0.$$

By Proposition 4.2, $g \in \mathsf{FP}$ and $L \in \mathsf{P}$.

(4 ⇒ 2): Let $g(x) = \lfloor f(x)/2^{p(|x|)} \rfloor$. By assumption, $g \in \mathsf{TotP}$, and for any $x \in \Sigma^*$:

$$x \in L \Leftrightarrow g(x) = 1,$$
$$x \notin L \Leftrightarrow g(x) = 0.$$

By Proposition 4.2, $g \in \mathsf{FP}$ and $L \in \mathsf{P}$.

(5 ⇒ 2): Let $g(x) = span(f(x), 2^{p(|x|)})$. By assumption, $g \in \mathsf{TotP}$, and for any $x \in \Sigma^*$:

$$x \in L \Leftrightarrow g(x) = 1,$$
$$x \notin L \Leftrightarrow g(x) = 2.$$

By Proposition 4.2, $g \in \mathsf{FP}$ and $L \in \mathsf{P}$.

(6 ⇒ 2): By assumption, there exists a function $g \in \mathsf{TotP}$ such that for any $x \in \Sigma^*$: $g(x) \in plu(0, 0, 2f(x), f(x) + 2^{p(|x|)}, 2^{p(|x|)+1})$,

$$x \in L \Leftrightarrow g(x) = 2^{p(|x|)+1},$$
$$x \notin L \Leftrightarrow g(x) = 0.$$

By Proposition 4.3, $L \in \mathsf{P}$. □

Acknowledgment. The author is grateful to anonymous reviewers for their valuable comments, to Alexander Shen for asking about a predicate based definition of TotP, and to Bruno Bauwens for his help in preparing of this paper. The work is supported by the HSE Basic Research Program.

References

1. Achilleos, A., Chalki, A.: Counting computations with formulae: logical characterisations of counting complexity classes. In: 48th International Symposium on Mathematical Foundations of Computer Science (2023)
2. Àlvarez, C., Jenner, B.: A very hard log-space counting class. Theoret. Comput. Sci. **107**(1), 3–30 (1993)
3. Antonopoulos, A., Bakali, E., Chalki, A., Pagourtzis, A., Pantavos, P., Zachos, S.: Completeness, approximability and exponential time results for counting problems with easy decision version. Theoret. Comput. Sci. **915**, 55–73 (2022)
4. Arenas, M., Muñoz, M., Riveros, C.: Descriptive complexity for counting complexity classes. Log. Methods Comput. Sci. **16** (2020)

5. Bakali, E., Chalki, A., Kanellopoulos, S., Pagourtzis, A., Zachos, S.: On the power of counting the total number of computation paths of NPTMs. In: Annual Conference on Theory and Applications of Models of Computation, pp. 209–220. Springer (2024)
6. Bakali, E., Chalki, A., Pagourtzis, A.: Characterizations and approximability of hard counting classes below #P. In: International Conference on Theory and Applications of Models of Computation, pp. 251–262. Springer (2020)
7. Bampas, E., Göbel, A.-N., Pagourtzis, A., Tentes, A.: On the connection between interval size functions and path counting. Comput. Complexity **26**(2), 421–467 (2016). https://doi.org/10.1007/s00037-016-0137-8
8. Beigel, R., Gill, J.: Counting classes: thresholds, parity, mods, and fewness. Theoret. Comput. Sci. **103**(1), 3–23 (1992)
9. Cook, S.A.: The complexity of theorem-proving procedures. In: Logic, automata, and computational complexity: The works of Stephen A. Cook, pp. 143–152 (2023)
10. Fenner, S.A., Fortnow, L.J., Kurtz, S.A.: Gap-definable counting classes. J. Comput. Syst. Sci. **48**(1), 116–148 (1994)
11. Fortnow, L.: Counting complexity. In: Hemaspaandra, L., Selman, A. (eds.) Complexity Theory Retrospective II, pp. 81–107 (1997)
12. Gill, J.: Computational complexity of probabilistic turing machines. SIAM J. Comput. **6**(4), 675–695 (1977)
13. Gupta, S.: The power of witness reduction. In: 1991 Proceedings of the Sixth Annual Structure in Complexity Theory Conference, pp. 43–44. IEEE Computer Society (1991)
14. Hemaspaandra, L.A., Vollmer, H.: The satanic notations: counting classes beyond #P and other definitional adventures. ACM SIGACT News **26**(1), 2–13 (1995)
15. Ivanashev, Y.: Low sets and closure properties of counting function classes. arXiv preprint arXiv:2507.04110 (2025)
16. Johnson, D.S., Yannakakis, M., Papadimitriou, C.H.: On generating all maximal independent sets. Inf. Process. Lett. **27**(3), 119–123 (1988)
17. Kiayias, A., Pagourtzis, A., Sharma, K., Zachos, S.: Acceptor-definable counting classes. In: Advances in Informatics: 8th Panhellenic Conference on Informatics, PCI 2001 Nicosia, Cyprus, 8–10 November 2001, Revised Selected Papers 8, pp. 453–463. Springer (2003)
18. Levin, L.: Universal sorting problems. Probl. Inform. Trammiss **9**, 265–266 (1973)
19. Ogiwara, M., Hemachandra, L.A.: A complexity theory for feasible closure properties. J. Comput. Syst. Sci. **46**(3), 295–325 (1993)
20. Pagourtzis, A., Zachos, S.: The complexity of counting functions with easy decision version. In: International Symposium on Mathematical Foundations of Computer Science, pp. 741–752. Springer (2006)
21. Simon, J.: On some central problems in computational complexity. Ph.D. thesis, Cornell University, Ithaca (1975)
22. Thierauf, T., Toda, S., Watanabe, O.: On closure properties of GapP. Comput. Complexity **4**, 242–261 (1994)
23. Valiant, L.G.: The complexity of computing the permanent. Theoret. Comput. Sci. **8**(2), 189–201 (1979)
24. Valiant, L.G.: The complexity of enumeration and reliability problems. SIAM J. Comput. **8**(3), 410–421 (1979)
25. Wagner, K.W.: The complexity of combinatorial problems with succinct input representation. Acta Inform. **23**, 325–356 (1986)

Fair k-Center Clustering with Minimum Representation Guarantees

Meiyun Lu and Longkun Guo(✉)

Fuzhou University, Fuzhou, China
lkguo@fzu.edu.cn

Abstract. We investigate the fair k-center clustering problem under the *Minimum Representation Fairness* (MR-fairness) criterion, which requires that a protected group constitutes at least half of the points in at least $\frac{k}{2}$ of the resulting clusters. This fairness notion is motivated by applications such as electoral districting and content recommendation, where group influence relies on achieving local majorities. We present a 9-approximation algorithm for the case of two-group data that guarantees this fairness constraint. Our approach integrates greedy center selection, all-or-nothing rounding, and a bipartite matching-based reassignment strategy to ensure both coverage quality and fair representation.

Keywords: Fair · k-center · Minimum representation · All-or-Nothing · Rounding

1 Introduction

Clustering is the basis of unsupervised learning, with the aim of dividing data into several clusters based on similarity. The k-center problem plays a crucial role in applications such as facility location, sensor placement, and emergency planning. The goal is to select k centers so that the maximum distance from any point to the nearest center is minimized. Classical algorithms, such as the 2-approximation by Gonzalez [23], Hochbaum and Shmoys [25], focus solely on geometric objectives, overlooking social and ethical considerations.

In recent years, fair clustering has received extensive attention, especially as it addresses the issues of bias and underrepresentation of protected groups in unsupervised learning outcomes [1,3,5–8,19,20,24,28,29,37]. In scenarios involving sensitive attributes such as race, gender or political affiliation, traditional clustering itself may lead to unfair treatment of vulnerable groups [9,34]. To mitigate such risks, fairness-aware clustering models have been developed, incorporating constraints such as balance [20], group bounded representation [6,24], individual fairness [33], and data summarization fairness [26,28].

In this work, we study the MR-fairness model, recently introduced by Lawless and Günlük [31], and integrate it into the k-center clustering paradigm. The MR-fairness model specifies that a protected group must constitute at least an

M. Li et al. (Eds.): TAMC 2025, LNCS 16084, pp. 25–36, 2026.
https://doi.org/10.1007/978-981-95-4839-2_3

α-fraction in at least β of the clusters, capturing representation-sensitive constraints without enforcing global balance.

Despite the expressiveness of MR-fairness, enforcing it under the k-center objective introduces significant computational challenges. Lawless and Günlük [31] prove that the problem is NP-hard even when center locations are fixed. To address this, we propose a novel algorithm that achieves a 9-approximation on the clustering objective while satisfying MR-fairness constraints via greedy center selection, an all-or-nothing assignment and rounding, and a bipartite matching strategy.

1.1 Related Work

Fair Clustering. Fair clustering has been studied under various fairness notions. Chierichetti et al. [20] proposed fairlets to enforce strict balance, requiring each cluster to reflect global group proportions. Later works [3,6,7,24] relaxed this to bounded representation fairness, allowing group ratios in clusters to lie within specified bounds.

Another line of work, motivated by fair data summarization, focuses on minimum representation guarantees, ensuring that each group is sufficiently represented among the selected centers or summaries [26,28]. This fairness model does not enforce intra-cluster balance, but rather group-level coverage, and is particularly suitable for summarization, recommendation, and public service scenarios where group visibility is essential.

MR-Fairness and Representation Constraints. The MR-fairness framework [31] captures realistic constraints by requiring that a protected group reach minimum representation in at least β clusters. This generalizes proportional fairness models [19] and group bounded representation [6,7,24], and aligns with applications like content diversity [22] and electoral fairness [35]. Celis et al. [11] discuss related fairness in top-k ranking and committee selection.

All-or-Nothing Supply Assignment. The all-or-nothing supply assignment strategy is a strategy where tasks are executed only if all conditions (or constraints) are fully met. This strategy plays a crucial role in various domains related to all-or-nothing supply assignment, such as the optimization of communication networks [4,12,32], as well as the all-or-nothing multi-commodity flow [12–15,27], the all-or-nothing generalized assignment problem [2,10,18,30], and the separable assignment problem (SAP) [21].

It is straightforward to verify that the all-or-nothing supply assignment problem is a special case of the all-or-nothing multi-commodity flow problem [14]. The problem is known to be NP-hard and APX-hard [16,36], even in the restricted setting of unit requirements and when the underlying graph is a tree [14,17].

1.2 Contribution

Our contributions are summarized as follows:

- We formalize the MR-fairness k-center problem for two protected groups.
- We develop a hybrid algorithm that combines greedy center selection, all-or-nothing rounding, and a bipartite matching-based reassignment technique to ensure fairness constraints are met. This approach balances theoretical guarantees with practical traceability.
- We achieve a constant-factor approximation in both fairness and clustering quality.

2 Preliminaries

In this section, we introduce the notation and formally define the problem setting. Let S be a finite set of n data points embedded in a metric space equipped with a distance function $d : S \times S \to \mathbb{R}_{\geq 0}$ that satisfies the triangle inequality. Given a parameter $k \in \mathbb{N}$, the classical **k-center clustering problem** seeks to select a subset $C \subseteq S$ of k centers such that the maximum distance from any point $s \in S$ to its nearest center in C is minimized:

$$\min_{C \subseteq S,\, |C|=k} \max_{s \in S} d(s, C), \quad \text{where } d(s, C) = \min_{c \in C} d(s, c).$$

The **fair k-center clustering problem** generalizes the standard formulation by incorporating fairness constraints based on sensitive group attributes. In this paper, we assume each point $s \in S$ belongs to one of two disjoint groups, encoded as colors: red or blue. Let $\chi : S \to \{\text{red}, \text{blue}\}$ denote the group membership function. For any subset $S' \subseteq S$, we define the red and blue subsets of S' as follows:

$$\text{red}(S'){:=}\{s \in S' \mid \chi(s) = \text{red}\}, \quad \text{blue}(S'){:=}\{s \in S' \mid \chi(s) = \text{blue}\}.$$

The fairness criterion we consider is called MR-Fairness, which is motivated by the idea that members of a protected group derive tangible benefits from a cluster only when they constitute a sufficiently large fraction of that cluster. We assume that red is the protected group requiring minimum representation.

Let $\mathcal{C} = \{C_1, \ldots, C_k\}$ be a clustering of S, and let α denote the minimum fraction required for the protected group to be consider red represented in a cluster. For any cluster $C_i \subseteq S$, we define the red and blue cardinalities as $\#\text{red}(C_i) = |\text{red}(C_i)|$ and $\#\text{blue}(C_i) = |\text{blue}(C_i)|$, respectively. We say that a cluster $C_i \in \mathcal{C}$ satisfies MR-Fairness (i.e., is red-majority) if the proportion of red points in C_i is at least α, that is,

$$\#\text{red}(C_i) \geq \alpha \cdot |C_i|.$$

Let $\Lambda(\mathcal{C}, \text{red}(S), \alpha)$ denote the number of red-majority clusters:

$$\Lambda(\mathcal{C}, \text{red}(S), \alpha) = \left|\{C_i \in \mathcal{C} \mid \#red(C_i) \geq \alpha|C_i|\}\right|.$$

We say that $\mathcal{C}$ is an (α, β)-**MR fair** clustering if

$$\Lambda(\mathcal{C}, \mathrm{red}(S), \alpha) \geq \beta,$$

where $\beta \in \mathbb{Z}_+$ specifies the minimum number of clusters in which group red must be fairly represented.

The **MR-fairness k-center problem** can now be stated as:

$$\min_{\mathcal{C}, c_1, \ldots, c_k} \max_{s_i \in C_j, j \in [k]} d(s_i, c_j)$$
$$\text{s.t.} \quad \Lambda(\mathcal{C}, \mathrm{red}(S), \alpha) \geq \beta.$$

In this paper, we study the $(\frac{1}{2}, \frac{k}{2})$-MR fair k-center clustering problem.

3 Algorithm

In this section, we present an approximation algorithm for the MR-fairness k-center clustering problem with two protected groups. The algorithm performs a binary search to find the minimum clustering radius that satisfies fairness constraints. For each radius guess, it constructs candidate centers and refines assignments through greedy selection, all-or-nothing supply rounding, and bipartite matching. The goal is to ensure both compactness and fairness by guaranteeing sufficient representation of each protected group. The full procedure is described in Algorithm 1.

3.1 Venn Diagram Regions and Joiners Regions

We begin by introducing geometric constructs that are used to describe the structure of cluster coverage regions and their overlaps in our algorithm.

Definition 1 (Venn Diagram Region). *Given a set C of k selected cluster centers and a radius parameter $\lambda > 0$, the λ-Venn diagram region induced by C is defined as:*

$$\mathcal{B} = \bigcup_{c \in C} B(c, \lambda),$$

where $B(c, \lambda)$ denotes the closed ball of radius λ centered at c.

We now define the concept of a joiner, which corresponds to the region of intersection of balls from a subset of centers while explicitly excluding overlaps with the rest.

Definition 2 (Joiner Region). *Let $C' \subseteq C$ be a nonempty subset of centers. The joiner region corresponding to C' is defined as:*

$$J_{C', \lambda} = \left(\bigcap_{c \in C'} B(c, \lambda) \right) \cap \left(\bigcap_{c \in C \setminus C'} \overline{B(c, \lambda)} \right),$$

where $\overline{B(c, \lambda)}$ denotes the complement of the ball $B(c, \lambda)$. We say a point $c \in \mathbb{R}^d$ belongs to the joiner $J_{C', \lambda}$ if $c \in J_{C', \lambda}$. A joiner is said to be empty if it contains no data points.

Algorithm 1 MR-fair k-center Algorithm

Input: Point set $S = \text{red}(S) \cup \text{blue}(S)$, number of clusters k, small convergence threshold $\epsilon > 0$.
Output: A clustering with at least $\frac{k}{2}$ red-majority clusters.

Initialize search bounds: $l = 0$, $r = 2D$, where $D = \max_{x,y \in S} d(x, y)$
while $r - l > \epsilon$ or the feasible radius λ has not been found **do**
  $\lambda = \frac{l+r}{2}$
  Select k centers C via greedy selection and ball $B(c, 3\lambda)$- supplementation
  Classify centers into P_R and P_B based on disjoint region majority, and identify redeemable centers $V \subseteq P_B$, compute their demands
  if $|P_B \setminus V| > \frac{k}{2}$ **then**
    $l = \lambda$ and **continue**
  end if
  Construct bipartite graph $G(U, V, E)$ from overlapping regions
  Solve LP for red supply to satisfy demands in V, and $Z \subseteq V$ denotes the set of demand points fully satisfied in a basic optimal solution
  if $\text{OPT}_{LP} + |P_R| < \frac{k}{2}$ **then**
    $l = \lambda$ and **continue**
  end if
  Perform All-or-Nothing-Rounding algorithm on $V \setminus Z$ as described in Sect. 3.3 and Algorithm 2
  Match remaining red and blue Points
  Reassign blue points and redistribute red points
  if number of red-majority clusters $\geq \frac{k}{2}$ **then**
    Record current clustering and $r = \lambda$
  else
    $l = \lambda$
  end if
end while
return A center set C and a best clustering

3.2 Description of the Algorithm

We begin with a binary search over the radius parameter λ in the range $[0, 2D]$, where D is the maximum pairwise distance in the dataset, to identify the smallest feasible value that allows formation of $\frac{k}{2}$ red-majority clusters. For each candidate radius, the algorithm proceeds through the following key steps:

Stage 1: Greedy Center Selection. The algorithm starts by arbitrarily choosing an initial center. At each subsequent step, it selects a point lying outside the 2λ-balls $B(c, 2\lambda)$ of all previously chosen centers $c \in C$ that is farthest from them. This process continues until no such point can be found.

If $|C| > k$, the guessed radius λ is too small, and a larger value should be tried. If $|C| < k$, the algorithm supplements the center set by selecting additional centers from the region $N_{\text{red}} = \bigcup_{c \in C_{\text{red}}} B(c, 3\lambda)$, where $C_{\text{red}} \subseteq C$ is the subset of current centers whose 3λ-balls contain at least one red point. At each step, the algorithm selects a point $i \in N_{\text{red}}$ that maximizes the objective $\sum_{j \in \text{red}(B(i,3\lambda))} \frac{1}{n_j}$, subject to the constraint that this sum is at least 1. Here,

$n_j = |\{c \in C \mid j \in B(c, 3\lambda)\}|$ denotes the number of current centers covering red point j.

This process continues until no such point remains in N_{red} satisfying the constraint. If $|C| < k$ after exhausting N_{red}, the remaining centers are selected one by one from the remaining points in $S \setminus C$, each time choosing the point that is farthest from all current centers in C, until $|C| = k$.

Stage 2: Ball Construction and Classification. For each selected center $c \in C$, construct a ball $B(c, 3\lambda)$ and compute the red and blue point counts in its disjoint region: $\#\text{red}(J_{\{c\},3\lambda})$ and $\#\text{blue}(J_{\{c\},3\lambda})$. If $\#\text{red}(J_{\{c\},3\lambda}) \geq \#\text{blue}(J_{\{c\},3\lambda})$, assign c to the red-majority set P_R; otherwise, assign it to the blue-majority set P_B. For each center $c \in P_B$, define its demand value as $\#\text{blue}(J_{\{c\},3\lambda}) - \#\text{red}(J_{\{c\},3\lambda})$. Check whether the number of red points in its overlapping regions is at least the demand value. If so, c is considered redeemable and added to the demand set V, with its demand value.

If the number of non-redeemable centers in P_B (i.e., $P_B \setminus V$) is strictly greater than $\frac{k}{2}$, then the current radius λ is too small to satisfy the fairness constraint. In this case, increase λ and restart the algorithm. Otherwise, represent each overlapping region corresponding to centers in V as a single point to form the supply set U. Each point in U has a supply equal to the number of red points in its overlapping region. The supply set U and demand set V will be used in the subsequent supply-demand matching and allocation steps.

Stage 3: Bipartite Graph Construction. Construct a bipartite graph $G(U, V, E)$. An edge $(u, v) \in E$ exists if the supply point $u \in U$ corresponds to an overlapping region that intersects with the demand point $v \in V$.

Stage 4: LP Relaxation and All-or-Nothing Supply Assignment. We formulate and solve the supply-demand LP relaxation on the bipartite graph $G(U, V, E)$, which models the assignment of supply to meet the demands. We denote the optimal value of LP as OPT_{LP}. The detailed formulation of LP will be presented in Sect. 3.3.

If $\text{OPT}_{LP} + |P_R| < \frac{k}{2}$, it indicates that the current radius λ is too small to satisfy the fairness constraint, and the algorithm should increase λ and restart the search. Otherwise, include in Z all demand points fully served in the basic optimal solution, and update residual supplies. Then, apply the All-Or-Nothing-Rounding subroutine on the residual graph to obtain an assignment that fully satisfies the demands of remaing centers. The details of this subroutine are presented in Sect. 3.3.

Stage 5: Matching Remaining Red and Blue Points. We construct a bipartite graph whose two partitions consist of red and blue points, respectively. For each center $i \in Z \cup Z'$, we begin by removing red points that were already assigned during Stage 4 from the overlapping region $B(i, 3\lambda)$. The remaining unassigned red and blue points within this region are included in the bipartite graph.

Next, for each center $i \in P_R$, we examine its disjoint region $J_{\{i\},3\lambda}$. If the number of red points exceeds the number of blue points in this region, i.e.,

$\#\text{red}(J_{\{i\},3\lambda}) - \#\text{blue}(J_{\{i\},3\lambda}) > 0$, then only $\#\text{blue}(J_{\{i\},3\lambda})$ red points are retained and assigned to i, while the surplus red points are added to the bipartite graph.

An edge is added between a red point and a blue point if there exists a center whose ball of radius 3λ contains both points. After collecting all red and blue points and constructing the edges as described, we compute a maximum matching in the resulting bipartite graph.

Stage 6: Reallocation of Red and Blue Points. We say that two centers c and c' overlap if their corresponding 3λ-balls intersect geometrically—that is, if the regions covered by the two balls geometrically intersect. This notion of overlap is purely geometric and does not depend on whether the two balls contain any common data points. Based on this definition, we construct the center-overlap graph $G = (C, E)$, where each node represents a center and an edge connects each pair of overlapping centers.

We assign a tentative majority color (red or blue) to each center to indicate the majority color of the cluster associated with it, according to the current matching results. That is, the color label reflects whether the center's corresponding cluster is expected to be red-majority or blue-majority. Specifically, all centers in the set $P_B \setminus Z \cup Z'$ are initially labeled as blue. For each remaining center, if there exists at least one unmatched blue point within its 3λ-ball, it is also labeled as blue; otherwise, it is labeled as red.

We count the number of red centers. If this number is at least $\frac{k}{2}$, we proceed to the next stage. Otherwise, we perform a greedy reallocation procedure to adjust the color labels:

For each center c_i, the score Δ_i is given by

$$\Delta_i = |\mathcal{N}_B^{\text{unproc}}(c_i)| + \alpha_i + \mathbb{1}_{\{c_i \in B\}}.$$

Here,

- $\mathcal{N}_B^{\text{unproc}}(c_i)$ is the set of unprocessed blue neighbor centers of c_i,
- $\mathbb{1}_{\{c_i \in B\}}$ is the indicator function that equals 1 if c_i is blue, and 0 otherwise,
- α_i equals 1 if there exists a neighbor that absorbs the blue points within the 3λ-radius ball centered at c_i.

At each iteration, select the center c_i with the highest score Δ_i. Then, c_i absorbs all blue points located within the 3λ-radius balls of its blue neighbor centers, excluding the neighbors that absorbed the blue points inside $B(c_i, 3\lambda)$, including the blue points that have been absorbed from c_i's 3λ-radius balls . After this absorption step, update the scores Δ_i accordingly.

Repeat this procedure until all centers satisfy $\Delta_i \leq 1$ and every blue point with one or more neighbors has at least one red neighbor. If the blue point remains without a red neighbor, redistribute all blue points within its 3λ-radius ball among its blue neighbors.

Finally, compute the total number of red points.

Stage 7: Fairness Check. If the number of red-majority clusters is at least $\frac{k}{2}$, the search proceeds to find a smaller valid λ. Otherwise, a larger λ is considered.

3.3 All-or-Nothing Supply Assignment and Rounding

In the All-or-Nothing supply assignment (AoNSA) problem, we are given a bipartite graph $G = (U, V, E)$. Each supply point $i \in U$ has a supply capacity $p_i \in \mathbb{Z}_0^+$, and each $j \in V$ has a demand $q_j \in \mathbb{Z}_0^+$. For each $w \in U \cup V$, let $N_G(w)$ be the set of its neighbors in G, i.e., $N_G(w) = \{w' \in\ U \cup V \mid (w, w') \in E\}$.

Let $x_{ij} \in \mathbb{Z}_{\geq 0}$ denote the amount of supply allocated from supply point i to demand point j for each edge $(i, j) \in E$. A binary variable $y_j \in \{0, 1\}$ is introduced for each j, where $y_j = 1$ indicates that j receives its full demand, and $y_j = 0$ otherwise. The objective of the AoNSA problem is to find a feasible assignment that maximizes the number of fully served demand points, i.e., $\sum_{j \in V} y_j$. The natural LP relaxation is given as follows:

$$\begin{aligned}
\max\ & \textstyle\sum_{j \in V} y_j \\
\text{s.t.}\ & \textstyle\sum_{j \in N_G(i)} x_{ij} \leq p_i, \quad \forall i \in U, \\
& \textstyle\sum_{i \in N_G(j)} x_{ij} = q_j y_j, \quad \forall j \in V, \\
& x_{ij} \geq 0, \quad \forall (i, j) \in E, \\
& 0 \leq y_j \leq 1, \quad \forall j \in V.
\end{aligned}$$

Given a basic optimal solution $\chi = (\boldsymbol{x}^*, \boldsymbol{y}^*)$, we define the basis graph and the full demand point as follows:

Definition 3 *(Basis Graph). The basis graph of G corresponding to χ is denoted by G_χ, which consists of the edges (i, j) such that $x_{ij}^* > 0$.*

Definition 4 *(Full Demand Point). A demand point $j \in V \cap G_\chi$ is called a full demand point if and only if $y_j^* = 1$. Let V_{full} denote the set of full demand points in G_χ.*

After solving the LP of $G(U, V, E)$ to obtain the optimal solution $(\boldsymbol{x}^*, \boldsymbol{y}^*)$, we add the demand points where $y_j^* = 1$ to the routable set Z. Next, we construct a residual graph G_{res} and perform the rounding on it. The residual graph G_{res} includes all demand points j with $y_j^* \in [0, 1)$ as V_{res}, together with their corresponding supply points as U_{res} and the associated edges as E_{res}. For each $i \in U_{res}$, the supply p_i is reduced by the total supply allocated from i to the full demand points V_{full}.

Upon constructing the residual graph G_{res}, we aim to maximize the number of satisfied demand points by reallocating the available supplies to round the demands in V_{res}. The detailed procedure is presented in Algorithm 2.

3.4 Theoretical Guarantees

We will now proceed to formally prove the approximation bound and performance of this algorithm in the following sections. This proof will demonstrate that the algorithm produces results that are within a constant factor of the optimal solution, thereby providing an effective approach to the MR-fairness k-center problem.

Algorithm 2 All-or-Nothing-Rounding Algorithm

Input: A residual graph G_{res} with supply and demand.
Output: A routable set Z'.

$J = V_{res}$, $Z' = \emptyset$
while $J \neq \emptyset$ **do**
  Select one demand point j in J
  $Z' = Z' \cup \{j\}$
  $(\boldsymbol{x}, \boldsymbol{y}) =$ solve LP(1) of G_{res} with $\{y_j = 1 \mid j \in Z'\}$
  if LP has a feasible solution **then**
    $J \leftarrow J \setminus \{j\}$
  else
    $J \leftarrow J \setminus \{j\}, Z' \leftarrow Z' \setminus \{j\}$
  end if
end while
return Z'

Lemma 1 *When the radius is at least $3R^*$, the set of centers C selected by our algorithm (as described in Sect. 3.2) ensures that every optimal cluster is entirely contained within at least one ball $B(c, 3R^*)$ for some $c \in C$, where R^* denotes the optimal radius of the instance.*

By leveraging Theorem 1 proved in [24], Lemma 1 can be obtained straightforwardly.

Lemma 2 *When the radius $\lambda \geq R^*$, we have*

$$|P_R| + \mathsf{OPT}_{LP} \geq \frac{k}{2},$$

where OPT_{LP} denotes the optimal value of LP(1) in Stage 4.

The proof is deferred to the Appendix. The key idea is to classify the optimal clusters based on their color majority and their intersection size with the selected center set C. By carefully counting how these clusters contribute either to the set P_R or to the LP objective through fully satisfied demand, we can show that the total contribution is at least $k/2$.

Theorem 1 *The MR-fairness k-center algorithm described in Sects. 3.2 is a 9-approximation algorithm.*

The proof of the theorem can be readily derived from Lemma 2 and the algorithm. The detailed proof is deferred to the Appendix.

4 Conclusion

This work advances the state-of-the-art of fair clustering by integrating MR-fairness into the k-center framework for two protected groups with equal representation. Our hybrid algorithm effectively bridges the gap between strong theoretical guarantees and practical scalability, leveraging key techniques such as the all-or-nothing supply assignment and rounding strategy to ensure fairness and efficiency. The MR-fairness notion addresses the critical challenge of balanced representation in clustering, which is essential for equitable decision-making in high-stakes applications. Future directions include extending our approach to handle variable representation ratios beyond the fixed $\frac{1}{2}$ setting, accommodating multiple protected groups, and adapting to dynamic data streams, thereby broadening the practical applicability and impact of fair clustering methodologies.

References

1. Abbasi, M., Bhaskara, A., Venkatasubramanian, S.: Fair clustering via equitable group representations. In: Proceedings of the 2021 ACM Conference on Fairness, Accountability, and Transparency, pp. 504–514 (2021)
2. Adany, R., et al.: All-or-nothing generalized assignment with application to scheduling advertising campaigns. ACM Trans. Algorithms (TALG) **12**(3), 1–25 (2016)
3. Ahmadian, S., Epasto, A., Kumar, R., Mahdian, M.: Clustering without over-representation. In: Proceedings of the 25th ACM SIGKDD International Conference on Knowledge Discovery & Data Mining, pp. 267–275 (2019)
4. Amzallag, D., Bar-Yehuda, R., Raz, D., Scalosub, G.: Cell selection in 4G cellular networks. IEEE Trans. Mob. Comput. **12**(7), 1443–1455 (2012)
5. Backurs, A., Indyk, P., Onak, K., Schieber, B., Vakilian, A., Wagner, T.: Scalable fair clustering. In: International Conference on Machine Learning, pp. 405–413 (2019). arXiv:1902.03519
6. Bera, S., Chakrabarty, D., Flores, N., Negahbani, M., Sidiropoulos, A., Svensson, O.: Fair algorithms for clustering. In: Advances in Neural Information Processing Systems, vol. 32 (2019). arXiv:1905.12777
7. Bercea, I.O., Groß, M., Khuller, S., Kumar, A., Rösner, C., Schmidt, D.R.: On the cost of essentially fair clusterings. arXiv preprint arXiv:1811.10319 (2019)
8. Brubach, B., Chakrabarti, D., Dickerson, J.P., Khuller, S., Srinivasan, A., Tsepenekas, L.: A pairwise fair and community-preserving approach to k-center clustering. In: International Conference on Machine Learning, pp. 1134–1144 (2020). arXiv:2007.07384
9. Buolamwini, J., Gebru, T.: Gender shades: intersectional accuracy disparities in commercial gender classification. In: Conference on Fairness, Accountability and Transparency, pp. 77–91. PMLR (2018)
10. Castillo-Zunino, F., Keskinocak, P.: Bi-criteria multiple knapsack problem with grouped items. J. Heuristics **27**(5), 747–789 (2021). https://doi.org/10.1007/s10732-021-09476-y
11. Celis, L.E., Stoyanovich, J., Vishnoi, N.K.: Ranking with fairness constraints. In: Proceedings of the 2018 Conference on Fairness, Accountability and Transparency, pp. 2–12 (2018). https://dl.acm.org/doi/10.1145/3287560.3287563
12. Chaturvedi, A., Chekuri, C., Richa, A.W., Rost, M., Schmid, S., Weber, J.: Improved throughput for all-or-nothing multicommodity flows with arbitrary demands. ACM SIGMETRICS Perform. Eval. Rev. **49**(3), 22–27 (2022)
13. Chekuri, C., Ene, A.: The all-or-nothing flow problem in directed graphs with symmetric demand pairs. Math. Program. **154**, 249–272 (2015)

14. Chekuri, C., Khanna, S., Shepherd, F.B.: The all-or-nothing multicommodity flow problem. In: Proceedings of the Thirty-Sixth Annual ACM Symposium on Theory of Computing, pp. 156–165 (2004)
15. Chekuri, C., Khanna, S., Shepherd, F.B.: Multicommodity flow, well-linked terminals, and routing problems. In: Proceedings of the Thirty-Seventh Annual ACM Symposium on Theory of Computing, pp. 183–192 (2005)
16. Chekuri, C., Khanna, S., Shepherd, F.B.: An $\mathcal{O}(\sqrt{n})$ approximation and integrality gap for disjoint paths and unsplittable flow. Theory Comput. **2**(1), 137–146 (2006)
17. Chekuri, C., Mydlarz, M., Shepherd, F.B.: Multicommodity demand flow in a tree and packing integer programs. ACM Trans. Algorithms (TALG) **3**(3), 27–es (2007)
18. Chen, L., Zhang, G.: Packing groups of items into multiple knapsacks. ACM Trans. Algorithms (TALG) **14**(4), 1–24 (2018)
19. Chen, X., Fain, B., Lyu, L., Munagala, K.: Proportionally fair clustering. In: International Conference on Machine Learning, pp. 1032–1041. PMLR (2019)
20. Chierichetti, F., Kumar, R., Lattanzi, S., Vassilvitskii, S.: Fair clustering through fairlets. In: Advances in Neural Information Processing Systems, vol. 30 (2017). arXiv:1802.05733
21. Fleischer, L., Goemans, M.X., Mirrokni, V.S., Sviridenko, M.: Tight approximation algorithms for maximum general assignment problems. In: SODA, vol. 6, pp. 611–620. Citeseer (2006)
22. Geyik, S.C., Ambler, S., Kenthapadi, K.: Fairness-aware ranking in search & recommendation systems with application to LinkedIn talent search. In: Proceedings of the 25th ACM SIGKDD International Conference on Knowledge Discovery & Data Mining, pp. 2221–2231 (2019). https://dl.acm.org/doi/10.1145/3292500.3330691
23. Gonzalez, T.F.: Clustering to minimize the maximum intercluster distance. Theor. Comput. Sci. **38**, 293–306 (1985). https://www.sciencedirect.com/science/article/pii/0304397585902245
24. Harb, E., Lam, H.S.: KFC: a scalable approximation algorithm for k- center fair clustering. Adv. Neural. Inf. Process. Syst. **33**, 14509–14519 (2020)
25. Hochbaum, D.S., Shmoys, D.B.: A best possible heuristic for the k-center problem. Math. Oper. Res. **10**(2), 180–184 (1985)
26. Jones, M., Nguyen, H., Nguyen, T.: Fair k-centers via maximum matching. In: International Conference on Machine Learning, pp. 4940–4949. PMLR (2020)
27. Kawarabayashi, K., Kobayashi, Y.: All-or-nothing multicommodity flow problem with bounded fractionality in planar graphs. In: 2013 IEEE 54th Annual Symposium on Foundations of Computer Science, pp. 187–196. IEEE (2013)
28. Kleindessner, M., Awasthi, P., Morgenstern, J.: Fair k-center clustering for data summarization. In: International Conference on Machine Learning, pp. 3448–3457. PMLR (2019)
29. Kleindessner, M., Samadi, S., Awasthi, P., Morgenstern, J.: Guarantees for spectral clustering with fairness constraints. In: International Conference on Machine Learning, pp. 3458–3467 (2019). arXiv:1906.08384
30. Kulik, A., Sarpatwar, K., Schieber, B., Shachnai, H.: Generalized assignment via submodular optimization with reserved capacity. In: 27th Annual European Symposium on Algorithms (ESA 2019). Schloss-Dagstuhl-Leibniz Zentrum für Informatik (2019)
31. Lawless, C., Günlük, O.: Fair minimum representation clustering via integer programming. arXiv preprint arXiv:2409.02963 (2024)
32. Liu, M., Richa, A.W., Rost, M., Schmid, S.: A constant approximation for maximum throughput multicommodity routing and its application to delay-tolerant

network scheduling. In: IEEE INFOCOM 2019-IEEE Conference on Computer Communications, pp. 46–54. IEEE (2019)
33. Mahabadi, S., Vakilian, A.: Individual fairness for k-clustering. In: International Conference on Machine Learning, pp. 6586–6596. PMLR (2020)
34. Mehrabi, N., Morstatter, F., Saxena, N., Lerman, K., Galstyan, A.: A survey on bias and fairness in machine learning. ACM Comput. Surv. (CSUR) **54**(6), 1–35 (2021)
35. Shrestha, Y.R., Yang, Y.: Fairness in algorithmic decision-making: applications in multi-winner voting, machine learning, and recommender systems. Algorithms **12**(9), 199 (2019)
36. Varadarajan, K., Venkataraman, G.: Graph decomposition and a greedy algorithm for edge-disjoint paths. In: Proceedings of the Fifteenth Annual ACM-SIAM Symposium on Discrete Algorithms, pp. 379–380 (2004)
37. Wu, X., Feng, Q., Huang, Z., Xu, J., Wang, J.: New algorithms for distributed fair k-center clustering: almost accurate as sequential algorithms. In: Proceedings of the 23rd International Conference on Autonomous Agents and Multiagent Systems, pp. 1938–1946 (2024)

Models of Computation

Approximate EFX and Efficient Allocations of Chores for Bounded Preferences

Hui Liu and Zhijie Zhang(✉)

Center for Applied Mathematics of Fujian Province, School of Mathematics and Statistics, Fuzhou University, Fuzhou, China
{230320053,zzhang}@fzu.edu.cn

Abstract. We study the problem of fairly allocating a set of indivisible chores among a set of agents, who possess heterogeneous costs over the chores. We assume that agents' costs are bounded and lie in the interval $[1, k]$. We aim to find both fair and efficient allocations under this setting. We first show that a k-EFX and fPO allocation can be computed in polynomial time. Next, we generalize the result to the weighted setting and show the existence of a k-WEFX and fPO allocation. Finally, for $k = 2$, we can compute an EFX allocation based on the well-known round-robin algorithm.

Keywords: Fair allocation · EFX · Pareto optimality

1 Introduction

Fair allocation of *indivisible chores* has recently garnered significant interest across multiple communities, including computer science, economics, and mathematics. The problem involves fairly distributing a set M of m indivisible chores among a set N of n agents, where agents possess heterogeneous costs over the chores. An *allocation* of the chores is an n-partition $X = (X_1, X_2, \ldots, X_n)$ of M, with bundle X_i assigned to agent i. The allocation is required to be fair according to some criterion.

In the literature, perhaps the most natural fairness notion is *envy-freeness* (EF) [8]. An allocation is envy-free if no agent prefers another agent's bundle over their own, i.e., $c_i(X_i) \leq c_i(X_j)$ for all $i \neq j$, where c_i is the cost function of agent i. Unfortunately, envy-free allocations are generally impossible to achieve for indivisible chores, e.g., distributing a single chore between two agents. This limitation has led researchers to focus on relaxed notions of envy-freeness. Two widely studied relaxations are *envy-freeness up to one item* (EF1) [12] and *envy-freeness up to any item* (EFX) [5]. Specifically, an EF1 allocation requires that the envy of an arbitrary agent could be eliminated by removing *some* chore from her bundle, whereas an EFX allocation requires that the envy should be eliminated after removing *any* chore from the agent's bundle. EFX is a stricter

M. Li et al. (Eds.): TAMC 2025, LNCS 16084, pp. 39–50, 2026.
https://doi.org/10.1007/978-981-95-4839-2_4

fairness guarantee than EF1. It is well-known that EF1 allocations exist for indivisible chores [3,4]. In contrast, whether EFX allocations exist remains a central open problem in the field. Researchers henceforth consider approximate EFX and an $O(n^2)$-EFX allocation was first found in [19]. The latest result showed that a 4-EFX allocation always exists [10].

Weighted Setting. In real-world scenarios, agents often undertake unequal obligations and possess distinct entitlements. To model this phenomenon, it is assumed in [6,11] that each agent $i \in N$ is associated with a weight $w_i > 0$ such that $\sum_{i \in N} w_i = 1$. In this weighted setting, all the aforementioned fairness notions transfer to the weighted versions. It was shown in [17,18] that WEF1 allocations always exist and can be computed in polynomial time. However, WEFX allocations may not exist [17].

Efficiency. In addition to fairness, researchers are also concerned with the efficiency of an allocation and are interested in finding both fair and efficient allocations. The common efficiency metric for chores allocation is *Pareto optimality* (PO) and *fractionally Pareto optimality* (fPO). An allocation is PO (fPO) if no alternative (fractional) allocation can reduce the cost for at least one agent without increasing the costs for the others. For chores allocation, it remains a major open problem whether EF1 and PO allocations exist. The latest results make progress toward this problem from two aspects. First, it was shown that $(n-1)$-EF1 and fPO allocations always exist [10]. Second, when considering bi-valued instances, where the cost of a chore is either 1 or k (with $k > 1$) to any agent, Garg et al. [9] and Ebadian et al. [7] show that EF1 and PO allocations exist and can be computed in polynomial time. This result was later generalized in [18] to show that WEF1 and PO allocations can also be computed in polynomial time.

1.1 Our Contributions

In this work, we consider instances where agents have bounded cost functions, that is, the cost of a chore to any agent lies within the interval $[1, k]$. Such instances are called $[1, k]$-instances and they generalize the bi-valued $\{1, k\}$-instances. We argue that this setting makes sense in reality, since a certain degree of consensus can generally be reached, despite differing individual preferences. We make progress in this setting in the following ways:

1. We present a polynomial time algorithm for computing a k-EFX and fPO allocation for $[1, k]$-instances.
2. We generalize the algorithm to show that k-WEFX and fPO allocations can also be computed for $[1, k]$-instances in polynomial time.
3. For $[1, 2]$-instances, we show that an EFX allocation can be computed in polynomial time.

1.2 Other Related Works

Fair allocation of goods (items with positive utilities to agents) is another popular topic in the literature. We thus list some of its results for comparison. Unlike chore division, Caragiannis et al. [5] showed that EF1 and PO allocations for goods always exist. Specifically, it was shown that the allocation that maximizes Nash social welfare is both EF1 and PO. Murhekar and Garg [14] extended the result by proposing a pseudo-polynomial time algorithm to compute an EF1 and fPO allocation. Murhekar and Garg [14] also considered the class of k-ary instances where each agent has at most k different values for goods. They show that for such instances, an EF1 and fPO allocation can be computed in polynomial time when k is constant. For the weighted setting, Bhaskar et al. [4] proved that WEF1 and PO allocations always exist.

For EFX fairness, it might be the most important open problem in the field to determine whether an EFX allocation exists. Previously, Amanatidis et al. [2] provided a polynomial time algorithm that returns a 0.618-EFX allocation, significantly improving upon an earlier result of 1/2-EFX guarantee [16]. Besides, a recent work [1] demonstrates that 2/3-EFX allocations can be achieved when there are at most seven agents. Neoh and Teh [15] recently showed that WEFX and PO allocations can be computed in polynomial time for binary valuations, where goods take values from $\{0, 1\}$.

2 Preliminaries

The fair allocation of chores aims to fairly distribute a set M of m indivisible chores among a set N of n agents. Each agent $i \in N$ has a non-negative *cost* function $c_i : 2^M \to \mathbb{R}_{\geq 0}$ that describes the agent's preferences over the chores. The cost functions are *additive*, that is, for any agent $i \in N$ and *bundle* $X \subseteq M$, $c_i(X) = \sum_{g \in X} c_i(g)$, where $c_i(g)$ denotes agent i's cost for chore g. In this paper, we further assume that the cost functions are bounded:

Definition 1. *The cost functions are k-bounded if for each $i \in N$ and $g \in M$, $c_i(g) \in [1, k_i]$ and $\max_{i \in N} k_i = k$. An instance of chores allocation with k-bounded cost functions is called $[1, k]$-instance.*

Recall that an n-partition $X = (X_1, X_2, \dots X_n)$ of a set M satisfies $X_i \cap X_j = \emptyset$ for all $i \neq j$ and $\cup_i X_i = M$. An *allocation* is an n-partition of the chores, with X_i the bundle assigned to agent i. We use the following concept to measure the fairness of an allocation.

Definition 2 (α-EFX). *An allocation X is said to be α-envy-free up to any chore (α-EFX) for some $\alpha \geq 1$ if for all $i, h \in N$ and $g \in X_i$,*

$$c_i(X_i \setminus \{g\}) \leq \alpha \cdot c_i(X_h).$$

Specifically, if $\alpha = 1$, we say that the allocation is EFX.

We also consider a weighted version of EFX fairness. Each agent i is also associated with a weight $w_i > 0$, reflecting her share of obligations. The weights are normalized so that $\sum_{i \in N} w_i = 1$. We introduce the following notion of fairness.

Definition 3 (α-WEFX). *An allocation X is α-weighted envy-free up to any chore (α-WEFX) if for all $i, h \in N$ and $g \in X_i$,*

$$\frac{c_i(X_i \setminus \{g\})}{w_i} \leq \alpha \cdot \frac{c_i(X_h)}{w_h}.$$

Note that when $w_i = 1/n$ for all $i \in N$, WEFX reduces to EFX.

Finally, we introduce the Pareto optimality (PO), a well-known criterion for efficiency.

Definition 4 (Pareto optimality (PO)). *An allocation X is* Pareto optimal *if there exists no allocation Y that dominates X. An allocation Y* dominates *allocation X if $c_i(Y_i) \leq c_i(X_i)$ for all $i \in N$, and there exists $h \in N$ such that $c_h(Y_h) < c_h(X_h)$. An allocation X is fractionally Pareto-optimal (fPO) if there is no fractional allocation that dominates X. An fPO allocation is PO, but not vice-versa.*

Fisher Market. In the Fisher market, there is a price vector $\mathbf{p}$ that assigns each chore $g \in M$ a positive price $p(g) > 0$. Given the price vector $\mathbf{p}$, we define the pain-per-buck ratio α_{ig} of chores g as $\alpha_{ig} = \frac{c_i(g)}{p_g}$. The minimum-pain-per-buck (MPB) ratio of agent i is given by $\alpha_i = \min_{g \in M} \alpha_{ig}$. For each agent i,we define $\text{MPB}_i = \{g \in M \mid \alpha_{ig} = \alpha_i\}$ as the set of chores with the minimum pain-per-buck ratio for agent i. An allocation $(X, \mathbf{p})$ is called an MPB allocation if for all $i \in N$ and $g \in M$, $X_{ig} > 0 \implies g \in \text{MPB}_i$, i.e. , agents are only assigned chores which are MPB for them. An MPB allocation forms a *Fisher market equilibrium* $(X, \mathbf{p})$.

Following the First Welfare Theorem [13], for any market equilibrium (X, p), the allocation X is PO. For completeness, we provide a brief proof here.

Lemma 1. *Let $(X, \mathbf{p})$ be an MPB allocation. Then X is fPO.*

Proof. Suppose the allocation X with price vector $\mathbf{p}$ constitutes a competitive equilibrium. Then, for any agents $i \neq j$ and chore $g \in X_i$, we have

$$\frac{c_i(g)}{\alpha_i} = p(g) = \frac{c_j(g)}{\alpha_{jg}} \leq \frac{c_j(g)}{\alpha_j}. \tag{1}$$

The allocation minimizes the objective$\sum_{i \in N} \frac{c_i(X_i)}{\alpha_i}$. Any Pareto improvement would strictly decrease this objective, which leads to a contradiction. Thus, the allocation X is PO.

3 Unweighted EFX Allocations

We first deal with the unweighted agents. In Sect. 3.1, we provide a polynomial time algorithm which computes a k-EFX and fPO allocation for $[1, k]$-instances. In Sect. 3.2, we investigate whether exact EFX can be achieved and show that for $[1, 2]$-instances, an EFX allocation can be computed in polynomial time.

3.1 k-EFX and fPO Allocation for $[1, k]$-Instances

In this section, we show that when agents' cost functions are k-bounded, a k-EFX and fPO allocation of chores can be computed in polynomial time. The key idea is to generate a *balanced* MPB allocation. An allocation is balanced if and only if the difference between the sizes of any two agent bundles is at most one. As we show below, such an allocation is k-EFX. Besides, an MPB allocation is fPO by Lemma 1. We maintain the MPB property by the following tool.

Definition 5 (MPB graph). *The augmented MPB graph* $G = (N, M, E)$ *associated with an integral MPB allocation* $(X, \mathbf{p})$ *is a directed bipartite graph with vertex set* $V(G) = N \cup M$*, and edge set* $E(G) = \{(i, g) : i \in N, g \in M, g \in X_i\} \cup \{(g, i) : i \in N, g \in M, g \in \mathrm{MPB}_i \setminus X_i\}$.

Our algorithm is inspired by [10], which is depicted as Algorithm 1. We now describe it in detail.

Algorithm 1 begins by assigning the complete set of chores M to an arbitrarily chosen agent h. The initial payments are set as $p_g = c_h(g)$ for all $g \in M$, ensuring the allocation satisfies the MPB condition. To balance the number of chores assigned to each agent, the algorithm iteratively redistributes chores from the overloaded agent h (holding the maximum number of chores) to the underloaded agent ℓ (holding the minimum number of chores). Let C_h denote the set of vertices reachable from h in the MPB graph. We consider two possible cases.

Case 1 ($\ell \in C_h$) (Chore-Transfer). We find a path P from h to ℓ. Assuming $P = (h = i_0, g_1, i_1, \ldots, i_{k-1}, g_k, i_k = \ell)$, we then transfer chore g_r from agent i_{r-1} to i_r for all $r \in [k]$. This adjustment reduces the number of chores of agent h by 1, increases that of agent ℓ by 1, and preserves that of the other agents. Since $j_r \in \mathrm{MPB}_{i_r}$ for all $r \in [k]$, the allocation remains MPB after the transfer.

Case 2 ($\ell \notin C_h$) (Payment-Raise). We uniformly increase payments for all chores in C_h until at least one chore $g \in C_h$ becomes MPB for an agent $i \notin C_h$. The process repeats until agent ℓ becomes reachable from agent h, so that we can again perform chore transfer. Note that the MPB condition is still preserved during the process.

We now give an analysis of Algorithm 1. We first show that Algorithm 1 runs in polynomial time.

Lemma 2. *Algorithm 1 runs in polynomial time.*

Proof. In any chore-transfer step, agent h loses exactly one chore, so $|X_h|$ decreases by one. Additionally, $|X_h|$ does not change in any payment raise iteration. Algorithm 1 therefore terminates after at most m chore-transfer steps. We claim that there can be at most n payment-raise iterations between chore transfer iterations. A payment-raise step results in an agent $i \notin C_h$ getting added to $|C_h|$, and does not remove any agents from $|C_h|$. Thus, after at most n payment-raise steps, it must be that ℓ is reachable from h, and Algorithm 1 performs a chore-transfer step. Thus, Algorithm 1 terminates after at most $mn + m$ iterations.

Algorithm 1: Balanced PO Allocation

Input : Chore allocation instance $(N, M, \{c_i\}_{i \in N})$
Output: A balanced PO allocation X

For some agent $h \in N$, set $X_h \leftarrow M$, and $X_i \leftarrow \emptyset$ for $i \neq h$
For each $g \in M$, set $p_g \leftarrow c_h(g)$
$\ell \leftarrow \arg\min_{i \in N} |X_i|$
while $|X_h| - 1 > |X_\ell|$ **do**
  $G_h \leftarrow$ vertices reachable from h in the MPB graph of $(X, \mathbf{p})$
  if $\ell \in C_h$ **then**
    Find a path $P = (h = i_0, g_1, i_1, \ldots, i_{k-1}, g_k, i_k = \ell)$ from h to ℓ
    for $1 \leq r \leq k$ **do**
      $X_{i_{r-1}} \leftarrow X_{i_{r-1}} \setminus \{g_r\}$
      $X_{i_r} \leftarrow X_{i_r} \cup \{g_r\}$
    Update $\ell \leftarrow \arg\min_{i \in N} |X_i|$
  else
    $\gamma \leftarrow \min_{i \in N \setminus C_h, g \in M \cap C_h} \frac{c_i(g)/p_g}{\alpha_i}$
    for $g \in C_h$ **do**
      $p_g \leftarrow \gamma \cdot p_g$
return X

We next show that the output allocation is balanced.

Lemma 3. *When Algorithm 1 terminates, we have $|X_i| - 1 \leq |X_j|$ for any $i, j \in N$.*

Proof. Let $(X, \mathbf{p})$ be an allocation in the run of Algorithm 1 before termination. We claim that the allocation excluding X_h always satisfies this lemma, i.e., for any $i, j \in N \setminus \{h\}$: $|X_i| - 1 \leq |X_j|$. We prove this by induction. Indeed, the initial allocation excluding h is trivially balanced. A chore-transfer step from h to ℓ along a path P only changes the number of chores assigned to h and ℓ, while the other agents remain unaffected. Since ℓ has the fewest chores and gains only one additional chore, the lemma excluding h remains satisfied. As the payment-raise step does not alter the allocation, the claim holds by induction. When Algorithm 1 terminates, the stopping criterion ensures that for any $j \in N$, $|X_h| - 1 \leq |X_\ell| \leq |X_j|$.

We next show that for any $i \in N$, $|X_i| - 1 \leq |X_h|$ when Algorithm 1 terminates. Before the last iteration is executed, $|X_h| - 1 > |X_\ell|$. By the above argument, $|X_\ell| \geq |X_i| - 1$ for all $i \neq h$. The two inequalities together imply that for all $i \neq h$,

$$|X_h| - 1 > |X_i| - 1. \tag{2}$$

After the last iteration, Algorithm 1 terminates, and we have $|X_h| - 1 \leq |X_\ell| \leq |X_i|$ for all $i \neq h$ by the stopping condition and the choice of ℓ. On the other hand, by Eq. (2), $|X_i|-1 < |X_h|-1$ for all $i \neq h$ before the last iteration. Thus, $|X_i| - 1 < |X_h|$ after the last iteration since $|X_h|$ decreases by 1.

The above lemma directly implies that Algorithm 1 return a k-EFX allocation:

Lemma 4. *The allocation returned by Algorithm 1 is k-EFX.*

Proof. It is easy to see that for any $i, j \in N$,

$$c(X_i \setminus \{g\}) \leq k(|X_i| - 1) \leq k|X_j| \leq k \cdot c_i(X_j).$$

The first inequality holds since agents are k-bounded, and the second follows from Lemma 3.

We conclude the analysis by the following theorem.

Theorem 1. *Algorithm 1 returns a k-EFX and fPO allocation in polynomial time.*

Proof. By Lemma 2, Algorithm 2 runs in polynomial time. By Lemma 4, the returned allocation is k-EFX. Finally, since the initial allocation is MPB, and every transfer step and payment raise preserves the MPB condition, the resulting allocation is MPB as well. Thus, Lemma 1 shows that the final allocation is fPO.

3.2 EFX Allocation for $[1, 2]$-Instances

In this section, we show that EFX allocations can be computed in polynomial time for $[1, 2]$-instances. Our algorithm is depicted as Algorithm 2. It is based on the well-known round-robin algorithm: chores are allocated to agents sequentially and in rounds according to a natural order, and agents always select a minimum-cost chore in their turn. Note that the algorithm will only allocate chores to the first $\ell = n \pmod{m}$ agents in the first round.

We analyze Algorithm 2 in the following theorem.

Theorem 2. *For $[1, 2]$-instances, Algorithm 2 returns an EFX allocation in polynomial time.*

Proof. It is easy to see that Algorithm 2 runs in $O(nm)$ time. We now turn to its fairness. Let $X = (X_1, \ldots, X_n)$ be the allocation produced by Algorithm 2. First observe that if $m \leq n$, the EFX property holds trivially. Hence, we assume that $m = kn + \ell$ for some $k \geq 1$ and $0 \leq \ell < n$. Let $g_{i0} \in X_i$ denote the chore allocated to agent $i \in [\ell]$ during the first **for** loop and $g_{ir} \in X_i$ the chore allocated to agent $i \in [n]$ in the r-th iteration of the outer **for** loop, where $r \in [k]$. Then, for all $r \in [k]$,

$$c_i(g_{i(r-1)}) \leq c_i(g_{ir}). \tag{3}$$

Below, by a case analysis, we show that any agent $i \in N$ is EFX toward any agent $j \neq i$. First, consider the case where agent i receives one more chore than

Algorithm 2: Modified Round-Robin Allocation

Input : Chore allocation instance $(N, M, \{c_i\}_{i \in N})$ with $n = |N|$ and $m = |M|$
Output: An EFX allocation $X = (X_1, \ldots, X_n)$

Compute $m = kn + \ell$, where $k \geq 0$, $0 \leq \ell < n$
for $i = 1$ **to** ℓ **do**
    $g \leftarrow \arg\min_{q \in M} c_i(q)$; `// Select item with minimum cost`
    $X_i \leftarrow X_i \cup \{g\}$; `// Allocate to agent i`
    $M \leftarrow M \setminus \{g\}$; `// Remove from unallocated set`
for $r = 1$ **to** k **do**
    for $i = 1$ **to** n **do**
        $g \leftarrow \arg\min_{q \in M} C_i(q)$; `// Select item with minimum cost`
        $X_i \leftarrow X_i \cup \{g\}$; `// Allocate to agent i`
        $M \leftarrow M \setminus \{g\}$; `// Remove from unallocated set`
return $X = (X_1, X_2, \ldots, X_n)$

agent j in Algorithm 2. This occurs only when $i \leq \ell$ and $j > \ell$. Thus, agent i always picks a chore before agent j in the same round. It follows that for all $r \in [k]$,

$$c_i(g_{ir}) \leq c_i(g_{jr}). \tag{4}$$

Then, we have

$$\max_{g \in X_i} c_i(X_i \setminus \{g\}) = c_i(X_i \setminus g_{i0}) = \sum_{r=1}^{k} c_i(g_{ir}) \leq \sum_{r=1}^{k} c_i(g_{jr}) = c_i(X_j).$$

The first equality is due to Eq. (3) and the inequality is due to Eq. (4). Consequently, agent i is EFX toward agent j.

Next, consider the other case where agent i receives no more chores than agent j. If both agents i and j receive $k+1$ chores, note that for all $r \in [k]$,

$$c_i(g_{i(r-1)}) \leq c_i(g_{jr}). \tag{5}$$

Besides, for $[1,2]$-instances, $c_i(g_{ik}), c_i(g_{j0}), c_i(g_{j1}) \in [1,2]$, which implies

$$c_i(g_{ik}) \leq c_i(g_{j0}) + c_i(g_{j1}). \tag{6}$$

Then, we have

$$\begin{aligned}\max_{g \in X_i} c_i(X_i \setminus \{g\}) = c_i(X_i \setminus g_{i0}) &= \sum_{r=1}^{k-1} c_i(g_{ir}) + c_i(g_{ik}) \\ &\leq \sum_{r=2}^{k} c_i(g_{jr}) + c_i(g_{j0}) + c_i(g_{j1}) = c_i(X_j).\end{aligned}$$

Algorithm 3: Balanced PO Allocation for Weighted Agents

Input : Chore allocation instance $(N, M, \{c_i\}_{i \in N})$
Output: A balanced PO allocation X

For some agent $h \in N$, set $X_h \leftarrow M$, and $X_i \leftarrow \emptyset$ for $i \neq h$
For each $g \in M$, set $p_g \leftarrow c_h(g)$
$\ell \leftarrow \arg\min_{i \in N} \frac{|X_i|}{w_i}$
while $\frac{|X_h|-1}{w_h} > \frac{|X_\ell|}{w_\ell}$ **do**
  $C_h \leftarrow$ vertices reachable from h in the MPB graph of $(X, \mathbf{p})$
  if $\ell \in C_h$ **then**
    Find a path $P = (h = i_0, g_1, i_1, \ldots, i_{k-1}, g_k, i_k = \ell)$ from h to ℓ
    for $1 \leq r \leq k$ **do**
      $X_{i_{r-1}} \leftarrow X_{i_{r-1}} \setminus \{g_r\}$
      $X_{i_r} \leftarrow X_{i_r} \cup \{g_r\}$
    Update $\ell \leftarrow \arg\min_{i \in N} |X_i|$
  else
    $\gamma \leftarrow \min_{i \in N \setminus C_h, g \in M \cap C_h} \frac{c_i(g)/p_g}{\alpha_i}$
    for $g \in C_h$ **do**
      $p_g \leftarrow \gamma \cdot p_g$
return X

The first equality is due to Eq. (3) and the inequality is due to Eq. (5) and Eq. (6). Furthermore, the same argument holds when both agents i and j receive k chores. It is also easy to see that a similar argument holds when agent i receives k chores but agent j receives $k + 1$ chores. In a word, agent i is EFX toward agent j in this case, which completes the proof.

4 Weighted EFX Allocations

In this section, we generalize the previous result to weighted agents and provide a polynomial algorithm for computing k-WEFX and fPO allocations for $[1, k]$-instances. The algorithm is depicted as Algorithm 3, which is almost the same as Algorithm 1 for the unweighted case, with a minor modification that the condition of the **while** loop is replaced by $\frac{|X_h|-1}{w_h} > \frac{|X_\ell|}{w_\ell}$. In this way, Algorithm 3 can also produce a "balanced" allocation.

We now give an analysis for Algorithm 3. We first show that the returned allocation is balanced.

Lemma 5. *When Algorithm 3 terminates, for any agents $i \neq j$, $\frac{|X_i|-1}{w_i} \leq \frac{|X_j|}{w_j}$.*

Proof. Let $X = (X_1, X_2, \cdots, X_n)$ be the allocation maintained during Algorithm 3. We have the following claims.

Claim. The allocation X satisfies the condition of the lemma for all $i, j \neq h$ throughout Algorithm 3.

Proof. The claim holds trivially before the algorithm starts. Assuming that it holds in some iteration, we show that it still holds in the next iteration. If the payment-raise step is executed, no chores are transferred, and hence the claim holds. If the chore-transfer step is executed, it is easy to see that $|X_\ell|/w_\ell$ increases by $1/w_\ell$ and $|X_i|/w_i$ remains unchanged for $i \notin \{h, \ell\}$. By induction, the claim holds for $i, j \notin \{h, \ell\}$, and $(|X_i| - 1)/w_i \leq |X_\ell|/w_\ell$ also holds after the step. Besides, since $|X_\ell|/w_\ell \leq |X_i|/w_i$ before the step, we have $(|X_\ell|-1)/w_\ell \leq |X_i|/w_i$ after the step. The claim is therefore proved.

Claim. The allocation X satisfies the condition of the lemma for h and all $i \neq h$ when Algorithm 3 terminates.

Proof. Before the last iteration is executed, $(|X_h| - 1)/w_h > |X_\ell|/w_\ell$. By the first claim, $|X_\ell|/w_\ell \geq (|X_i| - 1)/w_i$ for all $i \neq h$. The two inequalities together imply that for all $i \neq h$,

$$(|X_h| - 1)/w_h > (|X_i| - 1)/w_i. \tag{7}$$

After the last iteration, Algorithm 3 terminates, and we have $(|X_h| - 1)/w_h \leq |X_\ell|/w_\ell \leq |X_i|/w_i$ for all $i \neq h$ by the stopping condition and the choice of ℓ. On the other hand, by Eq. (7), $(|X_i|-1)/w_i < (|X_h|-1)/w_h$ for all $i \neq h$ before the last iteration. Thus, $(|X_i| - 1)/w_i < |X_h|/w_h$ after the last iteration since $|X_h|/w_h$ decreases by $1/w_h$. The claim is therefore proved.

The above claims together imply the lemma.

Lemma 6. *The allocation returned by Algorithm 3 is k-WEFX.*

Proof. It is easy to see that for any $i, j \in N$ and any $g \in X_i$,

$$\frac{c_i(X_i \setminus \{g\})}{w_i} \leq \frac{k(|X_i| - 1)}{w_i} \leq \frac{k|X_j|}{w_j} \leq k \cdot \frac{c_i(X_j)}{w_j}.$$

The first inequality holds since agents are k-bounded, and the second follows from Lemma 5.

We conclude the analysis by the following theorem.

Theorem 3. *Algorithm 3 returns a k-WEFX and fPO allocation in polynomial time.*

Proof. First, note that Lemma 2 for Algorithm 1 also applies to Algorithm 3. Therefore, Algorithm 3 runs in polynomial time. Next, by Lemma 6, the returned allocation is k-WEFX. Finally, since the initial allocation is MPB, and every transfer step and payment raise preserves the MPB condition, the resulting allocation is MPB as well. Thus, Lemma 1 shows that the final allocation is fPO.

5 Conclusion and Open Problems

In this work, we present polynomial-time algorithms that compute k-WEFX and PO allocations for $[1, k]$-instances. We also present a polynomial-time algorithm for computing EFX and PO allocations for $[1, 2]$-instances. Our results enhance people's understanding of (weighted) EFX allocations in chore allocation. Nonetheless, there remain a lot of open problems. We mention the two most related ones. First, improve the approximation ratio of k-WEFX and PO allocations for $[1, k]$-instances. Second, investigate whether constant-factor WEFX allocations exist (without PO property for $[1, k]$-instances).

References

1. Amanatidis, G., Filos-Ratsikas, A., Sgouritsa, A.: Pushing the frontier on approximate EFX allocations. In: Proceedings of the 25th ACM Conference on Economics and Computation, EC 2024, New Haven, CT, USA, 8–11 July 2024, pp. 1268–1286. ACM (2024)
2. Amanatidis, G., Markakis, E., Ntokos, A.: Multiple birds with one stone: beating 1/2 for EFX and GMMS via envy cycle elimination. In: The Thirty-Fourth AAAI Conference on Artificial Intelligence, AAAI 2020, New York, NY, USA, 7–12 February 2020, pp. 1790–1797. AAAI Press (2020)
3. Aziz, H., Caragiannis, I., Igarashi, A., Walsh, T.: Fair allocation of indivisible goods and chores. In: Proceedings of the Twenty-Eighth International Joint Conference on Artificial Intelligence, IJCAI 2019, Macao, China, 10–16 August 2019, pp. 53–59. ijcai.org (2019)
4. Bhaskar, U., Sricharan, A.R., Vaish, R.: On approximate envy-freeness for indivisible chores and mixed resources. In: Approximation, Randomization, and Combinatorial Optimization. Algorithms and Techniques, APPROX/RANDOM 2021, 16–18 August 2021, University of Washington, Seattle, Washington, USA (Virtual Conference). LIPIcs, vol. 207, pp. 1:1–1:23. Schloss Dagstuhl - Leibniz-Zentrum für Informatik (2021)
5. Caragiannis, I., Kurokawa, D., Moulin, H., Procaccia, A.D., Shah, N., Wang, J.: The unreasonable fairness of maximum nash welfare. In: Proceedings of the 2016 ACM Conference on Economics and Computation, EC 2016, Maastricht, The Netherlands, 24–28 July 2016, pp. 305–322. ACM (2016)
6. Chakraborty, M., Igarashi, A., Suksompong, W., Zick, Y.: Weighted envy-freeness in indivisible item allocation. In: Proceedings of the 19th International Conference on Autonomous Agents and Multiagent Systems, AAMAS 2020, Auckland, New Zealand, 9–13 May 2020, pp. 231–239. IFAAMAS (2020)
7. Ebadian, S., Peters, D., Shah, N.: How to fairly allocate easy and difficult chores. In: 21st International Conference on Autonomous Agents and Multiagent Systems, AAMAS 2022, Auckland, New Zealand, 9–13 May 2022, pp. 372–380. IFAAMAS (2022)
8. Foley, D.K.: Resource allocation and the public sector. Yale University (1966)
9. Garg, J., Murhekar, A., Qin, J.: Fair and efficient allocations of chores under bivalued preferences. In: Thirty-Sixth AAAI Conference on Artificial Intelligence, AAAI 2022, Virtual Event, 22 February–1 March 2022, pp. 5043–5050. AAAI Press (2022)

10. Garg, J., Murhekar, A., Qin, J.: Constant-factor EFX exists for chores. In: Proceedings of the 57th Annual ACM Symposium on Theory of Computing, STOC 2025, Prague, Czechia, 23–27 June 2025, pp. 1580–1589. ACM (2025)
11. Li, B., Li, Y., Wu, X.: Almost (weighted) proportional allocations for indivisible chores. In: WWW 2022: The ACM Web Conference 2022, Virtual Event, Lyon, France, 25–29 April 2022, pp. 122–131. ACM (2022)
12. Lipton, R.J., Markakis, E., Mossel, E., Saberi, A.: On approximately fair allocations of indivisible goods. In: Proceedings 5th ACM Conference on Electronic Commerce (EC-2004), New York, NY, USA, 17–20 May 2004, pp. 125–131. ACM (2004)
13. Mas-Colell, A., Whinston, M.D., Green, J.R.: Microeconomic theory. OUP Catalogue (1995)
14. Murhekar, A., Garg, J.: On fair and efficient allocations of indivisible goods. In: Thirty-Fifth AAAI Conference on Artificial Intelligence, AAAI 2021, Virtual Event, 2–9 February 2021, pp. 5595–5602. AAAI Press (2021)
15. Neoh, T.Y., Teh, N.: Understanding EFX allocations: counting and variants. In: AAAI-25, Sponsored by the Association for the Advancement of Artificial Intelligence, 25 February–4 March 2025, Philadelphia, PA, USA, pp. 14036–14044. AAAI Press (2025)
16. Plaut, B., Roughgarden, T.: Almost envy-freeness with general valuations. In: Proceedings of the Twenty-Ninth Annual ACM-SIAM Symposium on Discrete Algorithms, SODA 2018, New Orleans, LA, USA, 7–10 January 2018, pp. 2584–2603. SIAM (2018)
17. Springer, M., Hajiaghayi, M., Yami, H.: Almost envy-free allocations of indivisible goods or chores with entitlements. In: Thirty-Eighth AAAI Conference on Artificial Intelligence, AAAI 2024, 20–27 February 2024, Vancouver, Canada, pp. 9901–9908. AAAI Press (2024)
18. Wu, X., Zhang, C., Zhou, S.: Weighted EF1 allocations for indivisible chores. In: Proceedings of the 24th ACM Conference on Economics and Computation, EC 2023, London, United Kingdom, 9–12 July 2023, p. 1155. ACM (2023)
19. Zhou, S., Wu, X.: Approximately EFX allocations for indivisible chores. In: Proceedings of the Thirty-First International Joint Conference on Artificial Intelligence, IJCAI 2022, Vienna, Austria, 23–29 July 2022, pp. 783–789. ijcai.org (2022)

Strategies for Non-myopic Users in EIP-1559

Yiming Ding, Qi Qi(✉), and Bingzhe Wang

Gaoling School of Artificial Intelligence, Renmin University of China, Beijing, China
{dingym97,qi.qi,wbz2022}@ruc.edu.cn

Abstract. This paper analyzes strategic transaction submission behavior in Ethereum's EIP-1559, focusing on non-myopic users who optimize costs over finite horizons. While prior work assumes myopic users who schedule their transactions in the immediate block, we formalize the dynamic decision-making of non-myopic bidders facing stochastic base fee adjustments, variable block capacities, and transaction deferral risks. We model the evolving base fee and user interactions as a semi-Markov process (sMP) and frame the optimal bidding strategy as a finite-horizon optimal stopping problem. By establishing equivalence to a semi-Markov decision process (sMDP), we prove the existence of deterministic stationary optimal policies and characterize the value function as the minimal solution to a recursive optimality equation. A value iteration algorithm is proposed to compute ϵ-optimal stopping policies, with explicit convergence bounds dependent on the semi-Markov kernel's properties.

Keywords: Transaction Fee · Semi-Markov Process · Sequential Decision Making

1 Introduction

In blockchain systems, transaction records are stored on the blockchain, and each block contains a series of transactions [12]. To incentivize miners to participate actively in mining, the system typically offers two reward mechanisms: block rewards and transaction fees. Due to the nature of block rewards, they usually decrease over time. As a result, transaction fees are increasingly becoming a primary source of income for miners. Therefore, researches on the transaction fee mechanism have attracted growing interest.

Most of the cryptocurrency's transaction fee mechanisms are a simple first price auction mechanism based on greedy algorithm [4]. The first price auction mechanism is simple and easy to implement, but it is not dominant strategy incentive compatible(DSIC), meaning buyers can benefit by reporting a bid that is different from their true value [1]. With the first price auction being as the transaction fee mechanism, many problems have occurred during the operation

Y. Ding, Q. Qi and B. Wang—The authors are ordered alphabetically.

M. Li et al. (Eds.): TAMC 2025, LNCS 16084, pp. 51–62, 2026.
https://doi.org/10.1007/978-981-95-4839-2_5

of trading. For example, Basu et al. [2] indicated that the mechanism based on a first-price auction had a significantly high variance of the transaction fee, resulting in little stable payouts. So, there are numerous attempts to design a transaction fee mechanism beyond the first price auction. Among these, one of the first to be used in the real is the EIP-1559.

EIP-1559 was proposed by the Ethereum founding team and officially adopted by Ethereum [3]. EIP-1559 has introduced many significant changes compared to the original transaction fee mechanism, including the variable-size blocks, a time-changed reserve price, and the burning of a portion of the transaction fees [15].

Following the proposal of EIP-1559, numerous studies have been conducted on this topic, ranging from economic analysis to empirical analysis and other perspectives. However, there is little concentration on analyzing the strategic behavior of the users, especially the non-myopic users. Most existing studies have typically assumed that the participant users are myopic. The myopic users, also referred to be as impatient users in some other papers, typically schedule their transactions within the immediate block. In contrast, the non-myopic users are more strategic. They are not in a hurry to schedule the transaction into blocks. As a result, they will wait for the appropriate time to execute their transactions in response to changes in the reserve price. Current studies primarily focus on designing a transaction fee mechanism with desirable economic properties, with the assumption that users are myopic. Few attentions are paid to how non-myopic users should do to make a cost-minimizing bidding strategy. In this paper, we model a dynamic system and analyze the transaction strategies for non-myopic bidders under dynamic environments.

The main contributions of this paper can be summarized as follows: (1) We introduce a semi-Markov model to describe the behavior of non-myopic users and formulate finding the optimal bidding strategies as an optimal stopping problem in a semi-Markov process. (2) We prove the existence of finite-horizon optimal stopping time and characterize the value function as the minimal solution to a recursive optimality equation. (3) We construct an efficient value iteration algorithm to compute ϵ-optimal stopping policies, with explicit convergence bounds dependent on semi-Markov kernel properties.

2 Related Work

Since the introduction of EIP-1559, a growing body of literature has emerged analyzing its implications from various perspectives. Buterin et al. [3] proposed EIP-1559. They introduced a posted price mechanism based on the original first-price auction, alongside several other modifications. However, they did not provide a formal economic analysis of this mechanism. After that, numerous studies have concentrated on EIP-1559.

Roughgarden [14] established a comprehensive framework for designing transaction fee mechanisms, demonstrating its incentive compatibility for myopic miners and proving the provability of off-chain protocols. The analysis of transaction fees and waiting times concluded that EIP-1559 did not compromise system security. Leonardos et al. [9] investigated the stability of EIP-1559 within a dynamic systems framework, deriving threshold bounds for adjustment parameters. They identified that the base fee exhibited the Markov property, offering a significant modeling approach. Reijsbergen et al. [13] noted oscillations in block size following the London Hard Fork and proposed an additive increase/multiplicative decrease (AIMD) fee adjustment model. Complementing these theoretical analyses, Liu et al. [10] conducted an empirical investigation into the effect of EIP-1559 on the dynamics of the transaction fee, revealing that it enhanced the bidder experience.

Since demand, preferences, and supply may change over time during the process of trading on the blockchain, the dynamic model has been widely used in studies of the transaction fee mechanism.

Ferreira et al. [6] firstly introduced the dynamic model to design a transaction fee mechanism. They proposed a dynamic posted-price transaction fee mechanism. The mechanism computes a dynamic posted-price for every block, utilizing historical information. They proved that truthfully reporting their value was a Bayesian-Nash equilibrium for the users. The first work establishing the dynamic model of EIP-1559 comes from Leonardos et al. [9]. The Markov property of the base fee, as revealed in their paper, is helpful to our work. Besides, Nisan [11] considered the monopolistic miners that were consecutively granted the exclusive right to assemble transactions into the next block and studied the dynamic price chosen by the sequence of monopolists. Crapis et al. [5] proposed a comprehensive and practical framework to tackle the challenge of optimally designing dynamic fee mechanisms for various blockchain resources.

3 System Model and Dynamic Analysis

Under the EIP-1559 protocol, users submit transactions with parameterized bids (γ_{sn}, η_{sn}) to a public memory pool, where η_{sn} denotes the maximum fee cap (total payment ceiling) and γ_{sn} represents the priority fee (miner incentive per gas unit). The effective gas price for user n in block s is determined by:

$$p_{sn} = \min\{\phi_s + \gamma_{sn}, \eta_{sn}\} \tag{1}$$

where ϕ_s is the dynamically adjusted base fee (system-wide reserve price). EIP-1559 dynamically adjusts block sizes by doubling the maximum capacity to $\overline{A} =$ 30 million gas units while maintaining a target load of $\widetilde{A} = 15$ million gas. The base fee evolves via a Markov process dependent on the prior block state:

$$\phi_s = \phi_{s-1}\left(1 + \frac{A_{s-1} - \widetilde{A}}{\widetilde{A} \times (1+\delta)}\right), \tag{2}$$

where A_{s-1} is the actual gas consumption in block $s-1$, $\delta = 1/8$ stabilizes the adjustments. This creates a proportional-integral control rule where persistent congestion $(A_{s-1} > \widetilde{A})$ increases ϕ_s, while underutilization decreases it.

Users have independent private valuations $v_n \sim \mathcal{V}$, where each user n's valuation is drawn from a continuous strictly increasing distribution with support $\subset \mathbb{R}^+$. Transactions that exceed the maximum block capacity $\overline{A}$ and are therefore not included in block s roll over to $s+1$ but are permanently removed if either excluded from two consecutive blocks, or $\eta_{sn} < \phi_{s+1}$ at the subsequent auction. Users who prioritize the immediate submission of their transactions in the current block are termed *myopic users*. Lemma 1 implies that myopic users will bid truthfully in EIP-1559.

Lemma 1 [14]. *For myopic user n with value v_n, the optimal bid is $(\gamma^*_{sn}, \eta^*_{sn}) = (v_n - \phi_s, v_n)$. This bidding strategy constitutes a symmetric ex-post Nash equilibrium under the EIP-1559 mechanism.*

Extending beyond static mechanisms, we now analyze *non-myopic users* who can submit their transactions at any time within finite horizons $T \in \mathbb{Q}^+$. We now develop a dynamic stochastic framework to model strategic transaction submission behaviors on finite planning horizons $T \in \mathbb{Q}^+$. This extension incorporates temporal uncertainties in both block mining processes and transaction arrivals, providing a foundation for analyzing optimal bidding strategies under evolving network conditions. We assume that block mining times follow distribution $\mathcal{M}$ governing inter-block intervals, while transaction arrivals follow distribution $\mathcal{L}$ determining inter-arrival times between consecutive submissions. Let N_s denote the random variable representing the number of new transactions arriving between blocks s and $s+1$, which we assume to follow a Poisson distribution with parameter λ.

The evolving system state over this stochastic process can be formally characterized as a semi-Markov Process (sMP) $\{\mathcal{X}, K(\cdot,\cdot|x)\}$ operating in a measurable state space $\mathcal{X}$ (a Borel space with Borel σ-algebra $\mathcal{B}(\mathcal{X})$). Central to this formulation is the semi-Markov kernel $K(\cdot,\cdot|x) : \mathbb{R}^+ \times \mathcal{X} \to [0,1]$, which encodes transition probabilities between system states. This kernel satisfies three key properties: first, for any measurable set $\mathbf{X} \in \mathcal{B}(\mathcal{X})$ and state $x \in \mathcal{X}$, $K(t, \mathbf{X}|x)$ exhibits non-decreasing right-continuity in t with $K(0, \mathbf{X}|x) = 0$; second, for fixed $t \geq 0$, $K(t,\cdot|\cdot)$ operates as a sub-stochastic kernel on $\mathcal{X}$; and third, the long-term limit $\lim_{t\to\infty} K(t,\cdot|x)$ defines a proper stochastic kernel over $\mathcal{X}$. These properties ensure temporal consistency, transient behavioral modeling, and long-term probabilistic closure. Through canonical construction [8], we establish the measurable space $(\Omega, \mathscr{F})$, where $\Omega = \{(x_0, t_1, x_1, \ldots, t_s, x_s, \ldots)\}$, $x \in \mathcal{X}, t \in \mathbb{R}^+$ and $\mathscr{F}$ is the Borel σ-algebra product. Define the history of sMPs up to the s-th block as $h_{s+1} = (x_0, t_1, x_1, \ldots, t_{s+1}, x_{s+1}), h_0 = x_0$. H_s is the set of all histories h_s. For canonical coordinates $\omega = (x_0, \ldots, x_s, t_{s+1}, \ldots) \in \Omega$, canonical variables capturing block mining times $\Psi_s(\omega) = t_s$, system states $X_s(\omega) = x_s$, and cumulative mining time $\Upsilon_s(\omega) = \sum_{j=0}^{s} \Psi_j(\omega)$. The filtration $\{\mathscr{F}_s\}$ generated by the marked point process $\{(\Psi_s, X_s)\}$ formalizes the information available to decision-makers at each epoch.

We define a state vector $X_s = \left(s, \phi_s, t_s, N_s^{(0)}\right)$, where $N_s^{(0)}$ represents the number of transactions that have been deferred once. So that we can get the formula of the overall kernel K converting ϕ_s to ϕ_{s+1} as followings:

$$K = \int_0^{T-t_s} \sum_{N_s=0}^{\infty} \binom{N_s^{(0)}}{N_s^{surv}} p_0(u)^{N_s^{surv}} (1 - p_0(u))^{N_s^{(0)} - N_s^{surv}}$$

$$\cdot \frac{\lambda^{N_s}}{N_s!} \cdot e^{-\lambda} \cdot \binom{N_s}{N_s^{\text{eff}}} [1 - \mathcal{V}(\phi_s)]^{N_s - N_s^{\text{eff}}} [\mathcal{V}(\phi_s)]^{N_s^{\text{eff}}}$$

$$\cdot \phi_s \cdot \left(1 + \frac{A_s - \widetilde{A}}{\widetilde{A} \times (1 + \delta)}\right) d\mathcal{M}(u)$$

where $p_0(u) = (1 - \mathcal{D}(u))(1 - \mathcal{V}(\phi_s))$, $A_s = \min(\overline{A}, N_s^{surv} + N_s^{\text{eff}})$.

Non-myopic users choose times $\tau \in \Lambda$ to submit their transactions, where $\tau : \Omega \to \mathbb{N} \cup \{+\infty\}$ is called an $\mathscr{F}_s$*-stopping time* if $\{\tau = s\} \in \mathscr{F}_s, \quad \forall s \in \mathbb{N}$. We then define an underlying continuous-time process $\{X(t), t \in \mathbb{R}^+\}$ by $X(t) = X_s$ for $\Upsilon_s \leq t < \Upsilon_{s+1}$. Fix any planning horizon $T \in \mathbb{Q}^+$. For each stopping time $\tau \in \Lambda$, define the total cost as:

$$C_\tau^T = \begin{cases} \int_0^{\Upsilon_\tau} \zeta(X(t))\, dt + \nu(X(\Upsilon_\tau)), & \text{if } \Upsilon_\tau \leq T \\ \int_0^{\Upsilon_\tau} \zeta(X(t))\, dt, & \text{if } \Upsilon_\tau > T \end{cases} \tag{3}$$

where $\zeta(X(t))$ is a nonnegative measurable cost rate function $\zeta : \mathcal{X} \to \mathbb{R}^+$, which can be interpreted as a discount rate representing the loss incurred by the user as the transaction is delayed. $\nu(\cdot)$ is a nonnegative measurable terminal cost function $\nu : \mathcal{X} \to \mathbb{R}^+$. When $\Upsilon_\tau \leq T$, $\nu(X(\Upsilon_\tau)) = \phi_\tau$ represents the base fee the user needs to pay. When $\Upsilon_\tau > T$, $\int_0^{\Upsilon_\tau} \zeta(X(t))\, dt = v$ represents the loss incurred by the user due to the transaction not being processed, which is equivalent to the value v of the transaction for the user. The T-horizon expected cost for each stopping time $\tau \in \Lambda$ is given by:

$$V^\tau(T, x) = \mathbb{E}_x[C_\tau^T], \quad x \in \mathcal{X} \tag{4}$$

The optimal value function is defined as $V^*(T, x) = \inf_{\tau \in \Lambda} V^\tau(T, x)$. A rational non-myopic users seeking optimal transaction submission, the problem transforms into finding $\mathscr{F}_s$-stopping times $\tau \in \Lambda$ that minimize expected costs over the planning horizon. A stopping time $\tau^* \in \Lambda$ is called T-optimal if

$$V^*(T, x) = V^{\tau^*}(T, x) = \inf_{\tau \in \Lambda} V^\tau(T, x), \quad \forall x \in \mathcal{X}. \tag{5}$$

The main objective of the non-myopic user is to identify the T-optimal stopping times and provide an algorithm for computing the value function V^*. This is achieved by recognizing that finding the T-optimal stopping times in sMPs is equivalent to finding the T-optimal policies in a special class of semi-Markov decision processes (sMDPs). In the subsequent section, we will introduce the primary results pertaining to finite-horizon sMDPs.

4 Transaction Strategies in Dynamic Environments

In this section, we first equate the problem of a user selecting transaction times under dynamically changing base fees governed by sMPs with making sequential decisions in an sMDP. Here and throughout the paper, we use the notation '$\dot{}$' to distinguish sMDPs from sMPs. Subsequently, we demonstrate the existence of an optimal 'transaction submission time' strategy. At the end of this section, we present an approximate algorithm for determining the optimal transaction submission time.

4.1 Semi-Markov Decision Process Formulation

The sMDPs model is defined by the tuple

$$\left\langle \dot{\mathcal{X}}, \dot{\mathcal{A}}, \{\dot{\mathcal{A}}(x)\}_{x\in\dot{\mathcal{X}}}, \dot{K}(\cdot|\cdot,\cdot), \dot{\zeta}(x,a), \dot{\nu}(x,a) \right\rangle,$$

where $\dot{\mathcal{X}}$ is the state space. $\dot{\mathcal{A}}$ is the denumerable action set. $\dot{\mathcal{A}}(x) \subseteq \dot{\mathcal{A}}$ is the finite set of admissible actions at state $x \in \dot{\mathcal{X}}$. $\mathcal{J} = \{(x,a) \mid x \in \dot{\mathcal{X}}, a \in \dot{\mathcal{A}}(x)\}$ is the set of all admissible state-action pairs. $\dot{K}(\cdot|\cdot,\cdot)$ is the transition kernel (assumed but not explicitly defined here for brevity). $\dot{\zeta}(x,a)$ and $\dot{\nu}(x,a)$ are non-negative measurable cost functions on $\mathcal{J}$. Assume that $\mathcal{J} \in \mathcal{X}(\dot{\mathcal{X}}) \times \mathcal{X}(\dot{\mathcal{A}})$ and that there is a measurable mapping $f : \mathbb{R} \times \dot{\mathcal{X}} \to \dot{\mathcal{A}}$ such that $\forall (t,x) \in \mathbb{R}\times\dot{\mathcal{X}}, (x, f(t,x)) \in \mathcal{J}$. Assume that $\mathcal{J} \in \mathcal{X}(\dot{\mathcal{X}})\times\mathcal{X}(\dot{\mathcal{A}})$ and there is a measurable mapping $f : \mathbb{R} \times \dot{\mathcal{X}} \to \dot{\mathcal{A}}$ such that $\forall (t,x) \in \mathbb{R} \times \dot{\mathcal{X}}, (x, f(t,x)) \in \mathcal{J}$. At the s-th decision epoch, the non-myopic bidder obtains a remaining planning horizon $T - \sum_{k=1}^{s} t_k$ and an admissible history $\dot{h}_s = (x_0, a_0, t_1, x_1, \ldots, a_{s-1}, t_s, x_s)$, where $(x_m, a_m) \in \mathcal{J}, t_m \in \mathbb{R}^+, \forall m \in [s-1], x_s \in \dot{\mathcal{X}}$. Define the admissible histories equipped with the Borel σ-algebra as $\dot{H}_s$.

A policy $\pi = \{\pi_s\}_{s\geq 0}$ consists of stochastic kernels π_s on $\dot{\mathcal{A}}$ given $\mathbb{R} \times \dot{H}_s$ satisfying $\forall s \geq 0, T \in \mathbb{R}, \dot{h}_s \in \dot{H}_s, \pi_s(\dot{\mathcal{A}}(x_s)|z, \dot{h}_s) = 1$. The policy space is denoted by Π. We can classify policies as:

- *Markov Policy*: There exist stochastic kernels $\{\xi_s\} \in \Xi$ such that $\pi_s(\cdot|z,\dot{h}_s) = \xi_s(\cdot|z, x_s),\ \forall (z, \dot{h}_s) \in \mathbb{R} \times \dot{H}_s$. Denoted as $\pi = \{\xi_s\}$.
- *Stationary Markov Policy*: ξ_s is independent of s, denoted as $\pi = \xi$.
- *Deterministic Policy*: There exist measurable functions $\{\mathbf{d}_s\}$ with $\mathbf{d}_s : \mathbb{R} \times \dot{H}_s \to \dot{\mathcal{A}}$ satisfying $\pi_s(a|z,\dot{h}_s) = \delta_{\mathbf{d}_s(z,\dot{h}_s)}(a),\ \forall a \in \dot{\mathcal{A}}$. Denoted as $\pi = \{\mathbf{d}_s\}$.
- *Deterministic Markov Policy*: There exist $\{\mathbf{g}_s\} \in \mathscr{F}$ where $\xi_s(\cdot|z,x)$ is concentrated at $\mathbf{g}_s(z,x)$, denoted $\pi = \{\mathbf{g}_s\}$.
- *Deterministic Stationary Policy*: $\mathbf{g}_s$ is independent of s, denoted $\pi = \mathbf{g}$.

The sets of all Markov, stationary Markov, deterministic, deterministic Markov, and deterministic stationary policies are denoted by $\Pi_{\mathrm{M}}, \Pi_{\mathrm{S}}, \Pi_{\mathrm{D}}, \Pi_{\mathrm{DM}}, \Pi_{\mathrm{DS}}$, respectively. These sets have the following containment relationships: $\Pi_{\mathrm{DS}} = \mathbb{D} \subset \Pi_{\mathrm{DM}} \subset \Pi_{\mathrm{D}} \subset \Pi$ and $\mathbb{D} \subset \Xi = \Pi_{\mathrm{S}} \subset \Pi_{\mathrm{M}} \subset \Pi$.

Let $\dot{\Omega} = \mathbb{R} \times \dot{\mathcal{X}} \times \dot{\mathcal{A}}$ be a sample space and $\dot{\mathscr{F}}$ be the corresponding product σ-algebra. For $\dot{\omega} = (x_0, a_0, t_1, \ldots, x_s, a_s, t_{s+1}, \ldots) \in \dot{\Omega}$, define $\dot{X}_s(\dot{\omega}) = x_s$, $\dot{\Psi}_0(\dot{\omega}) = 0$, $\dot{\Psi}_s(\dot{\omega}) = t_s$, and $\dot{\mathcal{A}}_s(\dot{\omega}) = a_s$. For any s, let $\dot{\Upsilon}_s(\dot{\omega}) = \sum_{j=0}^{s} \dot{\Psi}_j(\dot{\omega})$. We then define $\{\dot{X}(t), \dot{\mathcal{A}}(t), t \in \mathbb{R}^+\}$ by:

$$\dot{X}(t) = \begin{cases} \dot{X}_s, & \dot{\Upsilon}_s \le t < \dot{\Upsilon}_{s+1}; \\ \partial_{\dot{\Upsilon}}, & t \ge \lim_{s\to\infty} \dot{\Upsilon}_s, \end{cases} \qquad \dot{\mathcal{A}}(t) = \begin{cases} \dot{\mathcal{A}}_s, & \dot{\Upsilon}_s \le t < \dot{\Upsilon}_{s+1}; \\ \partial_{\dot{\mathcal{A}}}, & t \ge \lim_{s\to\infty} \dot{\Upsilon}_s, \end{cases} \tag{6}$$

where $\partial_{\dot{\Upsilon}}$ and $\partial_{\dot{\mathcal{A}}}$ are extra states/actions joined to $\dot{\mathcal{X}}$ and $\dot{\mathcal{A}}$, respectively. Given $(z, x) \in \mathbb{R} \times \dot{\mathcal{X}}$ and $\pi = [\pi_s] \in \Pi$, by Tulcea's Theorem [7], there exists a unique probability measure $\dot{\mathbf{P}}^{\pi}_{(z,x)}$ on $(\dot{\Omega}, \dot{\mathscr{F}})$ satisfying, $\forall t \in \mathbb{R}^+, a \in \dot{\mathcal{A}}, \mathbf{B} \in \mathcal{X}(\dot{\mathcal{X}})$:

$$\dot{\mathbf{P}}^{\pi}_{(z,x)}(\dot{\Psi}_0 = 0,\ \dot{X}_0 = x) = 1, \tag{7}$$

$$\begin{aligned} \dot{\mathbf{P}}^{\pi}_{(z,x)}&(\dot{\mathcal{A}}_s = a | (\dot{X}_0, \dot{\mathcal{A}}_0, \dot{\Psi}_1, \dot{X}_1, \ldots, \dot{\mathcal{A}}_{s-1}, \dot{\Psi}_s, \dot{X}_s)) \\ &= \pi_s(a | (T - \dot{\Upsilon}_s), (\dot{X}_0, \ldots, \dot{\mathcal{A}}_{s-1}, \dot{\Psi}_s, \dot{X}_s)), \end{aligned} \tag{8}$$

$$\dot{\mathbf{P}}^{\pi}_{(z,x)}(\dot{\Psi}_{s+1} \le t,\ \dot{X}_{s+1} \in \mathbf{B} | (\dot{X}_0, \ldots, \dot{\mathcal{A}}_{s-1}, \dot{\Psi}_s, \dot{X}_s), \dot{\mathcal{A}}_s) = \dot{K}(t, \mathbf{B} | \dot{X}_s, \dot{\mathcal{A}}_s). \tag{9}$$

The expectation operator is denoted by $\dot{\mathbf{E}}^{\pi}_{(z,x)}$. Assume for all $(z, x) \in \mathbb{R} \times \dot{\mathcal{X}}$ and $\pi \in \Pi$:

$$\dot{\mathbf{P}}^{\pi}_{(z,x)}\left(\left\{\lim_{s\to\infty} \dot{\Upsilon}_s = \infty\right\}\right) = 1. \tag{10}$$

Given $(z, x) \in \mathbb{R}^+ \times \dot{\mathcal{X}}$, the T-horizon expected cost of a policy $\pi \in \Pi$ is defined as: $U^{\pi}(z, x) = \dot{\mathbf{E}}^{\pi}_{(z,x)}\left[\int_0^z \dot{\zeta}(\dot{X}(t), \dot{\mathcal{A}}(t))dt + \dot{\nu}(\dot{X}(z), \dot{\mathcal{A}}(z))\right]$, and the value function is: $U^*(z, x) = \inf_{\pi\in\Pi} U^{\pi}(z, x)$.

The definitions and assumptions provided in this section lay the groundwork for the subsequent lemmas and theorems. The different policy types define various ways in which decisions can be made based on the history and current state. The probability measures defined ensure that these policies can be evaluated in a stochastic setting. The assumption in Eq. 10 ensures that the defined processes behave well and do not terminate prematurely. The expected cost function and value function are critical in optimization problems, where the goal is to find the best policy that minimizes the cost over a given horizon.

4.2 Policy Equivalence and Optimality

A policy $\pi^* \in \Pi$ is *T-optimal* if $U^{\pi^*}(z, x) = U^*(z, x)$ for all $(z, x) \in [0, T] \times \mathcal{X}$, where $\mathcal{X}$ denotes the state space.

Lemma 2 *For any policy $\pi \in \Pi$ and $(z, x) \in \mathbb{R}^+ \times \mathcal{X}$, there exists a Markov policy $\dot{\pi} \in \Pi_M$ such that $U^{\pi}(z, x) = U^{\dot{\pi}}(z, x)$.*

This equivalence allows restricting T-optimal policy searches to Π_{M}. We proceed to characterize U^* through Bellman operators and demonstrate the existence of deterministic stationary T-optimal policies.

For deterministic stationary $f \in \mathbb{D}$, the expected cost $U^f(z,x)$ decomposes as: *No transition by time* z: Cost $\dot{\zeta}(x,a)z + \dot{\nu}(x,a)$ with probability $1 - \dot{K}(z,\mathcal{X} \mid x,a)$. And *Transition at* $t \in [0,z]$: Cost $\dot{\zeta}(x,a)t + U^f(z-t,y)$ integrated over $(t,y) \sim \dot{K}(\cdot \mid x,a)$. This yields the integral equation:

$$U^f(z,x) = \dot{\nu}(x,a)\left[1 - \dot{K}(z,\mathcal{X} \mid x,a)\right] + \int_0^z \int_{\mathcal{X}} U^f(z-t,y)\dot{K}(dt,dy \mid x,a) + \dot{\zeta}(x,a)\left[z\left(1 - \dot{K}(z,\mathcal{X} \mid x,a)\right) + \int_0^z t\dot{K}(dt,\mathcal{X} \mid x,a)\right]. \tag{11}$$

Let $\mathbb{B}$ be the Borel-measurable functions $v : [0,T]\times\mathcal{X} \to [0,\infty]$. For $a \in \mathcal{A}(x)$, define the Bellman operator:

$$\mathbb{O}^a v(z,x) = \begin{cases} \dot{\zeta}(x,a)\int_0^z \left[1 - \dot{K}(t,\mathcal{X} \mid x,a)\right]dt + \dot{\nu}(x,a)\left[1 - \dot{K}(z,\mathcal{X} \mid x,a)\right] \\ \quad + \int_0^z \int_{\mathcal{X}} v(z-t,y)\dot{K}(dt,dy \mid x,a), & a \in \mathcal{A}(x), \\ +\infty, & \text{otherwise.} \end{cases}$$

For randomized $\xi \in \Xi$, define $\mathbb{O}^\xi v = \sum_{a\in\mathcal{A}(x)} \xi(a \mid z,x)\mathbb{O}^a v$ and $\mathbb{O}v = \min_{a\in\mathcal{A}(x)} \mathbb{O}^a v$.

The value iteration sequence for $\pi \in \Pi_{\mathrm{M}}$ satisfies: $U^\pi_{s+1} = \mathbb{O}^{\xi_0} U^{\pi^1}_s$, $U^\pi = \mathbb{O}^{\xi_0} U^{\pi^1}$, where $\pi^1 = \{\xi_s\}_{s\geq 1}$. This recursion generalizes to arbitrary $\xi \in \Xi$ as $U^\xi_{s+1} = \mathbb{O}^\xi U^\xi_s$.

Theorem 1 *Under Eq. 10, the following hold:*

- *Value Iteration: Define the sequence* $\{U^*_s\}$: $U^*_0 = 0$ *and* $U^*_{s+1} = \mathbb{O}U^*_s$. *Then* $U^* = \lim_{s\to\infty} U^*_s \in \mathbb{B}$.
- *Optimality Equation: The value function* U^* *is the minimal solution in* $\mathbb{B}$ *satisfying* $U^* = \mathbb{O}U^*$. *Moreover, if* $u \in \mathbb{B}$ *satisfies* $u = \mathbb{O}u$, *then* $u \geq U^*$.
- *T-Optimal Policy: There exists* $f^* \in \mathbb{D}$ *such that* $U^* = \mathbb{O}^{f^*}U^*$. *Hence, the deterministic stationary policy* f^* *is* T*-optimal.*

4.3 Bidding Strategies for Non-Myopic Users

Then, we will establish an equivalence between optimal stopping problems on sMPs and sMDPs by constructing a policy-induced correspondence between stopping times and decision strategies. This framework allows application of Theorem 1 to sMPs through their sMDP formulations.

Define the augmented state space $\dot{\mathcal{X}} = \mathcal{X} \cup \{\Delta\}$ with absorbing state Δ. Actions $\mathcal{A}(x) = \{0,1\}$ for $x \in \mathcal{X}$ (0 = continue, 1 = stop) and $\mathcal{A}(\Delta) = \{1\}$ form a finite action space. The admissible state-action pairs $\mathcal{J} = (\mathcal{X}\times\{0,1\})\cup\{(\Delta,1)\}$ constitute a Borel subset of $\dot{\mathcal{X}} \times \mathcal{A}$. The semi-Markov kernel is defined as:

$$\dot{K}_T(t,\mathbf{B}|x,a) = \begin{cases} K(t,\mathbf{B}\setminus\{\Delta\}|x), & x \in \mathcal{X}, a = 0 \\ \mathbb{I}_{[T,+\infty)}(t)\delta_\Delta(\mathbf{B}), & x \in \dot{\mathcal{X}}, a = 1 \end{cases} \tag{12}$$

with holding costs $\dot{\zeta}(x,a) = \zeta(x)\mathbb{I}_{\{0\}}(a)$ and terminal costs $\dot{\nu}(x,a) = \nu(x)\mathbb{I}_{\{1\}}(a)$. Lemma 3 confirms this construction satisfies Eq. 10.

Next, we establish a bidirectional relationship between stopping times $\tau \in \Lambda$ and admissible policies $\pi \in \Pi_D$:

Lemma 3 *The constructed sMDP satisfies:*

1. *Regularity condition Eq. 10.*
2. *Induced stopping time validity: For any policy $\pi \in \Pi_D$, the stopping time $\tau_\pi^z(\omega) = \inf\left\{s \in \mathbb{N} \,\middle|\, \mathbf{d}_s(z - \Upsilon_s(\omega), \dot{h}_s) = 1\right\}$ is $\mathcal{F}$-measurable.*

This dual result ensures both formal compatibility with Theorem 1 and operational validity of the constructed stopping times.

Define the restricted policy space

$$\Pi_D^0 = \left\{\pi = \{\mathbf{d}_s\}_{s\geq 0} \in \Pi_D \,\middle|\, \mathbf{d}_s(0, \dot{h}_s) = 0,\ \forall s \geq 0,\ \dot{h}_s \in \dot{H}_s^0\right\} \subset \Pi_D$$

containing policies that never repeat 'continue' actions consecutively. The following equivalence holds:

Lemma 4 *For any $\tau \in \Lambda$ and $\pi \in \Pi_D^0$, the following hold uniformly over $(z,x) \in [0,T] \times \mathcal{X}$:*

1. *Cost equivalence: $U^\pi(z,x) = V^{\tau_\pi^z}(z,x)$*
2. *Policy containment: $\pi_\tau \in \Pi_D^0$ for any induced policy π_τ*
3. *Value preservation: $V^\tau(z,x) = U^{\pi_\tau}(z,x) = V^{\tau_{\pi_\tau}^z}(z,x)$*

This tripartite equivalence enables bidirectional analysis between stopping times and decision policies.

Then, the Bellman operator for the value function $V^* = \lim_{s\to\infty} V_s^*$ arises naturally from the sMDP formulation:

$$\mathbb{C}u(z,x) = \min\left\{\zeta(x)\int_0^z (1 - K(t,\mathcal{X}|x))dt + \int_{[0,z]}\int_{\mathcal{X}} u(z-t,y)K(dt,dy|x), \nu(x)\right\}$$

Lemma 5 *Define the sequence $\{V_s^*\}$ with $V_0^* = 0$ and $V_{s+1}^* = \mathbb{C}V_s^*$. Then, the optimal value function V^* satisfies $V^* = \lim_{s\to\infty} V_s^*$ and $V^* = \mathbb{C}V^*$.*

This lemma establishes that iterating the operator $\mathbb{C}$ starting from the zero function converges to the optimal value function V^*. This is crucial because it allows us to compute V^* numerically through repeated applications of $\mathbb{C}$.

Lemma 6 *Define the subset $R^* = \{(z,x) \in (0,T] \times \mathcal{X} \mid V^*(z,x) = \nu(x)\}$. For each $\omega \in \Omega$, define:*

$$\tau^*(\omega) = \inf\left\{s \in \mathbb{N} \,\middle|\, \left(T - \sum_{k=1}^{s} t_k, x_s\right) \in R^*\right\} \wedge \inf\left\{s \in \mathbb{N} \,\middle|\, T \leq \sum_{k=1}^{s} t_k\right\}.$$

Then, τ^* is T-optimal and satisfies $\forall x \in \mathcal{X}, \mathbf{P}_x(\tau^* = \infty) = 0$.

Algorithm 1. Value Iteration Algorithm for ϵ-T-optimal stopping times

1: **Input:** $\epsilon > 0, T, \mathcal{X}, \zeta(x), \nu(x), K(t, \mathcal{X}|x), \beta, \mathbb{C}, \{t_k\}, \{x_n\}, \Omega$.
2: **Output:** the ϵ-T-optimal stopping time $\tau^\epsilon(\omega)$.
3: **Initialization:** Set $V_0^*(z,x) = 0$ for all $(z,x) \in [0,T] \times \mathcal{X}$.
4: Let $s = 0$.
5: **repeat**
6: Update $V_{s+1}^*(z,x)$ using the value iteration formula:

$$V_{s+1}^*(z,x) = \min\left\{\zeta(x)\int_0^z (1 - Q(t,\mathcal{X}|x))dt + \int_{\mathcal{X}}\int_{[0,z]} V_s^*(s-t,y)Q(dt,dy|x), \nu(x)\right\}.$$

7: Increment $s = s+1$.
8: **until** $|V_{s+1}^*(z,x) - V_s^*(z,x)| \le \frac{1-\beta}{\beta}\epsilon$ for all $(z,x) \in [0,T] \times \mathcal{X}$
9: Compute the set D^ϵ:

$$D^\epsilon = \{(z,x) \in (0,T] \times \mathcal{X} |\ \nu(x) = \mathbb{C}V_s^*(z,x)\}.$$

10: Compute the ϵ-T-optimal stopping time $\tau^\epsilon(\omega)$:

$$\tau^\epsilon(\omega) = \inf\left\{j \in \mathbb{N} \mid (T - \sum_{k=1}^{j} t_k, x_n) \in D^\epsilon\right\} \wedge \inf\left\{j \in \mathbb{N} \mid T \le \sum_{k=1}^{j} t_k\right\}, \quad \forall \omega \in \Omega.$$

Theorem 2 *For a non-myopic user n, the cost-minimizing bidding strategy $(\gamma_{sn}^*, \eta_{sn}^*)$ is given by:*

$$\gamma_{sn}^* = 0, \quad \eta_{sn}^* = V^* - \int_0^{\Upsilon_{\tau^*}} \zeta(X(t))\,dt = \phi_{\tau^*}.$$

4.4 Approximation Algorithm for Practical Implementation

In practical applications, the optimal value function V^* required by Lemma 6 is often unknown. To address this, we can approximate V^* using the iterative sequence $\{V_s^*\}$ derived from Lemma 5. This approximation motivates the concept of ϵ-T-optimal stopping.

Given $\epsilon > 0$, a stopping time τ is defined as ϵ-T-optimal if for all $x \in \mathcal{X}$, the difference between the T-horizon expected cost $V^\tau(z,x)$ and the optimal value function $V^*(z,x)$ is bounded by ϵ. Mathematically, this is expressed as: $V^\tau(z,x) - V^*(z,x) \le \epsilon$, where $V^\tau(z,x)$ is the T-horizon expected cost defined in Eq. 4.

We will now demonstrate that for any $\epsilon > 0$, under certain conditions, we can iterate sufficiently many times to obtain an ϵ-T-optimal stopping time.

Theorem 3 *Assume that the semi-Markov kernel K satisfies $T\|\zeta\| + \|\nu\| = M < \infty$ and $\sup_{x\in\mathcal{X}} K(T,\mathcal{X}|x) = \beta < 1$. Define for each $s \ge 0$ and $\omega \in \Omega$:*

$$D_s = \{(z,x) \in (0,T] \times \mathcal{X} \mid \nu(x) = \mathbb{C}V_s^*(z,x)\},$$

and the stopping time

$$\tau_s(\omega) = \inf\left\{ s \in \mathbb{N} \;\middle|\; \left(T - \sum_{k=1}^{s} t_k, x_m\right) \in D_s \right\} \wedge \inf\left\{ s \in \mathbb{N} \;\middle|\; T \leq \sum_{k=1}^{s} t_k \right\}.$$

Then, the following statements hold:

- *For any* $\epsilon > 0$, *if* $\|V^*_{S+1} - V^*_S\| \leq \frac{1-\beta}{\beta}\epsilon$ *for some* $S \geq 0$, *then* τ_S *is* ϵ-*T*-*optimal.*
- *For all* $s \geq 0$: $\|V^* - V^*_s\| \leq \beta^s M$. *Defining the iteration count:*

$$s_\epsilon = \left\lceil \frac{\log(\epsilon(1-\beta)) - \log(M+1)}{\log \beta} \right\rceil,$$

then τ_{s_ϵ} *is* ϵ-*T*-*optimal.*

The theorem leverages the convergence properties of the value iteration algorithm. By bounding the difference between consecutive iterations of the value function, we can ensure that the stopping time derived from a sufficient number of iterations is close to the optimal stopping time within a specified error bound ϵ. The constant β and M relate to the properties of the semi-Markov kernel and help quantify the convergence rate.

The Value Iteration Algorithm 1 initializes the value function to zero and iteratively updates it using the value iteration formula until the difference between consecutive iterates is within the specified error bound. The stopping time is then computed based on the final approximated value function, ensuring that it is ϵ-T-optimal.

5 Conclusion

We model EIP-1559's dynamic transaction fee as a semi-Markov process, formulating non-myopic bidding as an optimal stopping problem. By equivalence to semi-Markov decision processes, we prove that finite-horizon deterministic optimal policies exist and propose a value iteration algorithm with explicit convergence bounds for ϵ-optimal transaction timing, enabling cost-minimizing strategies under stochastic fee adjustments.

References

1. Aggarwal, G., Hartline, J.D.: Knapsack auctions. In: SODA, vol. 6, pp. 1083–1092 (2006)
2. Basu, S., Easley, D., O'Hara, M., Sirer, E.G.: Stablefees: a predictable fee market for cryptocurrencies. Manage. Sci. **69**(11), 6508–6524 (2023)
3. Buterin, V.: EIP 1559 FAQ (2021)

4. Chung, H., Shi, E.: Foundations of transaction fee mechanism design. In: Proceedings of the 2023 Annual ACM-SIAM Symposium on Discrete Algorithms (SODA), pp. 3856–3899. SIAM (2023)
5. Crapis, D., Moallemi, C.C., Wang, S.: Optimal dynamic fees for blockchain resources. In: Clark, J., Shi, E. (eds.) FC 2024. LNCS, vol. 14744, pp. 271–291. Springer, Cham (2024). https://doi.org/10.1007/978-3-031-78676-1_15
6. Ferreira, M.V., Moroz, D.J., Parkes, D.C., Stern, M.: Dynamic posted-price mechanisms for the blockchain transaction-fee market. In: Proceedings of the 3rd ACM Conference on Advances in Financial Technologies, pp. 86–99 (2021)
7. Hernández-Lerma, O., Lasserre, J.B.: Discrete-Time Markov Control Processes: Basic Optimality Criteria, vol. 30. Springer, Cham (2012)
8. Kitayev, M.Y.: Semi-Markov and jump Markov controlled models: average cost criterion. Theory Probab. Appl. **30**(2), 272–288 (1986)
9. Leonardos, S., Monnot, B., Reijsbergen, D., Skoulakis, E., Piliouras, G.: Dynamical analysis of the EIP-1559 ethereum fee market. In: Proceedings of the 3rd ACM Conference on Advances in Financial Technologies, pp. 114–126 (2021)
10. Liu, Y., Lu, Y., Nayak, K., Zhang, F., Zhang, L., Zhao, Y.: Empirical analysis of EIP-1559: transaction fees, waiting times, and consensus security. In: Proceedings of the 2022 ACM SIGSAC Conference on Computer and Communications Security, pp. 2099–2113 (2022)
11. Nisan, N.: Serial monopoly on blockchains. arXiv preprint arXiv:2311.12731 (2023)
12. Nofer, M., Gomber, P., Hinz, O., Schiereck, D.: Blockchain. Bus. Inf. Syst. Eng. **59**, 183–187 (2017)
13. Reijsbergen, D., Sridhar, S., Monnot, B., Leonardos, S., Skoulakis, S., Piliouras, G.: Transaction fees on a honeymoon: ethereum's EIP-1559 one month later. In: 2021 IEEE International Conference on Blockchain (Blockchain), pp. 196–204. IEEE (2021)
14. Roughgarden, T.: Transaction fee mechanism design for the ethereum blockchain: an economic analysis of EIP-1559. arXiv preprint arXiv:2012.00854 (2020)
15. Roughgarden, T.: Transaction fee mechanism design. J. ACM **71**(4), 1–25 (2024)

A Lookup Table Design Method: Achieving $O(1)$-Query Complexity and No Memory Waste

Yu Wu[1(✉)] and Xiaoping Qiu[2,3]

[1] College of Management Science, Chengdu University of Technology, Chengdu 610059, China
wuyu14@mails.ucas.edu.cn
[2] School of Computing and Artificial Intelligence, Southwest Jiaotong University, Chengdu 611756, China
[3] Shanghai Dongtami Intelligent Technology Co., Ltd., Shanghai, China

Abstract. When solving complex Markov Decision Processes (MDPs), the state value function can be represented by a parameterized function or a lookup table. Parameterized functions are space-efficient but computationally complex, while lookup tables are efficient but struggle with large state spaces, leading to memory waste or high query complexity. This study proposes a lookup table design with $O(1)$ query complexity and no memory waste for complex MDPs commonly found in dynamic vehicle routing problems. We define a state key as a numerical triple and design two 3D arrays. Though initially memory-wasteful, we apply operations like mirroring, rotation, and translation to remove unused memory while maintaining $O(1)$ query complexity. Our contributions are: (1) making state aggregation optional in certain MDPs, and (2) efficiently managing memory for querying complex data. An illustrative example is used throughout to enhance clarity.

Keywords: Lookup Table · Query Complexity · Memory Waste · Markov Decision Process · Approximate Dynamic Programming/Reinforcement Learning

1 Introduction

In solving Markov Decision Processes (MDPs) with Approximate Dynamic Programming or Reinforcement Learning, two main methods are used to represent the state-value function. Lookup Table Representation (LTR) and Parametric Function Representation (PFR) [7]. PFR saves memory with parameterized functions, but requires domain knowledge and significant computational resources for function selection and parameter optimization [2]. Additionally, adjusting parameters is slower compared to learning under LTR. In contrast, LTR directly stores each

Supplementary Information The online version contains supplementary material available at https://doi.org/10.1007/978-981-95-4839-2_6.

M. Li et al. (Eds.): TAMC 2025, LNCS 16084, pp. 63–76, 2026.
https://doi.org/10.1007/978-981-95-4839-2_6

state's value function, enabling fast updates. However, for large state spaces, LTR can be memory-intensive and querying incurs time overhead. This study addresses these limitations by proposing a lookup table design with $O(1)$ query complexity and no memory waste, where all applied memory is used efficiently.

Many studies focus on LTR-relevant policies, but some do not address how to design lookup tables, despite their critical role in policy implementation. Instead, these studies discuss other aspects, such as the space complexity or querying efficiency of the lookup table, such as [10,15,16]. For example, [4] only mentions the space complexity of the lookup table but does not discuss its design or efficient querying. Other studies focus on either very simple state spaces [2] or significantly simplify the state space through aggregation [9,14], making lookup table design trivial. To date, no research has addressed the issue of designing lookup tables for complex state spaces. It is important to note that designing a lookup table is a crucial step in implementing tabular-based policies. Therefore, this paper specifically focuses on the design of lookup tables for complex state spaces, aiming to bridge the gap between theoretical tabular policies and their practical implementation.

We consider a finite-horizon MDP with a discrete state space, commonly found in dynamic vehicle routing problems. This MDP is represented as a quadruple (S, A, P, R), where S is the set of states, A is the set of actions, P is the transition model, and R is the reward function. Our focus is on the design of the lookup table, which only involves the first two elements S and A, while the transition model P and reward function R can take any form. Hence, we will specify only S and A. In particular, based on existing studies, such as [8,9,11–13], we define the MDP state at any decision epoch as consisting of four components: two sets representing customer types, a numerical component for the vehicle's location, and the remaining amount of a one-dimensional resource constraint (e.g., time [8,9] or vehicle capacity [11,12]). The action for each state is the destination of the vehicle's next journey, i.e., the customer to be visited. Thus, the MDP is expressed in Eq. (1), omitting the last two components.

$$\text{MDP} = \Big\{\{c \in \mathbb{N} \ \& \ c \leq k, loc, \mathcal{D} \subset \mathrm{D}, \mathcal{P} \subset \mathrm{P}\}; \quad a \in \mathcal{D} \cup \mathcal{P} \cup \{depot\}; \ \cdot \ ; \ \cdot \Big\} \tag{1}$$

Above, c denotes the remaining feasible constraint resource, and k represents the total amount of the limited resource. The use of a lookup table implies that the state space cannot be continuous, as the state-value function cannot be stored in a lookup table in memory. Therefore, we let c take integer values from 0 to k. loc refers to the vehicle's location, which could be either a customer's location or the depot. P and D represent two different customer sets, and $\mathcal{D}$ and $\mathcal{P}$ denote their respective subsets that need to be serviced later. For convenience, we use "D-elements" and "P-elements" to represent customers belonging to D and P, respectively[1]. a denotes the action corresponding to a given state, which is also either a customer's location or the depot. Additionally, the last two components of the model are omitted, as they are not relevant to this study. The considered MDPs are applicable to scenarios where a single vehicle, constrained by

[1] Note that "D" as well as "$\mathcal{D}$" refers to "deterministic" while "P" as well as "$\mathcal{P}$" refers to "probabilistic".

one-dimensional resources (such as capacity or travel time), serves two types of customers with uncertain demand. One type has predetermined service demand, while the other has stochastic service requests that arrive dynamically during the vehicle's shift. As illustrated by [8] and [12], such scenarios are commonly observed in city logistics services. To make our method more intuitive and easier to understand, we provide a small illustrative example, using this example throughout the paper to illustrate the evolution of our method. However, we emphasize that this example does not prove any results, and removing it would not affect the logical completeness of our proposed method.

Example : $\mathrm{D} = \{1,2\}, \mathrm{P} = \{3,4,5\}, k = 2.$

Our contributions are primarily manifested in: (1) at the strategic level, changing one step of implementing the lookup table representation method from necessary to optional, and (2) at the technical level, economically managing memory space for efficiently querying complex data. More details are stated as follows.

First, previous studies have simplified complex state spaces by aggregating them into discrete points in lower-dimensional spaces, avoiding the need for lookup tables. However, this approach has limitations. It is often difficult to determine a rational basis for state aggregation [1,3,6], and the granularity of aggregation affects the state space size [11]. If the granularity is too small, it fails to simplify the state space; if too large, it can cause significant discrepancies in the value functions of different original states. These discrepancies introduce errors in state-value function approximation that can't be mitigated by increasing iterations. This study offers a solution that removes the need for state aggregation in certain MDP problems, providing a flexible data structure that allows for more adaptable state aggregation and disaggregation, with varying granularity across different regions of the state space.

Second, with the rapid advancement of hardware, computer memory has reached unprecedented capacities, with desktop computers commonly equipped with 32GB of memory. As a result, memory capacity is no longer a concern for real-world problems, as long as memory is not idle. The main challenge now is balancing memory management and efficient data querying. This study contributes by designing a data structure with $O(1)$ query time and no memory waste. We define a state key as a numerical triple and create two 3D arrays indexed by the triple, initially achieving $O(1)$ query time but wasting memory. By exploring the structural properties of the triple, we apply operations like mirroring, rotation, and translation to update the arrays and eliminate idle memory, while maintaining $O(1)$ query complexity.

The remainder of this paper is organized as follows. State key is defined in Sect. 2. Using the defined key, the lookup table is designed in Sect. 3. Section 4 concludes this paper.

2 State Key

2.1 Key Definition

Clearly, querying entries in a lookup table involves two steps: first, determining the key, which may be a single value or a combination of values depending on the table's structure, and second, using the key to find the target entry. We associate D-elements with indexes from 1 to d, P-elements with indexes from $d+1$ to $d+p$. For simplicity, we directly use the associated indexes to represent these distinct elements, as shown in our illustrative *example*. First, we define a function denoted as $f_2(\mathrm{D}, \mathcal{D} \subset \mathrm{D})$ that outputs an index of a subset $\mathcal{D}$ through sorting subsets of D based on their binary representations.

Definition 1. *Give a basic set* D*, first arrange the elements of* D *in descending order. Then in response to a given subset* $\mathcal{D} \subset \mathrm{D}$*, represent* $\mathcal{D}$ *using a binary number of length* $|\mathrm{D}|$ *with each bit position corresponds to an element in* D*. If the bit is 1, it means the element is present in* $\mathcal{D}$*; else, the element is absent. Finally, convert this binary number into a decimal number, which serves as the output of* $f_2(\mathrm{D}, \mathcal{D})$.

Definition 2. *First, arrange all possible combinations of loc and* $\mathcal{P}$ *by a certain rule as follows: (i) combinations with* $loc \in \mathrm{D}$ *are ordered earlier than those with* $loc \in \mathrm{P}$*; (ii) combinations with a smaller size of* $\mathcal{P}$ *are indexed earlier; (iii) combinations with smaller loc are indexed earlier; and (iv) combinations with the same loc and the same size of* $\mathcal{P}$ *are indexed by applying alphabet order on their* $\mathcal{P}$*s. Then associate natural numbers starting from 0 to these combinations sequentially. Finally, for a specific combination of loc and* $\mathcal{P}$*,* $f_3(loc, \mathcal{P})$ *returns the number associated to this combination.*

Then, for a given state $(c,\ loc,\ \mathcal{D},\ \mathcal{P})$, we define a triple $i_1 \mid i_2 \mid i_3$ shown as Eq. (2) with $\sum_{\emptyset} = 0$.

$$i_1 = c,\ c \in \mathbb{N}\ \&\ c \le k \quad \Big| \quad i_2 = f_2(\mathrm{D}, \mathcal{D}) \quad \Big| \quad i_3 = f_3(loc, \mathcal{P}) \tag{2}$$

Above, i_1 takes the value of c directly, i_2 is obtained by $f_2(\mathrm{D}, \mathcal{D})$ (Definition 1), and i_3 takes the output of function $f_3(loc, \mathcal{P})$ with loc and $\mathcal{P}$ as its two independent variables. We define $f_3(loc, \mathcal{P})$ as Definition 2.

Lemma 1. (1) *For any two states whose* $\mathcal{D}$*s are disjoint and of which the union is* D*, the sum of their corresponding* i_2*s is* $2^d - 1$. (2) *There are totally* $\sum_{j=0}^{k} d \cdot \binom{p}{j} + \sum_{j=1}^{k} j \cdot \binom{p}{j}$[2] *different values for* i_3*, and* (3) *under the arrangement rule mentioned in Definition 2, these values vary from 0 to* $\sum_{j=0}^{k} d \cdot \binom{p}{j} + \sum_{j=1}^{k} j \cdot \binom{p}{j} - 1$.

Proof. (1) Based on Definition 1 and Eq. (2), this lemma holds obviously. (2) The first term $\sum_{j=0}^{k} d \cdot \binom{p}{j}$ corresponds to $loc \in \mathrm{D}$, where the component loc has totally d choices and the component $\mathcal{P}$ has totally $\binom{p}{j}$ choices. The second term $\sum_{j=1}^{k} j \cdot \binom{p}{j}$ corresponds to $loc \in \mathrm{P}$, which is derived from the fact that the number of total P-elements in combination of loc and $\mathcal{P}$ can not be more than k when $loc \in \mathrm{P}$. Specifically, in response to exactly j P-elements, there are totally $\binom{p}{j}$ choices for these P-elements and for each choice, any of these j P-elements can be taken as component loc. (3) According to the arrangement rule, this conclusion holds obviously.

[2] In case of $p < j$, $\binom{p}{j} = 0$.

Remark 1. Referring to Lemma 1, we can precompute an array of numbers 2^j for $j = 0, 1, \cdots, d-1$, in advance with time consumed $O(d)$. Consequently, during implementation, for any encountered state, its i_2 value can be obtained through no more than d addition operations.

Example : Values of i_2 and i_3 are respectively shown in Tables 1 and 2.

Table 1. The correspondence between i_2 and $\mathcal{D}$.

i_2	0	1	2	3
$\mathcal{D}$	$\emptyset$	{1}	{2}	{1, 2}

Table 2. The correspondence between i_3 and $(loc, \mathcal{P})$.

i_3	$(loc, \mathcal{P})$	i_3	$(loc, \mathcal{P})$	i_3	$(loc, \mathcal{P})$
0	(1, $\emptyset$)	8	(1, {3, 4})	16	(5, $\emptyset$)
1	(2, $\emptyset$)	9	(1, {3, 5})	17	(3, {4})
2	(1, {3})	10	(1, {4, 5})	18	(3, {5})
3	(1, {4})	11	(2, {3, 4})	19	(4, {3})
4	(1, {5})	12	(2, {3, 5})	20	(4, {5})
5	(2, {3})	13	(2, {4, 5})	21	(5, {3})
6	(2, {4})	14	(3, $\emptyset$)	22	(5, {4})
7	(2, {5})	15	(4, $\emptyset$)		

According to Definition 2 and Lemma 1, the number of $(loc, \mathcal{P})$s with $loc \in \mathrm{D}$ and $\mathcal{P}$ of size l is $d \cdot \binom{p}{l}$, and the number of states with $loc \in \mathrm{P}$ and $\mathcal{P}$ of size l is $(l+1) \cdot \binom{p}{l+1}$. Then, given a $(loc, \mathcal{P})$ as $(loc, \{j_1, j_2 \cdots, j_l\})$, its i_3's lower bound denoted by $i_{3,L}$ can be quickly computed through Eq. (3). Obviously, in case of $l = 0$ (i.e., $\mathcal{P} = \emptyset$), the lower bound obtained is the exact value of i_3.

$$i_{3,L} = \begin{cases} d\sum_{j=0}^{l-1} \binom{p}{j} + (loc-1) \cdot \binom{p}{l}, & loc \in \mathrm{D} \\ d\sum_{j=0}^{k} \binom{p}{j} + \sum_{j=1}^{l-1} j\binom{p}{j} + (loc-d-1)\binom{p-1}{l}, & loc \in \mathrm{P} \end{cases} \tag{3}$$

Combined with the arrangement rule mentioned in Definition 2, we get the exact i_3 through an iterative method, under a reasonable assumption that elements in $\mathcal{P}$ are arranged increasingly by their indices. First, with all $\binom{p}{l}$ combinations being lexicographically ordered, we compute the order for a specific combination by Algorithm 1 (A1) [5]. Then, we compute i_3 through Algorithm 2 (A2). In order to save computing time, similar to computation on i_2, we also compute values of $\binom{p-g}{j}, j = 0,1,\cdots, \min\{p-g, k\}$, $0 \le g < p$, in advance with time consumed $O(p \cdot k)$. Clearly, with these combinations being prepared in advance, Algorithm 1 is $O(p)$-time consumed, implying that Algorithm 2 is also $O(p)$-time consumed.

Algorithm 1:

Input: p, loc, and combination $(j_1, j_2, \cdots, j_l)$ in which $j_1 < j_2 < \cdots < j_l$

```
Let j_0 = 0 and r = 0 ;
for i from 1 to l do
    if j_{i-1} + 1 ≤ j_i − 1 then
        for g from j_{i-1} + 1 to j_i − 1 do
            r = r + C(p−g, l−i)
```

Output: r

Algorithm 2:

Input: $(i, [j_1, \cdots j_l])(l \neq 0)$, d, p

```
1 Set i_3 = i_{3,L}, P[l] = {j_1, j_2 ··· j_l} and P'[l] = ∅;
2 if i > d go to 4; else go to 3 ;
3 i_3 = i_3 + A1 ( p, l, P[l] ), go to 9;
4 for g from 0 to l do
5     if P[g] < i then
6         P'[g] = P[g]
7     else
8         P'[g] = P[g] − 1
9 i_3 = i_3 + A1 (p − 1, l, P'[l] );
```

Output: i_3.

2.2 Complexity and Memory Waste

We differentiate between the time spent preparing computations in advance for all state keys before problem-solving, and the time spent on each state key during problem-solving (e.g., on policy training or approximate value iteration). We term these time expenditures as offline time consumption and online time consumption, respectively. As mentioned before, offline time consumption is totally $O(d + p \cdot k)$.

Regard online time consumption, in response to an encountered state, computation on state's key values involves two parts. The first is to compute this encountered state's key such that, by using the key, its value function can be queried from the memory. From Sect. 2.1, its i_1 is trivial and i_2 can be obtained in $O(d)$. Further, with $i_{3,L}$ being obtained in $O(k)$, i_3 can be obtained in $O(p)$ $(k \leq p)$ by A2. Thus, this part of computation can be done with $O(d + p)$. The second is to compute keys of all possible subsequent states that could be resulted from an encountered state, making their value functions can be queried from the memory. Note that $i_{3,L}$ is only related with component loc and the size of $\mathcal{P}$. In response to each feasible action $j \in \mathcal{D} \cup \mathcal{P}$ of an encountered state $(c,\ loc,\ \mathcal{D},\ \mathcal{P})$, we compute all possible subsequent states' lower bound $i_{3,L}$s by Eq. (3). We use $(c',\ loc',\ \mathcal{D}',\ \mathcal{P}')$ to denote a possible subsequent state resulted from taking action j at $(c,\ loc,\ \mathcal{D},\ \mathcal{P})$. With all possible $(c',\ loc',\ \mathcal{D}',\ \mathcal{P}')$s being listed in a sequence where the set difference $\mathcal{D} \setminus \mathcal{D}'$s and $\mathcal{P} \setminus \mathcal{P}'$s are arranged lexicographically, the i_2 and i_3 of a $(c',\ loc',\ \mathcal{D}',\ \mathcal{P}')$ can directly obtained from its close precedent state within constant operations. Thus, clearly, for any given encountered state and one of its feasible actions, the computational complexity of computing these keys is $O(1)$.

So far, we have defined a state key consisting of three numerical values, on which the computation does not cause any deterioration in the time complexity of the MDP itself. It implies that the key can be obtained with $O(1)$ time consumed. Thus, clearly, if the key is used to query state value function in memory directly without any additional computation, an $O(1)$-query complexity method has already been obtained. We can achieve it by designing two 3-dimensional arrays with (i_1, i_2, i_3) and $\left(i_1, i_2, i_3 - \sum_{j=0}^{k} d \cdot \binom{p}{j}\right)$ respectively as indices in their three dimensions. This will result in two 3-dimensional arrays in the size of $\mathcal{S}_{o,1}^d \times \mathcal{S}_{o,2}^d \times \mathcal{S}_{o,3}^d$ and $\mathcal{S}_{o,1}^p \times \mathcal{S}_{o,2}^p \times \mathcal{S}_{o,3}^p$, with these dimensional sizes being specified respectively as Eq. (4). Note that here "o" denotes the array size is *original*, which will be updated later and marked by "u".

$$\begin{cases} \mathcal{S}_{o,1}^d = k+1, \quad \mathcal{S}_{o,2}^d = 2^d, \quad \mathcal{S}_{o,3}^d = d \cdot \sum_{j=0}^{k} \binom{p}{j} \\ \mathcal{S}_{o,1}^p = k+1, \quad \mathcal{S}_{o,2}^p = 2^d, \quad \mathcal{S}_{o,3}^p = \sum_{j=1}^{k} j \cdot \binom{p}{j} \end{cases} \tag{4}$$

Example: We use two arrays in the size of $3 \times 4 \times 14$ and $3 \times 4 \times 9$ to store D-triples and P-triple respectively. The storage result is shown as Fig. 1, with a cube representing an entry and a cube in white representing an applied-but-idle entry. We can observe that this storage method will result in wasted applied memory, accounting for totally 65.22% (i.e., the ratio of the number of white cubes to the total number of cubes).

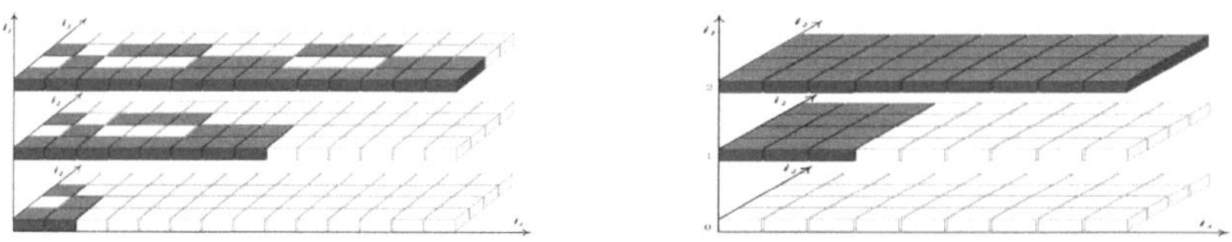

(a) D-array (b) P-array

Fig. 1. Arrays directly indexed by key triple in our *example*.

However, as illustrated by our *example* above, arrays directly indexed by key triple will result in a huge waste of memory. Therefore, in order to obtain a lookup table that does not waste memory while maintaining $O(1)$-query complexity, we need to design two specific data structures respectively for D-triples and P-triples. Specifically, such two structures will be derived through conducting a series of operations on arrays obtained above. For convenience, we will refer to the arrays obtained above as original D-array and original P-array later. We will detail these operations next section.

3 Lookup Table Design

In this section, by exploring properties of the state key, we design a lookup table with $O(1)$-query complexity and no memory waste.

3.1 Properties of State Key

Property 1. Each state is associated with a unique integer triple $i_1 \| i_2 \| i_3$.

By the definition of triple $i_1 \mid i_2 \mid i_3$, Property 1 is evident. It is just the reason why such triple can act as state key. Further, the key triples of all states forms a collection of integer triples. We first give a definition of an integer triple collection being consecutive as follows, and then propose Property 2.

Definition 3. *A collection of integer triples is consecutive if it includes all possible combinations of its components within their specified ranges.*

Property 2. The collection of all state key triple $i_1 \mid i_2 \mid i_3$s is not consecutive.

Proof. Our *example* can demonstrate this property. There, although i_1 can vary successively from 0 to 2, i_2 can vary successively from 0 to 3, and i_3 can vary successively from 0 to 22, there are no states corresponding to triples as $i_1 = 0 \mid i_2 \mid i_3 \geq 2$.

For one triple $i_1 \mid i_2 \mid i_3$, we regulate that it is valid if there is one state corresponding to it; otherwise, it is invalid. In addition, we define a triple as D-triple if it corresponds to a state with *loc* $\in$ D, while an one as P-triple if it corresponds to a state with *loc* $\in$ P.

Property 3. (1) For any two D-triples which have complimentary i_2 (the sum of their i_2s is $2^d - 1$) and the same i_1 and i_3, exactly one of them is valid. (2) For all P-triples with the same i_1 and i_3, they are either entirely valid or entirely invalid. (3) For those P-triples with same i_1, if one of them with $i_3 = i_3'$ is invalid, since i_3' denotes a specific value in the domain of i_3, all such triples satisfying $i_3 > i_3'$ are invalid. (4) All P-triples with $i_1 = 0$ are all invalid and all P-triples with $i_1 = k$ are all valid.

By the definition of triple $i_1 \mid i_2 \mid i_3$, the Property 3 obviously holds. We can also verify them through *example*.

Theorem 1. *The total number of valid* D*-triples with* $i_1 = i$ *and* $i_1 = k - 1 - i$, $0 \leq i \leq k - 1$, *is no more than the number of valid* D*-triples with* $i_1 = k$.

Proof. The number of all D-triples (valid and invalid) with $i_1 = k$ is $d \cdot 2^d \cdot \sum_{j=0}^{k} \binom{p}{j}$. In addition, the total number of all D-triples with $i_1 = i$ and $i_1 = k - 1 - i$ $(0 \leq i \leq k - 1)$ is $d \cdot 2^d \cdot \sum_{j=0}^{i} \binom{p}{j} + d \cdot 2^d \sum_{j=0}^{k-1-i} \binom{p}{j}$. Clearly, $\sum_{j=0}^{i} \binom{p}{j} + \sum_{j=0}^{k-1-i} \binom{p}{j} < \sum_{j=0}^{k} \binom{p}{j}$. Combined with Property 3, this theorem holds.

Theorem 2. *The total number of valid* P*-triples with* $i_1 = i$ *and* $i_1 = k - i$, $1 \leq i \leq k - 1$, *is no more than the number of valid* P*-triples with* $i_1 = k$.

Proof. The number of all P-triples (valid and invalid) with $i_1 = k$ is $\sum_{j=1}^{k} j \cdot \binom{p}{j}$. The total number of all P-triples with $i_1 = i$ and $i_1 = k - i$ $(1 \leq i \leq k - 1)$ is $\sum_{j=1}^{k-i} j \cdot \binom{p}{j} + \sum_{j=1}^{i} j \cdot \binom{p}{j}$. Clearly, $\sum_{j=1}^{k-i} j \cdot \binom{p}{j} + \sum_{j=1}^{i} j \cdot \binom{p}{j} \leq \sum_{j=1}^{k} j \cdot \binom{p}{j}$. Combined with Property 3, this theorem holds.

3.2 Lookup Table Design

D-triples and P-triples will be addressed respectively. Each are stored within a three-dimensional array in memory. Note that "store valid triple" always means "store the value function of the state indexed by this triple". We do not distinguish them later.

First, we design two 3-dimensional arrays in the sizes of $\mathcal{S}^d_{o,1} \times \mathcal{S}^d_{o,2} \times \mathcal{S}^d_{o,3}$ and $\mathcal{S}^p_{o,1} \times \mathcal{S}^p_{o,2} \times \mathcal{S}^p_{o,3}$ respectively to store D-triples and P-triples, referring to the original D-array and the original P-array. The values of $\mathcal{S}^d_{o,1}$, $\mathcal{S}^d_{o,2}$, $\mathcal{S}^d_{o,3}$, $\mathcal{S}^p_{o,1}$, $\mathcal{S}^p_{o,2}$ and $\mathcal{S}^p_{o,3}$ can be obtained through Eq. (4).

With notations being used a little confusingly, we index each entry in both D-array and P-array by (i_1, i_2, i_3) from now on. Clearly, one entry indices (i_1, i_2, i_3) in the D-array exactly corresponds to the state with the key $i_1 \mid i_2 \mid i_3$, but in the P-array, the indices (i_1, i_2, i_3) corresponds to the state with the key $i_1 \mid i_2 \mid \left(i_3 + \mathcal{S}^d_{o,3}\right)$. Then, we will derive our target arrays through performing a series of operations on these two original arrays. As a preliminary, we refer to entries with the same i_1 as an entry layer, and valid entries with the same i_1 as a valid entry layer. In addition, for the sake of conciseness later, we first define two functions related to k as Eq. (5).

$$h_d(k) = \begin{cases} 1 + k/2, & k \text{ is even} \\ 1 + (k+1)/2, & k \text{ is odd} \end{cases}, \quad h_p(k) = \begin{cases} 1 + k/2, & k \text{ is even} \\ (k+1)/2, & k \text{ is odd} \end{cases} \tag{5}$$

Operations on the Original D-Array. Operations performed on the original D-array include mirroring, rotation and translation. We will specify them in detail next.

Mirroring. According to the first point of Property 3, any two entries in the D-array that have the same i_1 and i_3 must have complimentary i_2s, that is, the sum of their i_2s is $2^p - 1$, and importantly, exactly one of them is valid. Combined with the fact that the number of entries with specific i_1 and i_3 is even, we focus on one specific entry layer that have same i_1 and i_3 and mirror valid entries of which the $i_2 \geq \mathcal{S}^d_{o,2}/2$ to their counterparts across the center line of the layer in the i_2 direction. In detail, we use a function denoted as $\mathscr{M}_\text{D}$ to describe the relationship between entry indices of the original D-array and the ones of the updated D-array by such mirroring as Eq. (6), where $\mathcal{S}^d_{o,2} - 1$ denotes the maximal index of entries in the i_2 direction.

$$\mathscr{M}_\text{D}(i_1, i_2, i_3) = \begin{cases} (i_1, i_2, i_3), & i_2 < \mathcal{S}^d_{o,2}/2 \\ (i_1, \mathcal{S}^d_{o,2} - 1 - i_2, i_3), & i_2 \geq \mathcal{S}^d_{o,2}/2 \end{cases} \tag{6}$$

Rotation. Continually, rotation will be performed on the updated array through mirroring. Specifically, we will rotate the valid entry layer with fixed i_1 satisfying $h_d(k) - 1 \leq i_1 < k$ in the plane formed by directions i_1 and i_3. The rotation is performed clockwise, transforming the corner entries of the valid entry array which are indexed by $(i_1, 0, 0)$, $(i_1, \mathcal{S}^d_{o,2}/2-1, 0)$, $\left(i_1,\ \mathcal{S}^d_{o,2}/2-1,\ d\sum_{j=0}^{i_1}\binom{p}{j}-1\right)$ and $\left(i_1,\ 0,\ d\sum_{j=0}^{i_1}\binom{p}{j}-1\right)$ before rotation respectively to the entries indexed by $(i_1, \mathcal{S}^d_{o,2}/2-1, \mathcal{S}^d_{o,3}-1)$, $(i_1, 0, \mathcal{S}^d_{o,3}-1)$, $(i_1, 0, 0)$ and $(i_1, \mathcal{S}^d_{o,2}/2-1, 0)$ after rotation. In detail, we use a function denoted as $\mathscr{R}_{\mathrm{D}}$ to describe the relationship between entry indices before and after such operation is performed as Eq. (7), where $h_d(k)$ is defined in Eq. (5).

$$\mathscr{R}_{\mathrm{D}}(i_1, i_2, i_3) = \left(i_1,\ \mathcal{S}^d_{o,2}/2-1-i_2,\ \mathcal{S}^d_{o,3}-1-i_3\right),\quad h_d(k)-1 \le i_1 < k \tag{7}$$

Translation. Focusing on the already-updated D-array so far, the translation operation primarily involves moving the entry layer with the same i_1 satisfying $h_d(k)-1 \le i_1 \le k$ along the i_1 direction. So such operation only makes i_1 values of the involved entries change while their i_2 and i_3 values remain unchanged. Specifically, we use a function denoted as $\mathscr{T}_{\mathrm{D}}$ to describe the relationship between entry indices before and after such operation as Eq. (8).

$$\mathscr{T}_{\mathrm{D}}(i_1, i_2, i_3) = \begin{cases} (k-1-i_1,\ i_2,\ i_3), h_d(k)-1 \le i_1 < k \\ (h_d(k)-1,\ i_2,\ i_3),\ i_1 = k \end{cases} \tag{8}$$

Operations on the Original P**-Array**

Mirroring. Given a specific entry layer with $i_1 \ge h_p(k)$, we mirror the valid entries in this layer to their counterparts across the entry layer's center line which is perpendicular to the i_3 direction. Specifically, we use a function denoted as $\mathscr{M}_{\mathrm{P}}$ to describe the relationship between entry indices before and after such mirroring operation as Eq. (9), where $\mathcal{S}^p_{o,3}-1$ denotes the maximal index of entries in the i_3 dimension.

$$\mathscr{M}_{\mathrm{P}}(i_1, i_2, i_3) = \left(i_1,\ i_2,\ \mathcal{S}^p_{o,3}-1-i_3\right),\quad i_1 \ge h_p(k) \tag{9}$$

Translation. Then, we preform translation operation on the updated P-array, similar as on the D-array. We translate valid entry layers with $i_1 \ge h_p(k)$ along the i_1 direction, making only i_1 values of the involved entries change while their i_2 and i_3 remain unchanged. We use a function denoted as $\mathscr{T}_{\mathrm{P}}$ to describe the relationship between entry indices before and after such operation as Eq. (10).

$$\mathscr{T}_{\mathrm{P}}(i_1, i_2, i_3) = \left(\mathcal{S}^p_{o,1}-1-i_1,\ i_2,\ i_3\right),\quad i_1 \ge h_p(k) \tag{10}$$

Array Size Reduction. Till now, after operations such as mirroring, rotation and translation have been performed, the already-updated two arrays have some properties shown as Property 4.

Property 4. (1) All entries in D-array with $i_1 \ge h_d(k)$ or $i_2 \ge \mathcal{S}^d_{o,2}/2$ are idle, and all entries in P-array with $i_1 \ge h_d(k)$ are idle. (2) Besides entries involved in (1) in both arrays, in the remaining entries those with variable i_2 but fixed i_1 and i_3 are either all valid or all idle. (3) In both arrays, no entry is reoccupied.

Proof. (1) For D-array, this property can be directly obtained from Eq. (6) and (8). For P-array this property can be obtained from Eq. (10). (2) This property is based on the key definition, especially the definition of i_2 as stated in Definition 1. (3) Property 3 and Theorem 2 guarantee there is no re-occupied entry in both arrays.

Based on the Property 4 above, we can reduce the original D-array by both changing $\mathcal{S}^d_{o,1}$ from $k+1$ to $h_d(k)$ and changing $\mathcal{S}^d_{o,2}$ from 2^d to 2^{d-1}. Also, we reduce the original P-array by changing $\mathcal{S}^p_{o,1}$ from $k+1$ to $h_p(k)$. However, although the reduction in both the i_1 and i_2 dimensions have released a large amount of idle entries, there are still some idle ones (as the white cubes shown in both the fourth step of Fig. 2 and the third step in Fig. 3. In order to achieve no memory being wasted completely, we set both $\mathcal{S}^d_{o,3}$ and $\mathcal{S}^d_{o,3}$ to be dynamic, which depend on the value of i_1. Thus, we treat the array size reduction as a new operation and denote the newly-updated array scale respectively as $\mathcal{S}^d_{u,1} \times \mathcal{S}^d_{u,2} \times \mathcal{S}^d_{u,3}$ and $\mathcal{S}^p_{u,1} \times \mathcal{S}^p_{u,2} \times \mathcal{S}^p_{u,3}$. The relationship between these new dimension sizes and the original ones is formally shown as Eq. (11–12).

$$\mathcal{S}^d_{u,1} = h_d(k),\ \mathcal{S}^d_{u,2} = \mathcal{S}^d_{o,2}/2,\quad \mathcal{S}^d_{u,3} = \begin{cases} d\sum\limits_{j=0}^{k}\binom{p}{j}, & i_1 = \mathcal{S}^d_{u,1} - 1 \\ d\sum\limits_{j=0}^{i_1}\binom{p}{j} + d\sum\limits_{j=0}^{k-1-i_1}\binom{p}{j}, & others \\ d\sum\limits_{j=0}^{i^d_1}\binom{p}{j}, & i^d_1 = (k-1)/2 \end{cases} \tag{11}$$

$$\mathcal{S}^p_{u,1} = h_p(k),\ \mathcal{S}^p_{u,2} = \mathcal{S}^p_{o,2}, \mathcal{S}^p_{u,3} = \begin{cases} \sum\limits_{j=1}^{i^p_1} j\binom{p}{j}, & i^p_1 = k/2 \\ \sum\limits_{j=1}^{k-i^p_1} j\cdot\binom{p}{j} + \sum\limits_{j=1}^{i^p_1} j\cdot\binom{p}{j}, & others \end{cases} \tag{12}$$

Summarize. To sum up, we give a theorem as follows.

Theorem 3. *With all involved combinations (i.e.,* $\binom{p}{j}, j = 0, 1, \cdots, k$ *) prepared in advance, for a valid triple* $i_1 \mid i_2 \mid i_3$*, we can query its state value function in a specific 3-dimensional array in* $O(1)$ *time, with no memory wasted.*

Proof. (1) Regarding no memory being wasted, relying on Theorems 1–2 and based on functions $\mathscr{M}_{\mathrm{D}}$ in (6), $\mathscr{R}_{\mathrm{D}}$ in (7), $\mathscr{T}_{\mathrm{D}}$ in (8),$\mathscr{M}_{\mathrm{P}}$ in (9) and $\mathscr{T}_{\mathrm{P}}$ in (10) as well as size reduction in (11–12), it is clear that in the finally-achieved two arrays, all idle entries have been deleted and no one is reoccupied. (2) With (i^u_1, i^u_2, i^u_3) denoting the index of one entry in the final arrays, we treat all the operations as a whole and use a function to denote the mapping from one valid entry indexed by (i_1, i_2, i_3) in the original array and the index (i_1^u, i_2^u, i_3^u) of its counterpart in the final array. Taking $\mathcal{I}_D(i_1, i_2, i_3)$ and $\mathcal{I}_P(i_1, i_2, i_3)$ respectively for D-array and P-array, they can be obtained as follows.

In the case of k being odd:

$$\mathcal{I}_D(i_1, i_2, i_3) = \begin{cases} (i_1, i_2, i_3),\ i_1 \le (k-1)/2 \ \& \ i_2 < 2^{d-1} \\ (i_1, 2^d - 1 - i_2, i_3), i_1 \le (k-1)/2 \ \& \ i_2 \ge 2^{d-1} \\ \left(k - 1 - i_1, 2^{d-1} - 1 - i_2, \mathcal{S}_{u,3}^d - 1 - i_3\right), \begin{cases} (k-1)/2 < i_1 < k \ \& \\ i_2 < 2^{d-1} \end{cases} \\ (k - 1 - i_1,\ i_2 - 2^{d-1}, \mathcal{S}_{u,3}^d - 1 - i_3), \begin{cases} (k-1)/2 < i_1 < k \ \& \\ i_2 \ge 2^{d-1} \end{cases} \\ ((k+1)/2, i_2,\ i_3), i_1 = k \ \& \ i_2 < 2^{d-1} \\ \left((k+1)/2, 2^d - 1 - i_2, i_3\right), i_1 = k \ \& \ i_2 \ge 2^{d-1} \end{cases} \tag{13}$$

$$\mathcal{I}_P(i_1, i_2, i_3) = \begin{cases} (i_1,\ i_2,\ i_3), & i_1 < (k+1)/2 \\ \left(k - i_1,\ i_2,\ \mathcal{S}_{u,3}^p - 1 - i_3\right), & i_1 \ge (k+1)/2 \end{cases} \tag{14}$$

In the case of k being even:

$$\mathcal{I}_D(i_1, i_2, i_3) = \begin{cases} (i_1, i_2, i_3),\ k/2 < i_1 \ \& \ i_2 < 2^{d-1} \\ \left(i_1, 2^d - 1 - i_2, i_3\right),\ k/2 < i_1 \ \& \ i_2 \ge 2^{d-1} \\ \left(k - 1 - i_1, 2^{d-1} - 1 - i_2, \mathcal{S}_{u,3}^d - 1 - i_3\right), k/2 \le i_1 < k \ \& \ i_2 < 2^{d-1} \\ \left(k - 1 - i_1, i_2 - 2^{d-1}, \mathcal{S}_{u,3}^d - 1 - i_3\right), k/2 \le i_1 < k \ \& \ i_2 \ge 2^{d-1} \\ (k/2,\ i_2,\ i_3), i_1 = k \ \& \ i_2 < 2^{d-1} \\ \left(k/2,\ 2^d - 1 - i_2,\ i_3\right), i_1 = k \ \& \ i_2 \ge 2^{d-1} \end{cases} \tag{15}$$

$$\mathcal{I}_P(i_1, i_2, i_3) = \begin{cases} (i_1,\ i_2,\ i_3), & i_1 \le k/2 \\ \left(k - i_1,\ i_2,\ \mathcal{S}_{u,3}^p - 1 - i_3\right), & i_1 > k/2 \end{cases} \tag{16}$$

Based on the calculation of $\mathcal{I}_D(i_1, i_2, i_3)$ and $\mathcal{I}_P(i_1, i_2, i_3)$ in both cases of k being even and odd shown as above, we can conclude that given one state's key triple $i_1 \mid i_2 \mid i_3$, its value function can be queried with $O(1)$-consumed time.

Corollary 1. *Since one state's key triple $i_1 \mid i_2 \mid i_3$ can be obtained with $O(1)$ time consumed, a lookup table design method with $O(1)$-query complexity and no memory waste has been obtained.*

Example: The generation of the final D-array, shown in Fig. 2, involves five sub-figures representing four steps: mirroring, rotation, translation, and size reduction. The first sub-figure depicts the original D-array (Fig. 1a), with different colors marking valid entries for various operations: gray for untouched, blue for mirroring, yellow for translation, pink for both mirroring and translation, green for rotation and translation, and red for all three operations. A similar process applies to the P-array, shown in Fig. 3, where gray marks untouched entries and pink indicates relocated entries. Four colors highlight the corner entries of the relocated layer. For detailed storage information of the original and final D-arrays and P-arrays, please refer to the supplement.

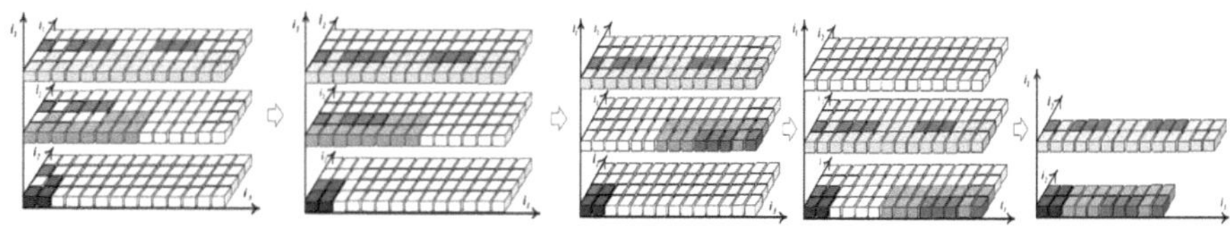

Fig. 2. The generation process of the final D-array in the *example*. (Color figure online)

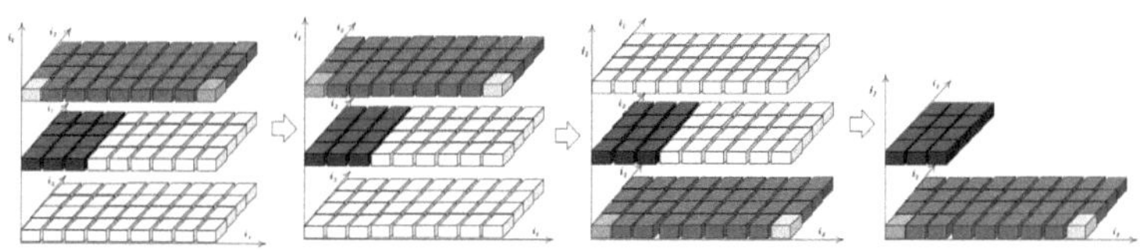

Fig. 3. The evolving process of the final P-array in the *example*. (Color figure online)

4 Conclusion

In solving complex MDPs, two common methods are used to represent the state value function: lookup table and parameterized function representation. The former is space-efficient but computationally expensive due to the need for optimizing function parameters. The latter is computationally efficient but struggles with large-scale state spaces, balancing between memory waste and query complexity. This study proposes a lookup table design method for dynamic vehicle routing problems, achieving $O(1)$ query complexity and no memory waste. Our contributions include: (1) making state aggregation optional in complex MDPs, and (2) efficiently managing memory for querying complex data. Future research will focus on applying this method to similar MDPs to improve solution efficiency and exploring methods for designing lookup tables that balance query efficiency and memory conservation for more complex MDPs.

References

1. Bertsekas, D., Tsitsiklis, J.N.: Neuro-dynamic programming. Athena Sci. (1996)
2. Corneil, D., Gerstner, W., Brea, J.: Efficient model-based deep reinforcement learning with variational state tabulation. In: International Conference on Machine Learning, pp. 1049–1058. PMLR (2018)
3. Duan, Y., Ke, T., Wang, M.: State aggregation learning from Markov transition data. In: Advances in Neural Information Processing Systems, vol. 32 (2019)
4. Jiang, N.: Notes on tabular methods (2020)
5. Kreher, D.L., Stinson, D.R.: Combinatorial algorithms: generation, enumeration, and search. ACM SIGACT News **30**(1), 33–35 (1999)
6. Rogers, D.F., Plante, R.D., Wong, R.T., Evans, J.R.: Aggregation and disaggregation techniques and methodology in optimization. Oper. Res. **39**(4), 553–582 (1991)
7. Sutton, R.S., Barto, A.G.: Reinforcement Learning: An Introduction. MIT Press, Cambridge (2018)
8. Ulmer, M.W., Goodson, J.C., Mattfeld, D.C., Hennig, M.: Offline-online approximate dynamic programming for dynamic vehicle routing with stochastic requests. Transp. Sci. **53**(1), 185–202 (2019)
9. Ulmer, M.W., Mattfeld, D.C., Köster, F.: Budgeting time for dynamic vehicle routing with stochastic customer requests. Transp. Sci. **52**(1), 20–37 (2018)
10. Wu, Y.: A dynamic and stochastic cumulative capacitated vehicle routing problem. Asia-Pac. J. Oper. Res. 2450008 (2024)
11. Wu, Y., Jian, M.: State aggregation and lower bound-based ADP approach for dynamic pick-up routing problem with capacity constraint. IEEE Trans. Intell. Transp. Syst. 1–13 (2024)
12. Wu, Y., Zeng, B.: Dynamic parcel pick-up routing problem with prioritized customers and constrained capacity via lower-bound-based rollout approach. Comput. Oper. Res. **154**, 106176 (2023)
13. Wu, Y., Zeng, B., Huang, S.: A dynamic strategy for home pick-up service with uncertain customer requests and its implementation. Sustainability **11**(7), 2060 (2019)
14. Wu, Y., Zeng, B., Jian, M.: ADP-and rollout-based dynamic vehicle routing for pick-up service via budgeting capacity. Flexible Serv. Manuf. J. 1–45 (2024)
15. Yin, M., Wang, Y.X.: Asymptotically efficient off-policy evaluation for tabular reinforcement learning. In: International Conference on Artificial Intelligence and Statistics, pp. 3948–3958. PMLR (2020)
16. Zhou, Z., Zhou, Z., Bai, Q., Qiu, L., Blanchet, J., Glynn, P.: Finite-sample regret bound for distributionally robust offline tabular reinforcement learning. In: International Conference on Artificial Intelligence and Statistics, pp. 3331–3339. PMLR (2021)

Logic and Proof Theory

On a Computability-Theoretic Approach to Boolean-Valued Models

Nikolay Bazhenov[1,2,3](✉) and Manat Mustafa[3]

[1] Novosibirsk State University, Novosibirsk, Russia
[2] Innopolis University, Innopolis, Russia
nickbazh@yandex.ru
[3] Nazarbayev University, Astana, Kazakhstan
manat.mustafa@nu.edu.kz

Abstract. The paper outlines a possible computability-theoretic approach to the theory of Boolean-valued models. We introduce the notion of a decidable Boolean-valued model **A**. This notion is a natural extension of the classical definition of a decidable structure. Intuitively speaking, in order to formally define **A**, one needs to (algorithmically) work only with countably many truth values taken from a computable Boolean algebra. This convention allows us to view such **A** through the lens of computable structure theory.

We show that the introduced approach provides a rich class of computable structures. We prove that any possible computable dimension can be realized by an appropriately constructed, decidable Boolean-valued model **A**. Consequently, there exists a model **A** having computable dimension 2 (that is, **A** possesses precisely two computable presentations, up to computable isomorphisms). In contrast, it is known that dimension 2 can be realized neither by a classical decidable structure nor by a computable Boolean algebra.

Keywords: Boolean-valued model · Computable structure · Decidable structure · Boolean algebra · Computable dimension

1 Introduction

The paper outlines a computability-theoretic approach to Boolean-valued models. Boolean-valued models constitute an integral part of the modern mathematical logic. The first groundbreaking application of Boolean-valued models was provided by Scott and Solovay (see [23]): they showed that the seminal Cohen's forcing could be fully re-cast in the terms of Boolean-valued models (a similar approach was independently developed by Vopěnka [25]). Nowadays, the theory of Boolean-valued models has become a standard tool in set theory and non-standard analysis. We refer to the monographs [4,15] for a detailed exposition of applications of Boolean-valued models.

On the other hand, to our best knowledge, computability-theoretic investigations of Boolean-valued models remain largely undeveloped. (We note that

M. Li et al. (Eds.): TAMC 2025, LNCS 16084, pp. 79–92, 2026.
https://doi.org/10.1007/978-981-95-4839-2_7

a somewhat related framework was considered in [21], where Boolean-valued models are used as a tool to develop the theory of fuzzy algebraic systems.) The main goal of this paper is to initiate systematic investigations of Boolean-valued models from the point of view of modern *computable structure theory.*

Computable structure theory studies relations between algorithmic and structural properties of various mathematical structures [2,5,8]. The core methods of this theory are based on Turing computable algorithms. Consider an algebraic structure $\mathcal{S} = (S; f_1, f_2, \ldots, f_m; P_1, P_2, \ldots, P_n)$—for example, one can think about the ordered field of rationals $(\mathbb{Q}; +, \times; \leq)$ or the ordered ring of integers $(\mathbb{Z}; +, \times; \leq)$. The structure $\mathcal{S}$ is *computable* if its domain S is a (Turing) computable set, the functions $f_1, f_2, \ldots, f_m$ are computable, and the predicates $P_1, P_2, \ldots, P_n$ are also computable. Recall that (as is customary in model theory) the collection $L = \{f_1, f_2, \ldots, f_m, P_1, P_2, \ldots, P_n\}$ is often called the *signature* of the structure $\mathcal{S}$.

Roughly speaking, the main obstacle in applying computable structure theory to Boolean-valued models is as follows. When one considers a Boolean-valued model $\mathbf{A}$, one has to take truth values t from a given complete Boolean algebra $\mathbb{B}$ (the formal details are discussed in Subsect. 2.1). If the algebra $\mathbb{B}$ is infinite, then $\mathbb{B}$ must be *uncountable* (see, e.g., [14]). However, most methods of computable structure theory are tailored to *countable structures.*

In order to overcome the discussed obstacle, we introduce the notion of a *decidable Boolean-valued model* (Definition 2 in Sect. 3). The basic intuition behind this notion is as follows. Fix a finite signature L for our Boolean-valued models. If one wants to define a Boolean-valued model $\mathbf{A}$, then it is sufficient to work only with (at most) *countably many* truth values $t = [\![\psi]\!]$ taken from a complete Boolean algebra $\mathbb{B}$. Thus, we can restrict our attention only to some countable subalgebra $\mathcal{C}$ of the algebra $\mathbb{B}$.

The described intuition is formalized in a 'streamlined' way as follows. The works [6,18] initiated systematic studies of *decidable structures.* We emphasize that the theory of decidable structures is one of the classical areas of computable structure theory, and we refer to, e.g., [8,11] for a discussion of known results in this area.

A computable structure $\mathcal{S}$ is called *decidable* iff there exists an algorithm which, given a first-order formula $\psi(a_1, a_2, \ldots, a_n)$ (where the elements a_i are taken from the domain $\mathrm{dom}(\mathcal{S})$), decides whether $\psi(a_1, a_2, \ldots, a_n)$ is (classically) true or false in $\mathcal{S}$. A decidable Boolean-valued model $\mathbf{A} = (A, \mathcal{C}, [\![\cdot]\!])$ must satisfy essentially the same condition, but in this case an algorithm must find the truth value $[\![\psi(a_1, a_2, \ldots, a_n)]\!]$ taken from the fixed computable Boolean algebra $\mathcal{C}$. The formal details are given in Sect. 3.

We note that the formal definitions (for Boolean-valued models) in Sects. 2 and 3 are given only for the case when L is a finite signature containing only unary predicates. Nevertheless, the introduced framework can be easily extended to arbitrary computable signatures. We choose to omit discussion of this extension, since the considered specific case is enough for our purposes.

After our working framework is formally established, we focus on the problem of *computable dimension* for decidable Boolean-valued models. We briefly discuss the necessary preliminaries.

If (algebraic) structures $\mathcal{M}$ and $\mathcal{N}$ are isomorphic, we write $\mathcal{M} \cong \mathcal{N}$. If $\mathcal{M}$ is an arbitrary structure and $\mathcal{S}$ is a computable structure such that $\mathcal{S} \cong \mathcal{M}$, then we say that $\mathcal{S}$ is a *computable presentation* of $\mathcal{M}$.

We say that two computable structures $\mathcal{S}$ and $\mathcal{T}$ are *computably isomorphic* if there exists an isomorphism f acting from $\mathcal{S}$ onto $\mathcal{T}$ such that f is a (Turing) computable function. Mal'tsev [17] first observed that the existence of an abstract isomorphism between structures $\mathcal{S}$ and $\mathcal{T}$ does not necessarily imply that $\mathcal{S}$ and $\mathcal{T}$ must be computably isomorphic. This phenomenon was the main motivating reason behind the notion of computable dimension.

Let $\mathcal{S}$ be a computable structure. The *computable dimension* of the structure $\mathcal{S}$ is the number of computable presentations of $\mathcal{S}$, counted up to computable isomorphism relation. For example, the fact that the computable dimension of $\mathcal{S}$ is equal to three means the following. There exist three computable structures $\mathcal{S}_1, \mathcal{S}_2, \mathcal{S}_3$ such that:

- each $\mathcal{S}_i$ is isomorphic to the original structure $\mathcal{S}$;
- for $i \neq j$, the structures $\mathcal{S}_i$ and $\mathcal{S}_j$ are *not computably isomorphic*;
- if $\mathcal{T}$ is a computable presentation of $\mathcal{S}$, then $\mathcal{T}$ must be *computably isomorphic* to some $\mathcal{S}_i$.

We denote the computable dimension of $\mathcal{S}$ by $\dim_c(\mathcal{S})$. Since there exist only countably many computable structures (in a fixed signature L), we always have $\dim_c(\mathcal{S}) \leq \omega$.

Goncharov [9] established that every cardinal k such that $1 \leq k \leq \omega$ can be realized as the computable dimension of a computable structure. In contrast, it is known that computable Boolean algebras have computable dimension either one or ω [10,22]. Nurtazin [19] obtained a similar result for decidable structures: if a structure $\mathcal{S}$ admits two decidable presentations which are not computably isomorphic, then $\dim_c(\mathcal{S}) = \omega$. We refer to, e.g., [3] for a survey of further known results on computable dimension.

The paper is arranged as follows. Section 2 contains the necessary preliminaries. In Sect. 3, we formally introduce our framework of decidable Boolean-valued models $\mathbf{A}$. Proposition 1 gives a useful method for constructing decidable Boolean-valued models.

In Sect. 4, we show how to 'constructively encode' a given computable undirected graph $\mathcal{G}$ into an appropriate decidable Boolean-valued model $\mathbf{A}_{\mathcal{G}}$. The proof of Theorem 1 gives this construction, and Theorem 1 itself says that the computable dimensions of $\mathbf{A}_{\mathcal{G}}$ and $\mathcal{G}$ coincide. This implies that decidable Boolean-valued models can realize any possible computable dimension (Corollary 1). This fact contrasts with the results on Boolean algebras and on decidable structures discussed above. In Sect. 5 we give our concluding remarks.

2 Preliminaries

For the preliminaries on computability theory, we refer to the monograph [20]. Background information on computable structure theory can be found in [2,5].

By $\langle\cdot,\cdot\rangle$ we denote the standard Cantor pairing function: that is, for $x, y \in \mathbb{N}$, we have $\langle x, y\rangle = (x+y)(x+y+1)/2 + x$.

Let L be a signature, and let A be a non-empty set. By $\mathrm{Sent}(L_A)$ we denote the set of all sentences (i.e., first-order formulas without free variables) in the extended signature $L \cup A$. Here we always assume that $L \cap A = \emptyset$, and an element $a \in A$ is treated as a constant symbol. In addition, the formulas from $\mathrm{Sent}(L_A)$ may contain the equality relation $=$.

Notice the following: if both L and A are at most countable, then the set $\mathrm{Sent}(L_A)$ is countable.

Convention 1. If L is a computable signature and A is a computable subset of $\mathbb{N}$, then we use the Gödel numbering γ of first-order formulas in $L \cup A$, and we identify a sentence $\psi \in \mathrm{Sent}(L_A)$ with its Gödel number $\gamma(\psi) \in \mathbb{N}$. Thus, in this case we may assume that $\mathrm{Sent}(L_A)$ is a computable subset of $\mathbb{N}$.

As is customary in computable structure theory, we may assume that every considered *infinite* computable structure $\mathcal{S}$ has domain $\mathbb{N}$ (if not explicitly stated otherwise).

If $\theta(x)$ is a first-order formula and $\mathcal{S}$ is a structure, then we use the following notation:

$$\theta[\mathcal{S}] = \{a \in \mathrm{dom}(\mathcal{S}) : \mathcal{S} \models \theta(a)\}.$$

Boolean algebras are viewed as structures in the signature $L_{BA} = \{\vee, \wedge, \overline{(\cdot)}, 0, 1\}$. We refer to the monograph [14] for preliminaries on Boolean algebras.

For a Boolean algebra $\mathcal{B}$, its standard ordering $\leq_{\mathcal{B}}$ is recovered as follows: $x \leq_{\mathcal{B}} y$ if and only if $x \vee y = y$. Recall that an element $a \in \mathcal{B}$ is an *atom* if a is a minimal non-zero element in $\mathcal{B}$, that is, $0 <_{\mathcal{B}} a$ and there is no b with $0 <_{\mathcal{B}} b <_{\mathcal{B}} a$.

If the considered algebra $\mathcal{B}$ is clear from the context, then we will write $\leq$ in place of $\leq_{\mathcal{B}}$. By $\mathrm{Atom}(\mathcal{B})$ we denote the set of atoms of $\mathcal{B}$.

A Boolean algebra $\mathcal{B}$ is *complete* if every subset $X \subseteq \mathrm{dom}(\mathcal{B})$ has supremum $\bigvee X$ and infimum $\bigwedge X$, with respect to the ordering $\leq_{\mathcal{B}}$. It is known that every infinite complete Boolean algebra has cardinality at least $2^{\aleph_0}$ (see, e.g., Corollary 3.5 in [14]).

In order to emphasize *completeness*, we denote complete Boolean algebras by blackboard bold letters: $\mathbb{A}, \mathbb{B}, \dots$. Meanwhile, arbitrary Boolean algebras are denoted by calligraphic letters: $\mathcal{A}, \mathcal{B}, \dots$.

2.1 Boolean-Valued Models

For background on Boolean-valued models, we refer to [4,24]. Fix a complete Boolean algebra $\mathbb{B}$. Here we provide the definition of a Boolean-valued model $\mathbf{A}$

only for the case when L is a finite signature comprised of unary predicates. Suppose that $L = \{P_0, P_1, \dots, P_n\}$.

A tuple $\mathbf{A} = (A, \xi_0, \xi_1, \dots, \xi_n)$ is a $\mathbb{B}$-*valued model of* L if A is a non-empty set and $\xi_i \colon A \to \mathbb{B}$ for $i \leq n$.

Given a sentence ψ in the extended signature $L \cup A$, one defines the truth value $[\![\psi]\!] \in \mathbb{B}$ recursively, as follows. Suppose that $a, b \in A$.

(A0) $[\![P_i(a)]\!] = \xi_i(a)$, for $i \leq n$.
(A1) $[\![a = b]\!] = 1_{\mathbb{B}}$ if $a = b$, and $[\![a = b]\!] = 0_{\mathbb{B}}$ if $a \neq b$. In other words, the equality relation in $\mathbf{A}$ is treated as in the classical predicate logic.
(A2) $[\![\psi_0 \vee \psi_1]\!] = [\![\psi_0]\!] \vee_{\mathbb{B}} [\![\psi_1]\!]$.
(A3) $[\![\psi_0 \& \psi_1]\!] = [\![\psi_0]\!] \wedge_{\mathbb{B}} [\![\psi_1]\!]$.
(A4) $[\![\neg\psi_0]\!] = \overline{[\![\psi_0]\!]}$.
(A5) $[\![\psi_0 \to \psi_1]\!] = \overline{[\![\psi_0]\!]} \vee_{\mathbb{B}} [\![\psi_1]\!]$.
(A6) $[\![\exists x \psi_0(x)]\!] = \bigvee\{[\![\psi_0(c)]\!] : c \in A\}$.
(A7) $[\![\forall x \psi_0(x)]\!] = \bigwedge\{[\![\psi_0(c)]\!] : c \in A\}$.

In other words, we have defined the map $[\![\cdot]\!] \colon \mathrm{Sent}(L_A) \to \mathbb{B}$.

In what follows, we will identify the $\mathbb{B}$-valued model $\mathbf{A} = (A, \xi_0, \xi_1, \dots, \xi_n)$ with the pair $(A, [\![\cdot]\!])$.

3 Decidable Boolean-Valued Models

In this section we formally introduce our approach to decidability in the setting of Boolean-valued models. Again, we fix a signature $L = \{P_0, P_1, \dots, P_n\}$ containing only unary predicates.

Our key observation (which was informally discussed in the introduction) is as follows. Suppose that A is a countable set, and $\mathbf{A} = (A, [\![\cdot]\!])$ is a (classical) $\mathbb{B}$-valued model of L. Then, since the set of sentences $\mathrm{Sent}(L_A)$ is countable, it is clear that the set $\mathrm{range}([\![\cdot]\!])$ is also (at most) countable. In addition, Axioms (A1)–(A4) given above ensure that $\mathrm{range}([\![\cdot]\!])$ forms a countable subalgebra $\mathcal{C}$ of the complete Boolean algebra $\mathbb{B}$.

More informally, one can say that for the case of a Boolean-valued model $\mathbf{A}$ with countable domain A, one needs to assign only countably many truth values $[\![\psi]\!]$, $\psi \in \mathrm{Sent}(L_A)$. In addition, these truth values form a Boolean algebra $\mathcal{C}$.

By applying natural computability-theoretic conditions, we introduce the following notion:

Definition 2. *A* decidable Boolean-valued model *(of the signature L) is an arbitrary triple $(A, \mathcal{C}, [\![\cdot]\!])$ such that:*

1. *A is a computable non-empty subset of the set of natural numbers,*
2. *$\mathcal{C}$ is a computable Boolean algebra,*
3. *$[\![\cdot]\!]$ is a computable function acting from* $\mathrm{Sent}(L_A)$ *into $\mathcal{C}$ such that $[\![\cdot]\!]$ satisfies Axioms* (A1)–(A7) *provided above.*

Recall that by Convention 1, we assume that $\mathrm{Sent}(L_A)$ is a computable subset of $\mathbb{N}$, hence the notion of a decidable Boolean-valued model is well-defined. We also note that it is straightforward to extend Definition 2 to the case of an arbitrary computable signature L. Nevertheless, we choose not to discuss this extension here, since our results will essentially use only unary predicates.

Intuitively speaking, Definition 2 provides an 'effectively arranged space' $\mathcal{C}$ of possible truth values, and it also gives an algorithm for effectively assigning truth values: $\psi \mapsto [\![\psi]\!]$.

We emphasize again that Definition 2 is a natural extension of the classical notion of *decidable structure* (or decidable model), developed in computable structure theory (see, e.g., [2,8,11]). Indeed, a decidable L-structure $\mathcal{S}$ with domain $A \subseteq \mathbb{N}$ is characterized by the following property: there exists an algorithm which tells whether a given sentence $\psi \in \mathrm{Sent}(L_A)$ is (classically) true or false. Notice that this characterization corresponds precisely to the version of Definition 2 restricted to the case when $\mathcal{C}$ is the two-element Boolean algebra $\{0_{\mathcal{C}}, 1_{\mathcal{C}}\}$.

If one wants to give an explicit construction for some decidable Boolean-valued model $(A, \mathcal{C}, [\![\cdot]\!])$, then the most 'tricky' part is how to ensure that Axioms (A6) and (A7) are satisfied. We illustrate the arising difficulty via an example. Suppose that we choose $\mathcal{C}$ as (a computable isomorphic copy of) the Boolean algebra of finite/cofinite subsets of $\mathbb{N}$. Let $A = \{c_i : i \in \mathbb{N}\}$. If we are not careful enough, then, e.g., we could choose the value $[\![P_0(c_i)]\!]$ as the element $\{2i\}$ from the algebra $\mathcal{C}$. But then the family $\{\{2i\} : i \in \mathbb{N}\}$ does not have a supremum in $\mathcal{C}$, and hence we fail to define the value $[\![\exists x P_0(x)]\!]$ satisfying Axiom (A6).

Our next result gives a method to alleviate the difficulty discussed above (for some specific cases).

Proposition 1. *Let A be a computable subset of $\mathbb{N}$, and let $\mathcal{C}$ be a computable Boolean algebra. For $i \leq n$, we choose computable maps ξ_i acting from A to $\mathcal{C}$ such that for every tuple $(\delta_0, \delta_1, \dots, \delta_n) \in \{0_{\mathcal{C}}, 1_{\mathcal{C}}\}^{n+1}$, the set*

$$\bigcap_{i \leq n} \xi_i^{-1}(\delta_i) \text{ is infinite.} \tag{1}$$

Then there exists a unique decidable Boolean-valued model $(A, \mathcal{C}, [\![\cdot]\!])$ such that $[\![P_i(a)]\!] = \xi_i(a)$ for all $i \leq n$ and $a \in A$.

Informally speaking, this result says that in the particular case described in the proposition, it is sufficient to effectively define only the truth values $[\![P_i(a)]\!]$ discussed in Axiom (A0), and then all the remaining truth values $[\![\psi]\!]$ will be recovered 'automatically'.

Proposition 1 allows to obtain a simple explicit example of a decidable Boolean-valued model $(\mathbb{N}, \mathcal{C}, [\![\cdot]\!])$ in the signature $\{P_0, P_1\}$. We choose $\mathcal{C}$ as a computable isomorphic copy of the free Boolean algebra with countably many generators g_i, $i \in \mathbb{N}$. For $k \in \mathbb{N}$, let

$$\mathrm{par}(k) = \begin{cases} 0, & \text{if } k \text{ is even,} \\ 1, & \text{if } k \text{ is odd.} \end{cases}$$

If $a = 2^k \cdot 3^\ell \in \mathbb{N}$, then we put $[\![P_0(a)]\!] = \mathrm{par}(k)$ and $[\![P_1(a)]\!] = \mathrm{par}(\ell)$. If $a \in \mathbb{N}$ does not have form $2^k \cdot 3^\ell$, then set $[\![P_0(a)]\!] = [\![P_1(a)]\!] = g_a$.

Proof (of Proposition 1*).* We recall the notion of a completion of a given Boolean algebra $\mathcal{C}$. A complete Boolean algebra $\mathbb{B}$ is called a *completion of* $\mathcal{C}$ if $\mathcal{C}$ is a dense subalgebra of $\mathbb{B}$ (that is, $\mathcal{C}$ is a subalgebra of $\mathbb{B}$ such that for any $b \in \mathbb{B} \setminus \{0_\mathbb{B}\}$, there exists an element $c \in \mathcal{C}$ such that $0_\mathbb{B} < c \leq b$). It is known that every Boolean algebra $\mathcal{C}$ has a unique completion up to isomorphism (see Theorem 4.19 in [14]).

We fix a completion $\mathbb{B}$ of our computable Boolean algebra $\mathcal{C}$. It is clear that the given maps ξ_i, $i \leq n$, induce a unique (classical) $\mathbb{B}$-valued model $\mathbf{A} = (A, [\![\cdot]\!])$ satisfying Axioms (A0)–(A7). Now we need to explain how to 'procure' a decidable Boolean-valued model from the classical Boolean-valued model $\mathbf{A}$.

Our proof essentially uses the fact that the signature L contains only unary predicates. We apply a classical result of Löwenheim (we refer to Chapter IV of [1] and to Sect. 5.2 of [7] for a more modern exposition of this result).

For $\delta \in \{0,1\}$ and $i \leq n$, we define the formula $P_i^\delta(x)$ as follows: $P_i^1(x) := P_i(x)$ and $P_i^0(x) := \neg P_i(x)$. We also introduce the following ancillary formulas:

Convention 3. Let $m \geq 1$, and let $\overrightarrow{\delta} = (\delta_0, \delta_1, \dots, \delta_n) \in \{0,1\}^{n+1}$. Then the existential sentence $\Theta(m, \overrightarrow{\delta})$ says that there exist pairwise different elements $y_1, y_2, \dots, y_m$ such that each y_j satisfies $P_0^{\delta_0}(y_j) \& P_1^{\delta_1}(y_j) \& \dots \& P_n^{\delta_n}(y_j)$.

Lemma 1 (Löwenheim [16]). *There exists an algorithm which, given a formula* $\theta(x_1, x_2, \dots, x_k)$ *in the signature* L*, produces an* L*-formula* $\theta^*(x_1, x_2, \dots, x_k)$ *such that the formulas* θ *and* θ^* *are logically equivalent, and* θ^* *is a Boolean combination of formulas having one of the following forms:*

- $\Theta(m, \overrightarrow{\delta})$ *for some* $m \geq 1$ *and some* $\overrightarrow{\delta} \in \{0,1\}^{n+1}$*; or*
- *an atomic formula (that is,* $(x_i = x_j)$ *or* $P_\ell(x_i)$*).*

We apply Lemma 1 to establish the following fact:

Lemma 2. (i) *For any sentence* $\psi \in \mathrm{Sent}(L_A)$*, its truth value* $[\![\psi]\!]$ *(taken in the classical* $\mathbb{B}$*-valued model* $\mathbf{A}$*) belongs to the subalgebra* $\mathcal{C}$*. In other words,* $\mathrm{range}([\![\cdot]\!]) \subseteq \mathrm{dom}(\mathcal{C})$.

(ii) *There exists an algorithm which, given a sentence* $\psi \in \mathrm{Sent}(L_A)$*, computes the corresponding truth value* $[\![\psi]\!] \in \mathcal{C}$*.*

Proof. **(i)** Consider a sentence $\psi = \theta(a_1, a_2, \dots, a_k)$, where $\theta(x_1, x_2, \dots, x_k)$ is a first-order L-formula, and $a_1, a_2, \dots, a_k \in A$. By applying Lemma 2, we algorithmically find the 'special' formula $\theta^*(x_1, x_2, \dots, x_k)$ which is logically equivalent to θ.

Notice that the original sentence $\psi = \theta(a_1, a_2, \dots, a_k)$ is logically equivalent to the sentence $\psi^* = \theta^*(a_1, a_2, \dots, a_k)$. This implies that the truth values $[\![\psi]\!]$ and $[\![\psi^*]\!]$ must be *equal* (see, e.g., Theorem 6.4 in [24]). Therefore, now it is sufficient to analyze the value $[\![\psi^*]\!]$.

We claim that for all $m \geq 1$ and all $\overrightarrow{\delta} = (\delta_0, \delta_1, \dots, \delta_n)$, we have $[\![\Theta(m, \overrightarrow{\delta})]\!] = 1_{\mathbb{B}}$. Indeed, by Eq. (1), we can choose m-many pairwise different elements $b_1, b_2, \dots, b_m$ satisfying:

- $\xi_i(b_j) = (\delta_i)_{\mathbb{B}}$ and $[\![P_i^{\delta_i}(b_j)]\!] = 1_{\mathbb{B}}$, for every $i \leq n$ and every $j \leq m$,
- $[\![b_j \neq b_r]\!] = 1_{\mathbb{B}}$ for all $j \neq r$.

Hence, a standard calculation shows that $[\![\Theta(m, \overrightarrow{\delta})]\!]$ is equal to $1_{\mathbb{B}}$.

If we consider an atomic formula, then we have $[\![a = b]\!] \in \{0_{\mathbb{B}}, 1_{\mathbb{B}}\}$, and $[\![P_i(a)]\!] = \xi_i(a) \in \mathcal{C}$ for all $a, b \in A$ and $i \leq n$. Therefore, we deduce that the truth value $[\![\psi^*]\!]$ is equal to a Boolean combination of elements from the countable subalgebra $\mathcal{C}$, and thus, the element $[\![\psi^*]\!] = [\![\psi]\!]$ itself lies in $\mathcal{C}$. We conclude that $\mathrm{range}([\![\cdot]\!]) \subseteq \mathrm{dom}(\mathcal{C})$.

(ii) An analysis of the proof of Item (i) shows the following. In order to compute the value $[\![\psi]\!]$, it is sufficient to have a computation algorithm for the function $(i, a) \mapsto [\![P_i(a)]\!]$. We notice that the existence of this algorithm follows from the computability of maps ξ_i, $i \leq n$. Lemma 2 is proven. □

Proof. Lemma 2 immediately implies that the structure $(A, \mathcal{C}, [\![\cdot]\!])$ is a well-defined decidable Boolean-valued model. In addition, Axioms (A1)–(A7) guarantee that $(A, \mathcal{C}, [\![\cdot]\!])$ must be a unique model satisfying $[\![P_i(a)]\!] = \xi_i(a)$ for all $i \leq n$ and all $a \in A$. This concludes the proof of Proposition 1. □

4 Encoding a Graph Into a Boolean-Valued Model

After our working framework has been established, we can investigate decidable Boolean-valued models from the point of view of computable structure theory. Recall that $L = \{P_0, P_1, \dots, P_n\}$, where each P_i is a unary predicate.

Convention 4. A decidable Boolean-valued model $\mathbf{A} = (A, \mathcal{C}, [\![\cdot]\!])$ can be viewed as a computable two-sorted structure $\mathbf{A}^{2\mathrm{s}} = (A, \mathcal{C}, \xi_0, \xi_1, \dots, \xi_n)$, where:

- The elements of the first sort are *basic elements* $a \in A$.
- The elements of the second sort are *truth values* $c \in \mathcal{C}$.
- The Boolean algebra operations $\vee, \wedge, \overline{(\cdot)}$ act on the truth values.
- For $i \leq n$, the function ξ_i maps A into $\mathrm{dom}(\mathcal{C})$, and $\xi_i(a) = [\![P_i(a)]\!]$.

We treat undirected graphs as structures in the signature $L_G = \{R\}$. In what follows, we assume that a binary relation R is symmetric and irreflexive.

In the proof of the next result, we (essentially) encode a given computable graph into an appropriately constructed, decidable Boolean-valued model.

Theorem 1. *Let $\mathcal{G} = (\mathbb{N}, R)$ be a computable undirected graph. There exists a decidable Boolean-valued model $\mathbf{A}_\mathcal{G}$ such that the computable dimensions of $\mathbf{A}_\mathcal{G}$ and $\mathcal{G}$ are the same.*

Recall that Goncharov [9] proved that for every n such that $1 \leq n \leq \omega$, there exists a computable undirected graph $\mathcal{G}_n$ with computable dimension n. We note that the graph $\mathcal{G}_n$ is not a decidable structure in the classical recursion-theoretic sense (see, e.g., the exposition in Theorem 12.10 of [2]).

The result of Goncharov and Theorem 1 imply the following:

Corollary 1. *For every n such that $1 \leq n \leq \omega$, there exists a decidable Boolean-valued model having computable dimension n.*

Proof (of Theorem 1). Without loss of generality, we may assume that the graph $\mathcal{G}$ has at least one isolated node. Indeed, if $\mathcal{G}$ has no isolated nodes, then we can add one 'external' isolated node to $\mathcal{G}$. It is well-known that the resulting new graph $\mathcal{G}_1$ has the same computable dimension as the original $\mathcal{G}$ (see, e.g., Sect. 3.3 in [13]). We will also assume that the node 0 is isolated in $\mathcal{G}$.

We choose a computable Boolean algebra $\mathcal{C}$ such that $\mathcal{C}$ is isomorphic to the Boolean algebra of finite/cofinite subsets of $\mathbb{N}$, and the set of atoms $\mathrm{Atom}(\mathcal{C})$ is computable. We fix a computable list $(c_i)_{i\in\mathbb{N}}$ which enumerates the atoms of $\mathcal{C}$ without repetitions.

The desired Boolean-valued model $\mathbf{A}_\mathcal{G}$ has signature $L = \{P_0, P_1\}$, that is, we will use only two unary predicates. We want to apply Proposition 1, and thus, for each $i \in \{0, 1\}$ and every $a \in \mathbb{N}$ we define the truth value $\xi_i(a) = [\![P_i(a)]\!]$ belonging to the Boolean algebra $\mathcal{C}$. For $k \in \mathbb{N}$ and for $\ell \neq m$, we put:

- $\xi_0(4\langle \ell, \ell\rangle) = \xi_1(4\langle \ell, \ell\rangle) = 0_\mathcal{C}$;
- $\xi_0(4k+1) = 0_\mathcal{C}$ and $\xi_1(4k+1) = 1_\mathcal{C}$;
- $\xi_0(4k+2) = 1_\mathcal{C}$ and $\xi_1(4k+2) = 0_\mathcal{C}$;
- $\xi_0(4k+3) = \xi_1(4k+3) = 1_\mathcal{C}$;
- if $\mathcal{G} \models \neg R(\ell, m)$, then $\xi_0(4\langle \ell, m\rangle) = c_\ell \vee c_m$ and $\xi_1(4\langle \ell, m\rangle) = 0_\mathcal{C}$;
- if $\mathcal{G} \models R(\ell, m)$, then $\xi_0(4\langle \ell, m\rangle) = 0_\mathcal{C}$ and $\xi_1(4\langle \ell, m\rangle) = c_\ell \vee c_m$.

Since the relation R is computable, the functions ξ_0 and ξ_1 are also computable. By Proposition 1, we obtain a (unique) decidable Boolean-valued model $\mathbf{A}_\mathcal{G} = (\mathbb{N}, \mathcal{C}, [\![\cdot]\!])$ such that $[\![P_i(a)]\!] = \xi_i(a)$ for $i \in \{0, 1\}$ and $a \in \mathbb{N}$.

Now we need to prove that the computable dimensions of $\mathcal{G}$ and of the constructed model $\mathbf{A}_\mathcal{G}$ are the same.

Beforehand, we notice the following properties of the model $\mathbf{A}_\mathcal{G}$:

(i) $\mathrm{range}(\xi_0) = \{c_\ell \vee c_m : \ell \neq m,\ \mathcal{G} \models \neg R(\ell, m)\} \cup \{0_\mathcal{C}, 1_\mathcal{C}\}$,

(ii) $\mathrm{range}(\xi_1) = \{c_\ell \vee c_m : \mathcal{G} \models R(\ell, m)\} \cup \{0_\mathcal{C}, 1_\mathcal{C}\}$.

We define several ancillary first-order formulas (these formulas will be used for defining some subsets of the computable two-sorted structure $\mathbf{A}_{\mathcal{G}}$, recall Convention 4).

(F.1) Since the node 0 is isolated in the graph $\mathcal{G}$, for each non-zero m, the element $c_0 \vee c_m$ belongs to range(ξ_0). Therefore, the set $\mathrm{Atom}(\mathcal{C}) = \{c_i : i \in \mathbb{N}\}$ is definable inside the two-sorted structure $\mathbf{A}_{\mathcal{G}}$ by the following existential formula:

$$\psi^{\exists}_{Atom}(x) := \exists u[\xi_0(u) \neq 1_{\mathcal{C}} \ \&\ 0_{\mathcal{C}} <_{\mathcal{C}} x <_{\mathcal{C}} \xi_0(u)].$$

On the other hand, the set $\mathrm{Atom}(\mathcal{C})$ is also definable by a universal formula:

$$\psi^{\forall}_{Atom}(x) := (0_{\mathcal{C}} <_{\mathcal{C}} x) \ \&\ \neg\exists y[0_{\mathcal{C}} <_{\mathcal{C}} y <_{\mathcal{C}} x].$$

(F.2) We introduce two existential formulas:

- $\psi_R(x,y) := \psi^{\exists}_{Atom}(x) \,\&\, \psi^{\exists}_{Atom}(y) \,\&\, \exists u[\xi_1(u) = x \vee y]$,
- $\psi_{\mathrm{not}R}(x,y) := \psi^{\exists}_{Atom}(x) \,\&\, \psi^{\exists}_{Atom}(y) \,\&\, \exists u[\xi_0(u) = x \vee y]$.

Notice the following: if one of the conditions $\psi_R(x,y)$ or $\psi_{\mathrm{not}R}(x,y)$ is true, then $x \neq y$ and $x, y \in \mathrm{Atom}(\mathcal{C}) = \{c_i : i \in \mathbb{N}\}$.

By applying Properties (i) and (ii) of the construction, we obtain the following:

Claim 3. *For natural numbers $i \neq j$, the following conditions are equivalent:*

$$\mathcal{G} \models R(i,j) \ \Leftrightarrow\ \mathbf{A}_{\mathcal{G}} \models \psi_R(c_i, c_j) \ \Leftrightarrow\ \mathbf{A}_{\mathcal{G}} \models \neg\psi_{\mathrm{not}R}(c_i, c_j).$$

The introduced ancillary formulas allow us to give the following definition:

Definition 5. *Let $\mathbf{D}$ be an arbitrary structure isomorphic to the two-sorted structure $\mathbf{A}_{\mathcal{G}}$. Then we define an undirected graph* $\mathrm{Gr}(\mathbf{D})$ *as follows:*

- *The domain of* $\mathrm{Gr}(\mathbf{D})$ *is equal to the set* $\psi^{\exists}_{Atom}[\mathbf{D}]$.
- *Consider two nodes $a \neq b$ from* $\mathrm{dom}(\mathrm{Gr}(\mathbf{D}))$. *The nodes a and b are connected by an edge if and only if $\mathbf{D} \models \psi_R(a,b)$.*

We give the main properties of the introduced graph $\mathrm{Gr}(\mathbf{D})$:

Lemma 4. *Let $\mathbf{D}$ and $\mathbf{E}$ be computable two-sorted structures which are isomorphic to $\mathbf{A}_{\mathcal{G}}$. Then the following hold:*

(a) The graph $\mathrm{Gr}(\mathbf{A}_{\mathcal{G}})$ *is computably isomorphic to the original graph $\mathcal{G}$.*
(b) The structure $\mathrm{Gr}(\mathbf{D})$ *is a computable graph which is isomorphic to $\mathcal{G}$. In addition, if g is a computable isomorphism between two-sorted structures $\mathbf{D}$ and $\mathbf{E}$, then the restriction of g to the set* $\mathrm{dom}(\mathrm{Gr}(\mathbf{D}))$ *is a computable isomorphism between the graphs* $\mathrm{Gr}(\mathbf{D})$ *and* $\mathrm{Gr}(\mathbf{E})$.
(c) If f is a computable isomorphism acting from the graph $\mathrm{Gr}(\mathbf{D})$ *onto the graph* $\mathrm{Gr}(\mathbf{E})$, *then there exists a computable isomorphism (of two-sorted structures) $F\colon \mathbf{D} \to \mathbf{E}$ such that $f \subset F$.*

Proof. **(a)** Notice that the domain $\mathrm{dom}(\mathrm{Gr}(\mathbf{A}_{\mathcal{G}}))$ is equal to the set $\mathrm{Atom}(\mathcal{C}) = \{c_i : i \in \mathbb{N}\}$. Hence, Claim 3 implies that the map $h\colon i \mapsto c_i$ is a computable isomorphism acting from $\mathcal{G}$ onto $\mathrm{Gr}(\mathbf{A}_{\mathcal{G}})$.

(b) Since $\mathbf{D} \cong \mathbf{A}_{\mathcal{G}}$ and the graph $\mathrm{Gr}(\mathbf{D})$ is defined via first-order formulas in $\mathbf{D}$, we apply Item (a) and we deduce that $\mathrm{Gr}(\mathbf{D}) \cong \mathrm{Gr}(\mathbf{A}_{\mathcal{G}}) \cong \mathcal{G}$.

Since $\mathbf{D} \cong \mathbf{A}_{\mathcal{G}}$, we obtain that the set $\mathrm{dom}(\mathrm{Gr}(\mathbf{D}))$ is definable both by the existential formula $\psi^{\exists}_{Atom}(x)$ and by the universal formula $\psi^{\forall}_{Atom}(x)$. Thus, by Post's Theorem, the domain $\mathrm{dom}(\mathrm{Gr}(\mathbf{D}))$ is a computable set.

By Claim 3, all elements $x \neq y$ taken from $\mathrm{Gr}(\mathbf{D})$ satisfy:

$$\mathbf{D} \models \psi_R(x, y) \ \Leftrightarrow\ \mathbf{D} \models \neg\psi_{\mathrm{not}R}(x, y).$$

Since both formulas ψ_R and $\psi_{\mathrm{not}R}$ are existential, we deduce (by Post's Theorem) that the edge relation in the undirected graph $\mathrm{Gr}(\mathbf{D})$ is computable. We conclude that $\mathrm{Gr}(\mathbf{D})$ is a computable graph isomorphic to $\mathcal{G}$.

Now suppose that g is a computable isomorphism between two-sorted structures $\mathbf{D}$ and $\mathbf{E}$. Define $g_0 := g \upharpoonright \mathrm{dom}(\mathrm{Gr}(\mathbf{D}))$. Since the set $\mathrm{dom}(\mathrm{Gr}(\mathbf{D}))$ is computable, the map g_0 is computable. In addition, since $\mathbf{D} \cong \mathbf{E}$ and the graphs $\mathrm{Gr}(\mathbf{D})$ and $\mathrm{Gr}(\mathbf{E})$ are definable by the same formulas (in $\mathbf{D}$ and $\mathbf{E}$, respectively), we deduce that g_0 is an isomorphism acting from $\mathrm{Gr}(\mathbf{D})$ onto $\mathrm{Gr}(\mathbf{E})$.

(c) Suppose that $\mathbf{D} = (D, \mathcal{U}, [\![\cdot]\!]_D)$ and $\mathbf{E} = (E, \mathcal{V}, [\![\cdot]\!]_E)$, and let f be a computable isomorphism acting from $\mathrm{Gr}(\mathbf{D})$ onto $\mathrm{Gr}(\mathbf{E})$.

We fix a computable list $(d_i)_{i\in\mathbb{N}}$ which enumerates (without repetitions) the domain $\mathrm{dom}(\mathrm{Gr}(\mathbf{D})) = \mathrm{Atom}(\mathcal{U})$. Then the isomorphism f is a computable bijection acting from $\{d_i : i \in \mathbb{N}\}$ onto the set $\mathrm{dom}(\mathrm{Gr}(\mathbf{E})) = \mathrm{Atom}(\mathcal{V})$.

Since $\mathcal{U}$ is isomorphic to the Boolean algebra of finite/cofinite subsets of $\mathbb{N}$, for a given element $a \in \mathcal{U} \setminus \{0_{\mathcal{U}}, 1_{\mathcal{U}}\}$, we can effectively find a non-empty finite set $\{d_{i_0}, d_{i_1}, \ldots, d_{i_r}\}$ such that:

- either $a = d_{i_0} \vee d_{i_1} \vee \cdots \vee d_{i_r}$ (that is, a is the finite sum of atoms d_{i_j}), or
- $\overline{a} = d_{i_0} \vee d_{i_1} \vee \cdots \vee d_{i_r}$ (that is, the complement of a is the finite sum of those atoms).

In the first case, we put $F(a) = f(d_{i_0}) \vee f(d_{i_1}) \vee \cdots \vee f(d_{i_r})$. In the second case, $F(a)$ is chosen as the complement of $f(d_{i_0}) \vee f(d_{i_1}) \vee \cdots \vee f(d_{i_r})$.

It is clear that the described definition provides a computable isomorphism between Boolean algebras $\mathcal{U}$ and $\mathcal{V}$. (Here, of course, we have to additionally set $F(0_{\mathcal{U}}) = 0_{\mathcal{V}}$ and $F(1_{\mathcal{U}}) = 1_{\mathcal{V}}$.)

For each $\overrightarrow{\delta} = (\delta_0, \delta_1) \in \{0,1\}^2$, we fix a computable bijection $g_{\overrightarrow{\delta}}$ acting from the infinite computable set $\{d \in D : [\![P_0(d)]\!]_D = \delta_0,\ [\![P_1(d)]\!]_D = \delta_1\}$ onto the set $\{e \in E : [\![P_0(e)]\!]_E = \delta_0,\ [\![P_1(e)]\!]_E = \delta_1\}$.

Now consider an arbitrary element $a \in D$. If $\{[\![P_0(a)]\!]_D, [\![P_1(a)]\!]_D\} \subseteq \{0_{\mathcal{U}}, 1_{\mathcal{U}}\}$, then we use an appropriate map $g_{\overrightarrow{\delta}}$ to define $F(a) = g_{\overrightarrow{\delta}}(a)$. (Intuitively, such elements a have been introduced to the structure $\mathbf{D} \cong \mathbf{A}_{\mathcal{G}}$ in order to ensure the condition given in Eq. (1)).

Otherwise, precisely one of the values $[\![P_0(a)]\!]_D, [\![P_1(a)]\!]_D$ is equal to a sum of two atoms (from the algebra $\mathcal{U}$), and the other value is equal to $0_{\mathcal{U}}$. Without loss of generality, we assume that

$$[\![P_0(a)]\!]_D = d_k \vee d_\ell, \quad [\![P_1(a)]\!]_D = 0_{\mathcal{U}},$$

where $k < \ell$. Then, since $\mathbf{E} \cong \mathbf{D} \cong \mathbf{A}_{\mathcal{G}}$ and $f\colon \mathrm{Gr}(\mathbf{D}) \cong \mathrm{Gr}(\mathbf{E})$, we can effectively find the (unique) element $b \in E$ satisfying

$$[\![P_0(b)]\!]_E = f(d_k) \vee f(d_\ell), \quad [\![P_1(b)]\!]_E = 0_{\mathcal{V}}.$$

We put $F(a) = b$.

For every element a from the two-sorted structure $(D, \mathcal{U}, [\![\cdot]\!]_D)$, we have described how to (computably) find the value $F(a)$ taken from $(E, \mathcal{V}, [\![\cdot]\!]_E)$. Notice that $f \subset F$. Furthermore, it is straightforward to verify that the map F is an isomorphism of two-sorted structures. Lemma 4 is proven. □

Lemma 4 gives us the necessary tools to deal with computable dimensions. Given a computable graph $\mathcal{H}$ which is isomorphic to $\mathcal{G}$, by $\mathbf{A}_{\mathcal{H}}$ we denote the two-sorted structure obtained by applying our construction (described above) to $\mathcal{H}$ in place of $\mathcal{G}$. It is clear that $\mathbf{A}_{\mathcal{H}}$ is a computable two-sorted structure isomorphic to $\mathbf{A}_{\mathcal{G}}$.

Recall that $\dim_c(\mathcal{S})$ denotes the computable dimension of a structure $\mathcal{S}$. We will establish two inequalities for computable dimensions:

$$\dim_c(\mathcal{G}) \leq \dim_c(\mathbf{A}_{\mathcal{G}}), \tag{2}$$

$$\dim_c(\mathbf{A}_{\mathcal{G}}) \leq \dim_c(\mathcal{G}). \tag{3}$$

(I.1) Suppose that $\dim_c(\mathcal{G}) = n$, where $1 \leq n \leq \omega$. We need to show that there exist at least n-many computable two-sorted structures $\mathbf{D}_i$ such that each $\mathbf{D}_i$ is isomorphic to $\mathbf{A}_{\mathcal{G}}$, but the structures $\mathbf{D}_i$ are pairwise not computably isomorphic. In order to obtain this fact, it is sufficient to establish the following claim: if $\mathcal{H}$ is a computable graph such that $\mathcal{H} \cong \mathcal{G}$ and $\mathcal{H}$ is not computably isomorphic to $\mathcal{G}$, then $\mathbf{A}_{\mathcal{H}}$ is not computably isomorphic to $\mathbf{A}_{\mathcal{G}}$.

Towards a contradiction, assume that $\mathbf{A}_{\mathcal{H}}$ and $\mathbf{A}_{\mathcal{G}}$ are computably isomorphic. Then by Lemma 4(b), the graphs $\mathrm{Gr}(\mathbf{A}_{\mathcal{G}})$ and $\mathrm{Gr}(\mathbf{A}_{\mathcal{H}})$ are also computably isomorphic. On the other hand, by Lemma 4(a), we have two additional computable isomorphisms: from $\mathcal{G}$ onto $\mathrm{Gr}(\mathbf{A}_{\mathcal{G}})$, and from $\mathcal{H}$ onto $\mathrm{Gr}(\mathbf{A}_{\mathcal{H}})$. This implies that $\mathcal{G}$ and $\mathcal{H}$ must be computably isomorphic, which contradicts our original choice of $\mathcal{G}$ and $\mathcal{H}$. Hence, Inequality (2) is proven.

(I.2) Similarly to Item (I.1), here it is sufficient to prove the following claim: if $\mathbf{D}$ and $\mathbf{E}$ are computable two-sorted structures such that $\mathbf{D} \cong \mathbf{E} \cong \mathbf{A}_{\mathcal{G}}$ and $\mathbf{D}$ is not computably isomorphic to $\mathbf{E}$, then the computable graphs $\mathrm{Gr}(\mathbf{D})$ and $\mathrm{Gr}(\mathbf{E})$ are not computably isomorphic.

Towards a contradiction, assume that there is a computable isomorphism from $\mathrm{Gr}(\mathbf{D})$ onto $\mathrm{Gr}(\mathbf{E})$. Then, by Lemma 4(c), the structures $\mathbf{D}$ and $\mathbf{E}$ must be computably isomorphic, which gives a contradiction. Inequality (3) is proven.

We deduce that the computable dimensions of the original graph $\mathcal{G}$ and the decidable Boolean-valued model $\mathbf{A}_{\mathcal{G}}$ coincide. This concludes the proof of Theorem 1. □

5 Concluding Remarks

The paper introduced a possible computability-theoretic approach to Boolean-valued models. The main notion of this approach is the notion of a decidable Boolean-valued model $\mathbf{A} = (A, \mathcal{C}, [\![\cdot]\!])$ (see Definition 2). This notion allows us to look at Boolean-valued models through the lens of modern computable structure theory. In Proposition 1, we gave a specific construction of decidable Boolean-valued models $\mathbf{A}$. Our main result (Theorem 1) showed that this construction allows us to produce a (computability-theoretically) rich class of structures: any possible computable dimension can be realized by an appropriately constructed $\mathbf{A}$ (Corollary 1). On the other hand, it is known that for (classical) decidable structures $\mathcal{S}$ such a behavior is impossible [19].

We note that the method developed in the proof of Theorem 1 allows us to transfer further computability-theoretic results into the setting of decidable Boolean-valued models: in particular, some theorems on $\mathbf{d}$-computable dimensions, degree spectra of relations, and categoricity spectra of structures. Nevertheless, we choose to omit discussion of these theorems here, since it involves cumbersome technical details which would blur the focus of the paper. We cite [3,12,13] for a detailed exposition of similar computability-theoretic results.

Acknowledgments. The work was supported by Nazarbayev University Faculty Development Competitive Research Grants 201223FD8823. The authors are grateful to the anonymous referees for their helpful comments.

Disclosure of Interests. The authors have no competing interests to declare that are relevant to the content of this article.

References

1. Ackermann, W.: Solvable Cases of the Decision Problem. North-Holland, Amsterdam (1954)
2. Ash, C.J., Knight, J.: Computable Structures and the Hyperarithmetical Hierarchy. Studies in Logic and the Foundations of Mathematics, vol. 144. Elsevier, Amsterdam (2000)
3. Bazhenov, N.A.: Categoricity spectra of computable structures. J. Math. Sci. **256**(1), 34–50 (2021). https://doi.org/10.1007/s10958-021-05419-x
4. Bell, J.L.: Set Theory: Boolean-Valued Models and Independence Proofs, 3rd edn. Oxford University Press, Oxford (2005)
5. Downey, R., Melnikov, A.: Computable Structure Theory: A Unified Approach. https://homepages.ecs.vuw.ac.nz/~melnikal/maindoc.pdf, to appear
6. Ershov, Y.L.: Constructive models. In: Selected Questions of Algebra and Logic, pp. 111–130. Nauka, Novosibirsk (1973). (in Russian)
7. Ershov, Y.L.: Decidability Problems and Constructive Models. Nauka, Moscow (1980). in Russian
8. Ershov, Y.L., Goncharov, S.S.: Constructive Models. Kluwer Academic, New York (2000)
9. Goncharov, S.S.: Problem of the number of non-self-equivalent constructivizations. Algebra Log. **19**(6), 401–414 (1980). https://doi.org/10.1007/BF01669323

10. Goncharov, S.S., Dzgoev, V.D.: Autostability of models. Algebra Log. **19**(1), 28–37 (1980). https://doi.org/10.1007/BF01669102
11. Harizanov, V.S.: Pure computable model theory. In: Handbook of Recursive Mathematics, vol. 1, Studies in Logic and the Foundations of Mathematics, vol. 138, pp. 3–114. Elsevier, Amsterdam (1998). https://doi.org/10.1016/S0049-237X(98)80002-5
12. Harrison-Trainor, M., Melnikov, A., Miller, R., Montalbán, A.: Computable functors and effective interpretability. J. Symb. Log. **82**(1), 77–97 (2017). https://doi.org/10.1017/jsl.2016.12
13. Hirschfeldt, D.R., Khoussainov, B., Shore, R.A., Slinko, A.M.: Degree spectra and computable dimensions in algebraic structures. Ann. Pure Appl. Logic **115**(1–3), 71–113 (2002). https://doi.org/10.1016/S0168-0072(01)00087-2
14. Koppelberg, S.: Handbook of Boolean Algebras, vol. 1. North-Holland, Amsterdam (1989)
15. Kusraev, A.G., Kutateladze, S.S.: Boolean Valued Analysis. Kluwer Academic, New York (1999)
16. Löwenheim, L.: Über Möglichkeiten im Relativkalkül. Math. Ann. **76**, 447–470 (1915). https://doi.org/10.1007/BF01458217
17. Mal'tsev, A.I.: Constructive algebras. I. Russ. Math. Surv. **16**(3), 77–129 (1961). https://doi.org/10.1070/RM1961v016n03ABEH001120
18. Morley, M.: Decidable models. Isr. J. Math. **25**(3–4), 233–240 (1976). https://doi.org/10.1007/BF02757002
19. Nurtazin, A.T.: Strong and weak constructivization and computable families. Algebra Log. **13**(3), 177–184 (1974). https://doi.org/10.1007/BF01463352
20. Odifreddi, P.: Classical Recursion Theory. Studies in Logic and the Foundations of Mathematics, vol. 125. Elsevier, Amsterdam (1992)
21. Pal'chunov, D.E., Yakhyaeva, G.E.: Interval fuzzy algebraic systems. In: Mathematical logic in Asia, pp. 191–202. World Scientific Publishing, Hackensack (2006). https://doi.org/10.1142/9789812772749_0014
22. Remmel, J.B.: Recursive isomorphism types of recursive Boolean algebras. J. Symb. Log. **46**(3), 572–594 (1981). https://doi.org/10.2307/2273757
23. Scott, D.: A proof of the independence of the continuum hypothesis. Math. Syst. Theory **1**, 89–111 (1967). https://doi.org/10.1007/BF01705520
24. Takeuti, G., Zaring, W.M.: Axiomatic Set Theory. Springer, New York (1973)
25. Vopěnka, P.: General theory of ∇-models. Comment. Math. Univ. Carolinae **8**, 145–170 (1967). http://eudml.org/doc/16203

Separation Between Walksat and DPLL

Tao Jiang[1,2], Shaowei Cai[1,2(✉)], Ziqun Li[1,2], Jiabao Lin[3], and Yijia Chen[4]

[1] Key Laboratory of System Software (Chinese Academy of Sciences) and State Key Laboratory of Computer Science, Institute of Software, Chinese Academy of Sciences, Beijing, China
{jiangt,caisw,liziqun}@ios.ac.cn

[2] School of Computer Science and Technology, University of Chinese Academy of Sciences, Beijing, China

[3] Huawei Taylor Lab, Shanghai, China

[4] School of Computer Science, Shanghai Jiao Tong University, Shanghai, China
yijia.chen@cs.sjtu.edu.cn

Abstract. Modern SAT solvers can be broadly classified into two categories. The first category comprises solvers based on the DPLL (Davis-Putnam-Logemann-Loveland) paradigm, which has achieved remarkable success with the combination of heuristics such as conflict-driven clause learning (CDCL) and restart mechanism. The second category consists of solvers employing the stochastic local search (SLS) approach, with Walksat being a prominent example. In this paper, we demonstrate that these two types of solvers can be distinguished by their performance on Tseitin formulas which have been extensively studied in proof complexity. Specifically, we show that Walksat can efficiently find a satisfying assignment for any satisfiable Tseitin formula with high probability in polynomial time. In contrast, we construct a family of Tseitin formulas on which any DPLL-based solver, even when enhanced with restarts, requires exponential time to find a solution. This analysis reveals a clear limitation of DPLL-based solvers compared to Walksat and suggests a potential direction for their improvements.

Keywords: SAT · SLS · DPLL

1 Introduction

The Boolean Satisfiability problem (SAT) asks whether a given propositional formula has a satisfying assignment. It is the first problem proved to be NP-complete, and thus unlikely to be decidable in polynomial time. Nevertheless, contemporary SAT solvers have demonstrated that SAT can in fact be solved efficiently for many real-life applications, including bounded model checking [6], program verification [7], and theorem proving [15].

M. Li et al. (Eds.): TAMC 2025, LNCS 16084, pp. 93–106, 2026.
https://doi.org/10.1007/978-981-95-4839-2_8

The existing SAT solvers generally fall into two broad categories. The first is derived from the classical Davis-Putnam-Logemann-Loveland (DPLL) algorithm [9], which systematically explores the space of assignments using branching heuristics and simplification rules. The most successful solvers of this category, known as conflict-driven clause learning (CDCL) solvers [27], extend DPLL by some advanced heuristics like clause learning and restarts. For an extensive survey on CDCL solvers, we refer the reader to [3]. The second category consists of solvers employing the stochastic local search (SLS) method, e.g., Walksat [25]. These solvers iteratively modify a candidate assignment, usually guided by some randomized heuristics, until a satisfying assignment is found or some resource limits are reached [16,21].

Empirical studies suggest SLS solvers excel on randomly generated satisfiable formulas [4,5,20,22], whereas DPLL-based ones, especially CDCL, tend to perform better on structured SAT instances. However, a theoretical analysis of this observation has been missing. In this paper, we make a first attempt to pinpoint the exact difference between SLS and DPLL by showing that on *Tseitin formulas* [28] there is an exponential time gap between Walksat and DPLL. In fact, the exponential gap persists even if DPLL is augmented with *restart*.

Given an undirected graph $G = (V, E)$ with a function $c : V \to \{0, 1\}$ which associates every vertex $v \in V$ with its *charge* $c(v)$, the Tseitin formula $T(G, c)$ asserts that each edge in E can be assigned a Boolean value such that for every vertex $v \in V$ the summation of the values over the edges adjacent to v is equal to the charge $c(v)$ mod 2. It is easy to see that $T(G, c)$ is satisfiable if and only if the sum of vertex charges in each connected component in G is even [29]. Tseitin formulas have been extensively studied in proof complexity [13], ?, ?. Particularly for proof systems like Resolution, it turns out that the structure of the underlying graph G plays a crucial role in determining the length of a resolution refutation of an unsatisfiable $T(G, c)$, particularly when G is a expander or a grid [29], ?, ?, ?. As DPLL is closely connected to Resolution, unsatisfiable Tseitin formulas must be hard for DPLL-based solvers. Nonetheless, to the best of our knowledge, there is no much work on Tseitin formulas from the perspective of SLS-based solvers.

Besides the above theoretical interest, Tseitin formulas have also found numerous applications in practical SAT solving. Instances from cryptography [23], Brent Equations over $\mathbb{Z}_2$ [24], and circuit-based CNF formulas [2] often contain Tseitin formulas as subformulas. As a consequence, benchmarks from recent SAT Competitions (2023 and 2024) include various instances based on Tseitin formulas.

1.1 Contributions

Our main contributions can be summarized as follows.

1. We show that for satisfiable Tseitin formula the Walksat algorithm finds a satisfying assignment in polynomial expected runtime.
2. We construct a family of graphs whose corresponding satisfiable Tseitin formulas are hard for the DPLL algorithm even when augmented with the restart heuristic.

3. We provide experimental results which confirm the above gap between Walksat and DPLL.

1.2 Related Work

To the best of our knowledge, our paper is the first attempt to understand the behavior of SAT solvers based on Stochastic Local Search with Tseitin formulas as inputs and compare it to that of DPLL. Most prior research investigated various proof systems suitable for Tseitin formulas [10,13,17], which does not apply to SAT solvers directly. On the other hand, Elffers et al. [12] examined the clause erasure of CDCL solvers on unsatisfiable Tseitin formulas. To study the efficiency of SAT solving, [30] constructed hard instances for some common branching strategies used in CDCL solvers. Building on this work, [19] explored the effectiveness of restarts as a heuristic in both DPLL and CDCL solvers.

1.3 Techniques

To establish the runtime upper bounds on the Walksat algorithm with Tseitin formulas $T(G, c)$ as inputs, we view its execution as a random procedure on the graph $G = (V, E)$, where several tokens are placed on some vertices in V and moved along the edges in E until they meet each other in pairs. This approach is inspired by the *random meet model* introduced by Coppersmith et al. [8]. As far as we can tell, it is the first time that such a model is applied to a SAT solver, which might be of some independent interest.

To construct Tseitin formulas hard for DPLL, we leverage graphs with *high treewidth*. Their corresponding Tseitin formulas, if unsatisfiable, are known to have only *tree-like* resolution of exponential length. By connecting these graphs by long paths we obtain the desired graphs G such that DPLL cannot run in polynomial time on satisfiable $T(G, c)$ in contrast to Walksat.

1.4 Organization

The rest of the paper is structured as follows. Section 2 provides necessary technical preliminaries including proof complexity, DPLL algorithms with restarts, Walksat, and relevant graph-theoretic concepts. Sections 3 and 4 contain our main theoretical results, i.e., polynomial-time analysis of Walksat and the construction of hard instances for DPLL. Section 5 explains our experiment results as a companion. We conclude in Sect. 6.

Due the space limitations, we defer some of the proofs to the appendices.

2 Preliminaries

2.1 Propositional Logic and Resolution

For our purposes, a Boolean formula of propositional logic is always expressed in *conjunctive normal form* (CNF), defined as a conjunction of clauses in the form of $F = C_1 \wedge C_2 \wedge \cdots \wedge C_m$. Each clause C_i is a disjunction of literals over distinct variables, i.e., $C_i = \ell_{i1} \vee \ell_{i2} \vee \cdots \vee \ell_{ik}$. A literal ℓ is either a Boolean variable x or its negation $\neg x$. In case $\ell = x$, its *opposite* literal, denoted by $\bar{\ell}$, is $\neg x$; while for $\ell = \neg x$ we have $\bar{\ell} = x$. We use $\mathrm{var}(F)$ to denote the set of variables in F.

Resolution is one of the central proof systems for propositional logic. For a unsatisfiable Boolean formula F in CNF, a resolution refutation for F is a sequence of clauses

$$\pi = D_1, D_2, \ldots, D_m$$

such that

- each clause D_i either belongs to F or is derived by applying the resolution rule
$$\frac{C \vee x \quad C' \vee \neg x}{C \vee C'},$$
where $C \vee x = D_{i_1}$ and $C' \vee \neg x = D_{i_2}$ for some variable x and $i_1, i_2 < i$. Thereby, D_i is derived from D_{i_1} and D_{i_2} by resolving the variable x.
- and D_m is the empty clause $\bot$.

The length of the resolution refutation π is the number of clauses it contains, i.e., $|\pi| = m$.

We can associate π with a directed acyclic graph (DAG) G_π consisting of $|\pi|$ nodes, each labeled with a clause D_i. In case D_i is derived from D_{i_1} and D_{i_2}, then there is an edge from D_{i_1} to D_i and also an edge from D_{i_2} to D_i. Therefore, the source nodes (also called leaves) of G_π are those labeled with the original clauses in F.

A resolution refutation π is said to be *tree-like* if the associated DAG G_π forms a tree. Moreover, a resolution refutation π is *regular* if no variable is resolved more than once along any path in G_π.

2.2 Basic Graph Theory and Tseitin Formulas

We only consider undirected graphs without self-loops and parallel edges. In particular, a graph $G = (V, E)$ consists of a set V of vertices and a set E of edges. Each edge $e \in E$ connecting vertices u and v is denoted by either (v, u) or (u, v). The set of edges incident to a vertex v is $E(v) = \{(v, u) \in E\}$, and the degree of v is $|E(v)|$. The degree of G, written $\Delta(G)$ is the maximum degree of a vertex in G. We use $\#G$ to denote the number of connected components in G.

A subset $V' \subseteq V$ *induces* a subgraph $G[V']$ of G with vertex set V' and edge set $E' = \{(u, v) \in E : u, v \in V'\}$. If $V' = \{v_1, v_2, \ldots, v_k\}$, we also refer to the induced graph $G[V']$ as $G[v_1, v_2, \ldots, v_k]$. On the other hand, the subgraph of G obtained by deleting the vertices in V', i.e., $G[V \setminus V']$, is often denoted by $G \setminus V'$. Similarly, the subgraph obtained by deleting a subset of edges $E' \subseteq E$ but keeping all the vertices in V is denoted by $G \setminus E'$.

For an edge $e = (u, v) \in E$, *contracting* e in G results in a graph G/e, where e is removed and vertices u and v are merged into a single vertex. A graph G' is a *minor* of G if it can be obtained from G by a sequence of vertex deletions, edge deletions, and edge contractions.

The *treewidth* of a graph G, denoted by $\mathrm{tw}(G)$, measures how close G is to being a tree. It is an important graph parameter with many algorithmic applications. For our presentation, we only need the following connection between treewidth and graph *minor*, thus skipping the precise definition of $\mathrm{tw}(G)$.

Lemma 1 [26]. *Let G be a graph and G' its minor. Then the treewidth of G' is less than or equal to the treewidth of G, i.e., $\mathrm{tw}(G') \leq \mathrm{tw}(G)$.*

A *Tseitin formula* $T(G, c)$ is a propositional formula defined from an undirected graph $G = (V, E)$ and a charge function $c : V \to \{0, 1\}$. For each edge $e \in E$, we introduce a propositional variable x_e. Then the parity condition at vertex $v \in V$ is defined as:

$$P_v = \left(\sum_{e \in E(v)} x_e \equiv c(v) \right) \mod 2, \tag{1}$$

where $E(v)$ is the set of edges incident to the vertex v. Recall that addition modulo 2 is the same as XOR. So equivalently, P_v is an *XOR equation* on all the variables x_e with $e \in E(v)$. Then the Tseitin formula $T(G, c)$ is the conjunction of these XOR equations over all vertices:

$$T(G, c) = \bigwedge_{v \in V} P_v.$$

We use the standard translation of XOR equations by CNF formulas. In particular, each XOR equation P_v is translated into $2^{|E(v)|-1}$ clauses, with each clause containing exactly $|E(v)|$ literals. For brevity, this is often referred to as the *canonical CNF encoding*.

Thus, $T(G, c)$ can be viewed as a CNF formula, with each P_v translated into its canonical set of clauses. We again use $T(G, c)$ to denote the resulting CNF formula. Whether $T(G, c)$ is viewed as a conjunction of XOR equations or as a CNF formula depends on the context. An illustrative example is shown in Fig. 1, where each vertex v is labeled by its charge value $c(v) \in \{0, 1\}$ and associated with its canonical clauses.

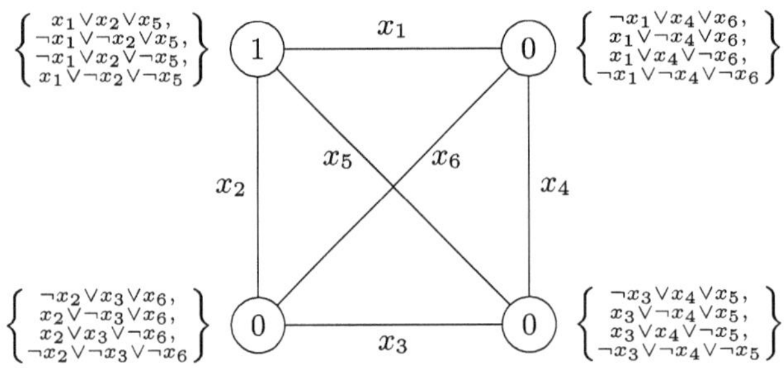

Fig. 1. An illustrative example of the canonical CNF encoding for a Tseitin formula. Each vertex v is labeled with its charge $c(v) \in \{0,1\}$ and associated with a set of canonical clauses derived from the parity condition at vertex v. Edges represent variables, and each parity condition translates into exactly $2^{|E(v)|-1}$ clauses, each containing exactly $|E(v)|$ literals.

When discussing assignments of variables in a Tseitin formula, variables are often identified with their corresponding edges. More specifically, for a Tseitin formulas $T(G,c)$, sometimes we say "assign an edge e in G" to indicate that we assign the variable associated with e in $T(G,c)$.

If the graph G consists of connected components $H_1, H_2, \ldots, H_t$, then the Tseitin formula $T(G,c)$ can equivalently be expressed as:

$$T(G,c) = \bigwedge_{i=1}^{t} T(H_i, c|_{V(H_i)}),$$

where $c|_{V(H_i)}$ is the restriction of the charge function c to the vertices in H_i. We say that H_i is satisfiable with respect to the charge function c if $T(H_i, c|_{V(H_i)})$ is satisfiable, and sometimes we just say that H_i is satisfiable if the charge function c is clear from the context. $T(G,c)$ is satisfiable if and only if H_i is satisfiable for $i = 1, \cdots, t$.

The following lemma is well-known.

Lemma 2 [29]. *Let $G = (V,E)$ be a connected graph and c a charge function on G. Then the Tseitin formula $T(G,c)$ is satisfiable if and only if the* total vertex charge of *c is even, i.e.,*

$$\sum_{v \in V} c(v) \equiv 0 \mod 2.$$

In particular, if the total vertex charge for c is odd, then $T(G,c)$ is unsatisfiable. For such Tseitin formulas, we have an exponential lower bound for their *regular* Resolution proofs in terms of the treewidth of G.

Lemma 3 [17]. *For any connected graph $G = (V,E)$ with $\Delta(G) \leq d$, i.e., the maximum degree of a vertex is G is at most d, and odd weight function c, the shortest regular resolution of $T(G,c)$ is of length $2^{\Omega(\mathrm{tw}(G))}$.*

2.3 The Walksat and DPLL Algorithms

Algorithm 1. The Walksat Algorithm

Input: A CNF formula F with variable set $\text{var}(F)$
Output: A satisfying assignment $\sigma : \text{var}(F) \to \{0, 1\}$

Select an initial assignment $\sigma : \text{var}(F) \to \{0, 1\}$ uniformly at random;
while F is not satisfied by σ **do**
    Select C: an unsatisfied clause in F chosen uniformly at random;
    Select x: a variable in C chosen uniformly at random;
    Flip the value of x and update $\sigma(x)$;
end while
Output σ.

The Walksat algorithm, initially introduced by Papadimitriou [25], stands as a foundational yet incomplete algorithm in the realm of heuristic optimization. It has inspired a plenty of subsequent heuristics that refine its fundamental processes. The operation of Walksat commences with the random selection of an unsatisfied clause within a CNF formula, followed by the random flipping of a variable within the clause to achieve a potentially more satisfactory assignment. A detailed description of this procedure is presented in Algorithm 1.

Algorithm 2. DPLL Algorithm

Input: CNF formula F.
Output: "Satisfiable" or "Unsatisfiable".

Simplify F using unit propagation.
if F is empty **then**
    return "Satisfiable".
end if
if F contains empty clause **then**
    return "Unsatisfiable".
end if
Pick a variable v and Boolean value ε (Heuristics A, B).
if $DPLL(F[v := \varepsilon])$ or $DPLL(F[v := \neg\varepsilon])$ returns "Satisfiable" **then**
    return "Satisfiable".
end if
return "Unsatisfiable".

The DPLL algorithm is a recursive procedure designed to determine the satisfiability of a propositional formula. At each step, the algorithm applies simplification rules to the input formula F, ensuring these transformations preserve the formula's satisfiability. Subsequently, The recursion continues until the formula is trivialized—either the formula becomes empty (indicating satisfiability) or a clause becomes empty (indicating unsatisfiability). A formal description of the algorithm is presented in Algorithm 2.

The specific behavior of the DPLL algorithm is governed by its simplification rules and two heuristics: **Heuristic A**, which determines the variable v to be chosen, and **Heuristic B**, which selects the polarity (i.e., whether to assign $v = 0$ or $v = 1$). It is worth noting that if P = NP, and if **Heuristic B** is nondeterministic, it could always select the correct variable assignments in every step, enabling the algorithm to terminate in polynomial time. Consequently, to establish unconditional lower bounds on the algorithm's runtime, it is necessary to impose restrictions on the simplification rules and heuristics, and then analyze the algorithm's performance under these constraints.

In this paper, we use unit propagation as the simplification rule. Heuristic A and B and the simplification rule (unit propagation) are set as:

Heuristic A: Randomly select an unassigned variable v from F.
Heuristic B: Randomly decide whether to branch on $v := 1$ or $v := 0$ first.
Unit propagation(UP): If the formula F contains a clause that consists of a single literal ℓ, replace F by $F[\ell := 1]$, where $\ell := 1$ denotes the elementary substitution that satisfies the literal ℓ.

Before DPLL (i.e., Algorithm 2) is intiated, an input formula goes through some preprocessing procedure which often significantly shortens the length of the formula. We consider two popular preprocessing techniques:

Self-subsuming resolution [11].. If we have clauses $C \vee \ell$ and $D \vee \bar{\ell}$, and $C \subseteq D$[1], then we remove $\bar{\ell}$ from D.
Blocked clause elimination [18]. If a clause C contains a literal ℓ such that resolving C with all clauses containing $\bar{\ell}$ only yields tautologies, then we remove C.

We use the following lemma to relate the length of the shortest tree-like resolution refutation and the running time of DPLL algorithms.

Lemma 4 [1]. *For each unsatisfiable formula, the shortest tree-like resolution proof is at most polynomially longer than the smallest recursion tree of a DPLL algorithm, and vice versa.*

Restart policies play a critical role in enhancing the practical performance of modern DPLL-based solvers. A restart policy periodically resets the algorithm's search state by abandoning the current search branch. This strategy prevents the search procedure from being trapped in difficult regions of the search space, thereby significantly improving efficiency. Common restart strategies include the *fixed restart*, occurring at regular intervals; the *exponential restart*, which exponentially increases intervals between restarts; and the *Luby restart*, using intervals defined by the Luby sequence to effectively balance exploration and exploitation. For a comprehensive overview of restart techniques, we refer readers to [3].

[1] Here, we identify a clause with the set of literals it contains.

3 Expected Runtime of Walksat on Tseitin Formulas

In this section, we analyze the expected runtime of Walksat algorithm when its inputs are Tseitin formulas, which turns out to be polynomial.

Theorem 1. *Consider a satisfiable Tseitin formula $T(G,c)$ associated with an n-vertex graph G and a charge function c. Then the expected runtime of Walksat on $T(G,c)$ is $O(n^4)$.*

Note on an unsatisfiable Tseitin formulas, Walksat does not halt. Given the above theorem, we can augment Walksat with an $\Theta(n^4)$-time clock, and thus ensure that on any Tseitin formula the algorithm always halts in time $O(n^4)$ and outputs a correct answer with high probability.

On a satisfiable Tseitin formula, the behavior of Walksat can be clearly understood as a random walk towards a satisfying assignment, which is exactly how we will analyze Walksat and prove Theorem 1. To that end, we first introduce the well-known notion of the *Hamming distance* between assignments.

Definition 1. *The* Hamming distance *between two Boolean assignments $x, y \in \{0,1\}^n$ is*

$$H(x,y) = \left|\{e \in E : x_e \neq y_e\}\right|,$$

i.e., the number of different bits of x and y.

The Hamming distance between an assignment x and a set S of assignments is the minimum distance from x to any assignment in S:

$$H(x,S) = \min_{y \in S} H(x,y).$$

The following lemma provides another key combinatorial property of Tseitin formulas, stating that the number of unsatisfied XOR equations (and clauses) for an unsatisfying assignment is always even.

Lemma 5. *For any bit assignment x to a satisfiable Tseitin formula $T(G,c)$ on a graph G, if x does not satisfy $T(G,c)$, then the number of unsatisfied XOR equations and the corresponding number of unsatisfied clauses in $T(G,c)$ are both even.*

Proof. Let $G=(V,E)$. By assumption $T(G,c)$ has a satisfying assignment $y \in \{0,1\}^{|E|}$. Since $x \in \{0,1\}^{|E|}$ does not satisfy $T(G,c)$, x and y must differ. We use the difference between x and y to define a subgraph $G_{\text{diff}} = (V_{\text{diff}}, E_{\text{diff}})$ of G by

$$V_{\text{diff}} = V \quad \text{and} \quad E_{\text{diff}} = \{e \in E \mid x_e \neq y_e\}.$$

Consider an XOR equation P_v (cf. (1)) that is *not* satisfied by x. As y must satisfy P_v, obviously there exist exactly an *odd* number of $e \in E(v)$ such that $x_e \neq y_e$. It is also easy to see that the converse is true. Therefore, any XOR equation P_v is not satisfied by x if and only if the degree of v in the graph G_{diff}

is odd. By the handshaking lemma, the number of vertices of odd degree in any graph is even. Hence, the number of unsatisfied XOR equations is even.

It remains to show that the number of unsatisfied clauses in $T(G, c)$ under x is also even. We observe that every XOR equation P_v is translated to a set of clauses each of which corresponds to exactly one unsatisfying (partial) assignment for the variables in P_v. Thus we have exactly one unsatisfied clause from P_v under x. It follows that the number of unsatisfied clauses is equal to the number of unsatisfied XOR equations, and we are done.

Our proof of Theorem 1 is built on the *random meet model* introduced by Coppersmith et al. in [8]. In this model, we are given a graph G and two tokens placed on two vertices in G. In each step, one of the tokens is selected and moved to an adjacent vertex. Both choices are made uniformly at random. The question is how many steps are needed so that the two tokens will eventually meet.

Lemma 6 [8]. *On a connected graph G with n vertices, the expected time for two tokens to meet each other is $O(n^3)$.*

We are now fully equipped with the necessary combinatorial and probabilistic tools for the analysis of the Walksat algorithm on satisfiable Tseitin formulas.

Proof (Proof of Theorem 1).

Let $G = (V, E)$ and c be a charge function. We assume that Walksat starts on $T(G, c)$ with an unsatisfying assignment x (otherwise it halts immediately and outputs x). Therefore,

$$U_x = \{v \in V : \text{the XOR equation } P_v \text{ is not satisfied by } x\}$$

is nonempty. And by Lemma 5, the size of U_x is even. of its variables (to flip). By the definition of $T(G, c)$ this is the same as choosing a vertex $u \in U_x$ and one adjacent edge $e \in E(u)$, say $e = (u, v)$. After flipping the variable x_e, the selected clause and hence also the XOR equation P_u is satisfied. Observe that the variable x_e only appears in P_u and P_v. Thus the flipping of x_e only affects the satisfiability of P_u, P_v, and their clauses. In particular, we have two cases concerning P_v.

- P_v is satisfied, which implies that it was *not* satisfied before the flipping. It also implies that the total number of unsatisfied XOR equations are reduced by 2.
- P_v is not satisfied. Hence, it was satisfied before the flipping, and that the total number of unsatisfied XOR equations does not change.

More generally, in each step indexed by some $t \in \mathbb{N}$, Walksat maintains an assignment x^t, which starts from $t = 0$ and $x^0 = x$, the original unsatisfying assignment. If x^t satisfies $T(G, c)$, then we are done. Otherwise, let U^t be the set of unsatisfied XOR equations, e.g., $U^0 = U_x$. Then Walksat chooses a $v \in U^t$ and flips a variable in P_v. By the above discussions, there are two possibilities of the size of U^{t+1}, i.e., the number of unsatisfied XOR equations under the new assignment x^{t+1}.

$|U^{t+1}| = |U^t| - 2$. It means that U^{t+1} contains two adjacent vertices $u, v \in V$ such that in Step t, the Walksat algorithm has chosen u and the edge (u, v).
$|U^{t+1}| = |U^t|$. That is, Walksat has selected some $u \in U^t$ and $e = (u, v) \in E$ with $v \notin U^t$. It follows that

$$U^{t+1} = (U^t \setminus \{u\}) \cup \{v\}.$$

Now we are ready to describe the behavior of Walksat by a (generalized) random meet model on the graph G. In the beginning, there are $|U_x|$ tokens placed on the vertex set U_x. Then in Step t for $t \geq 0$, one token is chosen and moved to a vertex adjacent to its current position, where both choices are made uniformly at random. If it meets another token on G, i.e., the above case of $|U^{t+1}| = |U^t| - 2$, then both tokens are removed from G. The process ends once all tokens are removed from G.

Therefore, we can divide the sequence of an execution of Walksat on x to $|U_x|/2$ subsequences, each viewed as a random procedure of two tokens moving along the edges in G until they meet each other. By Lemma 6, the expected length of such a subsequence is $O(|V|^3)$. So the overall expected runtime of Walksat is bounded by

$$\frac{|U_x|}{2} \cdot O(|V|^3) = O(|V|^4) = O(n^4).$$

4 Separation Instances in Tseitin Formulas

Having established that the Walksat algorithm can solve satisfiable Tseitin formulas in polynomial expected runtime, we now turn our attention to DPLL. While Walksat demonstrates efficiency on these instances, it remains unclear whether traditional backtracking-based DPLL algorithm exhibit similar performance. To rigorously explore this question, we construct a special family of Tseitin formulas explicitly designed to be challenging for DPLL algorithms. The analysis that follows will reveal a separation in performance between Walksat and DPLL.

Theorem 2. *There exists a family of Tseitin formulas T such that the DPLL algorithm with any restart strategy requires $2^{\Omega(\sqrt{\mathrm{var}(T)})}$ time to decide, with high probability.*

The proof of Theorem 2 can be found in Appendix ??. The desired family consists of Tseitin formulas L_n whose underlying graph G_n comprises a linear sequence of connected d-regular subgraphs $B^{(i)}$. Each $B^{(i)}$ is connected sequentially by paths $Q^{(i)}$. The structure ensures high treewidth and controlled propagation, creating a configuration that is provably difficult for DPLL-based solving strategies.

Definition 2 (L_n). *Fix a constant $d \geq 3$. For every $n \geq 1$ we construct a graph $G_n = (V_n, E_n)$ consisting of:*

- *Connected d-regular subgraphs* $B^{(1)}, B^{(2)}, \ldots, B^{(n)}$, *each containing exactly* n *vertices with treewidth* $\Theta(n)$. *The existence of such* n*-vertex graphs of treewidth* $\Theta(n)$ *is well known, see for example [14].*
- *Paths* $Q^{(1)}, Q^{(2)}, \ldots, Q^{(n-1)}$, *each of length* n *(thus containing* $n+1$ *vertices).*
- *Each* $Q^{(i)}$ *connects* $B^{(i)}$ *to* $B^{(i+1)}$ *by sharing exactly one vertex with* $B^{(i)}$ *and one with* $B^{(i+1)}$.

Figure 2 gives an illustration of G_n. *We denote the set of edges in the paths* $Q^{(i)}$ *by* $\mathcal{Q}$, *and the set of edges in the subgraphs* $B^{(i)}$ *by* $\mathcal{B}$. *Observe that* $|\mathcal{Q}| = (n-1)n$, $|\mathcal{B}| = \frac{dn^2}{2}$, *and the number of edges in* G_n *is* $|\mathcal{Q}| + |\mathcal{B}| = (\frac{dn}{2} + n - 1)n$.

We define a charge function $c : V_n \to \{0,1\}$ *ensuring that the resulting Tseitin formula* $L_n = T(G_n, c)$ *is satisfiable (as guaranteed by Lemma 2).*

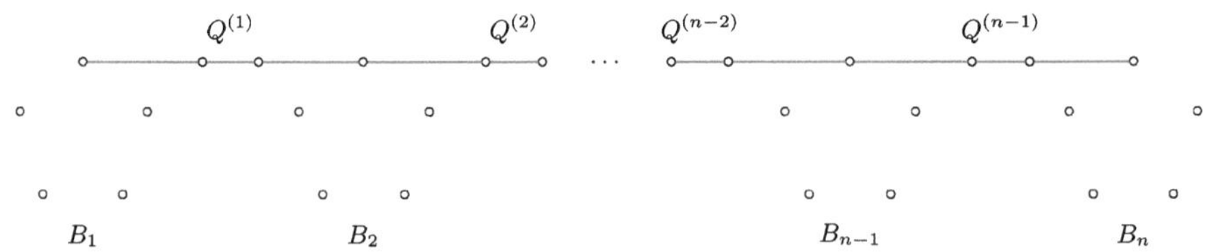

Fig. 2. The graph G_n contains subgraphs $B^{(i)}$ (orange edges) which are sequentially connected by paths $Q^{(i)}$ (purple edges). (Color figure online)

Recall that our DPLL algorithm begins with a set of preprocessing rules as described in Sect. 2.3. Interestingly, these preprocessing steps have no effect on the formulas L_n, which demonstrates the robustness of our construction.

Lemma 7. *The self-subsuming resolution and the blocked clause elimination do not apply to* L_n.

The proof of Lemma 7 is deferred to Appendix. Complementing our theoretical findings, we also provide experimental results in Appendix.

5 Conclusion and Future Work

We showed that Walksat runs in expected polynomial time on Tseitin formulas, while DPLL with restarts requires exponential time. It remains future work to understand whether techniques like clause learning and advanced branching heuristics (e.g., VSIDS), widely used in CDCL solvers, can change this behavior.

Acknowledgments. We thank the anonymous referees for their valuable comments and suggestions. This research is supported by NSFC 62532014. Yijia Chen is also supported by the National Natural Science Foundation of China (Project 62372291).

References

1. Alekhnovich, M., Hirsch, E.A., Itsykson, D.: Exponential lower bounds for the running time of DPLL algorithms on satisfiable formulas. In: Díaz, J., Karhumäki, J., Lepistö, A., Sannella, D. (eds.) ICALP 2004. LNCS, vol. 3142, pp. 84–96. Springer, Heidelberg (2004). https://doi.org/10.1007/978-3-540-27836-8_10
2. Beame, P., Liew, V.: Toward verifying nonlinear integer arithmetic. J. ACM (JACM) **66**(3), 1–30 (2019)
3. Biere, A., Heule, M., van Maaren, H., Walsh, T. (eds.) Handbook of Satisfiability. Frontiers in Artificial Intelligence and Applications, vol. 336, 2nd edn. IOS Press (2021). https://doi.org/10.3233/FAIA336
4. Cai, S., Luo, C., Su, K.: CCAnr: a configuration checking based local search solver for non-random satisfiability. In: Heule, M., Weaver, S. (eds.) SAT 2015. LNCS, vol. 9340, pp. 1–8. Springer, Cham (2015). https://doi.org/10.1007/978-3-319-24318-4_1
5. Cai, S., Luo, C., Zhang, X., Zhang, J.: Improving local search for structured SAT formulas via unit propagation based construct and cut initialization (short paper). In: Michel, L.D. (ed.) 27th International Conference on Principles and Practice of Constraint Programming, CP 2021, Montpellier, France (Virtual Conference), 25–29 October 2021. LIPIcs, vol. 210, pp. 5:1–5:10. Schloss Dagstuhl - Leibniz-Zentrum für Informatik (2021). https://doi.org/10.4230/LIPICS.CP.2021.5
6. Clarke, E., Biere, A., Raimi, R., Zhu, Y.: Bounded model checking using satisfiability solving. Formal Methods Syst. Des. **19**, 7–34 (2001)
7. Cook, B., Kroening, D., Sharygina, N.: Accurate theorem proving for program verification. In: Margaria, T., Steffen, B. (eds.) ISoLA 2004. LNCS, vol. 4313, pp. 96–114. Springer, Heidelberg (2006). https://doi.org/10.1007/11925040_7
8. Coppersmith, D., Tetali, P., Winkler, P.: Collisions among random walks on a graph. SIAM J. Discret. Math. **6**(3), 363–374 (1993)
9. Davis, M., Logemann, G., Loveland, D.: A machine program for theorem-proving. Commun. ACM **5**(7), 394–397 (1962)
10. de Colnet, A., Mengel, S.: Characterizing Tseitin-formulas with short regular resolution refutations. J. Artif. Intell. Res. **76**, 265–286 (2023)
11. Eén, N., Biere, A.: Effective preprocessing in SAT through variable and clause elimination. In: Bacchus, F., Walsh, T. (eds.) SAT 2005. LNCS, vol. 3569, pp. 61–75. Springer, Heidelberg (2005). https://doi.org/10.1007/11499107_5
12. Elffers, J., Giráldez-Cru, J., Gocht, S., Nordström, J., Simon, L.: Seeking practical CDCL insights from theoretical sat benchmarks. In: IJCAI, pp. 1300–1308 (2018)
13. Glinskih, L., Itsykson, D.: Satisfiable Tseitin formulas are hard for nondeterministic read-once branching programs. In: 42nd International Symposium on Mathematical Foundations of Computer Science (MFCS 2017). Schloss Dagstuhl-Leibniz-Zentrum fuer Informatik (2017)
14. Gruber, H.: On balanced separators, treewidth, and cycle rank. arXiv preprint arXiv:1012.1344 (2010)
15. Heule, M.J.H., Kullmann, O., Marek, V.W.: Solving and verifying the Boolean Pythagorean triples problem via cube-and-conquer. In: Creignou, N., Le Berre, D. (eds.) SAT 2016. LNCS, vol. 9710, pp. 228–245. Springer, Cham (2016). https://doi.org/10.1007/978-3-319-40970-2_15
16. Hoos, H.H., Stützle, T.: Stochastic Local Search: Foundations and Applications. Elsevier (2004)

17. Itsykson, D., Riazanov, A., Smirnov, P.: Tight bounds for Tseitin formulas. In: 25th International Conference on Theory and Applications of Satisfiability Testing (SAT 2022). Schloss Dagstuhl-Leibniz-Zentrum für Informatik (2022)
18. Järvisalo, M., Biere, A., Heule, M.: Blocked clause elimination. In: Esparza, J., Majumdar, R. (eds.) TACAS 2010. LNCS, vol. 6015, pp. 129–144. Springer, Heidelberg (2010). https://doi.org/10.1007/978-3-642-12002-2_10
19. Li, C., Fleming, N., Vinyals, M., Pitassi, T., Ganesh, V.: Towards a complexity-theoretic understanding of restarts in SAT solvers. In: Pulina, L., Seidl, M. (eds.) SAT 2020. LNCS, vol. 12178, pp. 233–249. Springer, Cham (2020). https://doi.org/10.1007/978-3-030-51825-7_17
20. Luo, C., Cai, S., Kaile, S., Wei, W.: Clause states based configuration checking in local search for satisfiability. IEEE Trans. Cybern. **45**(5), 1028–1041 (2014)
21. Luo, C., Cai, S., Wei, W., Jie, Z., Kaile, S.: CCLS: an efficient local search algorithm for weighted maximum satisfiability. IEEE Trans. Comput. **64**(7), 1830–1843 (2014)
22. Luo, C., Cai, S., Wu, W., Su, K.: Double configuration checking in stochastic local search for satisfiability. In: Proceedings of the AAAI Conference on Artificial Intelligence, vol. 28 (2014)
23. Massacci, F., Marraro, L.: Logical cryptanalysis as a sat problem. J. Autom. Reason. **24**, 165–203 (2000)
24. Nawrocki, W., Liu, Z., Fröhlich, A., Heule, M.J.H., Biere, A.: XOR local search for Boolean Brent equations. In: Li, C.-M., Manyà, F. (eds.) SAT 2021. LNCS, vol. 12831, pp. 417–435. Springer, Cham (2021). https://doi.org/10.1007/978-3-030-80223-3_29
25. Papadimitriou, C.H.: On selecting a satisfying truth assignment. In: Proceedings 32nd Annual Symposium of Foundations of Computer Science, pp. 163–169. IEEE Computer Society (1991)
26. Robertson, N., Seymour, P.D.: Graph minors. II. Algorithmic aspects of tree-width. J. Algorithms **7**(3), 309–322 (1986)
27. Marques Silva, J.P., Sakallah, K.A.: Grasp-a new search algorithm for satisfiability. In: Proceedings of International Conference on Computer Aided Design, pp. 220–227. IEEE (1996)
28. Tseitin, G.: On the complexity of derivation in propositional calculus. In: Studies in Constructive Mathematics and Mathematical Logic, pages 115–125 (1968)
29. Urquhart, A.: Hard examples for resolution. J. ACM (JACM) **34**(1), 209–219 (1987)
30. Vinyals, M.: Hard examples for common variable decision heuristics. In: Proceedings of the AAAI Conference on Artificial Intelligence, vol. 34, pp. 1652–1659 (2020)

Algorithmic Game Theory

How to Play Mastermind Without Game Master

Shota Ikeda[1(✉)] and Kazumasa Shinagawa[2,3(✉)]

[1] Ibaraki University, Hitachi, Japan
25nm709x@vc.ibaraki.ac.jp
[2] University of Tsukuba, Tsukuba, Japan
shinagawa@cs.tsukuba.ac.jp
[3] National Institute of Advanced Industrial Science and Technology, Tokyo, Japan

Abstract. Mastermind is a classic code-guessing game in which a game master secretly selects a multi-digit code with no duplicates and a codebreaker iteratively submits candidate codes. For each submission, the game master compares it to the secret code and tells the number of "hits" and "blows" as feedback. In this paper, we design a novel method to play Mastermind without a game master using card-based cryptographic techniques. In particular, we design feedback protocols and a code generation protocol, where the former outputs the number of hits and blows without revealing the code itself, and the latter outputs a sequence of cards representing a code chosen uniformly at random. By using these protocols, we can enjoy Mastermind without the need for a game master. All of the proposed protocols are extremely simple and can be executed in real gameplay. This enables people to learn fundamental principles of cryptography through playing Mastermind.

Keywords: Cryptography · Game Informatics · Mastermind

1 Introduction

1.1 Background

The fusion of game informatics and cryptography presents a rich and largely unexplored research field, offering both theoretical challenges and practical applications. In this context, we explore the classical deduction game *Mastermind*, a well-known game that requires logical reasoning and combinatorial analysis. Traditionally, it is played between a *game master* (also known as a *codemaker*), who selects a secret 4-digit number $(c_1, c_2, c_3, c_4) \in \{1, 2, \ldots, 6\}^4$, and a *player* (also known as a *codebreaker*), who attempts to guess this number by submitting queries $(g_1, g_2, g_3, g_4) \in \{1, 2, \ldots, 6\}^4$. For each query, the game master responds with two numbers: the *hit number*, which is the number of positions i for which $g_i = c_i$, and the *blow number*, which is the number of positions i for which g_i matches some c_j with $j \neq i$. Using this feedback, the player iteratively refines their guesses, aiming to deduce the secret number in as few attempts as possible.

M. Li et al. (Eds.): TAMC 2025, LNCS 16084, pp. 109–120, 2026.
https://doi.org/10.1007/978-981-95-4839-2_9

When played without electronic devices, this game inherently requires a trusted game master to manage the secret number and compute the correct feedback for each guess. However, this reliance on a central authority limits the autonomy and fairness of the gameplay.

This paper addresses the problem of how to play Mastermind without a game master by applying advanced cryptographic techniques, specifically secure multiparty computation. We propose a secure computation protocol that enables Mastermind to be played without a game master, using only physical cards. This approach falls within the field of *card-based cryptography*, which explores how cryptographic primitives such as secure computation and zero-knowledge proofs can be realized by using a deck of physical cards without electronic computers.

Research at the intersection of game informatics and cryptography carries significant value. Cryptography often relies on advanced mathematics, making it difficult to understand for people without a deep knowledge of mathematics. As a result, teaching cryptography and security to ordinary people can be quite challenging. By the fusion of game informatics and cryptography, we can create opportunities for people to learn fundamental ideas of cryptography through engaging gameplay. In particular, our protocols in this paper are extremely easy to implement, making it accessible and enjoyable to play. This simplicity enhances its potential as a powerful educational tool for teaching cryptography.

1.2 Mastermind

Mastermind (also known as *Bulls and Cows* or *Hit and Blow*) is a code-guessing game between two players called *codemaker* and *codebreaker*. The basic rule of Mastermind is given as follows.

1. The codemaker chooses a *code* privately. The code is a 4-digit number[1] $c = (c_1, c_2, c_3, c_4) \in \{1, 2, \dots, 6\}^4$. According to the basic rule, a duplicate code is not allowed, i.e., we impose the condition that all c_i's are distinct.
2. For $i = 1, 2, \dots, k$, where k is the maximum round, do the following:
   (a) The codebreaker makes a *guess* $g^{(i)} = (g_1^{(i)}, g_2^{(i)}, g_3^{(i)}, g_4^{(i)}) \in \{1, 2, \dots, 6\}^4$. Then he/she tells it to the codemaker.
   (b) The codemaker compares the code c with the guess $g^{(i)}$ and computes the number of *hits* $h(c, g^{(i)})$ and the number of *blows* $b(c, g^{(i)})$. If the same number exists in the same digit in c and $g^{(i)}$, it counts as a hit. If the same number exists in a different digit in c and $g^{(i)}$, it counts as a blow.
   (c) The codemaker tells the codebreaker $(h(c, g^{(i)}), b(c, g^{(i)}))$ as feedback. If $c = g^{(i)}$, the codebreaker wins.
3. If the codebreaker cannot guess the code within k rounds, he/she loses.

There are several variants of rules and settings. In this paper, we consider a generalized Mastermind with a code $c = (c_1, c_2, \dots, c_M) \in \{1, 2, \dots, N\}^M$.

[1] In the original Mastermind, the code is a sequence of 4 balls with 6 colors. In this paper, we deal with a generalized Mastermind of M balls with N colors.

Table 1. Summary of our protocols

Protocol	# of cards	# of shuffles	Remark
∘ Feedback protocol			
Section 3.1	NM	2 CS	public guess
Section 3.2	$2NM$	2 PSS	private guess
Section 3.3	$2NM$	2 CS + 1 PSS	public guess, code preserving
Section 3.3	$3NM$	3 PSS	private guess, code preserving
∘ Code generation protocol			
Section 3.4	kNM	1 PSS	k copies

Note: CS and PSS mean complete shuffle and pile-scramble shuffle.

1.3 Contribution

This paper presents a novel approach to playing the code-guessing game Mastermind without requiring a game master (also called a codemaker). Our first contribution is the introduction of a cryptographic perspective to the game of Mastermind. This game is highly compatible with card-based cryptography, and it can be played without a game master using extremely simple protocols. As technical contributions, we design several card-based protocols for playing Mastermind without a game master (Table 1). The key technical contributions are as follows:

- We design four types of *feedback protocols*, each of which securely computes the number of hits and blows given a secret code and a guess. These protocols differ in functionality:
  - A protocol with *public guess*, in which the guess is revealed but the code remains secret.
  - A protocol with *private guess*, where both the code and the guess remain hidden, and only the feedback is revealed.
  - A protocol with *preserving code*, which allows the code to persist in a hidden form even after multiple rounds.
- We propose a *code generation protocol* that generates a secret code in a verifiably random and concealed manner, without the need for any participant to learn the actual value of the code.
- All proposed protocols are realized as card-based cryptographic protocols. Our protocols are secure computation protocols with information-theoretic security, which guarantees that no information is leaked beyond the result.
- From an educational and practical standpoint, our protocols are exceptionally simple to implement. This makes them not only suitable for research on secure computation but also valuable as pedagogical tools in cryptography and computer science education.

1.4 Related Work

Card-based cryptography is the field to implement cryptographic protocols using a deck of physical cards [1,2,5]. Research in this field primarily follows two directions: secure computation and zero-knowledge proof.

Although card-based protocols usually implement cryptographic primitives, there are some works to apply card-based cryptography to the field of game informatics. Hashimoto et al. [3] proposed a protocol, which can be applied to the grouping of the Werewolf Game. Shinagawa et al. [9] introduced a card-based protocol that simulates virtual players in the game of Old Maid. Ruangwises and Shinagawa [8] designed a card-based protocol simulating virtual players for UNO. This paper focuses on removing the game master in Mastermind.

1.5 Outline of the Paper

This paper is organized as follows. Section 2 provides an overview of the cards, shuffling methods, and encoding techniques used throughout the paper. Section 3 describes several feedback protocols and code generation protocols for playing Mastermind. Section 4 presents multiple gameplay procedures for Mastermind based on the proposed protocols. Finally, Sect. 5 concludes the paper.

2 Preliminaries

Throughout this paper, we use $[N]$ to denote $[N] := \{0, 1, \ldots, N-1\}$.

2.1 Cards

In this paper, we use two types of decks of cards. The first one is called a *two-color deck*, which consists of cards with black $\boxed{\clubsuit}$ and red $\boxed{\heartsuit}$ on the front side and $\boxed{?}$ on the back side. The second one is called a *number deck*, which consists of cards with natural numbers $\boxed{1}\,\boxed{2}\cdots$ on the front side and $\boxed{?}$ on the back side. Here, the backs of the two-color and number decks can be common or different. For this reason, we will refer to both as $\boxed{?}$ in this paper.

2.2 Shuffle

In card-based cryptography, a shuffle operation is the operation of randomly rearranging a sequence of cards, where the corresponding permutation is chosen from a probability distribution $\mathcal{F}$ [5]. Here, we require that every player, including the player performing the shuffle, cannot guess which permutation is chosen beyond the fact that it is chosen by $\mathcal{F}$. In this paper, we use two types of shuffles: *complete shuffle* and *pile-scramble shuffle*.

A *complete shuffle* [6,7] is a shuffle that makes the order of cards completely random. For a sequence of 3 cards as an example, a complete shuffle, denoted by $[\cdot]$, generates one of 6 cyclic sequences uniformly at random as follows.

$$\left[\boxed{A}\boxed{B}\boxed{C}\right] \rightarrow \begin{cases} \boxed{A}\boxed{B}\boxed{C} & \text{with probability } 1/6; \\ \boxed{A}\boxed{C}\boxed{B} & \text{with probability } 1/6; \\ \boxed{B}\boxed{C}\boxed{A} & \text{with probability } 1/6; \\ \boxed{B}\boxed{A}\boxed{C} & \text{with probability } 1/6; \\ \boxed{C}\boxed{A}\boxed{B} & \text{with probability } 1/6; \\ \boxed{C}\boxed{B}\boxed{A} & \text{with probability } 1/6. \end{cases}$$

A *pile-scramble shuffle* [4] is a generalization of complete shuffles, which applies a uniform random permutation to a sequence of piles of cards. For a sequence of piles $(p_1, p_2, \ldots, p_n)$, a pile-scramble shuffle chooses a permutation $\pi \in S_n$ uniformly at random for the n-th symmetric group S_n, and then outputs $(p_{\pi^{-1}(1)}, p_{\pi^{-1}(2)}, \ldots, p_{\pi^{-1}(n)})$. Applying a pile-scramble shuffle to a sequence of cards is described by $[p_1|p_2|\cdots|p_n]$. An example of applying a pile-scramble shuffle to a sequence of 3 piles with 2 cards each is shown below.

$$\left[\begin{matrix}\boxed{A_1}\\\boxed{A_2}\end{matrix}\middle|\begin{matrix}\boxed{B_1}\\\boxed{B_2}\end{matrix}\middle|\begin{matrix}\boxed{C_1}\\\boxed{C_2}\end{matrix}\right] \rightarrow \begin{cases} \begin{matrix}\boxed{A_1}\,\boxed{B_1}\,\boxed{C_1}\\\boxed{A_2}\,\boxed{B_2}\,\boxed{C_2}\end{matrix} & \text{with probability } 1/6; \\ \begin{matrix}\boxed{A_1}\,\boxed{C_1}\,\boxed{B_1}\\\boxed{A_2}\,\boxed{C_2}\,\boxed{B_2}\end{matrix} & \text{with probability } 1/6; \\ \begin{matrix}\boxed{B_1}\,\boxed{C_1}\,\boxed{A_1}\\\boxed{B_2}\,\boxed{C_2}\,\boxed{A_2}\end{matrix} & \text{with probability } 1/6; \\ \begin{matrix}\boxed{B_1}\,\boxed{A_1}\,\boxed{C_1}\\\boxed{B_2}\,\boxed{A_2}\,\boxed{C_2}\end{matrix} & \text{with probability } 1/6; \\ \begin{matrix}\boxed{C_1}\,\boxed{A_1}\,\boxed{B_1}\\\boxed{C_2}\,\boxed{A_2}\,\boxed{B_2}\end{matrix} & \text{with probability } 1/6; \\ \begin{matrix}\boxed{C_1}\,\boxed{B_1}\,\boxed{A_1}\\\boxed{C_2}\,\boxed{B_2}\,\boxed{A_2}\end{matrix} & \text{with probability } 1/6. \end{cases}$$

2.3 Encoding for Integer

Using a set of N cards consisting of one $\boxed{\heartsuit}$ and $N-1$ $\boxed{\clubsuit}$ cards, the integer $i \in [N]$ is expressed as follows.

$$\overset{N-1}{\boxed{\clubsuit}}\ \overset{N-2}{\boxed{\clubsuit}} \cdots \overset{i+1}{\boxed{\clubsuit}}\overset{i}{\boxed{\heartsuit}}\overset{i-1}{\boxed{\clubsuit}} \cdots \overset{1}{\boxed{\clubsuit}}\overset{0}{\boxed{\clubsuit}}.$$

This encoding places $\boxed{\heartsuit}$ on the i-th card in a horizontal line with the rightmost card as the 0-th card, and $\boxed{\clubsuit}$ on all others. This type of encoding is called integer encoding [2]. If an integer $i \in [N]$ is encoded by one $\boxed{\heartsuit}$ and $N-1$ $\boxed{\clubsuit}$, it is written $E_N^{\heartsuit}(i)$, and if it is encoded by one $\boxed{\clubsuit}$ and $N-1$ $\boxed{\heartsuit}$, it is written $E_N^{\clubsuit}(i)$. This integer encoding is simple to encode integers, but requires many cards to represent large integers.

In this paper, we propose an *N-base M-digit encoding*, which is an extension of integer encoding. The N-base M-digit encoding represents multi-digit integers by applying integer encoding to each digit, providing an intuitive and easily interpretable representation. Denote by $E_{N,M}^{\heartsuit}(x)$ the sequence of NM face-down cards with M integer encodings of each digit for N-base M-digit integer x. As with integer encoding, the encoding in which the roles of $\boxed{\clubsuit}$ and $\boxed{\heartsuit}$ are exchanged is denoted by $E_{N,M}^{\clubsuit}(x)$. Our protocols use an N-base M-digit encoding $E_{N,M}^{\heartsuit}(x)$. The following is an example card sequence of 10-base 4-digit encoding.

$$\underbrace{\overset{9}{\boxed{?}}\overset{8}{\boxed{?}}\cdots\overset{1}{\boxed{?}}\overset{0}{\boxed{?}}}_{\text{1000's digit}}\;\underbrace{\overset{9}{\boxed{?}}\overset{8}{\boxed{?}}\cdots\overset{1}{\boxed{?}}\overset{0}{\boxed{?}}}_{\text{100's digit}}\;\underbrace{\overset{9}{\boxed{?}}\overset{8}{\boxed{?}}\cdots\overset{1}{\boxed{?}}\overset{0}{\boxed{?}}}_{\text{10's digit}}\;\underbrace{\overset{9}{\boxed{?}}\overset{8}{\boxed{?}}\cdots\overset{1}{\boxed{?}}\overset{0}{\boxed{?}}}_{\text{1's digit}}.$$

Using N-base M-digit encoding, various large integers can be represented with fewer cards. For example, to represent $0 \le x \le 9999$, the previous integer encoding uses 10000 cards but our 10-base 4-digit encoding uses only 40 cards.

3 Protocols for Simulating Codemaker

In this section, we design cryptographic protocols for simulating Codemaker. In particular, we propose *feedback protocols* and a *code generation protocol.* Given two sequences of face-down cards representing a code and a guess, respectively, a feedback protocol returns the result "x hits and y blows". We propose three types of feedback protocols, each with a different functionality. These protocols are used in each round to decide the number of hits and blows. On the other hand, the code generation protocol is used at the beginning of the game, which chooses a code privately and randomly, and generates a sequence of cards representing the code. In the following, we will refer to the code as $c = (c_M, c_{M-1}, \ldots, c_1) \in [N]^M$ and the guess as $g = (g_M, g_{M-1}, \ldots, g_1) \in [N]^M$, both are N-base M-digit numbers.

Throughout this paper, for simplicity, we assume that the codebreaker's guesses do not contain duplicate numbers. This is not an essential restriction. Indeed, our protocols with a minor modification can handle a guess with duplicate numbers. On the other hand, prohibiting duplicate numbers in the code is important for executing the code generation protocol described in Sect. 3.4.

3.1 Feedback Protocol with Public Guess

In this section, we propose a *feedback protocol with public guess*, which takes $E_{N,M}^{\heartsuit}(c)$ and a public value g as input, and outputs the number of hits $h(c, g)$ and the number of blows $b(c, g)$. It uses NM cards and 2 complete shuffles.

1. The input of the protocol is a value g and an encoding $E^{\heartsuit}_{N,M}(c)$ as follows:

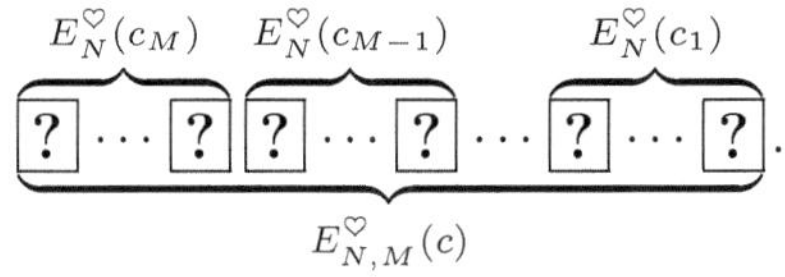

2. Let $\alpha_j^{(i)}$ be the j-th card of $E^{\heartsuit}_N(c_i)$, i.e., $E^{\heartsuit}_N(c_i) = (\alpha^{(i)}_{N-1}, \dots, \alpha^{(i)}_1, \alpha^{(i)}_0)$. For each $i \in \{1, 2, \dots, M\}$, take the g_i-th card $\alpha^{(i)}_{g_i}$. Then we have M face-down cards of $(\alpha^{(M)}_{g_M}, \dots, \alpha^{(2)}_{g_2}, \alpha^{(1)}_{g_1})$.
3. Apply a complete shuffle to the M face-down cards, and then open these cards. Suppose that the number of $\boxed{\heartsuit}$ cards is $k \in \{0, 1, \dots, M\}$ as follows:

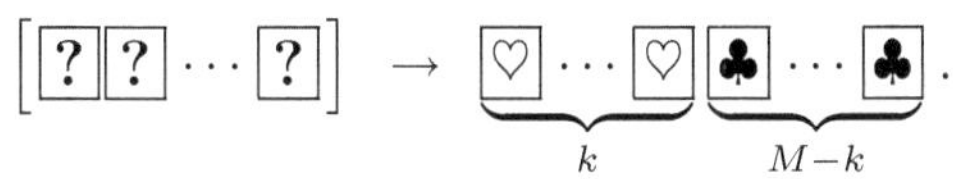

   Then the protocol concludes that the number of hits $h(c, g)$ equals k.
4. For each $i \in \{1, 2, \dots, M\}$, take $M-1$ cards $(\alpha^{(1)}_{g_i}, \alpha^{(2)}_{g_i}, \dots, \alpha^{(M)}_{g_i})$ except $\alpha^{(i)}_{g_i}$ used above. Then we have $M(M-1)$ face-down cards.
5. Apply a complete shuffle to the $M(M-1)$ cards, and then open these cards. Suppose that the number of $\boxed{\heartsuit}$ cards is $\ell \in \{0, 1, \dots, M(M-1)\}$ as follows:

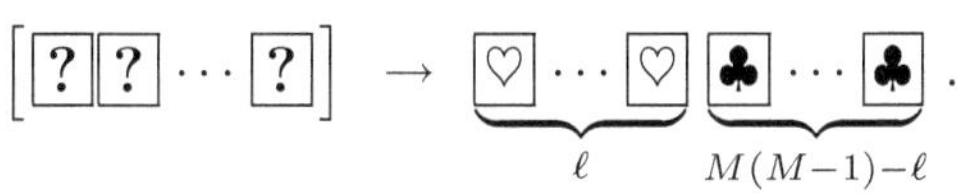

   Then the protocol concludes that the number of blows $b(c, g)$ equals ℓ.

3.2 Feedback Protocol with Private Guess

In this section, we propose a *feedback protocol with private guess*, which takes $E^{\heartsuit}_{N,M}(c)$ and $E^{\heartsuit}_{N,M}(g)$ as input, and outputs the number of hits $h(c, g)$ and the number of blows $b(c, g)$. It uses $2NM$ cards and 2 pile-scramble shuffles.

1. The input of the protocol is $E^{\heartsuit}_{N,M}(c)$ and $E^{\heartsuit}_{N,M}(g)$ as follows:

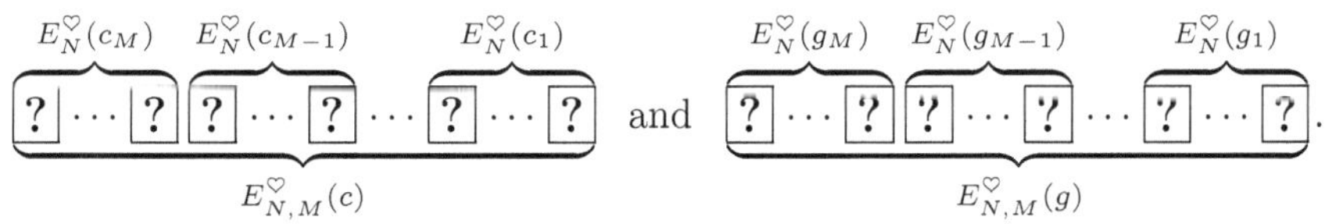

2. Make a $2M \times N$ matrix of cards as follows:

$$\begin{array}{ccccc l}
\boxed{?} & \boxed{?} & \boxed{?} & \cdots & \boxed{?} & E_N^{\heartsuit}(c_M) \\
\boxed{?} & \boxed{?} & \boxed{?} & \cdots & \boxed{?} & E_N^{\heartsuit}(c_{M-1}) \\
\vdots & \vdots & \vdots & \ddots & \vdots & \\
\boxed{?} & \boxed{?} & \boxed{?} & \cdots & \boxed{?} & E_N^{\heartsuit}(c_1) \\
\boxed{?} & \boxed{?} & \boxed{?} & \cdots & \boxed{?} & E_N^{\heartsuit}(g_M) \\
\boxed{?} & \boxed{?} & \boxed{?} & \cdots & \boxed{?} & E_N^{\heartsuit}(g_{M-1}) \\
\vdots & \vdots & \vdots & \ddots & \vdots & \\
\boxed{?} & \boxed{?} & \boxed{?} & \cdots & \boxed{?} & E_N^{\heartsuit}(g_1)
\end{array}.$$

Then apply a pile-scramble shuffle to the matrix as follows:

$$\left[\begin{array}{c|c|c|c|c}
\boxed{?} & \boxed{?} & \boxed{?} & \cdots & \boxed{?} \\
\boxed{?} & \boxed{?} & \boxed{?} & \cdots & \boxed{?} \\
\vdots & \vdots & \vdots & & \vdots \\
\boxed{?} & \boxed{?} & \boxed{?} & \cdots & \boxed{?}
\end{array}\right].$$

For $1 \le i \le M$, we refer to the i-th row (resp. $(M+i)$-th row) as C_i (resp. G_i), which is a randomized sequence of $E_N^{\heartsuit}(c_i)$ (resp. $E_N^{\heartsuit}(g_i)$).

3. Make a $2 \times MN$ matrix where the 1st row is $(C_M, C_{M-1}, \dots, C_1)$ and the 2nd row is $(G_M, G_{M-1}, \dots, G_1)$. Then apply a pile-scramble shuffle as follows:

$$\left[\begin{array}{c|c|c|c}
C_M & C_{M-1} & \cdots & C_1 \\
G_M & G_{M-1} & \cdots & G_1
\end{array}\right].$$

We refer to the first row as $(C'_M, C'_{M-1}, \dots, C'_1)$ and the second row as $(G'_M, G'_{M-1}, \dots, G'_1)$, where each C'_i and G'_i are sequences of N cards.

4. Turn over all cards in the $2 \times MN$ matrix. If there exists $i \in \{1, 2, \dots, M\}$ such that C'_i and G'_i have $\boxed{\heartsuit}$ in the same column, it counts as a "hit". If there exist $i, j \in \{1, 2, \dots, M\}$ such that C'_i and G'_j have $\boxed{\heartsuit}$ in the same position, it counts as a "blow" as follows:

$$\begin{array}{c ccccc}
 & {\scriptstyle N-1} & & {\scriptstyle x} & & {\scriptstyle 0} \\
C'_i & \boxed{\clubsuit} & \cdots & \boxed{\heartsuit} & \cdots & \boxed{\clubsuit} \\
G'_i & \boxed{\clubsuit} & \cdots & \boxed{\clubsuit} & \cdots & \boxed{\clubsuit}
\end{array}
\quad \text{and} \quad
\begin{array}{c ccccc}
 & {\scriptstyle N-1} & & {\scriptstyle x} & & {\scriptstyle 0} \\
C'_j & \boxed{\clubsuit} & \cdots & \boxed{\clubsuit} & \cdots & \boxed{\clubsuit} \\
G'_j & \boxed{\clubsuit} & \cdots & \boxed{\heartsuit} & \cdots & \boxed{\clubsuit}
\end{array}.$$

The protocol outputs the number of hits $h(c, g)$ and blows $b(c, g)$.

3.3 How to Preserve the Code

By slightly modifying the feedback protocols in Sects. 3.1 and 3.2, it is possible to preserve the sequence of face-down cards representing the code c at the end of the protocol, instead of having it disappear.

This can be achieved with a very simple idea. First, on top of each card in the encoding $E^{\heartsuit}_{N,M}(c)$, we attach number cards with 1 to $L := NM$ as follows:

$$\underbrace{\begin{array}{ccccc} \boxed{1} & \boxed{2} & \boxed{3} & \cdots & \boxed{L} \\ \boxed{?} & \boxed{?} & \boxed{?} & \cdots & \boxed{?} \end{array}}_{E^{\heartsuit}_{N,M}(c)}.$$

Then all number cards are turned face down. From now on, each column of two cards is treated as a single unit, and we execute the feedback protocols in Sects. 3.1 and 3.2. Here, even when the protocol involves turning over the cards of the code, we leave the number cards face down.

Suppose we execute a feedback protocol. At the end of the protocol, we then take all NM units corresponding to the code and the number cards, turn them face-down, perform a pile-scramble shuffle, turn only the number cards, and reorder the pairs so that the numbers appear in order from 1 to NM. Through this procedure, we can reconstruct the original sequence $E^{\heartsuit}_{N,M}(c)$.

3.4 Code Generation Protocol

In this section, we propose a *code generation protocol*, which generates k copies of $E^{\heartsuit}_{N,M}(c)$, where c is chosen uniformly at random. It uses kNM cards and 1 pile-scramble shuffle.

1. Make a $kM \times N$ matrix of cards as follows:

$$\begin{array}{cccccc|l}
\boxed{\clubsuit} & \boxed{\clubsuit} & \boxed{\clubsuit} & \cdots & \boxed{\clubsuit} & \boxed{\heartsuit} & E^{\heartsuit}_{N}(0) \\
\vdots & \vdots & \vdots & \ddots & \vdots & \vdots & \vdots \\
\boxed{\clubsuit} & \boxed{\clubsuit} & \boxed{\clubsuit} & \cdots & \boxed{\clubsuit} & \boxed{\heartsuit} & E^{\heartsuit}_{N}(0) \\
\boxed{\clubsuit} & \boxed{\clubsuit} & \boxed{\clubsuit} & \cdots & \boxed{\heartsuit} & \boxed{\clubsuit} & E^{\heartsuit}_{N}(1) \\
\vdots & \vdots & \vdots & \ddots & \vdots & \vdots & \vdots \\
\boxed{\clubsuit} & \boxed{\clubsuit} & \boxed{\clubsuit} & \cdots & \boxed{\heartsuit} & \boxed{\clubsuit} & E^{\heartsuit}_{N}(1) \\
\vdots & \vdots & \vdots & \ddots & \vdots & \vdots & \vdots \\
\boxed{\clubsuit} & \boxed{\clubsuit} & \boxed{\heartsuit} & \cdots & \boxed{\clubsuit} & \boxed{\clubsuit} & E^{\heartsuit}_{N}(M-1) \\
\vdots & \vdots & \vdots & \ddots & \vdots & \vdots & \vdots \\
\boxed{\clubsuit} & \boxed{\clubsuit} & \boxed{\heartsuit} & \cdots & \boxed{\clubsuit} & \boxed{\clubsuit} & E^{\heartsuit}_{N}(M-1)
\end{array}$$

Here, the first k rows are $E_N^{\heartsuit}(0)$, the next k rows are $E_N^{\heartsuit}(1)$, ..., and the last k rows are $E_N^{\heartsuit}(M-1)$.

2. Turn all cards face-down and apply a pile-scramble shuffle to the matrix as:

$$\left[\begin{array}{c|c|c|c|c} \boxed{?} & \boxed{?} & \boxed{?} & \cdots & \boxed{?} \\ \boxed{?} & \boxed{?} & \boxed{?} & \cdots & \boxed{?} \\ \vdots & \vdots & \vdots & & \vdots \\ \boxed{?} & \boxed{?} & \boxed{?} & \cdots & \boxed{?} \end{array}\right].$$

Suppose that the first k rows are $E_N^{\heartsuit}(c_1)$, the next k rows are $E_N^{\heartsuit}(c_2)$, ..., and the last k rows are $E_N^{\heartsuit}(c_M)$. Note that $c_1, c_2, \ldots, c_M \in [N]$ are all distinct.

3. By collecting $(E_N^{\heartsuit}(c_1), E_N^{\heartsuit}(c_2), \ldots, E_N^{\heartsuit}(c_M))$, we obtain k copies of $E_{N,M}^{\heartsuit}(c)$.

$$\underbrace{\overbrace{\boxed{?}\cdots\boxed{?}}^{E_N^{\heartsuit}(c_M)}\overbrace{\boxed{?}\cdots\boxed{?}}^{E_N^{\heartsuit}(c_{M-1})}\cdots\overbrace{\boxed{?}\cdots\boxed{?}}^{E_N^{\heartsuit}(c_1)}}_{E_{N,M}^{\heartsuit}(c)} \times k\,.$$

4 How to Play Mastermind Without Codemaker

In this section, we describe how to play Mastermind without a codemaker using our feedback protocols and our code generation protocol. Since there are several rules of Mastermind, we propose various methods tailored to each of these rules.

4.1 Playing by One Codebreaker

We can play Mastermind with only a single player as follows. The player's role is the codebreaker. If the player guesses the correct code within k rounds (for a predetermined number k), the player wins. Otherwise, the player loses.

1. Generate k copies of the code c using the code generation protocol.
2. For each round $i = 1, 2, \ldots, k$, the player decides the guess $g_i \in [N]^M$ and then executes the feedback protocol with public guess. Then the protocol outputs the number of hits $h(c, g_i)$ and blows $b(c, g_i)$ as feedback.
3. The player wins if the player can guess c within k rounds.

4.2 Playing by Two Codebreakers

We can play Mastermind with two players as follows. The players' roles are both codebreakers. The two players guess the same code alternatively, and the first to correctly guess the code wins. If both players cannot guess the correct code within k rounds (for a predetermined number k), the game is a tie.

1. Generate k copies of the code c using the code generation protocol.
2. For each round $i = 1, 2, \ldots, k$, Player 1 plays if i is odd, and Player 2 plays otherwise. The player decides the guess $g_i \in [N]^M$ and then executes the feedback protocol with a public guess or with a private guess. Then the protocol outputs the number of hits $h(c, g_i)$ and blows $b(c, g_i)$ as feedback.
3. If some player can guess the correct code within k rounds, the player wins. Otherwise, the game is a tie.

If the feedback protocol with public guesses is used, both players can make their own guesses on the other's previous guesses. On the other hand, if the feedback protocol with private guesses is used, each player makes their guesses without knowing the other's. In the latter style of play, if each player privately views only their own feedback and this information is kept hidden from the opponent, the game becomes a competition to determine who can make more intelligent guesses independently.

4.3 Playing with Preventing Incorrect Feedback

A variant rule of the Mastermind, also known as Bulls and Cows, involves two players each secretly choosing a code and taking turns to guess the other's code. The winner is the first to correctly guess the opponent's code. In this setting, there is a risk that the codemaker may accidentally provide incorrect feedback. By using our protocol, such mistakes can be prevented, ensuring that all feedback is computed correctly.

1. Player 1 and Player 2 set c_1 and c_2 as their code and make $E^{\heartsuit}_{N,M}(c_1)$ and $E^{\heartsuit}_{N,M}(c_2)$, respectively.
2. For each round $i = 1, 2, \ldots, k$, each player guesses the opponent's code and executes the feedback protocol with preserving the code.
3. The first player to guess the opponent's code wins.

5 Conclusion

In this paper, we propose card-based protocols for playing Mastermind without a codemaker. An open problem is to design more efficient protocols for Mastermind in terms of the number of cards and shuffles. Another research direction is to design protocols to play other games, which require a game master, without a game master.

Acknowledgments. This work was supported by JSPS KAKENHI Grant Numbers JP21K17702 and JP23H00479, and JST CREST Grant Number JPMJCR22M1, Japan. We would like to express our gratitude to Professor Kazuki Yoneyama for his valuable advice on improving the efficiency of our protocols presented in this paper.

References

1. Boer, B.: More efficient match-making and satisfiability *the five card trick*. In: Quisquater, J.-J., Vandewalle, J. (eds.) EUROCRYPT 1989. LNCS, vol. 434, pp. 208–217. Springer, Heidelberg (1990). https://doi.org/10.1007/3-540-46885-4_23
2. Crépeau, C., Kilian, J.: Discreet solitary games. In: Stinson, D.R. (ed.) CRYPTO 1993. LNCS, vol. 773, pp. 319–330. Springer, Heidelberg (1994). https://doi.org/10.1007/3-540-48329-2_27
3. Hashimoto, Y., Shinagawa, K., Nuida, K., Inamura, M., Hanaoka, G.: Secure grouping protocol using a deck of cards. IEICE Trans. Fundam. Electron. Commun. Comput. Sci. **E101.A**(9), 1512–1524 (2018)
4. Ishikawa, R., Chida, E., Mizuki, T.: Efficient card-based protocols for generating a hidden random permutation without fixed points. In: Calude, C.S., Dinneen, M.J. (eds.) UCNC 2015. LNCS, vol. 9252, pp. 215–226. Springer, Cham (2015). https://doi.org/10.1007/978-3-319-21819-9_16
5. Mizuki, T., Shizuya, H.: A formalization of card-based cryptographic protocols via abstract machine. Int. J. Inf. Secur. **13**(1), 15–23 (2014)
6. Niemi, V., Renvall, A.: Secure multiparty computations without computers. Theoret. Comput. Sci. **191**(1–2), 173–183 (1998)
7. Nuida, K.: Efficient card-based millionaires' protocols via non-binary input encoding. In: Shikata, J., Kuzuno, H. (eds.) IWSEC 2023. LNCS, vol. 14128, pp. 237–254. Springer, Cham (2023). https://doi.org/10.1007/978-3-031-41326-1_13
8. Ruangwises, S., Shinagawa, K.: Simulating virtual players for UNO without computers. In: Formenti, E., Manzoni, L. (eds.) UCNC 2025. LNCS. Springer, Cham (2025, to appear)
9. Shinagawa, K., Miyahara, D., Mizuki, T.: How to play old maid with virtual players. Theory Comput. Syst. **69**(1), 1–18 (2025)

Constrained Distributed Heterogeneous Two-Facility Location Problems with Max-Variant Cost

Xinru Xu[1], Wenjing Liu[1,2](✉), and Qizhi Fang[1]

[1] School of Mathematical Sciences, Ocean University of China, Qingdao 266100, China
xuxinru1207@stu.ouc.edu.cn, {liuwj,qfang}@ouc.edu.cn

[2] Laboratory of Marine Mathematics, Ocean University of China, Qingdao 266100, China

Abstract. We study a constrained distributed heterogeneous two-facility location problem, where a set of agents with private locations on the real line are divided into disjoint groups. The constraint means that the facilities can only be built in a given multiset of candidate locations and at most one facility can be built at each candidate location. Given the locations of the two facilities, the cost of an agent is the distance from her location to the farthest facility (referred to as max-variant). Our goal is to design strategyproof distributed mechanisms that can incentivize all agents to truthfully report their locations and approximately optimize some social objective. A distributed mechanism consists of two steps: for each group, the mechanism chooses two candidate locations as the representatives of the group based only on the locations reported by agents therein; then, it outputs two facility locations among all the representatives. We focus on a class of deterministic strategyproof distributed mechanisms and analyze upper and lower bounds on the distortion under the Average-of-Average cost (average of the average individual cost of agents in each group), the Max-of-Max cost (maximum individual cost among all agents), the Average-of-Max cost (average of the maximum individual cost among all agents in each group) and the Max-of-Average cost (maximum of the average individual cost of all agents in each group). Under four social objectives, we obtain constant upper and lower distortion bounds.

Keywords: Mechanism design without money · Facility location · Strategyproof · Distributed · Distortion

1 Introduction

The facility location problem is a classic combinatorial optimization problem. Its main goal is to select the optimal locations for facilities under given constraints to optimize some social objective. However, in many real-world scenarios, the information of agents (e.g., their residential addresses) may be private. Therefore,

M. Li et al. (Eds.): TAMC 2025, LNCS 16084, pp. 121–134, 2026.
https://doi.org/10.1007/978-981-95-4839-2_10

social planners can only locate facilities based on the information reported by agents. On the one hand, social planners aim to optimize some social objective. On the other hand, selfish agents may misreport their information in order to minimize their individual costs. Naturally, social planners want to design mechanisms that can ensure each agent's truthful report (e.g., strategyproof) while (approximately) optimizing some social objective. Procaccia and Tennenholtz [15] were the first to study approximate mechanism design without payment for facility location problems. Subsequently, various different facility location game models have been proposed; see the survey by Chan et al. [5].

In real-world scenarios, a collective decision-making process may be distributed, as follows: a set of agents are divided into groups, and each group makes a decision first (without considering the agents of the other groups), and then these decisions are aggregated into a collective decision. For example, in the selection of outstanding students at a university, first each faculty selects its outstanding student representatives, and then the university-level outstanding students are finally selected based on these representatives. To analyze these more complex problems, Filos-Ratsikas et al. [10] initiated the study of social choice problems in distributed settings, in which decisions are made by two-step mechanisms: for each group, the mechanism first selects a representative based on the local election with the agents therein, and then outputs one of the representatives as the winner. To quantify the inefficiency of distributed mechanisms, Filos-Ratsikas et al. [10] extended the notion of distortion, which is broadly used in social choice problems. Distortion of a mechanism refers to the worst-case ratio (over all instances) between the social objective value obtained by the mechanism and the optimal social objective value. In follow-up work, Filos-Ratsikas and Voudouris [11] studied a distributed single-facility location problem under minimizing the social cost objective (i.e., the sum of individual costs of all agents).

In the classic models, facilities can be built at any point in a metric space. However, in many real-world scenarios, the feasible regions for building facilities and the number of facilities that can be built at each location may be limited, due to land use restrictions, environmental protection, humanity factors, etc. Motivated by this, mechanism design for facility location problems with limited locations has been studied, which is termed as constrained facility location.

In this paper, we study a constrained distributed heterogeneous two-facility location problem. Here, we assume each agent approves the two facilities and the individual cost of each agent is the distance from her location to the farthest facility (referred to as max-variant cost). As for max-variant, consider a scenario where an express delivery outlet needs to transport regular packages to a ordinary distribution center and cold chain packages to a professional distribution center respectively. Assuming the outlet has multiple transport vehicles with the same speed, its waiting time depends on the distance to the farthest distribution center. We show upper and lower bounds on the distortion of strategyproof distributed mechanisms under four social objectives. More details are provided below.

1.1 Our Results

We study a constrained distributed heterogeneous two-facility location problem with max-variant, where the facilities can only be built in a given set of candidate locations and at most one facility can be built at each candidate location. A set of agents have private locations on the real line and they are divided into disjoint groups. Once the two facilities have been located, the individual cost of each agent is the distance from her location to the farthest facility. Our goal is to design strategyproof distributed mechanisms that take as input the locations reported by the agents and output the locations of the two facilities. A distributed mechanism consists of two steps: for each group, the mechanism chooses two candidate locations as the representatives of the group based only on the locations reported by agents therein; then, it outputs two facility locations among all the representatives.

Following the work of Anshelevich et al. [3], we focus on the following four social objectives: the average of the average individual cost of all agents in each group (i.e., Average-of-Average cost); the maximum individual cost among all agents (i.e., Max-of-Max cost); the maximum of the average individual cost of all agents in each group (i.e., Max-of-Average cost); and the average of the maximum individual cost among all agents in each group (i.e., Average-of-Max cost). While the Average-of-Average cost and the Max-of-Max cost are adaptations of objectives that have been considered in the classic setting (i.e., non-distributed setting), the Max-of-Average cost and the Average-of-Max cost are fairness-inspired objectives that are only meaningful for the distributed setting. Under these four social objectives, we show upper and lower bounds on the distortion of strategyproof distributed mechanisms. A summary of our results is shown in Table 1.

Table 1. Upper and lower bounds on the distortion of strategyproof distributed mechanisms.

Social objective	Upper bound	Lower bound
Average-of-Average cost	9	3
Max-of-Max cost	3	3
Max-of-Average cost	$2+\sqrt{5}$	$\frac{7}{2}$
Average-of-Max cost	$2+\sqrt{5}$	3

1.2 Related Work

Approximate mechanism design without money was initiated by Procaccia and Tennenholtz [15], who studied strategyproof mechanisms with constant approximation ratios for facility location problems on the line under the social cost objective and the maximum cost objective (i.e., the maximum individual cost among all agents). They considered single-facility location problems and homogeneous two-facility location problems where each agent's individual cost is the distance from her location to the closest facility (referred to as min-variant cost). Since then, numerous different truthful facility location problems have been well studied. For example, Alon et al. [1] studied single-facility location problems in circles and general graphs. Tang et al. [17] considered constrained single and two facility location problems where the facility can only be built in a given set of candidate locations of the line. Cheng et al. [7] discussed an obnoxious facility location game where every agent wants to stay far away from the facility. Cai et al. [4] studied facility location games under the minimax envy objective and Ding et al. [8] studied the envy ratio objective. We will highlight the work of heterogeneous facility location and distributed facility location problems respectively, which is most related to our work.

Heterogeneous Facility Location. Zou and Li [19] studied heterogeneous facility location problems with dual preferences where the facility can be desirable or obnoxious for every agent. Serafino and Ventre [16] considered heterogeneous facility location problems with optional preferences where each agent approves either one facility or both. In their setting, the agents' locations are public while agents' preferences are private and the individual cost of each agent is the sum of distances to her interested facilities (referred to as sum-variant cost). Later, Chen et al. [6] considered the optional preference model with max-variant and min-variant; and their results with min-variant cost were improved by Li et al. [14]. Anastasiadis and Deligkas [2] studied heterogeneous k-facility location problems with min-variant. When the locations of agents are private and the preferences of agents are public, Zhao et al. [18] studied the optional preference model with max-variant cost; Kanellopoulos et al. [13] studied the optional preference model with sum-variant cost. Fong et al. [12] proposed a fractional preference model where the preference of each agent for the facility is a number between 0 and 1.

Distributed Facility Location. Filos-Ratsikas et al. [10] initiated the study of the distortion in distributed social choice and then Anshelevich et al. [3] considered a distributed metric social choice setting where voters and alternatives can be considered as points in a metric space. Filos-Ratsikas and Voudouris [11] studied a distributed single-facility location problem under the social cost objective in the discrete setting (i.e., facilities can only be located in a finite set of candidate locations) and the continuous setting (i.e., facilities can be located at every point on the real line), respectively. For the discrete setting, they proved a tight bound of 7 for strategyproof distributed mechanism, and for the continuous setting, they proved a tight bound of 3 for strategyproof distributed mechanism. Further, Filos-Ratsikas et al. [9] considered a continuous distributed single-facility location problem under four social objectives: the average cost, the max cost, the average-of-max cost and the max-of-average cost. A summary of their results is shown in Table 2.

Table 2. Tight bounds on the distortion of strategyproof distributed mechanisms in Filos-Ratsikas et al. [9].

Social objective	Tight bounds
Average cost	3 [9]
Max cost	2 [9]
Max-of-Average cost	$1+\sqrt{2}$ [9]
Average-of-Max cost	$1+\sqrt{2}$ [9]

2 Preliminaries

Let $N=\{1,2,...,n\}$ be a set of agents. Each agent $i\in N$ has a private location $x_i\in\mathbb{R}$ and denote by $\mathbf{x}=(x_1,...,x_n)\in\mathbb{R}^n$ the location profile of all agents. The agents are divided into k (≥ 1) disjoint groups and let $D=\{1,...,k\}$ be the set of groups. For each group $d\in D$, let N_d be the set of agents that belong to d and $n_d=|N_d|$ the number of agents in group d.

Let $\mathcal{F}=\{F_1,F_2\}$ be the two heterogeneous facilities to be located and $A=\{a_1,...,a_m\}\in\mathbb{R}^m$ be a multiset of candidate locations. Assume that at most one facility can be located at each location in A. Denote by $I=(N,\mathbf{x},D,A)$ an instance.

Distributed Mechanism. A distributed mechanism M is a function that maps an instance I to a facility location profile $\mathbf{w}$ through two steps, i.e. $M(I)=\mathbf{w}=(w_1,w_2)$, where $w_1\in A$ is the location of F_1 and $w_2\in A\setminus\{w_1\}$ is the location of F_2. In detail, given an instance I, a distributed mechanism M consists of two steps:

- Step 1. For each group $d\in D$, M chooses two representative locations $y_{d(1)}\in A$, $y_{d(2)}\in A\setminus\{y_{d(1)}\}$ for group d based on the locations reported by agents in N_d.
- Step 2. M outputs a facility location profile $\mathbf{w}=(w_1,w_2)$ where $w_1\in A$, $w_2\in A\setminus\{w_1\}$ among all the representatives $\{y_{d(1)},y_{d(2)}\}_{d\in D}$.

Remark 1. Following the work of Filos-Ratsikas et al. [9], a distributed mechanisms should have the following properties. *(P1)*: For any two groups where the locations reported by agents are identical, the mechanism will output the same representative locations. *(P2)*: The selected representative locations for a group are independent of the locations reported by agents in other groups, as well as the number and sizes of the other groups. *(P3)*: The facility location profile $\mathbf{w}$ chosen by the mechanism is the same over all instances where the group representatives are identical. These properties are necessary for our work of the lower bounds.

For any two points $x,y\in\mathbb{R}$, let $\delta(x,y)=|x-y|$ be the distance between x and y. In our model, we assume that all agents approve the two heterogeneous facilities and the *individual cost* of each agent i for a facility location profile $\mathbf{w}$ is the distance from her location to the farthest facility:

$$c(x_i, \mathbf{w}) = \max\left\{\delta(x_i, w_1), \delta(x_i, w_2)\right\},$$

which is called the max-variant cost.

Social Objectives. We consider the following four cost-minimization social objectives:

(1) The *Average-of-Average cost* of a facility location profile $\mathbf{w}$ is the average of the average individual cost of agents in each group:

$$\text{AoA}(\mathbf{w}|I) = \frac{1}{k}\sum_{d\in D}\left\{\frac{1}{n_d}\sum_{i\in N_d} c(x_i, \mathbf{w})\right\}.$$

(2) The *Max-of-Max cost* of a facility location profile $\mathbf{w}$ is the maximum individual cost among all agents:

$$\text{MoM}(\mathbf{w}|I) = \max_{d\in D}\max_{i\in N_d} c(x_i, \mathbf{w}).$$

(3) The *Max-of-Average cost* of a facility location profile $\mathbf{w}$ is the maximum of the average individual cost of all agents in each group:

$$\text{MoA}(\mathbf{w}|I) = \max_{d\in D}\left\{\frac{1}{n_d}\sum_{i\in N_d} c(x_i, \mathbf{w})\right\}.$$

(4) The *Average-of-Max cost* of a facility location profile $\mathbf{w}$ is the average of the maximum individual cost among all agents in each group:

$$\text{AoM}(\mathbf{w}|I) = \frac{1}{k}\sum_{d\in D}\max_{i\in N_d} c(x_i, \mathbf{w}).$$

Strategyproofness. In order to minimize individual costs, selfish agents may misreport their locations. Thus, the strategyproofness of mechanisms should be taken into account. Formally, a mechanism M is *strategyproof* if each agent can never benefit by misreporting her location, regardless of the locations reported by the other agents, i.e., for every $i \in N$, for every $\mathbf{x} \in \mathbb{R}^n$, and for every $x_i' \in \mathbb{R}$, it must hold that

$$c(x_i, M(N, (x_i, \mathbf{x}_{-i}), D, A) \leq c(x_i, M(N, (x_i', \mathbf{x}_{-i}), D, A)),$$

where $\mathbf{x}_{-i} = (x_1, \ldots, x_{i-1}, x_{i+1}, \ldots, x_n)$ is the location profile of $N \setminus \{i\}$.

Distortion. In the distributed setting, due to lack of global information and the requirement for strategyproofness of a mechanism, the facility location profile chosen by a distributed mechanism may be suboptimal. Here, we adopt the notion of distortion to quantify the gap between a mechanism and the optimal mechanism. The *distortion* of a distributed mechanism M is the worst-case ratio between the social objective value obtained by M and the optimal social objective value over all possible instances:

$$\text{dist}(M) = \sup_I \frac{\text{cost}(M(I) \mid I)}{\text{cost}(\text{OPT}(I) \mid I)},$$

where $\text{OPT}(I)$ is the optimal solution for instance I and cost $\in \{AoA, MoM, MoA, AoM\}$. To simplify our notation, $\text{cost}(M(I)|I)$ is abbreviated as $\text{cost}(M|I)$.

For each agent $i \in N$, denote by $t(i)$ the closest candidate location to i in A and $s(i)$ the closest candidate location to i in $A \setminus \{t(i)\}$. Now, we give a class of distributed mechanisms which is called the (α, β)-Quantile mechanism ($\alpha, \beta \in [0, 1]$).

(α, β)-Quantile Mechanism[1]

- Step 1. For each group $d \in D$, denote by α_d the $\lceil \alpha \cdot n_d \rceil$-th leftmost agent in N_d. Let $y_{d(1)} = t(\alpha_d)$ and $y_{d(2)} = s(\alpha_d)$ be two representatives of group d.
- Step 2. Set $z_d := y_{d(1)}$ for each group $d \in D$ and return $w_1 :=$ the $\lceil \beta \cdot k \rceil$-th leftmost location in $\{z_d\}_{d \in D}$. For each group $d \in D$, update z_d as $y_{d(2)}$ if $y_{d(1)} = w_1$, and return $w_2 :=$ the $\lceil \beta \cdot k \rceil$-th leftmost location in $\{z_d\}_{d \in D}$.

Remark 2. Obviously, the (α, β)-Quantile mechanism has the following properties. For each group d, $y_{d(1)}$, $y_{d(2)}$ are two adjacent locations, which implies that $(y_{d(1)}, y_{d(2)})$ and $(y_{d'(1)}, y_{d'(2)})$ are never "interlaced "[2] for any two groups d and d'. Therefore, the locations w_1, w_2 output by the (α, β)-Quantile mechanism must serve as two representatives of some group d such that $w_1 = y_{d(1)}$, $w_2 = y_{d(2)}$.

We will prove that the (α, β)-Quantile mechanism is strategyproof.

Theorem 1. *The (α, β)-Quantile mechanism is strategyproof.*

Proof. Consider any instance I, and let $\mathbf{w} = (w_1, w_2)$ be the facility location profile output by the mechanism. According to the property of the mechanism, w_1 and w_2 must be two adjacent locations closest to some agent j, i.e., $w_1 = t(x_j)$, $w_2 = s(x_j)$. Let i be any agent belonging to some group d. Assume w.l.o.g. that $x_i \le x_{\alpha_d}$.

Case 1: $x_i \le x_{\alpha_d} \le x_j$. In order to affect the output of the mechanism, agent i must first become the $\lceil \alpha \cdot n_d \rceil$-th leftmost agent in N_d. So i has to report a location $x'_i > x_{\alpha_d}$. If $x_j \ge x'_i > x_{\alpha_d}$, then the output of the mechanism does not change, in which case i has no incentive to misreport. If $x'_i > x_j$, then the

[1] It is regulated that when $\alpha = 0$, the $\lceil \alpha \cdot n_d \rceil$-th leftmost location in N_d is the leftmost location in N_d. Similarly, when $\beta = 0$, the $\lceil \beta \cdot k \rceil$-th leftmost location is the leftmost location.

[2] Here, interlaced means the cases such as $y_{d(1)} < y_{d'(1)} < y_{d(2)} < y_{d'(2)}$.

output of the mechanism becomes two adjacent locations closest to some agent k with $x_k \geq x_j$, meaning that the cost of agent i does not decrease. So i has no incentive to misreport.

Case 2: $x_i < x_j < x_{\alpha_d}$. In order to affect the output of the mechanism, agent i must first become the $\lceil \alpha \cdot n_d \rceil$-th leftmost agent in N_d. So i has to report a location $x_i' > x_{\alpha_d}$. However, the output of the mechanism does not change, and then i has no incentive to misreport.

Case 3: $x_j < x_i \leq x_{\alpha_d}$. Similar to Case 1, i has no incentive to misreport.

Above all, the (α, β)-Quantile mechanism is strategyproof.

□

In the following sections, we analyze the upper bounds on the distortion under four social objectives by adjusting parameters α and β in the (α, β)-Quantile mechanism.

Notations. Denote by l, r, and m the leftmost, rightmost, and median, respectively, agent in N and l_d, r_d, m_d the leftmost, rightmost, and median, respectively, agent in N_d. For a facility location profile $\mathbf{w} = (w_1, w_2)$ and any $i \in N$, denote by $w(x_i)$ the farthest one to agent i in $\{w_1, w_2\}$.

3 Average-of-Average Cost

In this section, we consider the Average-of-Average cost objective, which is the average of the average individual cost of agents in each group. For the upper bound, we consider the $(\frac{1}{2}, \frac{1}{2})$-Quantile mechanism,[3] which achieves a distortion of at most 9. For the lower bound, the distortion of any strategyproof mechanism is at least $3 - \epsilon$, for any $\epsilon > 0$.

$(\frac{1}{2}, \frac{1}{2})$-Quantile Mechanism

- Step 1. For each group $d \in D$, let $y_{d(1)} = t(m_d)$, $y_{d(2)} = s(m_d)$.
- Step 2. Set $z_d := y_{d(1)}$ for each group $d \in D$ and return $w_1 :=$ the median location in $\{z_d\}_{d \in D}$. For each group $d \in D$, update z_d as $y_{d(2)}$ if $y_{d(1)} = w_1$, and return $w_2 :=$ the median location in $\{z_d\}_{d \in D}$.

Theorem 2. *For the Average-of-Average cost,[4] the distortion of the $(\frac{1}{2}, \frac{1}{2})$-Quantile mechanism is at most 9.*

Theorem 3. *For the Average-of-Average cost, the distortion of any strategyproof mechanism is at least $3 - \epsilon$, for any $\epsilon > 0$.*

---

[3] Here, we use the method of undetermined coefficients to conclude that the $(\frac{1}{2}, \frac{1}{2})$-Quantile mechanism achieves a minimum upper bound on the distortion in (α, β)-Quantile.

[4] Due to space constrains, all missing proofs can be found in the appendix.

Proof. Assume for contradiction that there exists a strategyproof mechanism M that has a distortion strictly smaller than $3-\epsilon$, for some $\epsilon > 0$. Let $\theta > 0$ be an infinitesimal. We will reach contradictions through considering the following several instances with the set of candidate locations $A = \{0, 0, 1, 1\}$.

Instance I_1: Consider an instance I_1 with one agent in a single group, where $x_1^1 = 0$. In this case, the representative chosen by mechanism M must be $(0, 0)$, i.e., $M(I_1) = (0, 0)$. Otherwise, the distortion of M is infinite.

Instance I_2: Consider an instance I_2 with one agent in a single group, where $x_1^2 = \frac{1}{2} - \theta$. If $M(I_2) \neq (0, 0)$, then $c(x_1^2, M(I_2)) = \frac{1}{2} + \theta$. Considering $c(x_1^2, M(I_1)) = \frac{1}{2} - \theta$, agent 1 at x_1^2 can decrease her cost by misreporting her location as 0, in contradiction to strategyproofness. Thus, $M(I_2) = (0, 0)$.

Instance I_3: Consider an instance I_3 with one agent in a single group, where $x_1^3 = 1$. Similar to Instance I_1, mechanism M must output $(1, 1)$ as the representative of the group, i.e., $M(I_3) = (1, 1)$.

Instance I_4: Consider an instance I_4 with one agent in a single group, where $x_1^4 = \frac{1}{2} + \theta$. Similar to Instance I_2, $M(I_4) = (1, 1)$.

Instance I_5: Consider an Instance I_5 with two groups:

- In group 1, there is one agent at $\frac{1}{2} - \theta$.
- In group 2, there is one agent at 1.

Considering $M(I_2) = (0, 0)$ and $M(I_3) = (1, 1)$, by *(P1)* and *(P2)*, the representatives of group 1 and group 2 in Instance I_5 are $(0, 0)$ and $(1, 1)$, respectively. Since $AoA(0, 0) = \frac{3}{4} - \frac{\theta}{2}$, $AoA(0, 1) = AoA(1, 0) = \frac{3}{4} + \frac{\theta}{2}$ and $AoA(1, 1) = \frac{1}{4} + \frac{\theta}{2}$, M must output $(1, 1)$ as the overall facility location profile, i.e., $M(I_5) = (1, 1)$. Otherwise, the distortion of M is at least $\frac{\frac{3}{2} - \theta}{\frac{1}{2} + \theta} \geq 3 - \epsilon$, for $\theta \leq \frac{\epsilon}{8 - 2\epsilon}$.

Now, we can reach a contradiction by considering the following instance I_6 with two groups:

- In group 1, there is one agent at $\frac{1}{2} + \theta$.
- In group 2, there is one agent at 0.

Considering $M(I_4) = (1, 1)$ and $M(I_1) = (0, 0)$, by *(P1)* and *(P2)*, the representatives of group 1 and group 2 in Instance I_6 are $(1, 1)$ and $(0, 0)$, respectively. Since the group representatives in Instance I_5 and Instance I_6 are identical, according to *(P3)*, we have that $M(I_6) = (1, 1)$. Since $AoA(0, 0) = \frac{1}{4} + \frac{\theta}{2}$, $AoA(0, 1) = AoA(1, 0) = \frac{3}{4} + \frac{\theta}{2}$ and $AoA(1, 1) = \frac{3}{4} - \frac{\theta}{2}$, the distortion of M is at least $\frac{\frac{3}{2} - \theta}{\frac{1}{2} + \theta} \geq 3 - \epsilon$, for $\theta \leq \frac{\epsilon}{8 - 2\epsilon}$. This contradicts the assumption that M has a distortion strictly smaller than $3 - \epsilon$. □

Remark 3. In the facility location problem, the most natural social objective is the social cost. In fact, our results under the average-of-average cost can be easily generalized to the social cost.

4 Max-of-Max Cost

In this section, we study the Max-of-Max cost objective. For the upper bound, we consider the $(1,1)$-Quantile mechanism,[5] which achieves a distortion of at most 3.

$(1,1)$-Quantile Mechanism

- Step 1. For each group $d \in D$, let $y_{d(1)} = t(r_d)$, $y_{d(2)} = s(r_d)$.
- Step 2. Set $z_d := y_{d(1)}$ for each group $d \in D$ and return $w_1 :=$ the rightmost location in $\{z_d\}_{d \in D}$. For each group $d \in D$, update z_d as $y_{d(2)}$ if $y_{d(1)} = w_1$, and return $w_2 :=$ the rightmost location in $\{z_d\}_{d \in D}$.

Theorem 4. *For the Max-of-Max cost, the distortion of the $(1,1)$-Quantile mechanism is at most 3.*

Proof. Given any instance I, let $\mathbf{w} = (w_1, w_2)$ be the location profile chosen by the mechanism and $\mathbf{o} = (o_1, o_2)$ be an optimal solution. According to the property of this mechanism, we have that $w_1 = t(r)$ and $w_2 = s(r)$. Denote by i^* the agent such that $MoM(\mathbf{w}) = \max\{\delta(x_{i^*}, w_1), \delta(x_{i^*}, w_2)\} = \delta(x_{i^*}, w(x_{i^*}))$. Since $w_1 = t(r)$ and $w_2 = s(r)$ are the closest candidate locations to r and using the triangle inequality, we have that

$$\begin{aligned} MoM(\mathbf{w}) &= \delta(x_{i^*}, w(x_{i^*})) \\ &\le \delta(x_{i^*}, o_1) + \delta(o_1, x_r) + \delta(x_r, w(x_{i^*})) \\ &\le \delta(x_{i^*}, o(x_{i^*})) + \delta(x_r, o(x_r)) + \delta(x_r, w(x_r)) \\ &\le \delta(x_{i^*}, o(x_{i^*})) + 2 \cdot \delta(x_r, o(x_r)). \end{aligned} \tag{1}$$

For the optimal solution $\mathbf{o}$, we have $MoM(\mathbf{o}) \ge \delta(x_i, o(x_i))$, for any agent $i \in N$. Therefore, we obtain

$$MoM(\mathbf{w}) \le \delta(x_{i^*}, o(x_{i^*})) + 2 \cdot \delta(x_r, o(x_r)) \le 3 \cdot MoM(\mathbf{o}). \tag{2}$$

□

Next, we show a lower bound on the distortion of all strategyproof mechanisms.

Theorem 5. *For the Max-of-Max cost, the distortion of any strategyproof mechanism is at least $3 - \varepsilon$, for any $\varepsilon > 0$.*

Proof. Assume for contradiction that there exists a strategyproof mechanism M that has a distortion strictly smaller than $3 - \varepsilon$, for some $\varepsilon > 0$. Let $\theta > 0$ be an infinitesimal. We will reach a contradiction through considering the following several instances with the set of candidate locations $A = \{0, 0, 2, 2, 4, 4\}$.

[5] In fact, whatever values α and β take, the (α, β)-Quantile mechanism would achieve a distortion of at most 3 under the max-of-max cost.

Instance I_1: Consider an instance I_1 with two agents in a single group, with the location profile $\mathbf{x^1}$, where $x_1^1 = -1$ and $x_2^1 = 1 - \theta$. We can obtain $MoM(0,0) = 1, MoM(0,2) = MoM(2,0) = MoM(2,2) = 3, MoM(4,4) = MoM(0,4) = MoM(4,0) = MoM(2,4) = MoM(4,2) = 5$. Clearly, $OPT(I_1) = (0,0)$, so we claim that M must choose $(0,0)$ as the representative of this group, i.e., $M(I_1) = (0,0)$. Otherwise, the distortion of M is at least 3, which contradicts the assumption.

Instance I_2: Consider an instance I_2 with two agents in a single group, with the location profile $\mathbf{x^2}$, where $x_1^2 = 1 - \theta$ and $x_2^2 = 1 - \theta$. If $M(I_2) \neq (0,0)$, then $c(x_1^2, M(I_2)) \geq 1 + \theta$. Considering $c(x_1^2, M(I_1)) = 1 - \theta$, agent 1 at x_1^2 can decrease her cost by misreporting her location as -1, in contradiction to strategyproofness. Thus, $M(I_2) = (0,0)$.

Instance I_3: Consider an instance I_3 with two agents in a single group, with the location profile $\mathbf{x^3}$, where $x_1^3 = 5$ and $x_2^3 = 3 + \theta$. Since Instance I_3 and Instance I_1 have symmetry with respect to A, we have that $M(I_3) = (4,4)$.

Instance I_4: Consider an instance I_4 with two agents in a single group, with the location profile $\mathbf{x^4}$, where $x_1^4 = 3 + \theta$ and $x_2^4 = 3 + \theta$. Similar to Instance I_2, we have that $M(I_4) = (4,4)$.

Finally, to reach a contradiction, consider the following instance I_5 with two groups:

- In group 1, there are two agents at $1 - \theta$.
- In group 2, there are two agents at $3 + \theta$.

Considering $M(I_2) = (0,0)$ and $M(I_4) = (4,4)$, by *(P1)* and *(P2)*, the representatives of group 1 and group 2 are $(0,0)$ and $(4,4)$, respectively. Therefore, $M(I_5) = (0,0), (0,4), (4,0)$ or $(4,4)$ and $MoM(M|I_5) = 3 + \theta$. However, $OPT(I_5) = (2,2)$ and $MoM(OPT|I_5) = 1 + \theta$.

Thus, the distortion of mechanism M is at least $\frac{3+\theta}{1+\theta} \geq 3 - \varepsilon$, for $\theta \leq \frac{\epsilon}{2-\epsilon}$; which is a contradiction. □

5 Max-of-Average Cost

In this section, we focus on the Max-of-Average cost objective, which is the maximum of the average individual cost of all agents in each group. For the upper bound, we consider the $(\frac{3-\sqrt{5}}{2}, 1)$-Quantile mechanism, which achieves a distortion of at most $2 + \sqrt{5}$. For the lower bound, the distortion of any strategyproof mechanism is at least $\frac{7}{2} - \epsilon$, for any $\epsilon > 0$.

We first consider the $(\alpha, 1)$-Quantile mechanism and analyze its performance.

$(\alpha, 1)$-Quantile Mechanism

- Step 1. For each group $d \in D$, denote by α_d the $\lceil \alpha \cdot n_d \rceil$-th leftmost agent in N_d. Let $y_{d(1)} = t(\alpha_d)$, $y_{d(2)} = s(\alpha_d)$.
- Step 2. Set $z_d := y_{d(1)}$ for each group $d \in D$ and return $w_1 :=$ the rightmost location in $\{z_d\}_{d \in D}$. For each group $d \in D$, update z_d as $y_{d(2)}$ if $y_{d(1)} = w_1$, and return $w_2 :=$ the rightmost location in $\{z_d\}_{d \in D}$.

Lemma 1. *For the Max-of-Average cost, the distortion of the* $(\alpha, 1)$*-Quantile mechanism is at most* $\max\left\{1 + \frac{2(1-\alpha)}{\alpha}, 1 + \frac{2}{1-\alpha}\right\}$.

By solving the equation $1 + \frac{2(1-\alpha)}{\alpha} = 1 + \frac{2}{1-\alpha}$, we can obtain the following theorem.

Theorem 6. *For the Max-of-Average cost, the distortion of the* $(\frac{3-\sqrt{5}}{2}, 1)$*-Quantile mechanism is at most* $2 + \sqrt{5}$.

Theorem 7. *For the Max-of-Average cost, the distortion of any strategyproof mechanism is at least* $\frac{7}{2} - \varepsilon$*, for any* $\varepsilon > 0$.

6 Average-of-Max Cost

In this section, we turn our attention to the last social objective, Average-of-Max cost objective, which is the average of the maximum individual cost among all agents in each group. For the upper bound, we consider the $(1, \frac{3-\sqrt{5}}{2})$-Quantile mechanism, which achieves a distortion of at most $2+\sqrt{5}$. For the lower bound, the distortion of any strategyproof mechanism is at least $3 - \epsilon$, for any $\epsilon > 0$.

We first consider the $(1, \beta)$-Quantile mechanism and analyze its performance.

$(1, \beta)$-Quantile Mechanism

- Step 1. For each group $d \in D$, let $y_{d(1)} = t(r_d)$, $y_{d(2)} = s(r_d)$.
- Step 2. Set $z_d := y_{d(1)}$ for each group $d \in D$ and return $w_1 :=$ the $\lceil \beta \cdot k \rceil$-th leftmost location in $\{z_d\}_{d \in D}$. For each group $d \in D$, update z_d as $y_{d(2)}$ if $y_{d(1)} = w_1$, and return $w_2 :=$ the $\lceil \beta \cdot k \rceil$-th leftmost location in $\{z_d\}_{d \in D}$.

Lemma 2. *For the Average-of-Max cost, the distortion of the* $(1, \beta)$*-Quantile mechanism is at most* $\max\left\{1 + \frac{2(1-\beta)}{\beta}, 1 + \frac{2}{1-\beta}\right\}$.

By solving the equation $1 + \frac{2(1-\beta)}{\beta} = 1 + \frac{2}{1-\beta}$, we can obtain the following theorem.

Theorem 8. *For the Average-of-Max cost, the distortion of the* $(1, \frac{3-\sqrt{5}}{2})$*-Quantile mechanism is at most* $2 + \sqrt{5}$.

Theorem 9. *For the Average-of-Max cost, the distortion of any strategyproof mechanism is at least* $3 - \epsilon$*, for any* $\epsilon > 0$.

7 Conclusions and Future Work

In this paper, we studied a constrained distributed heterogeneous two-facility location problem and showed upper and lower bounds on the distortion of strategyproof mechanisms under four social objectives. There are at least three directions for future work. First, it would be interesting to close the gaps between our lower and upper bounds. And in our model, the candidate location set A is a multiset, and the lower bounds on the distortion under four social objectives rely on this condition; when A is a simple set, do the same lower bounds still hold? Second, it would be meaningful to consider more than just two facilities to be built and more general metric spaces than just the real line. Third, one could try some other social objectives in distributed settings (such as minimax envy), and consider settings where agents have other preferences over the facilities (such as the dual preferences).

Acknowledgments. This research was supported in part by the National Natural Science Foundation of China (12201590, 12171444) and Natural Science Foundation of Shandong Province (ZR2024MA031).

Disclosure of Interests. The authors declare that they have no conflict of interest.

References

1. Alon, N., Feldman, M., Procaccia, A.D., Tennenholtz, M.: Strategyproof approximation of the minimax on networks. Math. Oper. Res. **35**(3), 513–526 (2010)
2. Anastasiadis, E., Deligkas, A.: Heterogeneous facility location games. In: Proceedings of the 17th International Conference on Autonomous Agents and MultiAgent Systems, pp. 623–631 (2018)
3. Anshelevich, E., Filos-Ratsikas, A., Voudouris, A.A.: The distortion of distributed metric social choice. Artif. Intell. **308**, 103713 (2022)
4. Cai, Q., Filos-Ratsikas, A., Tang, P.: Facility location with minimax envy. In: AAAI Press/International Joint Conferences on Artificial Intelligence (2016)
5. Chan, H., Filos-Ratsikas, A., Li, B., Li, M., Wang, C.: Mechanism design for facility location problems: a survey. In: Proceedings of the Thirtieth International Joint Conference on Artificial Intelligence, pp. 4356–4365. International Joint Conferences on Artificial Intelligence Organization (2021)
6. Chen, Z., Fong, K.C., Li, M., Wang, K., Yuan, H., Zhang, Y.: Facility location games with optional preference. Theoret. Comput. Sci. **847**, 185–197 (2020)
7. Cheng, Y., Yu, W., Zhang, G.: Strategy-proof approximation mechanisms for an obnoxious facility game on networks. Theoret. Comput. Sci. **497**, 154–163 (2013)
8. Ding, Y., Liu, W., Chen, X., Fang, Q., Nong, Q.: Facility location game with envy ratio. Comput. Ind. Eng. **148**, 106710 (2020)
9. Filos-Ratsikas, A., Kanellopoulos, P., Voudouris, A.A., Zhang, R.: The distortion of distributed facility location. Artif. Intell. **328**, 104066 (2024)
10. Filos-Ratsikas, A., Micha, E., Voudouris, A.A.: The distortion of distributed voting. Artif. Intell. **286**, 103343 (2020)
11. Filos-Ratsikas, A., Voudouris, A.A.: Approximate mechanism design for distributed facility location. In: International Symposium on Algorithmic Game Theory, pp. 49–63. Springer, Cham (2021)

12. Fong, K.C., Li, M., Lu, P., Todo, T., Yokoo, M.: Facility location games with fractional preferences. In: 32nd AAAI Conference on Artificial Intelligence (AAAI 2018), pp. 1039–1046. AAAI Press (2018)
13. Kanellopoulos, P., Voudouris, A.A., Zhang, R.: Truthful two-facility location with candidate locations. Theoret. Comput. Sci. **1024**, 114913 (2025)
14. Li, M., Lu, P., Yao, Y., Zhang, J.: Strategyproof mechanism for two heterogeneous facilities with constant approximation ratio. In: Proceedings of the Twenty-Ninth International Conference on International Joint Conferences on Artificial Intelligence, pp. 238–245 (2021)
15. Procaccia, A.D., Tennenholtz, M.: Approximate mechanism design without money. ACM Trans. Econ. Comput. (TEAC) **1**(4), 1–26 (2013)
16. Serafino, P., Ventre, C.: Heterogeneous facility location without money. Theoret. Comput. Sci. **636**, 27–46 (2016)
17. Tang, Z., Wang, C., Zhang, M., Zhao, Y.: Mechanism design for facility location games with candidate locations. In: International Conference on Combinatorial Optimization and Applications, pp. 440–452. Springer, Cham (2020)
18. Zhao, Q., Liu, W., Nong, Q., Fang, Q.: Constrained heterogeneous facility location games with max-variant cost. J. Comb. Optim. **45**(3), 90 (2023)
19. Zou, S., Li, M.: Facility location games with dual preference. In: Proceedings of the 2015 International Conference on Autonomous Agents and Multiagent Systems, pp. 615–623 (2015)

Truthful Two-Obnoxious-Facility Location Games with Optional Preferences and Minimum Distance Constraint

Xiaojia Han[1], Wenjing Liu[1,2(✉)], and Qizhi Fang[1]

[1] School of Mathematical Sciences, Ocean University of China, Qingdao 266100, China
hxj1069@stu.ouc.edu.cn, {liuwj,qfang}@ouc.edu.cn

[2] Laboratory of Marine Mathematics, Ocean University of China, Qingdao 266100, China

Abstract. In this paper, we study a truthful two-obnoxious-facility location problem, in which each agent has a private location in $[0, 1]$ and a public optional preference over two obnoxious facilities, and there is a minimum distance constraint $d \in [0, 1]$ between the two facilities. Each agent wants to be as far away as possible from the facilities that affect her, and the utility of each agent is the total distance from her to these facilities. The goal is to decide how to place the facilities in $[0, 1]$ so as to incentivize agents to report their private locations truthfully as well as maximize the social utility. First, we consider the special setting where $d = 0$, that is, the two facilities can be located at any point in $[0, 1]$. We propose a deterministic strategyproof mechanism with approximation ratio of at most 4 and a randomized strategyproof mechanism with approximation ratio of at most 2, respectively. Then we study the general setting where $d \in [0, 1]$. We propose a deterministic strategyproof mechanism with approximation ratio of at most 8 and a randomized strategyproof mechanism with approximation ratio of at most 4, respectively. Furthermore, we provide lower bounds of 2 and $\frac{14}{13}$ on the approximation ratio for any deterministic and any randomized strategyproof mechanism, respectively.

Keywords: Facility location game · Mechanism design · Strategyproof · Optional preferences · Minimum distance constraint

1 Introduction

In the classic truthful facility location problems, there is a set of agents who have private locations on the real line, and there is a facility (for example, a park or a school) that will be located somewhere on that line. The government aims to design a mechanism to determine the facility location that approximately optimizes some social objective. Since the mechanism operates based on the information reported by agents, each self-interested and rational agent

M. Li et al. (Eds.): TAMC 2025, LNCS 16084, pp. 135–147, 2026.
https://doi.org/10.1007/978-981-95-4839-2_11

may strategically misreport her location to manipulate the facility location, thus reducing her own cost or increasing her own utility. Therefore, the government's goal is to design mechanisms that guarantee truthful reporting by agents (e.g., strategyproof) while approximately optimizing the social objective. The inconsistency between individual and collective interests implies that, in mechanism design, strategyproof mechanisms generally cannot guarantee the optimization of social objectives. Therefore, Procaccia and Tennenholtz [13] introduced the approximation ratio, which is the worst-case ratio (over all possible instances) of the social objective value obtained by the mechanism to the optimal social objective value[1], to measure the performance of strategyproof mechanisms.

In this paper, we study two-obnoxious-facility location games with optional preferences and minimum distance constraint with respect to two facilities. As a stylized example of our setting, consider locating a bar and a prison along a linear street. Both facilities are generally perceived as obnoxious by residents: the bar due to noise pollution and the prison due to potential psychological distress. However, some residents may be indifferent to whether these facilities are located close to their houses. Critically, due to the concern that former inmates congregating in nearby bars may cause disturbances, these facilities must maintain a minimum separation distance.

1.1 Our Results

Assuming that a set of agents have private locations in $[0, 1]$ and public optional preferences over two obnoxious facilities. Specifically, some agents prefer to be far from exactly one facility (with utility defined as the distance to that facility), while the others prefer to be far from both facilities (with utility defined as the sum of distances to both facilities). There is a minimum distance constraint $d \in [0, 1]$ between the two facilities. Our goal is to design mechanisms that take the agents' reported locations and the public preferences as input, and decide how to place the two facilities, so that (a) incentivize agents to report their locations truthfully, and (b) (approximately) maximize the social utility (i.e., the sum of the utilities of all agents).

In Sect. 3, we study the two-obnoxious-facility-location problem with optional preferences but without any distance constraint (i.e. $d = 0$). We propose a deterministic strategyproof mechanism (Mechanism 1) with approximation ratio of at most 4, and a randomized strategyproof mechanism (Mechanism 2) with approximation ratio of at most 2.

In Sect. 4, we study the two-obnoxious-facility-location problem with optional preferences and minimum distance constraint (i.e. $d \in [0, 1]$). We design a deterministic strategyproof mechanism (Mechanism 3) with approximation ratio of at most 8, which switches between two cases depending on the cardinalities of the

[1] This definition applies to cost-minimization social objectives. For utility-maximization social objectives, the approximation ratio is defined as the worst-case ratio (over all instances) of the optimal social objective value to the social objective value obtained by the mechanism.

sets of agents with different preferences, and a randomized strategyproof mechanism (Mechanism 4) with approximation ratio of at most 2. Furthermore, we derive lower bounds of 2 and $\frac{14}{13}$ on the approximation ratio for any deterministic and randomized strategyproof mechanism, respectively.

Our results are summarized in Table 1, where UB and LB indicate the upper bound and the lower bound of the approximation ratio, respectively.

Table 1. A summary of our results.

Minimum distance constraint	Deterministic	Randomized
Special setting	UB: 4 (Theorem 2)	UB: 2 (Theorem 3)
	LB: 2 (Theorem 6)	LB: 1.077 (Theorem 8)
General setting	UB: 8 (Theorem 5)	UB: 2 (Theorem 7)
	LB: 2 (Theorem 6)	LB: 1.077 (Theorem 8)

1.2 Related Work

Mechanism design without money for facility location games has been widely studied in recent years. Moulin [12] presented a complete characterization of strategyproof mechanisms when agents are located on the line and have single-peaked preferences. Procaccia and Tennenholtz [13] were the first to propose approximate mechanism design without money through facility location problems. Fotakis and Tzamos [4] investigated deterministic strategyproof mechanisms for two-facility location games on the line and proved a tight lower bound of $n-2$ for minimizing the social cost. Cheng et al. [2] discussed obnoxious facility location games, where each agent hopes that the facility will be built as far away from her as possible. Since then, researchers have studied various facility location problems from different perspectives. In what follows, we focus on reviewing two research settings closely related to our study: (a) agents have optional preferences over facilities, and (b) there are distance constraints between facilities.

Facility Location Games with Optional Preferences. Serafino and Ventre [14] introduced heterogeneous two-facility location games with optional preferences where facilities are built in a set of feasible alternative locations, and each agent has a public location and a private preference on facilities, with her cost being the sum of the distances to her preferred facilities (referred to as sum-variant). This model was later extended by Chen et al. [1] from a discrete set to the continuous line, and the cost of each agent is the minimum distance or the maximum distance to the facilities they are interested in (referred to as min-variant or max-variant, respectively). Li et al. [10] improved upper bounds of the approximation ratio when the individual cost is min-variant. Kanellopoulos et al. [7] improved upper and lower bounds of the approximation ratio when the individual cost is sum-variant.

When agents' locations are private information and preferences on facilities are public information, Zhao et al. [19] studied the two-facility location problem on a given set of candidate locations, where the cost of each agent is max-variant. Subsequently, Lofti et al. [11] proposed a new strategyproof mechanism that improves upon Zhao et al. [19]'s approximation ratio. Kanellopoulos et al. [8] considered deterministic strategyproof mechanisms, where the cost of agent is sum-variant. The facilities mentioned above are all desirable facilities. Kanellopoulos and Voudouris [6] studied the two-obnoxious-facility location problem with optional preferences where facilities are built in a set of feasible alternative locations.

Facility Location Games with Distance Constraints. For the case where there is a maximum distance constraint between two facilities, Zou and Li [20] studied the dual preference model where each agent considers a single facility either desirable or obnoxious, but different agents have potentially different opinions. Gai et al. [5] designed strategyproof mechanisms for two homogeneous facilities. For the case where there is a minimum distance constraint between two facilities, Xu et al. [16] proposed deterministic strategyproof mechanisms for both desirable and obnoxious facilities. Wu et al. [15] studied two homogeneous and heterogeneous obnoxious facilities on a circle. Duan et al. [3] and Xu et al. [17] studied the two-desirable-facility location problem with fractional preferences and optional preferences, respectively.

Kanellopoulos and Voudouris [6], Xu et al. [16], and Xu et al. [17] are the most related to our work among all researches on the heterogeneous facility location problem. However, there are several main differences between our work: (1) we study two obnoxious facility location problem with optional preferences and minimum distance constraint, whereas Kanellopoulos and Voudouris [6] only consider each agent has optional preference, and Xu et al. [16] only consider the minimum distance constraint; (2) each agent has a private location and a public preference on facilities in our model while the locations is public information and preferences is private information in Xu et al. [17], and Xu et al. [17] consider two desirable facilities.

2 Preliminaries

Let $N = \{1, 2, \ldots, n\}$ be a set of agents located in an interval $I = [0, 1]$ and $F = \{F_1, F_2\}$ be a set of two obnoxious facilities to be located in $[0, 1]$. Each agent $i \in N$ has a *private location* $x_i \in I$ and a public facility preference $p_i = (p_{i1}, p_{i2}) \in \{0, 1\}^2$ over the two facilities F_1 and F_2 with $p_{i1} + p_{i2} \geq 1$: indicating whether i is affected by facility F_j ($p_{ij} = 1$) or not ($p_{ij} = 0$), which is referred to as *optional preferences*. Let N_j ($j \in [2]$) be the set of agents that are affected by facility F_j. To be concise, denote an instance as $\mathbf{c} = (\mathbf{x}, \mathbf{p})$, where $\mathbf{x} = (x_1, x_2, \cdots, x_n)$ is the *location profile* and $\mathbf{p} = (p_1, p_2, \cdots, p_n)$ is the *preference profile.*

Suppose that there is a minimum distance constraint $d \in [0, 1]$ with respect to the two facilities. That is, the facility location profile $\mathbf{y} = (y_1, y_2)$ (y_j for F_j)

must satisfy $|y_1 - y_2| \geq d$. The social planner selects the locations of the two obnoxious facilities based on the locations reported by the agents and the public preferences.

A deterministic mechanism $f(\mathbf{c}) = (y_1, y_2) \in [0,1]^2$ is a function which maps the reported location profile $\mathbf{x}$ and the public preference profile $\mathbf{p}$ to a facility location profile $\mathbf{y}$. A randomized mechanism f is a function which maps the reported location profile $\mathbf{x}$ and the public preference profile $\mathbf{p}$ to a probability distribution over $[0,1]^2$.

Given a deterministic outcome $f(\mathbf{c}) = (y_1, y_2)$, the utility of agent i is defined as the sum of her distances to the facilities that affect her

$$u_i((x_i, p_i), f(\mathbf{c})) = \sum_{j \in [2]} p_{ij} \cdot |x_i - y_j|.$$

If $f(\mathbf{c})$ is a probability distribution returned by a randomized mechanism f, then the utility of agent i is defined as the expected total distance to all facilities that affect her

$$u_i((x_i, p_i), f(\mathbf{c})) = \mathbb{E}_{(y_1, y_2) \sim f(\mathbf{c})}[\sum_{j \in [2]} p_{ij} \cdot |x_i - y_j|].$$

The *social utility* of a mechanism f with respect to an instance $\mathbf{c}$ is defined as the total utility of all agents

$$SU(f|\mathbf{c}) = \sum_{i \in N} u_i((x_i, p_i), f(\mathbf{c})).$$

We aim to design mechanisms so that (a) the agents are incentivized to truthfully report their private locations, and (b) the social utility objective is (approximately) optimized.

A mechanism f is said to be *strategyproof* in two-obnoxious-facility location games if no agent can increase her utility by misreporting her location, regardless of the others' strategies. Formally, for every $i \in N$, every location profile $\mathbf{x} \in I^n$, and every $x_i' \in I$, it holds that

$$u_i((x_i, p_i), f(\mathbf{x}, \mathbf{p})) \geq u_i((x_i, p_i), f((x_i', \mathbf{x}_{-i}), \mathbf{p})),$$

where $\mathbf{x}_{-i} = (x_1, \cdots, x_{i-1}, x_{i+1}, \cdots, x_n)$ is the location profile of $N \setminus \{i\}$.

A mechanism f is *group strategyproof* in the two-obnoxious-facility location games if for any nonempty subset of agents, at least one of them cannot benefit if they misreport their locations jointly. Formally, for every nonempty set $S \subseteq N$ of agents, every location profile $\mathbf{x} = (\mathbf{x}_S, \mathbf{x}_{-S}) \in I^n$, and every $\mathbf{x}_S' \in I^{|S|}$, there exists $i \in S$, satisfying

$$u_i((x_i, p_i), f((\mathbf{x}_S, \mathbf{x}_{-S}), \mathbf{p})) \geq u_i((x_i, p_i), f((\mathbf{x}_S', \mathbf{x}_{-S}), \mathbf{p})).$$

We denote $OPT(\mathbf{c})$ to be the optimal social utility with respect to instance $\mathbf{c} = (\mathbf{x}, \mathbf{p})$. The *approximation ratio* of a mechanism f under the social utility objective is the worst-case ratio (over all possible instances) between the optimal social utility and the (expected) f-value, that is

$$\sup_{\mathbf{c}} \frac{SU(OPT|\mathbf{c})}{SU(f|\mathbf{c})}.$$

In this paper, we are interested in deterministic and randomized (group) strategyproof mechanisms with small approximation ratios under the social utility objective. First, we consider the special setting where $d = 0$, that is, the two facilities can be located at any point within $[0, 1]$. Then we study the general setting where $d \in [0, 1]$.

3 Special Setting

In this section, we consider the special setting where there is no distance constraint between two facilities (i.e., $d = 0$). We present a deterministic strategyproof mechanism with approximation ratio of at most 4 and a randomized group strategyproof mechanism with approximation ratio of at most 2, respectively. We provide lower bounds of 2 and $\frac{14}{13}$ on the approximation ratio for any deterministic and randomized strategyproof mechanism, respectively, as detailed in Sect. 4.

3.1 Deterministic Mechanism

In this subsection, we propose a deterministic strategyproof mechanism with approximation ratio of at most 4.

Mechanism 1: Given an instance $\mathbf{c} = (\mathbf{x}, \mathbf{p})$, the facility location profile $\mathbf{y} = (y_1, y_2)$ is given as follows: if more than $\frac{|N_1|}{2}$ agents in N_1 are located in $[0, 1/2]$, output $y_1 = 1$; otherwise, output $y_1 = 0$. If more than $\frac{|N_2|}{2}$ agents in N_2 are located in $[0, 1/2]$, output $y_2 = 1$; otherwise, output $y_2 = 0$.

Theorem 1. *Mechanism 1 is strategyproof.*

Due to space constraints, the proof of Theorem 1 can be found in the appendix.[2]

Lemma 1. *In the special setting, for any instance* $\mathbf{c}$, *the optimal social utility cannot exceed* $n + |N_1 \cap N_2|$.

Theorem 2. *The approximation ratio of Mechanism 1 is at most 4.*

[2] Due to space constraints, all missing proofs can be found in the appendix.

Proof. Let f denote Mechanism 1. For any instance $\mathbf{c} = (\mathbf{x}, \mathbf{p})$, we now discuss four cases with respect to $f(\mathbf{c}) = (y_1, y_2)$.

Case 1. $(y_1, y_2) = (1, 1)$. In this case, there are more than $\frac{|N_1|}{2}$ N_1-agents and more than $\frac{|N_2|}{2}$ N_2-agents located in $[0, 1/2]$. Thus, the social utility of Mechanism 1 is

$$\begin{aligned} SU\left((1,1) \mid \mathbf{c}\right) &= \sum_{i \in N} u_i((x_i, p_i), f(\mathbf{c})) = \sum_{i \in N_1} (1 - x_i) + \sum_{i \in N_2} (1 - x_i) \\ &\geq \frac{|N_1|}{2} \cdot \left(1 - \frac{1}{2}\right) + \frac{|N_2|}{2} \cdot \left(1 - \frac{1}{2}\right) \\ &= \frac{|N_1| + |N_2|}{4} \\ &= \frac{1}{4}(n + |N_1 \cap N_2|). \end{aligned}$$

Combining Lemma 1, we have $\frac{SU(OPT|\mathbf{c})}{SU(f|\mathbf{c})} \leq 4$.

Case 2. $(y_1, y_2) = (0, 0)$. This case is similar to Case 1.

Case 3. $(y_1, y_2) = (1, 0)$. In this case, there are more than $\frac{|N_1|}{2}$ N_1-agents located in $[0, \frac{1}{2}]$ and more than $\frac{|N_2|}{2}$ N_2-agents located in $[\frac{1}{2}, 1]$. Thus, the social utility of Mechanism 1 is

$$\begin{aligned} SU\left((1,0) \mid \mathbf{c}\right) &= \sum_{i \in N} u_i((x_i, p_i), f(\mathbf{c})) = \sum_{i \in N_1} (1 - x_i) + \sum_{i \in N_2} x_i \\ &\geq \frac{|N_1|}{2} \cdot \left(1 - \frac{1}{2}\right) + \frac{|N_2|}{2} \cdot \frac{1}{2} \\ &= \frac{|N_1| + |N_2|}{4} \\ &= \frac{1}{4}(n + |N_1 \cap N_2|). \end{aligned}$$

Combining Lemma 1, we have $\frac{SU(OPT|\mathbf{c})}{SU(f|\mathbf{c})} \leq 4$.

Case 4. $(y_1, y_2) = (0, 1)$. This case is similar to Case 3.

Therefore, the approximation ratio of Mechanism 1 is at most 4. □

3.2 Randomized Mechanism

In this subsection, we propose a randomized group strategyproof mechanism with approximation ratio of at most 2.

Mechanism 2: Given an instance $\mathbf{c} = (\mathbf{x}, \mathbf{p})$, return the facility location profile $(0, 0)$,$(0, 1)$, $(1, 0)$, and $(1, 1)$ with probability of $\frac{1}{4}$, respectively.

Theorem 3. *Mechanism 2 is a group strategyproof mechanism with approximation ratio of at most 2.*

Proof. Mechanism 2 is group strategyproof since it does not depend on the agents' reported locations.

Let f denote Mechanism 2. For any instance $\mathbf{c} = (\mathbf{x}, \mathbf{p})$, the social utility of Mechanism 2 is

$$\begin{aligned} SU(f|\mathbf{c}) &= \frac{1}{4}(\sum_{i \in N_1} (x_i + x_i + (1 - x_i) + (1 - x_i)) + \sum_{i \in N_2} (x_i + (1 - x_i) + x_i + (1 - x_i))) \\ &= \frac{1}{4}(\sum_{i \in N_1} 2 + \sum_{i \in N_2} 2) \\ &= \frac{1}{2}(|N_1| + |N_2|) \\ &= \frac{1}{2}(n + |N_1 \cap N_2|). \end{aligned}$$

Combining Lemma 1, the approximation ratio of Mechanism 2 is at most 2. □

4 General Setting

In this section, we study the two-obnoxious-facility location problem with optional preferences and minimum distance constraint $d \in [0, 1]$ between the two facilities.

4.1 Deterministic Mechanisms

In the following, we first present a deterministic strategyproof mechanism with approximation ratio of at most 8. Furthermore, we provide a lower bound of 2 on the approximation ratio for any deterministic strategyproof mechanism.

Mechanism 3: Given an instance $\mathbf{c} = (\mathbf{x}, \mathbf{p})$, let $j^* = argmax_{j \in [2]} |N_j \setminus N_{3-j}|$. The two facilities are located as follows:

- **Case 1.** If $|N_1 \cap N_2| \geq |N_{j^*} \setminus N_{3-j^*}|$, let n_L and n_R denote the number of agents in $N_1 \cap N_2$ located in $[0, \frac{1}{2}]$ and $(\frac{1}{2}, 1]$, respectively. If $n_L \geq n_R$, output $(y_{j^*}, y_{3-j^*}) = (1, 1-d)$; otherwise, output $(y_{j^*}, y_{3-j^*}) = (0, d)$.
- **Case 2.** If $|N_{j^*} \setminus N_{3-j^*}| > |N_1 \cap N_2|$, let n'_L and n'_R denote the number of agents in $N_{j^*} \setminus N_{3-j^*}$ located in $[0, \frac{1}{2}]$ and $(\frac{1}{2}, 1]$, respectively. If $n'_L \geq n'_R$, output $(y_{j^*}, y_{3-j^*}) = (1, 0)$; otherwise, output $(y_{j^*}, y_{3-j^*}) = (0, 1)$.

Theorem 4. *Mechanism 3 is strategyproof.*

Proof. For any instance $\mathbf{c} = (\mathbf{x}, \mathbf{p})$, without loss of generality, assume that $j^* = 1$. Note that the two cases in Mechanism 3 depend only on agent's preferences, which are public information. Thus, N_1, N_2 are public. Hence, no agent can induce the mechanism to shift from (Case 1) to (Case 2) and vice versa by misreporting. We will discuss each case separately.

Case 1. $|N_1 \cap N_2| \geq |N_1 \setminus N_2|$. Without loss of generality, assume that $(y_1, y_2) = (1, 1-d)$ is the facility location profile that Mechanism 3 outputs. Since all $N_1 \cap N_2$-agents located in $[\frac{1}{2}, 1]$ and $(N_1 \backslash N_2) \cup (N_2 \backslash N_1)$-agents cannot change the locations of F_1 and F_2 by misreporting, they have no incentive to misreport. Consider agent $i \in N_1 \cap N_2$ in $[0, \frac{1}{2}]$. For $d \in [0, \frac{1}{2}]$,

$$u_i((x_i, p_i), (0, d)) = x_i + |d - x_i| \leq (1 - x_i) + (1 - d - x_i) = u_i((x_i, p_i), (1, 1 - d)),$$

and for $d \in (\frac{1}{2}, 1]$,

$$u_i((x_i, p_i), (0, d)) = x_i + (d - x_i) \leq (1 - x_i) + |1 - d - x_i| = u_i((x_i, p_i), (1, 1 - d)).$$

Thus, any agent $i \in N_1 \cap N_2$ located in $[0, \frac{1}{2}]$ has no incentive to misreport.

Case 2. $|N_1 \setminus N_2| > |N_1 \cap N_2|$. Without loss of generality, assume that $(y_1, y_2) = (1, 0)$ is the facility location profile that Mechanism 3 outputs. Consider agent $i \in N_1 \cap N_2$, we have

$$u_i((x_i, p_i), (0, 1)) = x_i + (1 - x_i) = u_i((x_i, p_i), (1, 0)),$$

thus all $N_1 \cap N_2$-agents have no incentive to misreport.

Agents in $N_2 \setminus N_1$ cannot affect the output of Mechanism 3 in this case, and have no incentive to misreport.

The utility of agent $i \in N_1 \backslash N_2$ located in $[0, \frac{1}{2}]$ is

$$u_i((x_i, p_i), (0, 1)) = x_i \leq 1 - x_i = u_i((x_i, p_i), (1, 0)),$$

so she has no incentive to misreport. For any agent $i \in N_1 \backslash N_2$ located in $(\frac{1}{2}, 1]$, misreporting can only possibly increase n'_L, and have no way to change the output. □

Lemma 2. *In the general setting, for any instance* $\mathbf{c} = (\mathbf{x}, \mathbf{p})$, *the optimal social utility cannot exceed* $n + (1 - d) \cdot |N_1 \cap N_2|$.

Proposition 1. *For instances with* $|N_1 \cap N_2| \geq |N_{j^*} \setminus N_{3-j^*}|$, *the worst-case ratio between the optimal social utility and the social utility of Mechanism 3 is at most* $2(4 - d)$ *for* $d \in [0, \frac{1}{2}]$ *and at most* $\frac{4-d}{d}$ *for* $d \in (\frac{1}{2}, 1]$.

Proof. Without loss of generality, assume that $j^* = 1$, and $\mathbf{y} = (1, 1 - d)$ is the facility location profile that Mechanism 3 outputs.

(1) If $d \in [0, \frac{1}{2}]$, the social utility of Mechanism 3 is

$$\begin{aligned} SU(\mathbf{y}|\mathbf{c}) &= \sum_{i \in N} u_i((x_i, p_i), \mathbf{y}) \\ &= \sum_{i \in N_1 \cap N_2} (|1 - x_i| + |1 - d - x_i|) + \sum_{i \in N_1 \backslash N_2} |1 - x_i| + \sum_{i \in N_2 \backslash N_1} |1 - d - x_i| \\ &\geq \sum_{i \in N_1 \cap N_2, x_i \in [0, \frac{1}{2}]} (1 - d) + \sum_{i \in N_1 \cap N_2, x_i \in (\frac{1}{2}, 1]} d \\ &\geq \frac{|N_1 \cap N_2|}{2}. \end{aligned}$$

Note that $|N_1 \cap N_2| \geq \frac{n}{3}$, since $|N_1 \cap N_2| + |N_1 \setminus N_2| + |N_2 \setminus N_1| = n$, $|N_1| \geq |N_2|$, and $|N_1 \cap N_2| \geq |N_1 \setminus N_2|$.

Combining Lemma 2, we have

$$\frac{SU(OPT|\mathbf{c})}{SU(\mathbf{y}|\mathbf{c})} \leq 2 \cdot \frac{n + (1-d) \cdot |N_1 \cap N_2|}{|N_1 \cap N_2|} \leq 2(4-d) \in [7, 8].$$

(2) If $d \in (\frac{1}{2}, 1]$, the social utility of Mechanism 3 is

$$\begin{aligned} SU(\mathbf{y}|\mathbf{c}) &= \sum_{i \in N} u_i((x_i, p_i), \mathbf{y}) \\ &= \sum_{i \in N_1 \cap N_2} (|1 - x_i| + |1 - d - x_i|) + \sum_{i \in N_1 \setminus N_2} |1 - x_i| + \sum_{i \in N_2 \setminus N_1} |1 - d - x_i| \\ &\geq d \cdot |N_1 \cap N_2|. \end{aligned}$$

Combining Lemma 2, we have

$$\frac{SU(OPT|\mathbf{c})}{SU((\mathbf{y}|\mathbf{c})} \leq \frac{n + (1-d) \cdot |N_1 \cap N_2|}{d \cdot |N_1 \cap N_2|} \leq \frac{4-d}{d} \in [3, 7).$$

□

Proposition 2. *For instances with* $|N_{j^*} \setminus N_{3-j^*}| \geq |N_1 \cap N_2|$, *the worst-case ratio between the optimal social utility and the social utility of Mechanism 3 is at most* 8.

Proof. Without loss of generality, assume that $j^* = 1$, and $\mathbf{y} = (1, 0)$ is the facility location profile that Mechanism 3 outputs.

The social utility of Mechanism 3 is

$$\begin{aligned} SU(\mathbf{y}|\mathbf{c}) &= \sum_{i \in N} u_i((x_i, p_i), \mathbf{y}) \\ &= \sum_{i \in N_1 \setminus N_2} (1 - x_i) + \sum_{i \in N_2 \setminus N_1} x_i + \sum_{i \in N_1 \cap N_2} ((1 - x_i) + x_i) \\ &\geq \frac{|N_1 \setminus N_2|}{2} \cdot \frac{1}{2} + |N_1 \cap N_2| \\ &= \frac{|N_1|}{4} + \frac{3}{4}|N_1 \cap N_2|. \end{aligned}$$

Combining Lemma 2, we have

$$\frac{SU(OPT|\mathbf{c})}{SU(\mathbf{y}|\mathbf{c})} \leq 4 \cdot \frac{n + (1-d) \cdot |N_1 \cap N_2|}{|N_1| + 3|N_1 \cap N_2|}. \tag{1}$$

The expression on the right-hand side of Eq. (1) is monotonically decreasing with both $|N_1|$ and $|N_1 \cap N_2|$. Note that $|N_1| \geq n/2$, since $|N_1 \setminus N_2| \geq |N_2 \setminus N_1|$. Therefore

$$\frac{SU(OPT|\mathbf{c})}{SU(\mathbf{y}|\mathbf{c})} \leq 8.$$

□

Combining Proposition 1 and Proposition 2, we obtain the following result.

Theorem 5. *The approximation ratio of Mechanism 3 is at most* 8.

Theorem 6. *Under the social utility objective, the approximation ratio of any deterministic strategyproof mechanism is at least 2.*

4.2 Randomized Mechanisms

In this subsection, we consider randomized mechanisms. First, by setting an appropriate probability distribution, a randomized group strategyproof mechanism with approximation ratio of at most 2 can be achieved. Then, we provide a lower bound of $\frac{14}{13}$ on the approximation ratio of any randomized strategyproof mechanism.

Mechanism 4: Given an instance $\mathbf{c} = (\mathbf{x}, \mathbf{p})$, return the facility location profile $(0, 1)$ and $(1, 0)$ with probability of $\frac{1}{2}$, respectively.

Theorem 7. *Mechanism 4 is a group strategyproof mechanism with approximation ratio of at most* 2.

Proof. Mechanism 4 is group strategyproof since it does not depend on the agents' reported locations and the known preferences.

Let f denote Mechanism 4. For any instance $\mathbf{c} = (\mathbf{x}, \mathbf{p})$, the social utility of Mechanism 4 is

$$\begin{aligned} SU(f|\mathbf{c}) &= \frac{1}{2}(\sum_{i \in N_1} x_i + \sum_{i \in N_2} (1 - x_i)) + \frac{1}{2}(\sum_{i \in N_1} (1 - x_i) + \sum_{i \in N_2} x_i) \\ &= \frac{1}{2}(\sum_{i \in N_1} x_i + \sum_{i \in N_1} (1 - x_i)) + \frac{1}{2}(\sum_{i \in N_2} (1 - x_i) + \sum_{i \in N_2} x_i) \\ &= \frac{1}{2}(|N_1| + |N_2|) \\ &= \frac{1}{2}(n + |N_1 \cap N_2|). \end{aligned}$$

Combining Lemma 2, we have

$$\frac{SU(OPT|\mathbf{c})}{SU(f|\mathbf{c})} \leq 2 \cdot \frac{n + (1 - d) \cdot |N_1 \cap N_2|}{n + |N_1 \cap N_2|}. \tag{2}$$

The expression on the right-hand side of Eq. (2) is monotonically decreasing with $|N_1 \cap N_2|$. Note that $|N_1 \setminus N_2| \geq 0$. Therefore

$$\frac{SU(OPT|\mathbf{c})}{SU(f|\mathbf{c})} \leq 2.$$

□

Theorem 8. *Under the social utility objective, the approximation ratio of any randomized strategyproof mechanism is at least* $14/13 \approx 1.077$.

5 Conclusions and Future Work

In this paper, we studied mechanism design for the two-obnoxious-facility-location problem with optional preferences and minimum distance constraint. For the objective of maximizing the social utility, we designed deterministic and randomized strategyproof mechanisms with proven approximation ratios, and derived lower bounds of 2 and $\frac{14}{13}$ on the approximation ratios for any deterministic and randomized strategyproof mechanism, respectively.

There are many other interesting future research directions. For example, narrow the gap between the upper and the lower bounds for deterministic and randomized mechanisms in our settings. We can also extend to other preference settings where agents may have triple preferences or dual preferences. In addition, it would be interesting to change the assumption about which information is public or private, for example, assuming locations are public information and preferences are private information. Finally, it might make sense to consider other social objective functions, such as the sum of square distance [18] or the egalitarian welfare [9].

Acknowledgments. This research was supported in part by the National Natural Science Foundation of China (12201590, 12171444) and Natural Science Foundation of Shandong Province (ZR2024MA031).

Disclosure of Interests. The authors declare that they have no conflict of interest.

References

1. Chen, Z., Fong, K.C., Li, M., Wang, K., Yuan, H., Zhang, Y.: Facility location games with optional preference. Theoret. Comput. Sci. **847**, 185–197 (2020)
2. Cheng, Y., Yu, W., Zhang, G.: Strategy-proof approximation mechanisms for an obnoxious facility game on networks. Theoret. Comput. Sci. **497**, 154–163 (2013)
3. Duan, L., Gong, Z., Li, M., Wang, C., Wu, X.: Mechanism design for facility location with fractional preferences and minimum distance. In: Computing and Combinatorics: 27th International Conference, COCOON 2021, Tainan, Taiwan, October 24–26, 2021, Proceedings 27, pp. 499–511. Springer (2021)
4. Fotakis, D., Tzamos, C.: On the power of deterministic mechanisms for facility location games. ACM Trans. Econo.Comput.(TEAC) **2**(4), 1–37 (2014)
5. Gai, L., Qian, D., Wu, C.: Two-facility location games with distance requirement. In: International Workshop on Frontiers in Algorithmics, pp. 15–24. Springer (2022)
6. Kanellopoulos, P., Voudouris, A.A.: Constrained truthful obnoxious two-facility location with optional preferences. arXiv preprint arXiv:2410.16131 (2024)
7. Kanellopoulos, P., Voudouris, A.A., Zhang, R.: On discrete truthful heterogeneous two-facility location. SIAM J. Discret. Math. **37**(2), 779–799 (2023)
8. Kanellopoulos, P., Voudouris, A.A., Zhang, R.: Truthful two-facility location with candidate locations. Theoret. Comput. Sci. **1024**, 114913 (2025)
9. Li, F., Plaxton, C.G., Sinha, V.B.: The obnoxious facility location game with dichotomous preferences. Theoret. Comput. Sci. **961**, 113930 (2023)

10. Li, M., Lu, P., Yao, Y., Zhang, J.: Strategyproof mechanism for two heterogeneous facilities with constant approximation ratio. In: Proceedings of the Twenty-Ninth International Conference on International Joint Conferences on Artificial Intelligence, pp. 238–245 (2021)
11. Lotfi, M., Voudouris, A.A.: On truthful constrained heterogeneous facility location with max-variant cost. Oper. Res. Lett. **52**, 107060 (2024)
12. Moulin, H.: On strategy-proofness and single peakedness. Public Choice **35**(4), 437–455 (1980)
13. Procaccia, A.D., Tennenholtz, M.: Approximate mechanism design without money. ACM Trans.Econ.Comput.(TEAC) **1**(4), 1–26 (2013)
14. Serafino, P., Ventre, C.: Heterogeneous facility location without money. Theoret. Comput. Sci. **636**, 27–46 (2016)
15. Wu, X., Mei, L., Zhang, G.: Two-facility location games with a minimum distance requirement on a circle. In: Combinatorial Optimization and Applications: 15th International Conference, COCOA 2021, Tianjin, China, December 17–19, 2021, Proceedings 15, pp. 497–511. Springer (2021)
16. Xu, X., Li, B., Li, M., Duan, L.: Two-facility location games with minimum distance requirement. J. Artif. Intell. Res. **70**, 719–756 (2021)
17. Xu, X., Zhang, J., Xie, L.: Minmax for facility location game with optional preference under minimum distance requirement. J. Comb. Optim. **46**(4), 23 (2023)
18. Ye, D., Mei, L., Zhang, Y.: Strategy-proof mechanism for obnoxious facility location on a line. In: International Computing And Combinatorics Conference, pp. 45–56. Springer (2015)
19. Zhao, Q., Liu, W., Nong, Q., Fang, Q.: Constrained heterogeneous facility location games with max-variant cost. J. Comb. Optim. **45**(3), 90 (2023)
20. Zou, S., Li, M.: Facility location games with dual preference. In: Proceedings of the 2015 international conference on autonomous agents and multiagent systems, pp. 615–623 (2015)

Approximating EFX Through a New Notion of Fairness

Rui Dai[1(✉)], Yuxuan Wang[1], Zhengyang Liu[1], and Zihe Wang[2]

[1] Beijing Institute of Technology, Beijing, China
dairuircs@gmail.com
[2] Renmin University of China, Beijing, China

Abstract. This paper addresses the fair division of indivisible chores in the additive setting. We propose a polynomial-time algorithm that guarantees either a 3.69-EFX or 1.24-proportional allocation for three agents. Additionally, we introduce a new fairness concept called discrete perfect (DP), which applies to chore allocations. We prove the existence of DP for two agents by providing an efficient algorithm and extend this by proposing an algorithm that approximates DP for the general case with n agents. We believe the DP concept holds independent value and could have broader applications.

Keywords: Fair Division · Game Theory · Approximation Algorithm

1 Introduction

With the rise of e-commerce, the topic of fair division has been extensively studied over the past few decades. In the regime of fair division, we study how to distribute a bundle of items among multiple agents in a way that is considered fair according to certain criteria. Note that agents may have totally different valuations to the items. Besides, the items are named *goods* when the valuation functions are positive and *chores* otherwise. Many works have introduced various notions to delineate the concept of fairness.

Among all the notions, *envy-freeness* (EF) [21] is the most original and well-studied one. An allocation is called envy-free if each agent believes she has received at least as much value as anyone else. *Proportionality* (PROP) [36] is another criteria from fair division, similar to envy-freeness, but with a distinct focus on ensuring that each agent receives at least their "fair share" of the total value. When the items are infinitely divisible, [3] guaranteed the existence of EF and PROP allocations. For indivisible items, however, it is not the case, even for two agents with one item. Therefore, [30] made a relaxation named *envy-freeness up to one item* (EF1), which means that the envy between agents can be eliminated by removing a specified item. Surprisingly, the EF1 allocations can be found for goods [30], chores and even the mixture of goods and chores [13], respectively.

M. Li et al. (Eds.): TAMC 2025, LNCS 16084, pp. 148–162, 2026.
https://doi.org/10.1007/978-981-95-4839-2_12

Another stronger relaxation to EF is *envy-freeness up to any item* (EFX) [15, 24]. We say an allocation is EFX if the envy between agents can be eliminated by removing any item, which can reduce some extreme unfairness that EF1 leads to. Unlike the universal existence of EF1, whether EFX exists in general is still a challenging problem. For goods, many works had positive results under specific instances or with some approximations (see Sect. 1.2). For chores, Christoforidis [20] gave a counter example that an EFX allocation for chores need not exist under general valuation functions. Exploring specific cases of EFX for chores and seeking improved approximate EFX allocations are promising avenues for future research. Recently, for the general additive valuations, Zhou [38] obtained a $(2+\sqrt{6})$-approximate EFX allocation for three agents and a $(3n^2-n)$ approximate EFX allocation for n agents. In this paper, we aim to improve the corresponding ratios.

1.1 Main Results

In this paper, we first introduce a new notion of relaxation named Discrete Perfect (DP) Partition to ensure a more comprehensive fair division status. This notion concentrates on the global fairness of an allocation that makes all the agents suppose the allocation is fair, which is much more comprehensive than the other notions where fairness only happens when agents compare themselves to the others. For the case with two agents, we propose a novel polynomial-time algorithm to obtain DP. For cases with more than three agents, we prove that there exists a $2n(n-1)$-DP.

Result 1. (Theorem 1). For any instance $\langle [2], M, C\rangle$ whose disutility functions are all additive, there is a polynomial algorithm that computes a DP partition.

Result 2. (Theorem 2). For any instance $\langle [n], M, C\rangle$ whose disutility functions are all additive, there is an algorithm that computes a $2n(n-1)$-DP partition.

Moreover, we propose a novel algorithm to obtain a $\frac{1+4\beta}{1-5\beta}$-EF allocation of indivisible chores under the case with three agents, where $\beta = \max_{i\in[n],e\in M}\frac{c_i(e)}{c_i(M)}$. And thus, we come up with another novel polynomial-time algorithm that computes either a $\frac{3+\sqrt{19}}{2}\approx 3.69$-EFX allocation or $\frac{2\sqrt{19}-5}{3}\approx$ 1.239-prop allocation, which is an improvement of the result of $(\sqrt{6}+2)$-EFX in [38]. Note that although we make a slight relaxation on pure 3.69-EFX allocation, the guarantees in this result are actually desirable, for that proportionality is a very desirable property which is stronger than MMS, and the current best MMS approximation when there are three agents is 1.4-MMS allocation in [7] which is weaker than our 1.239-prop.

Another result is a combination of EFX and PROP, as we believe that PROP is sufficiently robust to serve as a complement to EFX. A similar approach is also demonstrated in Theorem 2 of [2], where the authors introduced a polynomial-time algorithm that computes either a tEFX allocation or a proportional allocation for any instance involving three agents.

Result 3. (Theorem 3). For any instance $\langle [3], M, C\rangle$ whose disutility functions are all additive, and $\frac{c_i(e)}{c_i(M)} \leq \beta$ for all $i \in [n], e \in M$, there is a polynomial algorithm that computes a $\frac{1+4\beta}{1-5\beta}$-EF allocation.

Result 4. (Theorem 4). For any instance $\langle [3], M, C\rangle$ whose disutility functions are all additive, there is a polynomial algorithm that computes either a $\frac{3+\sqrt{19}}{2} \approx 3.69$-EFX allocation or $\frac{2\sqrt{19}-5}{3} \approx 1.239$-prop allocation.

1.2 Related Work

EFX for Goods. For goods, Plaut [35] showed that EFX allocations always exist for two agents with general valuation functions and any number of agents with identical ordering (IDO) valuation functions. As a breakthrough work, Chaudhury [17] demonstrated the existence of EFX allocations on goods among three additive agents. Akrami [1] generalized this result to nearly general valuations. For the cases with more agents under additive valuation functions, there are multiple works that study partial EFX allocations. Chaudhury [18] showed the existence of an EFX allocation while leaving at most $n-1$ items unallocated. Berger [12] constructed an EFX allocation of four agents exist while leaving one item unallocated. Chan [16] and plaut [35] independently found a $\frac{1}{2}$-EFX allocation and [6] showed a 0.618-EFX one. Very recently, [5] improved the approximation ratio to $\frac{2}{3}$ under specific instances. There are also some other studies concentrating on special instances like small numbers of items [32], two types of items [23], lexicographic preference [26], two valuation types [31], EFX in graphs [19] and binary valuations [4,11,25].

EFX for Chores. The first major concern for chores is whether EFX exists in a general setting. Christoforidis [20] answered this question with a negative result by giving a counter example. However, there are still a few studies concentrating on some special instances, e.g., inctances with small number of chores [28], IDO instances [29], binary instances [37], instances with two types of chores [8], instances with leveled preferences [22] and dichotomous valuation instances [28]. The work most closely related to ours is [38]. For general additive valuations, they obtained a $(2+\sqrt{6})$-approximate EFX allocation for three agents and a $(3n^2-n)$ approximate EFX allocation for more agents. In [38], they classify all items into two categories: large items and small items. They then apply different algorithms to address the allocation of large and small items separately. In this paper, we propose a new algorithm that achieves a better EFX allocation than [38], especially when the items are relatively small.

Other Motions of Relaxation. Apart from EFX, there are also several notions of fair division. The notion of EF1 originates in the work of [30]. For chores, Bhaskar [13] proved EF1 allocations are known to exist and are able to be computed in polynomial-time. The concept of Maximin Share (MMS) was first introduced in [14]. [9] showed that MMS allocations are not guaranteed to exist for indivisible chores, and [27] showed the best approximation ratio at $\frac{13}{11}$. There are also many other fairness notions. [15] introduced the notion of PMMS, [34] introduced the notion of PROPX for chores, and [10] introduced two new fairness notions named GMMS and EFL. It remains unknown that whether there are other notions that provide better approximations to delineate fair division.

1.3 Paper Outline

The remaining parts of this paper are outlined below. We will introduce the formal model and relevant definitions in Sect. 2. Section 3 proposes our definition of the new relaxation notion, together with some approximation results we obtained through a novel algorithm. Section 4 shows an important intermediate result. And in Sect. 5 we show improved approximations under additive valuation functions.

2 Preliminaries

In the rest of this paper, we will focus exclusively on fair division for chores. An instance of fair division with indivisible chores can be represented by a tuple $\langle N, M, C\rangle$, where $N = [n]$ represents the set of agents, $M = \{e_1, e_2, \ldots, e_m\}$ represents the set of chores, and $C = (c_1(\cdot), c_2(\cdot), c_3(\cdot), \ldots, c_n(\cdot))$ is the set of disutility functions, where $c_i : 2^M \rightarrow R_{\geq 0}$ is the disutility function for agent i. A disutility function f is said to be additive if $f(S) = \sum\limits_{e \in S} f(e), \forall S \subseteq M$. For ease of notation, we use e instead of $\{e\}$ for an item e. In order to allocate items to agents fairly, we first introduce some well-known fariness notions. An allocation is represented by a n-partition $X = (X_1, X_2, \ldots, X_n)$ where $X_i \cap X_j = \emptyset, \forall i \neq j$ and $\bigcup\limits_{i \in [n]} X_i = M$.

Envy-freeness might be the first concept of fairness that comes to mind. That is, every agent feels that she receives at most as much burden as any other one.

Definition 1 (EF). For an instance $\langle N, M, C\rangle$, an allocation X is called *envy-free* (EF) if for any $i, j \in [n], i \neq j, c_i(X_i) \leq c_i(X_j)$.

However, since envy-freeness is too strong to always be satisfiable, we relax it to obtain other fairness notions. EF1 is such a weaker form of envy-freeness.

Definition 2 (EF1). For an instance $\langle N, M, C\rangle$, an allocation X is called *envy-free up to one item* (EF1) if for any $i, j \in [n], i \neq j$, there exists an item e in X_i such that $c_i(X_i \setminus e) \leq c_i(X_j)$.

For any general instance $\langle N, M, C\rangle$, there always exists an EF1 allocation [33]. Therefore, we investigate the EFX allocation, which is stronger than EF1 but weaker than EF.

Definition 3 (EFX). For an instance $\langle N, M, C\rangle$, an allocation X is called envy-free up to any item (EFX) if for any $i, j \in [n], i \neq j$ and any item $e \in X_i$, we still have that $c_i(X_i \setminus e) \leq c_i(X_j)$.

It remains an open question whether EFX allocations exist in general settings with additive disutilities. Therefore, a promising approach we can do is to introduce an approximate solution concept to EFX.

Definition 4 (α-EFX)**.** For an instance $\langle N, M, C\rangle$, an allocation X is called α-EFX if for any $i, j \in [n], i \neq j$ and any item $e \in X_i$, we have $c_i(X_i \setminus e) \leq \alpha c_i(X_j)$. If for a pair of $i, j \in [n], i \neq j$, for all $e \in X_i, c_i(X_i \setminus e) \leq \alpha c_i(X_j)$, agent i is said to be α-EFX towards agent j.

Similarly, we can also define approximate envy-freeness. Another crucial property related to our work is proportionality.

Definition 5 (PROP)**.** For an instance $\langle N, M, C\rangle$, an allocation X is called *proportional* if for any agent i, we have that $c_i(X_i) \leq \frac{c_i(M)}{n}$.

However, as with envy-freeness, proportionality may not always be achievable, so we also consider the approximated version.

Definition 6 (α-PROP)**.** For an instance $\langle N, M, C\rangle$, an allocation X is called α-proportional iff. for all agent i, $c_i(X_i) \leq \alpha \cdot \frac{c_i(M)}{n}$.

For those approximation ratio α, the smaller α indicates better fair allocation.

3 Discrete Perfect Partition

In this section, we present a new notion of fairness: *discrete perfect partition* (DP partition).

Definition 7 (DP partition)**.** For an instance $\langle N, M, C\rangle$, let $b_i = \max\{c_i(e) : e \in M\}$, a partition $X = (X_1, X_2, \ldots, X_n)$ is called *discrete perfect* if for any $i, j, k \in [n]$, we always have that $|c_i(X_j) - c_i(X_k)| \leq b_i$.

Unlike the envy-freeness and proportionality, this notion focuses on global fairness. That is, each bundle should be allocated as equally as possible. In Sects. 4 and 5, we will show that this notion could help us find an EF or prop allocation. However EF is not stronger than DP, that is, we cannot convert to a DP allocation from an EF one. Given an instance described as Table 1, an EF allocation would assign item c to agent 1 and items a and b to agent 2, but it is not a DP partition.

Similarly with envy-freeness, *discrete perfect* is also too strong to be satisfied in some instances, and approximated α-DP comes.

Definition 8 (α-DP)**.** For an instance $\langle N, M, C\rangle$, let $b_i = \max\{c_i(e) : e \in M\}$, a partition $X = (X_1, X_2, \ldots, X_n)$ is called α-DP if for any $i, j, k \in [n]$, we have $|c_i(X_j) - c_i(X_k)| \leq \alpha b_i$.

Table 1. An example: two agents with three chores.

items	$c_1(\cdot)$	$c_2(\cdot)$
a	5	0
b	5	0
c	1	1

Here, we present our first result:

Theorem 1. *For any instance $\langle [2], M, C\rangle$ whose disutility functions are all additive, there is a polynomial algorithm that computes a DP partition.*

To find this valid partition, we propose a greedy algorithm by sorting all the items according to $f_1(\cdot)$, preserving an interval from left to right until we find a valid partition.

Equal Divide Algorithm. Given 2 disutility functions f_1 and f_2, the output is a partition of S, i.e. two sets U, V where $U \cap V = \emptyset, U \cup V = S$, satisfying the condition below:

$$f_1(U) \in \left[\frac{f_1(S) - b_1}{2}, \frac{f_1(S) + b_1}{2}\right],$$
$$f_2(V) \in \left[\frac{f_2(S) - b_2}{2}, \frac{f_2(S) + b_2}{2}\right],$$

where b_i is the maximum value of $f_i(e)$ for all $e \in S$.

Equal divide algorithm is presented in Algorithm 1. And we show that this algorithm gives a valid partition in Appendix A.

Then, Theorem 2 holds easily when we let $f_1 = c_1, f_2 = c_2$. Since $c_i(X_1) - c_2(X_2) \le \frac{b_i}{2} \cdot 2 = b_i$, the output is a valid partition satisfying the condition of DP partition.

However, can we obtain DP partition for general $n > 2$?

Theorem 2. *For any instance $\langle [n], M, C\rangle$ whose disutility functions are all additive, there is an algorithm that computes a $2n(n-1)$-DP partition.*

Proof. Theorem 2 holds intuitively with the help of Theorem 1.2 in [3]. $\mu_1, \mu_2, \ldots, \mu_n$ are n continuous probability measures where $\mu_i(x) = \sum_{j\in[m]} c_i(e_j) \cdot ||[\frac{j-1}{m}, \frac{j}{m}] \cap [0, x]||$ (let $||[l, r]||$ be the length of interval $[l, r]$ i.e. $r - l$). Actually, we just place all the m items sequentially in the interval $[0, 1]$, and $\mu[l, r]$ is the value of all the items covered in $[l, r]$ (for those items who are not completely covered, it just contributes the portion of the covered length). Then, Theorem 1.2 in [3] can cut the interval $[0, 1]$ in $n(n-1)$ places and partition the resulting

Algorithm 1 Equal divide

Input: two additive disutility functions f_1 and f_2, a set of items $S = (e_1, e_2, \dots, e_m)$ where $\forall i \in [2], e \in S, f_i(e) \le b_i$

Output: a partition of S, i.e. two sets U, V where $U \cap V = \emptyset, U \cup V = S$

sort all the items according to $f_1(e)$, i.e. $f_1(e_1) \le f_1(e_2) \le \dots \le f_1(e_m)$
for $j = 1 \to m$ **do**
  if $f_1(\bigcup_{t=1}^{j} e_t) \ge \frac{f_1(S)-b_1}{2}$ **then**
    break
  end if
end for
Let $i = 1$
while True **do**
  if $f_2(\bigcup_{t=i}^{j} e_t) \in [\frac{f_2(S)-b_2}{2}, \frac{f_2(S)+b_2}{2}]$ **then**
    return $U = \bigcup_{t=i}^{j} e_t, V = S \setminus U$
  end if
  if $f_1(\bigcup_{t=i+1}^{j} e_t) < \frac{f_1(S)-b_1}{2}$ **then**
    Let $j = j + 1$
  end if
  Let $i = i + 1$
end while

$n(n-1)+1$ pieces into n families $F_1, F_2, \dots, F_n$. The final partition is computed as follows: X_i contains all the items that are completely covered in F_i. For those unallocated items, allocate then to arbitary F_i covering it. Therefore, $c_i(X_j) = \frac{1}{k} + t$ where t is the value gained or lost caused by partially covered items. Note that we cut at most $n(n-1)$places and in one cut, the gained or lost value is at most b_i, hence $c_i(X_j) \in [\frac{1}{k} - n(n-1) \cdot b_i, \frac{1}{k} + n(n-1) \cdot b_i]$, and $|c_i(X_j) - c_i(X_k)| \le 2n(n-1)b_i$

Theorem 2 is quite natural according to [3]. When $n = 2$, Theorem 1 gives a 4-DP partition. However, a discrete perfect partition exists when $n = 2$. Unfortunately, a DP partition does not always exist for any instances when $n > 2$. We present an instance with $n = 3$ that has no DP partition in Table 2. Hence, we are curious about the approximation ratio of the DP partition when $n \ge 3$.

4 $\frac{1+4\beta}{1-5\beta}$-EF Allocation

Then we present our third result. In this result, we discuss the existence of approximation EF when the items are all relatively small with the help of the existence of *discrete perfect partition*.

Table 2. an instance that has no DP partition

items	$c_1(\cdot)$	$c_2(\cdot)$	$c_3(\cdot)$
e_1	1	0	1
e_2	1	0	1
e_3	1	0	1
e_4	1	0	1
e_5	1	0	1
e_6	0	1	0
e_7	0	1	1
e_8	1	1	0

Theorem 3. *For any instance $\langle [3], M, C\rangle$ whose disutility functions are all additive, and $\frac{c_i(e)}{c_i(M)} \leq \beta$ for all $i \in [n], e \in M$, there is a polynomial algorithm that computes a $\frac{1+4\beta}{1-5\beta}$-EF allocation.*

Since approximate envy-freeness is scale free, we can normalize all disutility function, which is replacing every $c_i(e)$ with $\frac{c_i(e)}{c_i(M)}$. After normalization, for all $i \in [n]$, $c_i(M) = 1$.

Although we are discussing chores scenario, we are going to utilize *Round Robin* algorithm to generate EF1 allocation for goods [33].

Algorithm 2 Round Robin
Input: 3 value functions v_1, v_2, v_3, goods set M
Output: EF1 allocation

```
i = 1
while M is not ∅ do
    X_i = X_i ∪ arg max_{e∈M} v_i(e)
    i = i mod n +1
end while
return X
```

The idea that we use goods scenario is that EF1 for goods help agents to select those items with large disutility, and those selected items should be allocated to other agents to increase others' disutility. However, we want to allocate the only one bundle to other two agents. Then we use the DP partition we proposed in the previous section.

Here, we present Algorithm 3 for computing a $\frac{1+4\beta}{1-5\beta}$-EF allocation in this case. The correctness of this algorithm is shown in Appendix B.

5 Fair Division For Three Agents

In this section, we use the result above to obtain an almost EFX allocation.

Algorithm 3 Compute $\frac{1+4\beta}{1-5\beta}$-EF for no large agent
Input: 3 disutility functions c_1, c_2, c_3, chores set M
Output: An α-EFX allocation.

$A_1, A_2, A_3 = RoundRobin(c_1, c_2, c_3, M)$
$S_2, S_1 = equaldivide(A_1, c_2, c_3)$
if $c_2(S_1) < c_2(S_2)$ **then**
    swap(S_1,S_2)
end if
$S_4, S_3 = equaldivide(A_2, c_2, c_1)$
if $c_3(S_3) < c_3(S_4)$ **then**
    swap(S_3,S_4)
end if
$S_6, S_5 = equaldivide(A_3, c_3, c_2)$
if $c_1(S_5) < c_1(S_6)$ **then**
    swap(S_5,S_6)
end if
return Agent 1:$S_3 \cup S_6$, Agent 2: $S_2 \cup S_5$, Agent 3: $S_1 \cup S_4$

Theorem 4. *For any instance $\langle [3], M, C \rangle$ whose disutility functions are all additive, there is a polynomial algorithm that computes either a $\frac{3+\sqrt{19}}{2} \approx 3.69$-EFX allocation or $\frac{2\sqrt{19}-5}{3} \approx 1.239$-prop allocation.*

Similar to the previous section, we can assume $c_i(M) = 1$. We introduce another notion $\sigma_i()$, where $\sigma_i(j)$ represents the j-th largest item for $c_i()$. Hence, $c_i(\sigma_i(1)) \geq c_i(\sigma_i(2)) \geq \cdots \geq c_i(\sigma_i(m))$.

Now, we give some indispensable definition to distinguish different types of items.

Definition 9. If $c_i(e) \geq \beta$, e is said to be large for agent i.

Definition 10. If $c_i(\sigma_i(1)) \geq \beta$, agent i is said to be a large agent, otherwise, agent i is said to be a small agent.

Here, $\beta = \frac{\sqrt{19}-4}{3}$, and we also set $\alpha = \frac{3+\sqrt{19}}{2}$ in this section. Actually, we have $\frac{1+4\beta}{1-5\beta} = \frac{1-\beta}{2\beta} = \alpha$.

Then for any instance $\langle [3], M, C \rangle$, there are three cases:

- There is no large agent.
- There is only one or two large agents.
- All the agents are large agents.

For the situations above, we will present the algorithm respectively.

There is no Large Agent. In this case, $c_i(e) \leq \beta$ for all agents and items. Then, in the previous section, we can obtain a $\frac{1+4\beta}{1-5\beta} = \alpha$ allocation.

There is One or Two Large Agent. W.l.o.g. let agent 1 be the small agent and agent 2 the large one.

[29] proved that when agents have identical disutility functions, an EFX allocation exists and can be computed in polynomial time. The algorithm is presented in Algorithm 4, and in the beginning of Algorithm 5, we will use Algorithm 4 to obtain an allocation that is EFX towards agent 1.

Algorithm 4 EFX-Identical
Input: one single disutility function c, chores set M, agents $[n]$
Output: EFX allocation

while $M \neq \emptyset$ **do**
  $e = \arg\max_{e \in M} c(e)$
  $i = \arg\min_{i \in [n]} c(X_i)$
  $X_i = X_i \cup \{e\}$
end while
return X

If there is an item e such that $c_2(e) \geq \beta$ and $c_3(e) \geq \beta$, then we can obtain a $\frac{1-\beta}{2\beta}$-EFX allocation easily by algorithm 5. [38] has shown that this algorithm yields a $\frac{1}{2\beta}$-EFX allocation, but we show that it is actually a $\frac{1-\beta}{2\beta}$-EFX allocation. After this algorithm, agent 1 is EFX towards other 2 agents because he has got only one item. For agent 3, $c_3(X_3) \leq \frac{c_3(M')}{2} \leq \frac{1-\beta}{2}$, since he picked favorite bundle between S, T, he doesn't envy agent 2. As for agent 1, since $\frac{c_3(X_3)}{c_3(X_1)} \leq \frac{\frac{1-\beta}{2}}{\beta} = \frac{1-\beta}{2\beta}$, then agent 3 is $\frac{1-\beta}{2\beta}$-EFX towards agent 1.

For agent 2, since S, T are EFX allocation according to his disutility function, $c_2(X_2 \setminus e') \leq c_2(X_3)$ where $e' = \arg\min_{e' \in X_2} c_2(e')$, which means agent 2 is EFX towards agent 3. Therefore $c_2(X_2) \leq \frac{c_2(M')}{2} = \frac{1-\beta}{2}$, and $\frac{c_2(X_2)}{c_2(X_1)} \leq \frac{1-\beta}{2\beta}$, hence agent 2 is also $\frac{1-\beta}{2\beta}$-EFX towards agent 1. Since $\alpha = \frac{1-\beta}{2\beta}$, we obatin an α-EFX allocation in this case.

Algorithm 5 $\frac{1-\beta}{2\beta}$-EFX allocation for special case

$X_1 = e$
$M' = M \setminus e$
$S, T = \text{EFX-Identical}(c_2(), M', [2])$
Let X_3 pick his favorite bundle in S, T
return X

Now, we can assume that for all items that are large for agent 2, they are small to agent 3. We propose Algorithm 6 for this case to obtain either an α-EFX or $1 + 2\beta$-prop allocation.

The correctness of this algorithm is shown in Appendix C.

Algorithm 6 α-EFX or $1+2\beta$-prop allocation
Input: an instance $\langle [3], M, C \rangle$
Output: α-EFX allocation or $(1+2\beta)$-prop allocation

```
S_1, S_2, S_3 = EFX-Identical(c_1, M)
Let u = argmin_u (c_2(S_u))
Let v = argmin_v (c_3(S_v))
if u ≠ v then
    Let X_1 = M \ (S_u ∪ S_v)
    Let X_2 = S_u
    Let X_3 = S_v
    return X
else
    if c_2(S_u) ≥ β then
        Let X_2 = argmin_{S ∈ S_1 ∪ S_2 ∪ S_3 \ S_u} c_2(S)
        Let X_3 = S_u
        Let X_1 = M \ (X_2 ∪ X_3)
        return X
    else
        Let e be the large item for agent 2, and let S' be the bundle e belongs to
        Let S_t = argmin_{S ∈ {S_1, S_2, S_3} \ {S_u}} c_3(S)
        if c_3(S_u) + c_3(e) ≥ 3(c_3(S_t) − c_3(e)) then
            Let X_2 = S_u
            Let X_3 = S_t
            Let X_1 = M \ (S_u ∪ S_t)
            return X
        else
            S' = S' \ e
            S_u = S_u ∪ e
            Let X_2 = argmin_{S ∈ {S_1, S_2, S_3} \ {S_u}} c_2(S)
            Let X_3 = S_u
            Let X_1 = M \ (X_2 ∪ X_3)
            return X
        end if
    end if
end if
```

6 Conclusion

In this paper, we propose an innovative notion of fairness, termed the *discrete perfect partition*, which we use to achieve an almost envy-free allocation. We demonstrate the existence of a DP partition for two agents and extend this to show the existence of a $2n(n-1)$-DP partition in the general case involving n agents. Based on the existence of a DP partition for two agents, we propose an efficient algorithm that provides an almost envy-free allocation for three agents when the items are relatively small. Furthermore, we leverage this result to improve upon the $(2+\sqrt{6})$-EFX allocation presented in [38], achieving either a 3.69-EFX allocation or a 1.24-proportional allocation. For the future work, improving the ratio is an obvious open question. We also wonder if the discrete perfect partition can be further used in other problems.

Acknowledgments. This work was supported by the National Natural Science Foundation of China (Grant Nos. 62472029 and 62172422) and Key Laboratory of Interdisciplinary Research of Computation and Economics (Shanghai University of Finance and Economics), Ministry of Education. Zhengyang Liu is the corresponding author.

A The Correctness of Algorithm 1

Proof. In the second loop (Line 8–16 of **Algorithm 1**), we claim that

$$f_1(\{e_i, e_{i+1}, \dots, e_j\}) \in [\frac{f_1(S) - b_1}{2}, \frac{f_1(S) + b_1}{2}]$$

.

If $f_1(\bigcup_{t=i+1}^{j} e_t) < \frac{f_1(S)-b_1}{2}$, then $f_1(\bigcup_{t=i+1}^{j+1} e_t) = f_1(\bigcup_{t=i}^{j} e_t) + f_1(e_{j+1}) - f_1(e_i)$.

Since $f_1(e_{j+1}) - f_1(e_i) \in [0, b_1]$, on the one hand, $f_1(\bigcup_{t=i+1}^{j+1} e_t) > f_1(\bigcup_{t=i}^{j} e_t) \geq \frac{f_1(S)-b_1}{2}$, on the other hand, $f_1(\bigcup_{t=i+1}^{j+1} e_t) < f_1(\bigcup_{t=i}^{j} e_t) + b_1 \leq \frac{f_1(S)+b_1}{2}$, hence, segment $[i+1, j+1]$ is still valid for f_1.

Then consider the value of j after the first loop. Since $f_2(\bigcup_{t=1}^{j} e_t) + f_2(\bigcup_{t=j+1}^{m} e_t) = f_2(S)$, then $(f_2(\bigcup_{t=1}^{j} e_t) - \frac{f_2(S)}{2}) \cdot (f_2(\bigcup_{t=j+1}^{m} e_t) - \frac{f_2(S)}{2}) \leq 0$. Note that in every iteration of the second loop, the value of $f_2(\bigcup_{t=i}^{j} e_t)$ will not be changed over b_2, because there is at most one item changed in every iteration, hence, there must be one iteration segment$[i, j]$ satisfying $f_2(\bigcup_{t=i}^{j} e_t) \in [\frac{f_2(S)-b_2}{2}, \frac{f_2(S)+b_2}{2}]$

B The Correctness of Algorithm 3

Since for every agent, $c_i(e) \leq \beta$, A_1, A_2 and A_3 are generated by goods-EF1 Round Robin algorithm, we have $c_i(A_i) \geq c_i(A_j) - \beta$ for all $i = 1, 2, 3, j \neq i$

Then, in the step 2, since all the agents are symmetric, we only need to analyze $\frac{c_1(S_3 \cup S_6)}{c_1(S_2 \cup S_5)}$. Firstly, according to the swap operation, $c_1(S_6) \leq c_1(S_5)$. According to the result of step 1, $c_1(S_1 \cup S_2) \geq c_1(S_3 \cup S_4) - \beta$. Meanwhile, according to the property of equal divide,

$$c_1(S_2) \geq \frac{c_1(S_1 \cup S_2) - \beta}{2},$$

$$c_1(S_3) \leq \frac{c_1(S_3 \cup S_4) + \beta}{2} \leq \frac{c_1(S_1 \cup S_2) + 2\beta}{2}.$$

Since $c_1(A_1) \geq c_1(A_2) - \beta$ and $c_1(A_1) \geq c_1(A_3) - \beta$, we have $c_1(A_1) \geq \frac{1-2\beta}{3}$, and $c_1(S_2 \cup S_5) \geq c_1(S_2) \geq \frac{\frac{1-2\beta}{3} - \beta}{2}$.

Hence, we have that $c_1(S_2) + \frac{3\beta}{2} \geq c_1(S_3)$. That is, $\frac{c_1(S_3 \cup S_6)}{c_1(S_2 \cup S_5)} \leq \frac{c_1(S_2 \cup S_5) + \frac{3\beta}{2}}{c_1(S_2 \cup S_5)} = 1 + \frac{\frac{3\beta}{2}}{c_1(S_2 \cup S_5)} < 1 + \frac{\frac{3\beta}{2}}{\frac{\frac{1-2\beta}{3} - \beta}{2}} = 1 + \frac{9\beta}{1-5\beta} = \frac{1+4\beta}{1-5\beta}$. Hence, the result holds.

C The Correctness of Algorithm 6

Case 1: $u \neq v$. agent 2 and agent 3 both get their favorite bundle, and agent 1 is EFX towards S_1, S_2, S_3 because of EFX-Identical algorithm. When $u = v$, which means agent 2's favorite bundle is the same as agent 3's, we need to do more delicate adjustment. *Case 2: $c_2(S_u) \geq \beta$.* We analyze the envy rate for each agent. For agent 2, since S_t is one of the bundle in S_1, S_2, S_3 other than S_u which minimize $c_3(S_t)$, $c_2(X_2) \leq \frac{c_2(M \setminus S_u))}{2} \leq \frac{1-\beta}{2}$. Moreover, he is envy free towards another bundle besides S_u, and $\frac{c_2(X_2)}{c_2(S_u)} \leq \frac{1-\beta}{2\beta} = \alpha$, which means agent 2 is α-EFX towards the other 2 agents. Agent 3 gets her favorite bundle, so he will not envy anyone. For agent 1, she is EFX towards S_1, S_2, S_3 due to the EFX-Identical algorithm.

Case 3: $c_2(S_u) < \beta$.

Case 3.1: $c_3(S_u) + c_3(e) \geq 3(c_3(S_t) - c_3(e))$. We have

$$4c_3(e) \geq 3c_3(S_t) - c_3(S_u)$$

meanwhile $c_3(S_t) \geq c_3(S_u)$ therefore

$$2c_3(e) \geq c_3(S_t)$$

$$c_3(S_t) < 2\beta < \frac{1}{3}$$

Agent 3 get a bundle that is proportional. Agent 2 get his favorite bundle so he is proportional as well. For agent 1, since S_1, S_2, S_3 is generated by EFX-Identical algorithm, $c_3(S_i) \in [\frac{1-2\beta}{3}, \frac{1+2\beta}{3}]$, then whichever bundle agent 1 get, he is invariably $(1+2\beta)$-proportional. Therefore, we obtain a $1+2\beta$-prop allocation.

Case 3.2: $c_3(S_u \cup e) < 3c_3(S_t \setminus e)$. This indicates that $\frac{c_3(S_u \cup e)}{c_3(S_t \setminus e)} < 3$. In the last branch (Line 24–29), we also have $c_3(S') \geq c_3(S_t)$, hence $\frac{c_3(S_u \cup e)}{c_3(S' \setminus e)} \leq \frac{c_3(S_u \cup e)}{c_3(S_t \setminus e)} < 3$, which means agent 3 is 3-EFX towards other agents. For agent 2, since he got his favorite bundle other than S_u, and $c_2(S_u) > \beta$, the envy rate for him is still $\frac{1-\beta}{2\beta} = \alpha$. For agent 1, we can notice that after transferring e to S_u, $c_1(S_i) \geq \frac{1-2\beta}{3} - \beta = \frac{1-5\beta}{3}$. Besides, for any bundle S other than S_u, $c_1(S) \leq \frac{1+2\beta}{3}$, since agent 1 will not get S_u, the envy rate of him is at most $\frac{1+2\beta}{1-5\beta} < \frac{1+4\beta}{1-5\beta} = \alpha$. We obtain an α-EFX allocation.

All the Agents are Large Agent. In this case, we follow Algorithm 3 in [38]. Since $\alpha > 3.561$, we can still obtain an α-EFX allocation.

References

1. Akrami, H., Alon, N., Chaudhury, B.R., Garg, J., Mehlhorn, K., Mehta, R.: Efx: a simpler approach and an (almost) optimal guarantee via rainbow cycle number. In: 24th ACM Conference on Economics and Computation, EC 2023, p. 61. Association for Computing Machinery (2023)
2. Akrami, H., Ray Chaudhury, B., Garg, J., Mehlhorn, K., Mehta, R.: Fair and efficient allocation of indivisible chores with surplus. In: Thirty-Second International Joint Conference on Artificial Intelligence, pp. 2494–2502. IJCAI (2023)
3. Alon, N.: Splitting necklaces. Adv. Math. **63**(3), 247–253 (1987)
4. Amanatidis, G., Birmpas, G., Filos-Ratsikas, A., Hollender, A., Voudouris, A.A.: Maximum Nash welfare and other stories about EFX. Theoret. Comput. Sci. **863**, 69–85 (2021)
5. Amanatidis, G., Filos-Ratsikas, A., Sgouritsa, A.: Pushing the frontier on approximate EFX allocations. arXiv preprint arXiv:2406.12413 (2024)
6. Amanatidis, G., Markakis, E., Ntokos, A.: Multiple birds with one stone: beating 1/2 for EFX and GMMS via envy cycle elimination. Theoret. Comput. Sci. **841**, 94–109 (2020)
7. Aziz, H., Li, B., Wu, X.: Approximate and strategyproof maximin share allocation of chores with ordinal preferences. Math. Program. **203**(1), 319–345 (2024)
8. Aziz, H., Lindsay, J., Ritossa, A., Suzuki, M.: Fair allocation of two types of chores. In: 2023 International Conference on Autonomous Agents and Multiagent Systems, pp. 143–151 (2023)
9. Aziz, H., Rauchecker, G., Schryen, G., Walsh, T.: Algorithms for max-min share fair allocation of indivisible chores. In: Proceedings of the AAAI Conference on Artificial Intelligence, vol. 31 (2017)
10. Barman, S., Biswas, A., Krishnamurthy, S., Narahari, Y.: Groupwise maximin fair allocation of indivisible goods. In: 32nd AAAI Conference on Artificial Intelligence (2018)
11. Benabbou, N., Chakraborty, M., Igarashi, A., Zick, Y.: Finding fair and efficient allocations for matroid rank valuations. ACM Trans. Econ. Comput. **9**(4), 1–41 (2021)
12. Berger, B., Cohen, A., Feldman, M., Fiat, A.: Almost full EFX exists for four agents. In: Proceedings of the AAAI Conference on Artificial Intelligence, vol. 36, pp. 4826–4833 (2022)
13. Bhaskar, U., Sricharan, A., Vaish, R.: On approximate envy-freeness for indivisible chores and mixed resources. Randomization, and Combinatorial Optimization. Algorithms and Techniques, Approximation (2021)
14. Budish, E.: The combinatorial assignment problem: approximate competitive equilibrium from equal incomes. J. Polit. Econ. **119**(6), 1061–1103 (2011)
15. Caragiannis, I., Kurokawa, D., Moulin, H., Procaccia, A.D., Shah, N., Wang, J.: The unreasonable fairness of maximum Nash welfare. ACM Trans. Econ. Comput. (TEAC) **7**(3), 1–32 (2019)
16. Chan, H., Chen, J., Li, B., Wu, X.: Maximin-aware allocations of indivisible goods. In: 18th International Conference on Autonomous Agents and MultiAgent Systems, pp. 1871–1873 (2019)
17. Chaudhury, B.R., Garg, J., Mehlhorn, K.: EFX exists for three agents. J. ACM **71**(1), 1–27 (2024)
18. Chaudhury, B.R., Kavitha, T., Mehlhorn, K., Sgouritsa, A.: A little charity guarantees almost envy-freeness. SIAM J. Comput. **50**(4), 1336–1358 (2021). https://doi.org/10.1137/20M1359134

19. Christodoulou, G., Fiat, A., Koutsoupias, E., Sgouritsa, A.: Fair allocation in graphs. In: Proceedings of the 24th ACM Conference on Economics and Computation, pp. 473–488 (2023)
20. Christoforidis, V., Santorinaios, C.: On the pursuit of EFX for chores: non-existence and approximations. arXiv preprint arXiv:2406.10752 (2024)
21. Foley, D.K.: Resource allocation and the public sector. Yale University (1966)
22. Gafni, Y., Huang, X., Lavi, R., Talgam-Cohen, I.: Unified fair allocation of goods and chores via copies. ACM Trans. Econ. Comput. **11**(3–4), 1–27 (2023)
23. Gorantla, P., Marwaha, K., Velusamy, S.: Fair allocation of a multiset of indivisible items. In: Proceedings of the 2023 Annual ACM-SIAM Symposium on Discrete Algorithms (SODA), pp. 304–331. SIAM (2023)
24. Gourvès, L., Monnot, J., Tlilane, L.: Near fairness in matroids. In: ECAI 2014, pp. 393–398. IOS Press (2014)
25. Halpern, D., Procaccia, A.D., Psomas, A., Shah, N.: Fair division with binary valuations: one rule to rule them all. In: Chen, X., Gravin, N., Hoefer, M., Mehta, R. (eds.) WINE 2020. LNCS, vol. 12495, pp. 370–383. Springer, Cham (2020). https://doi.org/10.1007/978-3-030-64946-3_26
26. Hosseini, H., Sikdar, S., Vaish, R., Xia, L.: Fair and efficient allocations under lexicographic preferences. In: Proceedings of the AAAI Conference on Artificial Intelligence, vol. 35, pp. 5472–5480 (2021)
27. Huang, X., Segal-Halevi, E.: A reduction from chores allocation to job scheduling. In: Proceedings of the 24th ACM Conference on Economics and Computation, pp. 908–908 (2023)
28. Kobayashi, Y., Mahara, R., Sakamoto, S.: EFX allocations for indivisible chores: matching-based approach. In: International Symposium on Algorithmic Game Theory, pp. 257–270. Springer (2023)
29. Li, B., Li, Y., Wu, X.: Almost (weighted) proportional allocations for indivisible chores. In: 31st ACM World Wide Web Conference. WWW 2022, pp. 122–131. Inc, Association for Computing Machinery (2022)
30. Lipton, R.J., Markakis, E., Mossel, E., Saberi, A.: On approximately fair allocations of indivisible goods. In: Proceedings of the 5th ACM Conference on Electronic Commerce. p. 125–131. EC '04, Association for Computing Machinery, New York, NY, USA (2004). https://doi.org/10.1145/988772.988792
31. Mahara, R.: Existence of EFX for two additive valuations. Discret. Appl. Math. **340**, 115–122 (2023)
32. Mahara, R.: Extension of additive valuations to general valuations on the existence of EFX. Math. Oper. Res. **49**(2), 1263–1277 (2024)
33. Markakis, E.K.: Approximation algorithms and hardness results for fair division with indivisible goods (2017). https://api.semanticscholar.org/CorpusID:32862104
34. Moulin, H.: Fair division in the internet age. Ann. Rev. Econ. **11**(1), 407–441 (2019)
35. Plaut, B., Roughgarden, T.: Almost envy-freeness with general valuations. SIAM J. Discret. Math. **34**(2), 1039–1068 (2020)
36. Steinhaus, H.: The problem of fair division. Econometrica **16**, 101–104 (1948)
37. Tao, B., Wu, X., Yu, Z., Zhou, S.: On the existence of EFX (and pareto-optimal) allocations for binary chores. arXiv preprint arXiv:2308.12177 (2023)
38. Zhou, S., Wu, X.: Approximately EFX allocations for indivisible chores. Artif. Intell. **326**, 104037 (2024)

Parallel and Distributed Computing

A Distributed Algorithm for Robust Sequential Submodular Optimization in Multi-robot Systems

Bo Zhao[1], Fengmin Wang[2], Dachuan Xu[1], Ruiqi Yang[1(✉)], and Yifei Zou[3]

[1] Institute of Operations Research and Information Engineering, Beijing University of Technology, Beijing, P.R. China
bo.zhao07@outlook.com, {xudc,yangruiqi}@bjut.edu.cn

[2] Beijing Jinghang Research Institute of Computing and Communication, Beijing, P.R. China

[3] School of Computer Science and Technology, Institute of Intelligent Computing, Shandong University, Qingdao, P.R. China
yfzou@sdu.edu.cn

Abstract. With the integration of next-generation mobile communication systems and artificial intelligence, multi-robot systems utilize integrated sensing and communication (ISAC) technologies to meet the perception and communication needs for high performance, stability, and security. We introduce distributed communication strategies to enhance multi-robot cooperation, especially in sequential decision-making and complex adversarial environments. In this paper, we model these problems as a robust sequential submodular optimization framework, which restricts the multiple robots to satisfy the partition matroid constraints. We adopt the divide-and-conquer strategy to design the Distributed Robust Sequential Submodular Maximization (DRSSM) algorithm to enhance the information sharing and cooperative perception of robots. The DRSSM algorithm is proven to obtain curvature-dependent approximation guarantees. The numerical experiments demonstrate that the algorithm outperforms the centralized greedy algorithm and the centralized robust algorithm in solution quality, runtime, and robustness under attack.

Keywords: Approximation algorithms · Fault-tolerant systems · Autonomous aerial vehicles · Mobile robots · Multirobot systems

1 Introduction

In contemporary society, artificial intelligence increasingly empowers various fields, driving the integration of the physical and digital worlds. Intelligent transformation is vital for urban development, traffic management, and industrial production. The next generation of mobile communication systems places higher demands on the communication and perception capabilities of robotic systems

M. Li et al. (Eds.): TAMC 2025, LNCS 16084, pp. 165–177, 2026.
https://doi.org/10.1007/978-981-95-4839-2_13

and sensor networks, while emphasizing the efficiency, stability, and security of services [2,3]. Next-generation mobile communication systems integrating artificial intelligence could drive innovation and play a crucial role in scenarios such as smart cities [3], smart factories [13], and disaster response [7]. In these application scenarios, various mobile robots are equipped with sensors and communication modules.

In the real world, multi-robot systems are required to consistently and sequentially perform real-time target tracking [15], autonomous navigation [4], and environmental monitoring [4,8,14] tasks in complex dynamic scenarios where sensors can be attacked or failed at any time; for example, UAVs may be electronically disrupted during flight tracking, and the perception capabilities of mobile robots in post-disaster earthquakes or fires may be impaired by dense smoke and landslides. Existing work [12] introduces a constraint on the total number of activated sensors while neglecting the complementary advantages of multi-modal perception systems, such as cameras, infrared imagers, and LiDAR. To this end, we refine the resource constraints from a single total-sensor limitation to individual number constraints for different types of decisions, which not only reduce the resource consumption but also enhance the diversity of decision types.

In particular, these applications involve sequentially selecting a subset of elements from a finite set to optimize solution performance over multiple consecutive time steps, subject to various resource constraints. For instance, in target tracking applications, a subset of sensors in a wireless sensor network must be activated over time to minimize estimation error while adhering to resource limitations. To address performance degradation caused by potential failures or adversarial attacks, we consider a robust formulation of this problem. In this setting, the selected solution at any time step may be subject to worst-case element removals, constrained by an upper bound on the number of deletions. The objective is to maximize the overall value of the objective function across all time steps, accounting for the potential impact of such removals. We formalize the research problem as a robust sequential submodular optimization framework with constraints. This type of problem, originally proposed in [12] under cardinality constraints is computationally intractable. In contrast to previous studies that are limited to cardinality constraints, we introduce partition matroid constraints which more accurately reflect the resource class constraints of multi-robot systems in resource allocation.

Formally, we consider a system consisting R robots operating over a sequential decision-making process spanning T time steps. Let $\mathcal{X}_i^{(t)}$ denote the candidate set of decision options of various types that robot t can activate at time t, and let $D^{(t)}$ represent the subset of decisions removed by an adversary at time t. At each time step, we sequentially select a decision set $S^{(t)}$ to maximize an objective function f. We define the index sets $[R] = \{1, 2, \cdots, R\}$ and

$[T] = \{1, 2, \cdots, T\}$. The problem is formally defined as follows:

$$\begin{aligned} &\max_{S^{(1)} \subseteq \mathcal{X}^{(1)}} \min_{D^{(1)} \subseteq S^{(1)}} \cdots \max_{S^{(T)} \subseteq \mathcal{X}^{(T)}} \min_{D^{(T)} \subseteq S^{(T)}} f(S^{(1)} \backslash D^{(1)}, \cdots, S^{(T)} \backslash D^{(T)}) \\ &\text{s.t.} \ \left| S^{(t)} \cap \mathcal{X}_i^{(t)} \right| = 1, \quad i \in [R], t \in [T], \\ &\qquad \left| D^{(t)} \right| \leq d^{(t)}, \quad t \in [T], \end{aligned}$$

where $d^{(t)}$ is a given number of attacks at time step t. The objective function evaluates the quality of the selected action sets and typically satisfies submodularity and monotonicity. For example, in target tracking, the objective function can be used to measure the number of targets continuously tracked. In sensor scheduling, it can be used to assess the cumulative batch state estimation error. For ease of representation later, we define the following notation $\mathcal{I}^{(t)} := \{S \mid S \subseteq \mathcal{X}^{(t)}, |S \cap \mathcal{X}_i^{(t)}| = 1, \forall i \in [R]\} (\forall t \in [T])$, $\mathcal{J}^{(t)} := \{D \mid D \subseteq S^{(t)}, |D| \leq d^{(t)}\} (\forall t \in [T])$. Then, the optimal value of this problem is defined as

$$f_\star \triangleq \max_{S^{(1)} \in \mathcal{I}^{(1)}} \min_{D^{(1)} \in \mathcal{J}^{(1)}} \cdots \max_{S^{(T)} \in \mathcal{I}^{(T)}} \min_{D^{(T)} \in \mathcal{J}^{(T)}} f(S^{(1)} \backslash D^{(1)}, \cdots, S^{(T)} \backslash D^{(T)}).$$

Related Work. Tzoumas et al. [10] studied the resilience selection problem, modeling submodular optimization under adversarial denial-of-service (DoS) attacks. They proposed the first scalable approximation algorithm that defends against any number of attacks, with improved approximation performance for functions with lower curvature. Tzoumas et al. [11] addressed resilient maximization under matroid constraints, proposing the first approximation algorithm to handle arbitrary DoS attacks or failures, and providing curvature-based approximation guarantees. Schlotfeldt et al. [5] addressed the resilient active information gathering by mobile robots, presenting an algorithm with an approximation ratio and bounded communication rounds for monotone submodular or monotone set function. Zhou et al. [15], extending [11], tackled the resilient active target tracking problem against robot failures. Later, Tzoumas et al. [12] proposed the Robust Sequential Submodular Maximization problem to address failures in adversarial settings, designing the first online, adaptive algorithm resilient to any number of attacks—though it assumes a centralized architecture. However, the algorithm does not apply to distributed network architectures. To overcome this limitation, Zhou et al. [16] proposed a robust submodular maximization problem for distributed multi-robot path planning, introducing a distributed approximation algorithm. Shi et al. [6] studied robust maximization for sequential network submodular functions, developing a resilient greedy algorithm with guarantees dependent on deletions and network topology. Their study focused on directed acyclic networks, where the submodular function is defined over the edges. Despite these advances, most prior work is limited to static settings; addressing sequential decision making in adversarial, distributed communication scenarios remains to be explored.

Our Contribution. First, we address the problem background of multi-robot systems making multi-stage sequential decisions in an adversarial environment.

Specifically, we formulate it as a robust sequential submodular optimization problem with partition matroid constraints. Next, we propose the first distributed approximation algorithm for this framework. The algorithm is based on a divide-and-conquer strategy in which robots exchange information collaboratively in a distributed manner. Then, we derive approximation guarantees based on the total curvature of the objective function, supported by rigorous theoretical analysis. Finally, we conduct numerical experiments on multi-UAV target tracking, comparing our algorithm with centralized non-robust and robust algorithms. The results show that our proposed algorithm outperforms the compared algorithms in terms of objective function, running time, and number of calls to Oracle.

2 Preliminaries

In this paper, multi-robot distributed communication is constructed using a communication graph. Robots can only communicate with other robots within a specified communication range. We assume that all robots have the same communication radius r. Based on this, a communication graph can be constructed, where the nodes represent the set of robots, and the edges connect pairs of robots whose distance does not exceed the communication radius. Consequently, each robot can only exchange information with neighboring robots that are connected by an edge in the communication graph.

Submodular Function. We define a set function $f : 2^{\mathcal{N}} \to \mathbb{R}$ to be normalized if $f(\emptyset) = 0$; non-decreasing if $f(S_1) \leq f(S_2)$ for every $S_1 \subseteq S_2 \subseteq \mathcal{N}$; and submodular if $f(S_1)+f(S_2) \geq f(S_1 \cup S_2)+f(S_1 \cap S_2)$ for all $S_1, S_2 \subseteq \mathcal{N}$. For a set function, the definition of submodularity is equivalent to the diminishing returns property, which states that $f(S_1 \cup \{\nu\}) - f(S_1) \geq f(S_2 \cup \{\nu\}) - f(S_2)$ for any $S_1 \subseteq S_2 \subseteq \mathcal{N}$ and $\nu \in \mathcal{N} \setminus S_2$. The marginal gain of the function f with respect to an element ν and a set $S \subseteq \mathcal{N}$ is defined as $f(\nu \mid S) := f(S \cup \{\nu\}) - f(S)$. We define the symbol $\Delta f_T(S)$ to marginal returns concerning the objective function f, which is denoted as

$$\Delta f_T(S) \triangleq f(S \cup T) - f(S), \forall S, T \subseteq V. \tag{1}$$

Matroid. Given a finite set $\mathcal{N}$ and a collection $\mathcal{I}$ of subsets of $\mathcal{N}$, a matroid $\mathcal{M} := (\mathcal{N}, \mathcal{I})$ is an independent set system and satisfies the following properties: (i) $\emptyset \in \mathcal{I}$; (ii) If $S' \subseteq S$, $S \in \mathcal{I}$, then $S' \in \mathcal{I}$; (iii) If two sets S', S hold $|S'| > |S|$, then $\exists \nu \in S' \setminus S$ s.t. $S \cup \{\nu\} \in \mathcal{I}$.

Total Curvature. The total curvature of a monotone submodular function $f : 2^{\mathcal{N}} \to \mathbb{R}$ is denoted by

$$\kappa_f \triangleq 1 - \min_{\nu \in \mathcal{X}} \frac{f(\mathcal{N}) - f(\mathcal{N} \backslash \nu)}{f(\nu)} \tag{2}$$

where $\mathcal{X} = \{\nu \in \mathcal{N} \mid f(\nu) > 0\}$, and $\mathcal{N}$ represents the ground set [1]. The value of this curvature lies in the closed interval $[0, 1]$. The total curvature κ_f represents how close the function $f(\cdot)$ is to being modular.

Given a monotone set function $f : 2^{\mathcal{N}} \rightarrow \mathbb{R}$, the generalized curvature [9] is defined as

$$c_f \triangleq 1 - \min_{\nu \in \mathcal{X}} \min_{A,B \subseteq \mathcal{N} \setminus \{\nu\}} \frac{f(A \cup \{\nu\}) - f(A)}{f(B \cup \{\nu\}) - f(B)}. \tag{3}$$

Similarly to the total curvature, the generalized curvature also takes values in the interval $[0, 1]$. If f is submodular, the generalized curvature is equivalent to the total curvature.

3 Algorithm

Before presenting the algorithm, we define the necessary notation and essential symbols, as detailed in Table 1.

Table 1. Notation

Symbols	Descriptions
P	positions of robots
$\mathcal{C}$	partition groups at current moment
C_k	k-th partition group at current moment
$\mathcal{X}_k$	actions set of all robots in k-th partition group at current moment
$S_k^{(t)}$	solution of k-th partition group at time t
$S^{(t)}$	solutions set of all partition groups at time t
N_i	neighbor set of robot i at current moment
$A_k^{(t)}$	desired set of attacks of k-th partition group at time t
$B_k^{(t)}$	greedy set of k-th partition group at time t

The DRSSM algorithm operates as follows: First, the DGD algorithm partitions the robot swarm into groups, enabling one-hop communication within each group. Then, for each group, the following distributed operations are performed: viable decision sets are constructed; based on these sets, the objective function, and the number of robots to be omitted, the LRO algorithm computes each team's decision solution. Finally, a global decision is made for the entire swarm and the status of each robot is updated accordingly.

The DGD algorithm, described in Algorithm 2, takes as input the robots' position states, communication radius, and the maximum group size, and outputs a set of robot groups. The algorithm proceeds as follows: in a distributed manner, each robot determines its neighbors, shares its neighbor information with them, and receives information from others. Using this collective information along with its own, each robot computes the intersection of additional number constraints to determine its group assignment. Each robot then communicates its group affiliation to all neighbors.

Table 2. Abbreviations

Symbol Abbreviations
$S_k^{(t)} \triangleq A_k^{(t)} \cup B_k^{(t)}$
$A^{(t)} \triangleq \bigcup_{j=1}^{K} A_k^{(t)}$
$B^{(t)} \triangleq \bigcup_{j=1}^{K} B_k^{(t)}$
$A_{[1:k]}^{(t)} \triangleq (A_1^{(t)}, \cdots, A_k^{(t)}) \equiv A_1^{(t)} \cup \cdots \cup A_k^{(t)}$
$B_{[1:k]}^{(t)} \triangleq (B_1^{(t)}, \cdots, B_k^{(t)}) \equiv B_1^{(t)} \cup \cdots \cup B_k^{(t)}$
$S^{(t)} \triangleq \bigcup_{j=1}^{K} S_k^{(t)}$
$S_{[1:k]}^{(t)} \triangleq (S_1^{(t)}, \cdots, S_k^{(t)})$
$S^{[1:t]} \triangleq (S^{(1)}, \cdots, S^{(t)}) \equiv S^{(1)} \cup \cdots \cup S^{(t)}$
$S_{[1:k]}^{[1:t]} \triangleq \left(\bigcup_{j=1}^{k} S_j^{(1)}, \cdots, \bigcup_{j=1}^{k} S_j^{(t)}\right)$
$f(S^{(1)}, \cdots, S^{(T)}) \triangleq f(S^{(1)} \cup \cdots \cup S^{(T)})$
$D_\star^{(t)}$ is an optimal (worst-case) removal from $S^{(t)}$
$D_\star^{[1:T]} = (D_\star^{(1)}, \cdots, D_\star^{(T)})$
$S^{[1:T]} \setminus D_\star^{[1:T]} = (S^{(1)} \setminus_\star^{(1)}, \cdots, S^{(T)} \setminus D_\star^{(T)})$

Algorithm 1: DRSSM: Distributed Robust Sequential Submodular Maximization

Input: : f, R, $\{\mathcal{X}_i\}_{i=1}^{|R|}$, $P^{(1)}$, r, $\{d^{(t)}\}_{t=1}^{T}$, δ
Output: : $\{S^{(t)}\}_{t=1}^{T}$
for $t = 1, \cdots, T$ **do**
    $\{C_k\}_{k=1}^{K(t)} \leftarrow \mathrm{DGD}(P^{(t)}, r, \delta \cdot |R|)$, $\mathcal{C} \leftarrow \{C_k\}_{k=1}^{K(t)}$
    for ***each*** $C_k \in \mathcal{C}$ ***parallelly*** **do**
        $\mathcal{X}_k \leftarrow \bigcup_{i \in C_k} \mathcal{X}_i$
        $S_k^{(t)} \leftarrow \mathrm{LRO}(\mathcal{X}_k, f, d^{(t)})$
    $S^{(t)} \leftarrow \bigcup_{k=1}^{K(t)} S_k^{(t)}$
    Update information $P^{(t+1)}$ utilizing every robots' status after executing $S^{(t)}$

Algorithm 2: DGD: Distributed Group Division

Input: : P, r, n
Output: : Group information $\{C_i\}_{i=1}^{K(t)}$
for *each i distributively* **do**
    Identify i's neighbor set N_i within r_c
    Send information $I_i = \{i, N_i\}$ to robot i's neighbors, and reveice information $I_j = \{j, N_j\}$ from its neighbors($j \in N_i$)
    Compute the intersection set between I_i and each I_j, and the size of this set not exceeding n, and denote it as i's group C_i
    Send information C_i to i's neighbors

Algorithm 3: LRO: Local Robust Optimization

Input: : $\mathcal{X}_k$, f, $d^{(t)}$
Output: : $S_k^{(t)}$
$A_k^{(t)} \leftarrow \emptyset,\ U_1^{(t)} \leftarrow \mathcal{X}_k$
while $U_1^{(t)} \neq \emptyset$ **do**
    $a \in \arg\max_{s \in U_1^{(t)}} f(s)$
    if $|A_k^{(t)} \cup \{a\}| \leq d^{(t)}$ *and* $|(A_k^{(t)} \cup \{a\}) \cap \mathcal{X}_i^{(t)}| \leq 1$ *for all* $i \in [R]$ **then**
        $A_k^{(t)} \leftarrow A_k^{(t)} \cup \{a\}$
    $U_1^{(t)} \leftarrow U_1^{(t)} \setminus \{a\}$
$B_k^{(t)} \leftarrow \emptyset,\ U_2^{(t)} \leftarrow \mathcal{X}_k \setminus A_k^{(t)}$
while $U_2^{(t)} \neq \emptyset$ **do**
    $b \in \arg\max_{s \in U_2^{(t)}} \Delta f_s(S^{(1)} \setminus D^{(1)}, \cdots, S^{(t-1)} \setminus D^{(t-1)}, B_k^{(t)})$
    if $|(A_k^{(t)} \cup B_k^{(t)} \cup \{b\}) \cap \mathcal{X}_i| \leq 1$ *for all* $i \in [R]$ **then**
        $B_k^{(t)} \leftarrow B_k^{(t)} \cup \{b\}$
    $U_2^{(t)} \leftarrow U_2^{(t)} \setminus \{b\}$
$S_k^{(t)} \leftarrow A_k^{(t)} \cup B_k^{(t)}$

The LRO algorithm, detailed in Algorithm 3, takes as input robot decisions, a target function oracle, and the number of robots to be removed, and outputs a decision set for the current time step. The algorithm consists of two phases: one simulates adversarial attacks, and the other performs greedy selection. In the first phase, a solution set is constructed by selecting the decision with the highest individual value, verifying its feasibility, and ensuring that the attack-induced solution does not exceed the number of allowed attacks. Feasible decisions are added to the set, while others are discarded. The process repeats until no decisions remain. In the second phase, a greedy solution set is built by selecting decisions with the highest marginal gain. Each candidate is tested for feasibility and included if valid. The process continues until all decisions have been evaluated. Finally, the two sets from both phases are merged to form the decision set for the current time step.

4 Analysis

In the following section, we present the approximation guarantee and analyze the computational complexity of Algorithm 1. Due to space constraints, the full proofs of these lemmas and theorem 1 appear in the full version of this paper.

Theorem 1. *Algorithm 1 output $S^{[1:T]}$ such that $|S^{(t)} \cap \mathcal{X}_i^{(t)}| = 1, \forall i \in [R]$, for $t \in [T]$, and $D^{[1:T]}$ is an optimal deletion from $S^{[1:T]}$. If f has the generalized curvature c_f, then*

$$f(S^{[1:T]} \setminus D_\star^{[1:T]}) \geq (1 - c_f)^5 \cdot f_\star \ (T \geq 1); \tag{4}$$

If f is submodular, then

$$f(S^{[1:T]} \setminus D_\star^{[1:T]}) \geq (1-\kappa_f)^4 \cdot f_\star \ (T > 1); \tag{5}$$

$$f(S^{[1:T]} \setminus D_\star^{[1:T]}) \geq \frac{(1-\kappa_f)^2}{1+\kappa_f} \cdot f_\star \ (T = 1). \tag{6}$$

We compare our algorithm with an existing method [16] under varying curvature conditions and highlight its general applicability to dynamic settings.

Remark 1. When $T = 1$, the problem reduces to a multi-target tracking task with static targets. The algorithm proposed in [16] addresses this specific case. Our algorithm is similar to online approaches for dynamic multi-target tracking over time but is designed to handle more general cases. We compare the approximation ratios of the two algorithms when the objective function is submodular, as shown in Fig. 1. In the figure, the blue and yellow lines represent the approximation ratios of the two algorithms, respectively. The horizontal axis denotes the curvature of the objective function, while the vertical axis indicates the corresponding theoretical approximation ratios. When the curvature equals 1/3, both algorithms offer the same approximation guarantee. For curvature values below 1/3, our algorithm demonstrates significantly better performance. However, when the curvature lies in the interval (1/3, 1), the algorithm from [16] achieves better approximation results. The figure also illustrates approximation ratio results for both submodular and non-submodular cases when $T > 1$. These findings indicate that while our algorithm may not always outperform existing methods in specific scenarios, it applies to more general problems. Moreover, its approximation ratios exhibit a structured behavior.

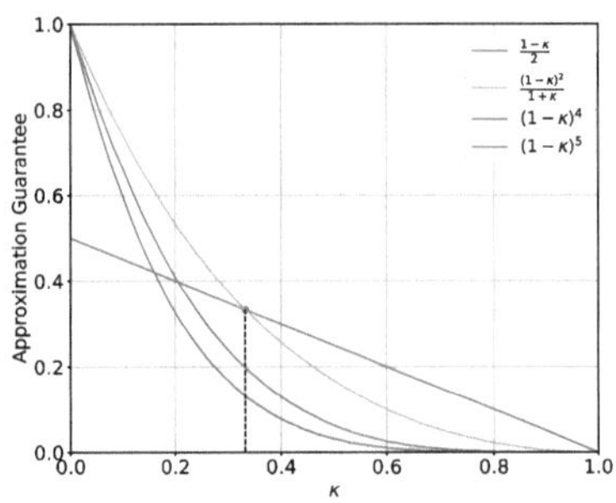

Fig. 1. Plot of value functions regarding approximation guarantees.

Remark 2. As shown in Fig. 1, for objective functions with low curvature ($\kappa < 1/3$), our algorithm achieves strong approximation performance, improving as κ approaches zero. However, for high-curvature functions, the curvature-dependent approximation ratio deteriorates significantly. In offline, centralized settings, the algorithm from [1] achieves a tight approximation guarantee of $(1/(1+\kappa))$, where κ is the curvature of a non-decreasing submodular function. When the function becomes modular, this algorithm yields the optimal solution. Similarly, the algorithm from [9] achieves a tight bound of $1-c$ in polynomial time, where c is the total curvature. When $T=1$, (6) does not reach the tight bound in the offline centralized setting. The tightness and optimality of (4) and (5) remain open questions. Specifically, (6) approaches 1 as $\kappa_f \to 0$, indicating the algorithm may find the optimal solution when the objective function becomes modular. While this is theoretically possible, achieving optimality in a distributed setting requires careful consideration of local optimization and communication strategies. A well-designed distributed protocol could enable the algorithm to match centralized performance. However, practical challenges such as dynamic environments, inconsistent information, and communication delays can degrade performance. Thus, although optimality is theoretically attainable, robust algorithms are essential to address real-world constraints.

At each time step t, the solution produced by our algorithm consists of two components. Given the partial solution $A^{(T)}$, we establish an inequality relation between the partial solution $B^{(T)}$ and any feasible solution that satisfies the same constraints.

Lemma 1. *Let $A^{(1)},\cdots,A^{(T)}$ be the sets selected by Algorithm 1. We denote any subset of remaining elements $\mathcal{X}^{(t)}\setminus A^{(t)}$ by $U^{(t)}$ such that $|(U^{(t)}\cup A^{(t)})\cap \mathcal{X}_i^{(t)}|\leq 1$ for each $i\in[R]$, where $t=1,\cdots,T$. Then, we get*

$$\sum_{t=1}^{T}\sum_{k=1}^{K}\sum_{s\in B_k^{(t)}} f(s)\geq (1-c_f)^3 f(U^{[1:T]}).$$

If f is submodular, then we have

$$\sum_{t=1}^{T}\sum_{k=1}^{K}\sum_{s\in B_k^{(t)}} f(s)\geq (1-\kappa_f)^2 f(U^{[1:T]}).$$

Lemma 2. *Let $A^{(1)},\cdots,A^{(T)}$ denote the sets selected by Algorithm 1. We consider the ground set to be $G^{(t)}=\mathcal{X}^{(t)}\setminus A^{(t)}$ at time t and $\bar{\mathcal{I}}^{(t)}=\{S\mid |(S\cup A^{(t)})\cap\mathcal{X}_i^{(t)}|\leq 1,\forall i\in[R]\}$ at time t. Also, we denote*

$$O_\star^{[1:T]}\in\arg\max_{\substack{O^{(1)}\subseteq G^{(1)},\\ O^{(1)}\in\bar{\mathcal{I}}}}\cdots\max_{\substack{O^{(T)}\subseteq G^{(T)},\\ O^{(T)}\in\bar{\mathcal{I}}^{(T)}}} f(O^{[1:T]}),\tag{7}$$

such that $O_\star^{(t)}\subseteq G^{(t)}$ and $O_\star^{(t)}\in\bar{\mathcal{I}}^{(t)}$. Let $O_g^{[1:T]}$ be the solution of the Online Greedy algorithm which is shown in the full version of this paper. Then, we have

$$f(O_g^{[1:T]})\geq(1-c_f)f(O_\star^{[1:T]}).\tag{8}$$

If f is submodular, then we have

$$f(O_g^{[1:T]})\geq(1-\kappa_f)f(O_\star^{[1:T]}).\tag{9}$$

Lemma 3. *Let $A^{(1)}, \cdots, A^{(T)}$ and $B^{(1)}, \cdots, B^{(T)}$ be the sets selected by Algorithm 1. We consider the ground set to be $G^{(t)} = \mathcal{X}^{(t)} \setminus A^{(t)}$ at time t and $\bar{\mathcal{I}}^{(t)} = \{S \mid |(S \cup A^{(t)}) \cap \mathcal{X}_i^{(t)}| \leq 1, \forall i \in [R]\}$ at time t.*

$$\sum_{t=1}^{T} \sum_{k=1}^{K} \sum_{s \in B_k^{(t)}} f(s) \geq (1 - c_f)^4 f(O_\star^{[1:T]}). \tag{10}$$

If f is submodular, then we have

$$\sum_{t=1}^{T} \sum_{k=1}^{K} \sum_{s \in B_k^{(t)}} f(s) \geq (1 - \kappa_f)^3 f(O_\star^{[1:T]}). \tag{11}$$

Lemma 4. *We continue with the notation from the above lemma. We consider the ground set to be $G^{(t)} = \mathcal{X}^{(t)} \setminus A^{(t)}$ at time t and $\bar{\mathcal{I}}^{(t)} = \{S \mid |(S \cup A^{(t)}) \cap \mathcal{X}_i^{(t)}| \leq 1, \forall i \in [R]\}$ at time t. Thus, it holds*

$$f(O_\star^{[1:T]}) \geq f_\star. \tag{12}$$

Remark 3. (Computational and communications performance). At any time $t \in [T]$ of the execution of the algorithm DRSSM, the computational complexity of the algorithm, i.e., the time required for evaluations of the objective function, is $O(|\mathcal{X}^{(t)}|^2)$, where $\mathcal{X}^{(t)} = \bigcup_{i=1}^{|R|} \mathcal{X}_i^{(t)}$. At each execution time $t \in [T]$, every robot $i \in [R]$ has five-round communications, with three rounds occurring in the algorithm DGD and two rounds in the algorithm LRO. Within C_k, each robot $i \in C_k$ exchanges $3|N_i| + 2(|C_k| - 1)$ messages through distributed communication.

5 Experiments

Among the real-world applications of the studied mathematical model, this section focuses on evaluating a representative scenario in which a group of UAVs performs multi-target tracking tasks. In the active target tracking scenario, the objective is to maximize the number of targets covered, and the submodularity of the target function for coverage naturally holds. The proposed algorithm is implemented in Python. Due to limited computational resources, we approximate the runtime by dividing the total execution time by the number of groups, assuming parallel operation. The runtime includes both communication and computation costs. We compare the centralized greedy algorithm, the centralized robust algorithm, and our algorithm. The primary difference between the first two classical algorithms is that the former is not resistant to attacks, whereas the latter is. The comparison is made with these two algorithms to show the difference in their algorithmic performance and runtime.

We consider N_r UAVs tasked with tracking N_t ground-moving targets. UAVs may encounter attacks or malfunctions, with N_a representing the number of UAVs encountering potential risks. Each UAV has a communication range r_c

and moves in a fixed plane, avoiding collisions. To simplify complex motion interference and focus on validating algorithm performance, we assume that at any given moment, a UAV can choose from four movement strategies: forward, backward, left, or right. Each UAV has a square field of vision of $c \times c$, which extends to $b = c + a$ in the chosen direction, where a is the distance moved. We assume that UAVs can sense and estimate target positions, and the objective function f is based on the expected number of targets covered. Over T time steps, some UAVs may fail or be attacked; affected UAVs cannot operate at that moment but can be reactivated and resume tasks in the next time step.

Our experiments investigate worst-case attacks. Due to the large problem scale, brute-force search for identifying worst-case attack scenarios is computationally infeasible. Specifically, given the trajectory allocation of all UAVs, computing the worst-case attack strategy corresponds to minimizing a non-increasing submodular function—an NP-hard problem. Therefore, we assume that attackers employ a greedy heuristic approach instead of executing true worst-case attacks. The greedy heuristic algorithm can effectively simulate the core impact of attacks. This approach may overestimate the robustness of the results to some extent, but the experimental objective is to compare and validate computational efficiency and relative robustness, focusing on the relative performance evaluation of algorithms, with the greedy attack serving as a consistent evaluation benchmark.

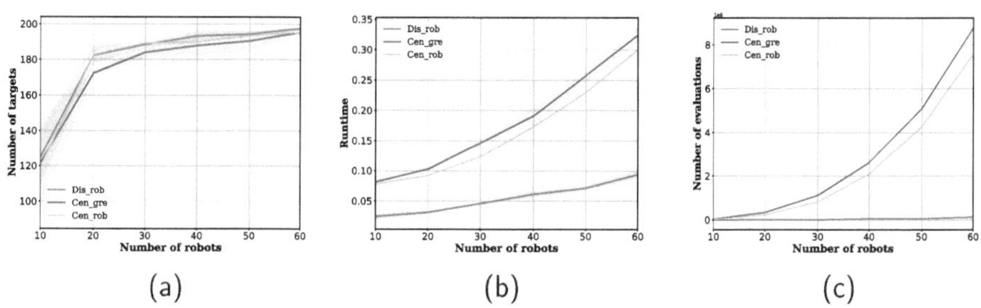

Fig. 2. Comparison when the number of robots changes.

Figure 2 shows the relationship between the number of robots and the number of tracking targets. As the number of robots increases, all three algorithms track more targets. A sharp increase is observed between 10 and 20 robots, followed by a slower growth. Both "Cen_rob" and "Dis_rob" consistently outperform "Cen_gre" in target count, with "Dis_rob" slightly exceeding "Cen_rob". Notably, "Dis_rob" exhibits a smaller confidence interval across multiple runs, indicating greater stability. Figure 2 bshows the runtime versus the number of robots. "Dis_rob" achieves the lowest runtime and scales more gradually with robot count compared to the other two methods. Figure 2 cpresents the number of function evaluations. "Dis_rob" requires significantly fewer evaluations and shows slower growth as the number of robots increases.

Due to space constraints, the results and analysis of the remaining three experiments are provided in the full version of this paper. In summary, we conduct a comprehensive performance evaluation of the proposed distributed algorithm through a series of numerical experiments. The results demonstrate that it significantly outperforms both centralized non-robust and centralized robust algorithms in tracking accuracy, computational efficiency, and the number of objective function evaluations.

6 Conclusion

This study addresses the challenge of sequential decision-making by multiple robots under adversarial attacks or failures, formulating it as a robust sequential submodular maximization problem with constraints. To solve this, we propose a distributed approximation algorithm based on a divide-and-conquer strategy, where each subproblem follows a robust selection process. Theoretical analysis demonstrates that the algorithm has approximation guarantees that depend on the total curvature of the objective function. Experimental validation on multi-target tracking demonstrates the algorithm's superior solution quality, faster runtime, and enhanced stability compared to centralized robust and non-robust methods. These results provide a strong theoretical and empirical foundation for distributed multi-stage robust decision-making.

Acknowledgements. This work was partially supported by Beijing Natural Science Foundation (Nos. Z220004, IS24001) and National Natural Science Foundation of China (No. 12131003).

References

1. Conforti, M., Cornuéjols, G.: Submodular set functions, matroids and the greedy algorithm: tight worst-case bounds and some generalizations of the rado-edmonds theorem. Discret. Appl. Math. **7**(3), 251–274 (1984)
2. Gera, B., Raghuvanshi, Y.S., Rawlley, O., Gupta, S., Dua, A., Sharma, P.: Leveraging AIâĂŘenabled 6GâĂŘdriven IoT for sustainable smart cities. Int. J. Commun. Syst. **36**(16) (2023). https://doi.org/10.1002/dac.5588
3. Kamruzzaman, M.M.: Key technologies, applications and trends of internet of things for energy-efficient 6G wireless communication in smart cities. Energies **15**(15), 5608 (2022). https://doi.org/10.3390/en15155608
4. Li, P., Fan, J., Wu, J.: Exploring the key technologies and applications of 6g wireless communication network. iScience **28**(5) (2025)
5. Schlotfeldt, B., Tzoumas, V., Thakur, D., Pappas, G.: Resilient active information gathering with mobile robots. In: Proc. IEEE Int. Conf. Intell. Rob. Syst., pp. 4309–4316 (2018)
6. Shi, Q., et al.: Robust sequence networked submodular maximization. In: Proc. 37th AAAI Conf. Artif. Intell. vol. 37, pp. 15100–15108 (2023). https://doi.org/10.1609/aaai.v37i12.26762

7. Sun, H., Hu, L., Shou, W., Wang, J.: Self-organized crowd dynamics: research on earthquake emergency response patterns of drill-trained individuals based on GIS and multi-agent systems methodology. Sensors **21**(4), 1353 (2021). https://doi.org/10.3390/s21041353
8. Sun, S., Li, W., Wang, Y., Liao, W., Yu, P.S.: Continuous monitoring of maximum clique over dynamic graphs. IEEE Trans. Knowl. Data Eng. **34**(4), 1667–1683 (2020)
9. Sviridenko, M., Vondrák, J., Ward, J.: Optimal approximation for submodular and supermodular optimization with bounded curvature. Math. Oper. Res. **42**(4), 1197–1218 (2017). https://doi.org/10.1287/moor.2016.0842
10. Tzoumas, V., Gatsis, K., Jadbabaie, A., Pappas, G.: Resilient monotone submodular function maximization. In: Proc. 56th IEEE Conf. Decis. Control, pp. 1362–1367 (2017)
11. Tzoumas, V., Jadbabaie, A., Pappas, G.: Resilient non-submodular maximization over matroid constraints. arXiv preprint arXiv:1804.01013 (2018). https://doi.org/10.48550/arXiv.1804.01013
12. Tzoumas, V., Jadbabaie, A., Pappas, G.J.: Robust and adaptive sequential submodular optimization. IEEE Trans. Autom. Control **67**(1), 89–104 (2022). https://doi.org/10.1109/tac.2020.3046222
13. Wete, E., Greenyer, J., Kudenko, D., Nejdl, W., Flegel, O., Eisner, D.: A tool for the automation of efficient multi-robot choreography planning and execution. In: Proc. 25th ACM/IEEE Int. Conf. Model Driven Eng. Lang. Syst., MODELS : Companion Proc, pp. 37–41 (2022). https://doi.org/10.1145/3550356.3559090
14. Xu, W., et al.: Reward maximization for disaster zone monitoring with heterogeneous uavs. IEEE/ACM Trans. Networking **32**(1), 890–903 (2023)
15. Zhou, L., Tzoumas, V., Pappas, G., Tokekar, P.: Resilient active target tracking with multiple robots. IEEE Robot. Autom. Lett. **4**(1), 129–136 (2018)
16. Zhou, L., Tzoumas, V., Pappas, G., Tokekar, P.: Distributed attack-robust submodular maximization for multirobot planning. IEEE Trans. Rob. **38**(5), 3097–3112 (2022)

Approximation Algorithms for the Parallel-Machine Customer Order Scheduling with Weighted and Submodular Rejection Penalties

Wencheng Wang[1], Tianjin Lu[1], and Xiaofei Liu[2,3](✉)

[1] School of Mathematics and Computer Science, Yunnan Minzu University, No.2929 Yuehua Street, Kunming 650504, Yunnan, China
wencheng@ymu.edu.cn

[2] School of Information and Engineering, Yunnan University, East Outer Ring South Road, Kunming 650500, Yunnan, China

[3] Yunnan Key Laboratory of Intelligent Systems and Computing, Kunming 650500, Yunnan, China
lxfjl2016@163.com

Abstract. In this paper, we consider the parallel-machine customer order scheduling with weighted and submodular rejection penalties, which is formally defined as follows. We are given a set $\mathcal{O}$ of n customer orders and a set of m dedicated machines. Each order consists of m different product types and has an associated weight, where each product type must be processed on a corresponding dedicated machine. Each order can either be rejected, in which case a rejection penalty must be paid, or accepted and processed on the m dedicated machines. If an order is accepted, the weighted completion time is computed as the product of its completion time and its weight. The completion time of an order is defined as the maximum completion time among all its product types. The objective is to select a rejected order set $\mathcal{R} \subseteq \mathcal{O}$ and schedule the remaining orders $\mathcal{A} = \mathcal{O} \setminus \mathcal{R}$ such that the sum of the maximum weighted completion time $W_{\max}(\mathcal{A})$ of the accepted orders in $\mathcal{A}$ and the penalty $\pi(\mathcal{R})$ is minimized, where the penalty is determined by a nondecreasing submodular function $\pi(\cdot)$. We present a deterministic polynomial-time $\frac{e}{e-1}$-approximation algorithm based on the randomized (α, β)-rounding algorithm.

Keywords: Order scheduling · Submodular function · Approximation algorithms

1 Introduction

Given a set of n customer orders $\mathcal{O} = \{O_1, O_2, \cdots, O_n\}$ and a set of m dedicated machines $\mathcal{M} = \{M_1, M_2, \cdots, M_m\}$ in parallel, each order $O_j \in \mathcal{O}$ consists of m distinct types of products, denoted as $O_{1j}, O_{2j}, \ldots, O_{mj}$, where the i-th type of product O_{ij} can only be processed non-preemptively on the dedicated parallel machine M_i,

M. Li et al. (Eds.): TAMC 2025, LNCS 16084, pp. 178–191, 2026.
https://doi.org/10.1007/978-981-95-4839-2_14

and the processing time of O_{ij} on machine $M_i \in \mathcal{M}$ is denoted by p_{ij}, *i.e.*, the time required for processing O_j on M_i. Each order $O_j \in \mathcal{O}$ has a associated weight w_j. The completion time of order $O_j \in \mathcal{O}$ is denoted by C_j and is the time when the processing of all the products for order O_j have been completed. That is $C_j = \max_{i:1\leq i\leq m}\{C_{ij}\}$, where C_{ij} denote the finish time of product O_{ij} of order O_j. With this notation, various objective functions of C_j can be defined similarly to those in classical scheduling models. A typical objective is to find a schedule for $\mathcal{O}$ that minimizes the total weighted completion time, i.e., $\sum w_j C_j$. Following the notation in [11], the problem is denoted as $PD||\sum w_j C_j$ when the number of machines m is arbitrary, and as $PD_m||\sum w_j C_j$ when m is fixed.

The total weighted completion time order scheduling problem was originally introduced by Ahmadi and Bagchi [1]. The same authors [2] showed that $PD_2||\sum w_j C_j$ is strongly NP-hard. Leung et al. [11] further demonstrated that $PD_m||\sum C_j$ remains strongly NP-hard for $m \geq 3$. Ahmadi et al. [3] presented four m-approximation algorithms for the $PD_m||\sum w_j C_j$ problem. Wang and Cheng [25] presented an LP-based algorithm which has an approximation ratio of $16/3$. Recently, Leung et al. [13] obtained a 2-approximation LP-based algorithm for the $PD_m||\sum w_j C_j$ problem. Lu et al. [24] considered the single machine scheduling problem with rejection to minimize the weighted makespan. Based on the relaxed integral programming, they propose a 2-approximation algorithm for this problem. This model has many applications, including component assembly in electronics manufacturing and sintering in electric-ceramic production [12, 15, 30].

In many practical scenarios, constrained production capacity or strict delivery deadlines force manufacturers to accept only a portion of jobs, assigning the selected ones to machines for processing. Chen and Li [4] in 2020 studied the customer order scheduling problem with delivery time and rejection, in which each customer order can be either rejected and paid a rejection penalty, or accepted and processed on dedicated machines. It is asked to find a rejected order set and schedule the remaining orders such that the sum of the maximum delivery completion time of accepted orders and the total penalty of rejected orders is minimized, where delivery completion time is defined as the sum of its complete processing time and its delivery time. Chen and Li showed that this problem is *NP*-hard by a transformation from the Exact Cover by 3-sets problem [8] and proposed three approximation algorithms to solve this problem.

Generally, the relationship between the set of rejected objects and the rejection penalties tends to be a nondecreasing set function rather than linear. Submodular optimization has been extensively studied due to its wide range of applications and theoretical significance. Yang et al. [31] considered the regularized two-stage submodular maximization problem in streams. Liu and Li [18] studied multiprocessor scheduling with submodular penalties. Wang and Liu [26] studied parallel-machine scheduling with release times and submodular penalties. Du et al. [6] considered the facility location problem with submodular penalties and proposed a primal-dual combinatorial algorithm with a 3-approximation guarantee. Liu et al. [20] addressed the minimum power cover problem with both submodular and linear penalties, proposing a combinatorial approximation algorithm based on a primal-dual scheme [14, 28]. Zheng et al. [32] considered parallel-machine customer order scheduling with delivery time and submodular rejection penalties problem. They proposed an LP-rounding algorithm achieving an approximation ratio of $n + 1$ for the scheduling problem. Recently, Zheng et al. [10] presented a noncombinatorial 2-approximation algorithm for this problem. Xu et al. [29] consider the submodular vertex cover problems with linear and submodular penalties. Liu and Li [19] considered the K-prize-collecting multicut problem in trees with submodular penalties and presented a combinatorial approximation algorithm.

Motivated by these optimization problems mentioned above, Wang et al. [27] considered the parallel-machine customer order scheduling with weighted and submodular rejection penalties. Given a set $\mathcal{O}$ of n customer orders and a set of m dedicated machines, each order consists of m different product types and has an associated weight, where each product type must be processed on a corresponding dedicated machine. Each order can either be rejected, in which case a rejection penalty must be paid, or accepted and processed on the m dedicated machines. If an order is accepted, the weighted completion time is computed as the product of its completion time and its weight. The completion time of an order is defined as the maximum completion time among all its product types. The objective is to select a rejected order set $\mathcal{R} \subseteq \mathcal{O}$ and schedule the remaining orders $\mathcal{A} = \mathcal{O} \setminus \mathcal{R}$ such that the sum of the maximum weighted completion time $W_{\max}(\mathcal{A})$ of the accepted orders in $\mathcal{A}$ and the penalty $\pi(\mathcal{R})$ is minimized, where the penalty is determined by a nondecreasing submodular function $\pi(\cdot)$ with $\pi(\emptyset) = 0$. Wang et al. [27] presented a 2-approximation algorithm for this problem. In this paper, we propose a deterministic polynomial-time $\frac{e}{e-1}$-approximation algorithm for the problem by combining Lov sz rounding [22] with the randomized (α, β)-rounding algorithm [9, 17]. This approximation ratio is better than that obtained in [27].

The remaining sections of this paper are organized as follows. In Sect. 2, we introduce parallel-machine customer order scheduling with weighted and submodular rejection penalties and formulate this problem as an integer linear program. In Sect. 4, we present both a randomized (α, β)-rounding algorithm and a deterministic polynomial-time approximation algorithm. Finally, we provide a brief conclusion.

2 Preliminaries

Given a finite set $\mathcal{O}$, let $\pi : 2^{\mathcal{O}} \rightarrow \mathbb{R}$ be a real-valued function defined on the power set of $\mathcal{O}$ with $\pi(\emptyset) = 0$. We assume that $\pi(\cdot)$ is given as an evaluation oracle, which returns the value of $\pi(\mathcal{R})$ for any $\mathcal{R} \subseteq \mathcal{O}$ in polynomial time. If

$$\pi(\mathcal{R}_1 \cup \mathcal{R}_2) + \pi(\mathcal{R}_1 \cap \mathcal{R}_2) \leqslant \pi(\mathcal{R}_1) + \pi(\mathcal{R}_2), \ \ \forall \mathcal{R}_1, \mathcal{R}_2 \subseteq \mathcal{O},$$

function $\pi(\cdot)$ is called **submodular**. If

$$\pi(\mathcal{R}_1) \leqslant \pi(\mathcal{R}_2), \forall \mathcal{R}_1 \subseteq \mathcal{R}_2 \subseteq \mathcal{O},$$

function $\pi(\cdot)$ is called **nondecreasing**.

We are given a set of n customer orders $\mathcal{O} = \{O_1, O_2, \cdots, O_n\}$ and a set of m dedicated machines $\mathcal{M} = \{M_1, M_2, \cdots, M_m\}$ in parallel. Each order $O_j \in \mathcal{O}$ consists of m distinct types of products, denoted as $O_{1j}, O_{2j}, \ldots, O_{mj}$, where the i-th type of product O_{ij} can only be processed non-preemptively on the dedicated parallel machine M_i, and the processing time of O_{ij} on machine $M_i \in \mathcal{M}$ is denoted by p_{ij}, *i.e.*, the time required for processing O_j on M_i. Each order $O_j \in \mathcal{O}$ can be either rejected and paid a rejection penalty, or accepted and processed on the m dedicated machines. If order O_j is accepted, its completion time c_j is equal to the maximum completion time of all types of O_j, and its weighted completion time W_j is equal to $w_j \cdot c_j$, where $w_j > 0$ is a weight of O_j which is a given constant. The objective is to select a rejected order set $\mathcal{R} \subseteq \mathcal{O}$ and schedule the remaining orders $\mathcal{A} = \mathcal{O} \setminus \mathcal{R}$ such that the sum of the maximum weighted completion time $W_{\max}(\mathcal{A})$ of the accepted orders in $\mathcal{A}$ and the penalty $\pi(\mathcal{R})$ is minimized, where the penalty is determined by a nondecreasing submodular function $\pi(\cdot)$. We refer to the problem as the parallel-machine customer order scheduling with weighted and submodular rejection penalties. Using the notation introduced by Leung et al. [11], our problem can be denoted by $PD_m \mid w_j \mid W_{\max}(\mathcal{A}) + \pi(\mathcal{R})$.

Then, we introduce the largest weight (LW) rule inspired by [16,24], where the LW rule for any machine is the next processed customer order with the longest weight.

Lemma 1. *For the $PD_m \mid w_j \mid W_{\max}(\mathcal{A}) + \pi(\mathcal{R})$ problem, there exists an optimal solution in which the accepted orders are scheduled according to the LW rule.*

Proof. Suppose that for any optimal solution $(\mathcal{A}^*, \mathcal{R}^*)$, the LW rule is not an optimal sequence for $\mathcal{A}^*$. For convenience, let O_j be the j-th processed order in an optimal sequence for $\mathcal{A}^*$. Then, for any $O_j \in \mathcal{R}^*$, we have $j >\mid \mathcal{A}^* \mid$.

Since an optimal sequence for $\mathcal{A}^*$ does not satisfy the LW rule, there must exist at least one pair of adjacent orders that violate this rule. Let O_f and O_{f+1} be the first pair of adjacent orders that are processed in a sequence that does not satisfy the LW *i.e.*, order O_f is processed before order O_{f+1} and $w_f < w_{f+1}$, and let c_j^* be defined as the complete processing time of order O_j based on the processing sequence of the optimal solution.

We interchange the processing sequences of orders O_f and O_{f+1} to obtain a new set of processing sequences. Let c_j' denote the complete processing time of order O_j on the basis of the new processing sequence. Since the processing sequence of the first $f-1$ orders remains unchanged, for any $j < f$, we have $c_j' = c_j^*$. As O_{f+1} is processed earlier, we have $c_{f+1}' < c_{f+1}^*$. Owing to the non-preemptive and non-interruptive nature of the machine processing, the complete processing time of order O_f in the new sequence, c_f', equals c_{f+1}^*. Based on this analysis, for any $j > f+1$, we also have $c_j' = c_j^*$. Thus, we have the following relationships:

$$\begin{cases} w_j \cdot c_j' = w_j \cdot c_j^*, & j < f; \\ w_f \cdot c_f' = w_f \cdot c_{f+1}^* < w_{f+1} \cdot c_{f+1}^* & ; \\ w_{f+1} \cdot c_{f+1}' \leq w_{f+1} \cdot c_{f+1}^* & ; \\ w_j \cdot c_j' = w_j \cdot c_j^*, & j > f+1. \end{cases}$$

According to the above formula, interchanging the processing sequence of orders O_f and O_{f+1} obviously does not increase the maximum weighted completion time. Since the set of rejected orders remains unchanged, this adjustment does not increase the objective function value, *i.e.*, the adjusted processing sequence is still an optimal processing sequence for $\mathcal{A}^*$.

By repeatedly applying the operations mentioned above, we can obtain another optimal processing sequence. Clearly, this sequence conforms to the largest weight first (LW) rule. This conclusion contradicts the initial assumption, and therefore, the theorem holds. □

For convenience, we assume that the orders satisfy

$$w_1 \geqslant w_2 \geqslant \cdots \geqslant w_n. \tag{1}$$

To model the $PD_m \mid w_j \mid W_{\max}(\mathcal{A}) + \pi(\mathcal{R})$problem as an integer program, we define a binary variable x_j for each customer order $O_j \in \mathcal{O}$, where $x_j = 1$ if order O_j is accepted, and $x_j = 0$ otherwise. Similarly, for each subset S, we define a binary variable z_S, such that $z_S = 1$ if S is a rejection set, and $z_S = 0$ otherwise. If order O_j is accepted and processed on the m dedicated machines, based on inequality (1) and the LW rule, its completion time is $c_j = \max_{i:1\leq i\leq m} \sum_{k=1}^{j} p_{ik} \cdot x_k$, and its weighted completion time is

$$W_j = w_j \cdot c_j = w_j \cdot \max_{i:1\leq i\leq m} \sum_{k=1}^{j} p_{ik} \cdot x_k.$$

Thus, $PD_m \mid w_j \mid W_{\max}(\mathcal{A}) + \pi(\mathcal{R})$ can be formulated as the following integer linear program.

$$\begin{aligned} \min \max_{j:O_j \in \mathcal{O}} & \left(w_j \cdot \max_{i:1\leq i\leq m} \sum_{k=1}^{j} p_{ik} \cdot x_k \right) + \sum_{S:S\subseteq \mathcal{O}} \pi(S) \cdot z_S \\ \text{s.t.} \quad & x_j + \sum_{S:O_j \in S \subseteq \mathcal{O}} z_S \geq 1, \quad \forall O_j \in \mathcal{O} \\ & x_j \in \{0,1\}, \quad z_S \in \{0,1\}, \quad \forall O_j \in \mathcal{O}, \quad \forall S \subseteq \mathcal{O}, \end{aligned} \tag{2}$$

where the first constraint ensures that each order is either accepted or in the rejected order set. Note that the total number of variables in the integer linear programming formulation is $n + 2^n$.

3 An $\frac{e}{e-1}$-Approximation Algorithm

In this section, we present a deterministic polynomial-time $\frac{e}{e-1}$-approximation algorithm for this problem. Note that when $\pi(\cdot)$ is a submodular function, we can solve a related convex relaxation of the integer programming formulation (2) in polynomial time using the Lovĺész extension, instead of solving the relaxed integer program (2) with an exponential number of variables.

The Lovĺész extension of $\pi(\cdot)$ is defined as the function $\widehat{\pi}\colon \mathbb{R}^n{}_{\geqslant 0} \to \mathbb{R}_{\geqslant 0}$ satisfying:

$$\begin{aligned} \widehat{\pi}(\boldsymbol{z}) &= \max \sum_{j=1}^{n} z_j \cdot \alpha_j \\ s.t. \quad & \sum_{j:O_j \in S} \alpha_j \leqslant \pi(S), \forall S \subseteq \mathcal{O}, \\ & \alpha_j \geqslant 0. \end{aligned} \tag{3}$$

Property 1. For each subset $S \subseteq \mathcal{O}$, $\widehat{\pi}(I_S)$=$\pi(S)$, where I_S is the characteristic vector of subset S, *i.e.*, $I_S \in \{0,1\}^n$ and if $O_j \in S$, the jth component of I_S is 1; otherwise, the jth component I_S is 0.

Property 2. For any number γ and any vector $\boldsymbol{z}$, $\widehat{\pi}(\gamma \cdot \boldsymbol{z}) = \gamma \cdot \widehat{\pi}(\boldsymbol{z})$.

Property 3. $\widehat{\pi}(\cdot)$ is a monotonically increasing convex function.

Note that for a given vector $\boldsymbol{z} = (z_1, z_2, \ldots, z_n) \in \mathbb{R}^n_{\geq 0}$, the linear program (3) involves an exponentially large number of constraints. However, it is not necessary to solve this linear program directly in order to obtain its optimal solution. We can consider a permutation $L = \{i_1, i_2, \ldots, i_n\}$ such that the inequality $z_{i_1} \geqslant z_{i_2} \geqslant \cdots \geqslant z_{i_n}$ holds. Define $S_j = \{O_{i_1}, O_{i_2}, \ldots, O_{i_j}\}$ for $j = 1, 2, \ldots, n$, where $S_0 = \emptyset$. According to [[7], Lemma 3.1], the optimal solution $\boldsymbol{\alpha}^* = (\alpha_1^*, \alpha_2^*, \ldots, \alpha_n^*)$ to the linear program (3) can be constructed, where $\alpha_j^* = \pi(S_j) - \pi(S_{j-1})$.

Based on $\widehat{\pi}(\cdot)$, we can construct the following convex program and prove that its optimal value is equal to the optimal value of the integer programming problem (2).

$$\begin{aligned} \min \quad & \max_{j:O_j \in \mathcal{O}} \big(w_j \cdot \max_{i:1 \leqslant i \leqslant m} \sum_{k=1}^{j} p_{ik} \cdot x_k\big) + \widehat{\pi}(\boldsymbol{z}) \\ s.t. \quad & x_j + z_j \geqslant 1, \ \forall O_j \in \mathcal{O}, \\ & x_j, z_j \in \{0,1\}, \ \forall O_j \in \mathcal{O}. \end{aligned} \tag{4}$$

Lemma 2.

$$OPT_{CP} = OPT,$$

where OPT_{CP} and OPT denote the optimal values of the convex program (4) and the integer programming problem (2), respectively.

Proof. Let $(\boldsymbol{x}, \boldsymbol{z})$ be an arbitrary feasible solution to the convex program (4), and let its objective function value be $OPT_{CP}(\boldsymbol{x}, \boldsymbol{z})$. On the basis of $(\boldsymbol{x}, \boldsymbol{z})$, we construct a feasible solution $(\boldsymbol{x}', \boldsymbol{z}')$ for the integer programming problem (2) via the following strategy and denote its objective function value as $OPT(\boldsymbol{x}', \boldsymbol{z}')$. We will now prove that $OPT_{CP}(\boldsymbol{x}, \boldsymbol{z}) = OPT(\boldsymbol{x}', \boldsymbol{z}')$.

We define $\boldsymbol{x} = \boldsymbol{x}'$, and for each $S \subseteq \mathcal{O}$, if $I_S = \boldsymbol{z}$, the corresponding component of $\boldsymbol{z}'$ is set to 1, that is $z_S = 1$; otherwise, it is set to 0.

Since $\boldsymbol{x} = \boldsymbol{x}'$, and $\widehat{\pi}(I_S)$=$\widehat{\pi}(\boldsymbol{z}) = z_S \cdot \pi(S) = \pi(S)$ according to Property 1, we have $OPT_{CP}(\boldsymbol{x}, \boldsymbol{z}) = OPT(\boldsymbol{x}', \boldsymbol{z}')$.

Conversely, let $(\boldsymbol{x}', \boldsymbol{z}')$ be a feasible solution to the integer programming problem (2). Similarly, we can construct a feasible solution $(\boldsymbol{x}, \boldsymbol{z})$ to the convex program (4) and prove that $OPT_{CP}(\boldsymbol{x}, \boldsymbol{z}) = OPT(\boldsymbol{x}', \boldsymbol{z}')$.

Specifically, we define $\boldsymbol{x} = \boldsymbol{x}'$, and for the $S \subseteq \mathcal{O}$ with $z_S = 1$, we set $\boldsymbol{z} = I_S$. Clearly, $OPT_{CP}(\boldsymbol{x}, \boldsymbol{z}) = OPT(\boldsymbol{x}', \boldsymbol{z}')$.

This completes a proof of the lemma. □

Using the ellipsoid method, we can find an optimal solution $(\boldsymbol{x}^*, \boldsymbol{z}^*)$ of the relaxation of convex program (4) in polynomial-time. Inspired by the rounding framework in [5,9,17], we first design a randomized rounding algorithm based on $(\boldsymbol{x}^*, \boldsymbol{z}^*)$.

(1) Initialize a parameter $\alpha \in (0, 1)$.

(2) Find an optimal solution $(\boldsymbol{x}^*, \boldsymbol{z}^*) = (x_1^*, x_2^*, \ldots, x_n^*, z_1^*, z_2^*, \ldots, z_n^*)$ of the convex relaxation of (4).

(3) Select a parameter β uniformly at random from the interval $(0, \alpha]$.

(4) Output a feasible solution $(\mathcal{A}', \mathcal{R}')$, where $\mathcal{R}' = \{O_j \in \mathcal{O} \mid x_j^* < 1 - \beta\}$ is the rejected order set, and where $\mathcal{A}' = \mathcal{O} \setminus \mathcal{R}$ is the remaining order set, in which the orders in $\mathcal{A}'$ are processed on the dedicated machines according to the LW rule.

For the randomized rounding algorithm, we can determine the following results.

Lemma 3.

$$\mathbb{E}(W_{\max}(\mathcal{A})) \leqslant \frac{1}{\alpha} \ln(\frac{1}{1-\alpha}) W_{\max}(\boldsymbol{x}^*).$$

Proof. To obtain this inequality, we define $I_{\mathcal{A}}$ as the characteristic vector of the subset $\mathcal{A}$, where $I_{\mathcal{A}} \in \{0, 1\}^n$. The j-th entry of $I_{\mathcal{A}}$, denoted as $I_{\mathcal{A}_j}$, is 1 if $O_j \in \mathcal{A}$ and 0 otherwise, for each $O_j \in \mathcal{O}$. Based on the randomized rounding algorithm, if $O_j \in \mathcal{A}$, then $x_j^* \geqslant 1 - \beta$, and we have

$$I_{\mathcal{A}_j} = 1 \leq \frac{x_j^*}{1 - \beta}.$$

Since the processing sequence of the orders in $\mathcal{A}$ is determined based on the LW rule, we have

$$\begin{aligned} W_{\max}(\mathcal{A}) &= \max_{j:O_j \in \mathcal{A}} \max_{1 \leqslant i \leqslant m} \{w_j c_j\} \\ &= \max_{j:O_j \in \mathcal{A}} \max_{1 \leqslant i \leqslant m} \{w_j \sum_{k:\substack{1 \leqslant k \leqslant j; \\ O_k \in \mathcal{A}}} p_{ik}\} \\ &= \max_{j:O_j \in \mathcal{A}} \max_{1 \leqslant i \leqslant m} \{w_j \sum_{k:\substack{1 \leqslant k \leqslant j; \\ O_k \in \mathcal{A}}} p_{ik} I_{\mathcal{A}_k}\} \end{aligned}$$

$$
\begin{aligned}
&\leqslant \max_{j:O_j\in\mathcal{A}}\max_{1\leqslant i\leqslant m}\{w_j \sum_{\substack{k:\ 1\leqslant k\leqslant j;\\ O_k\in\mathcal{A}}} p_{ik}\cdot\frac{x_k^*}{1-\beta}\} \\
&\leqslant \max_{j:O_j\in\mathcal{O}}\max_{1\leqslant i\leqslant m}\{w_j \sum_{k=1}^{j} p_{ik}x_k^*\}\cdot\frac{1}{1-\beta} \\
&= W_{\max}(\boldsymbol{x}^*)\cdot\frac{1}{1-\beta}.
\end{aligned}
$$

Then, we have

$$
\begin{aligned}
\mathbb{E}(W_{\max}(\mathcal{A})) &\leqslant W_{\max}(\boldsymbol{x}^*)\cdot\mathbb{E}\left(\frac{1}{1-\beta}\right) \\
&= W_{\max}(\boldsymbol{x}^*)\cdot\int_0^{\alpha}\frac{1}{1-\beta}\cdot\frac{1}{\alpha}\,d\beta \\
&= W_{\max}(\boldsymbol{x}^*)\cdot\frac{1}{\alpha}\cdot\ln\left(\frac{1}{1-\alpha}\right).
\end{aligned}
$$

This completes a proof of the lemma. □

Lemma 4.

$$\mathbb{E}(\pi(\mathcal{R}))\leqslant\frac{1}{\alpha}\hat{\pi}(\boldsymbol{z}^*).$$

Proof. Without loss of generality, we assume that $\boldsymbol{z}^*=(z_1^*,z_2^*,\ldots,z_n^*)$, where $z_1^*\geqslant z_2^*\geqslant\cdots\geqslant z_n^*$. Furthermore, we assume that there are exactly l distinct, nonzero values in $\{z_1^*,z_2^*,\ldots,z_n^*\}$, which satisfy $z_{q_1}^*>z_{q_2}^*>\cdots>z_{q_l}^*$, and that $z_{q_{l+1}}^*=0$.

Note that for each $j=1,2,\ldots,n$, there exists an index $j'\in\{1,2,\ldots,l\}$ such that $z_j^*=z_{q_{j'}}^*$. For convenience, we define $q_{j'}=\max\{k\mid z_k^*=z_j^*\}$, which represents the largest index k for which $z_k^*=z_j^*$.

Additionally, for each $j=0,1,2,\ldots,n$, we define $\mathcal{O}_j=\{O_0,O_1,O_2,\ldots,O_j\}$, where $\mathcal{O}_0=\emptyset$. By the comments of the definition of the Lovász extension, we define $q_0=0$. Then, we have

$$
\begin{aligned}
\widehat{\pi}(\boldsymbol{z}^*) &= \sum_{j=1}^{n} z_j^*\alpha_j^* \\
&= \sum_{j=1}^{n} z_j^*\left(\pi(\mathcal{O}_j)-\pi(\mathcal{O}_{j-1})\right) \\
&= \sum_{j=1}^{l}\sum_{j'=q_{j-1}+1}^{q_j} z_{q_{j'}}^*\cdot\left(\pi(\mathcal{O}_{j'})-\pi(\mathcal{O}_{j'-1})\right) \\
&= \sum_{j=1}^{l} z_{q_j}^*\cdot\left(\pi\left(\mathcal{O}_{q_j}\right)-\pi\left(\mathcal{O}_{q_{j-1}}\right)\right).
\end{aligned}
$$

Clearly, for the given constant $\alpha \in (0,1)$, we can find an integer k' satisfying $z^*_{q_{k'+1}} < \alpha \leqslant z^*_{q_{k'}}$ (as shown in Fig. 1). For each $j = 1, 2, \ldots, l$, we define $R_{q_j} = \{O_{j'} \in \mathcal{O} \mid z^*_{j'} \geq z^*_{q_j}\}$ as shown in Fig. 1. These sets satisfy the inclusion relationship $R_{q_1} \subset R_{q_2} \subset \cdots \subset R_{q_l}$.

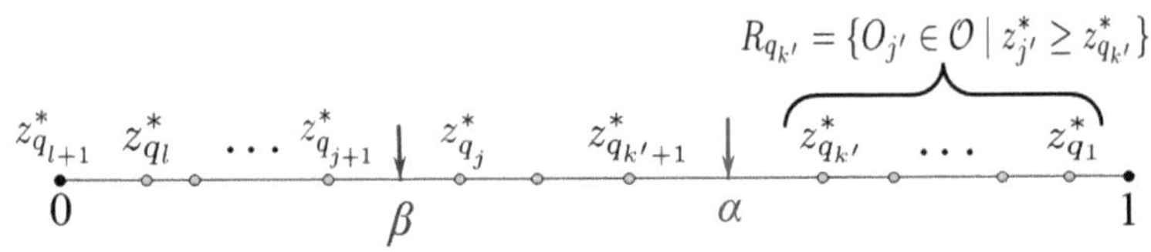

Fig. 1. The process of the proof for Lemma 4.

Now, for each $j = 1, 2, \ldots, l$, we estimate the rejecting probability $\mathbf{Pr}(R = R_{q_j})$. We consider the following three cases.

case 1 $j < k'$.
In this case, since $\beta \leqslant \alpha$, we have $\mathbf{Pr}(R = R_{q_j}) = 0$.

case 2 $j = k'$.
In this case, since $\beta \sim U(0, \alpha)$, R_{q_j} is a rejection set only when $\beta \in (z^*_{q_{j+1}}, \alpha]$. Then, the probability that $R = R_{q_{k'}}$ is given by:

$$\mathbf{Pr}(R = R_{q_{k'}}) = \frac{\alpha - z^*_{q_{k'+1}}}{\alpha}.$$

case 3 $j > k'$.
For this case, since $\beta \sim U(0, \alpha)$, R_{q_j} is a rejection set only when $\beta \in (z^*_{q_{j+1}}, z^*_{q_j}]$. Then, the probability that $R = R_{q_j}$ is given by:

$$\mathbf{Pr}(R = R_{q_j}) = \frac{z^*_{q_j} - z^*_{q_{j+1}}}{\alpha}.$$

For each $j = 1, 2, \ldots, l$, we assume that $I_{R_{q_j}}$ is the characteristic vector of the subset R_{q_j}. Then, we have that

$$\pi(R_{q_j}) = \widehat{\pi}(I_{R_{q_j}}) = \sum_{k=1}^{j} (\pi(\mathcal{O}_{q_k}) - \pi(\mathcal{O}_{q_{k-1}})) = \pi(\mathcal{O}_{q_j}).$$

Thus, we have

$$\begin{aligned}
\mathbb{E}(\pi(R)) &= \sum_{j=1}^{l} \mathbf{Pr}(R = R_{q_j}) \cdot \pi(R_{q_j}) \\
&= \sum_{j=1}^{k'-1} 0 \cdot \pi(R_{q_j}) + \frac{\alpha - z^*_{q_{k'+1}}}{\alpha} \cdot \pi(R_{q_{k'}}) + \sum_{j=k'+1}^{l} \frac{z^*_{q_j} - z^*_{q_{j+1}}}{\alpha} \cdot \pi(R_{q_j}) \\
&= \frac{\alpha - z^*_{q_{k'+1}}}{\alpha} \cdot \widehat{\pi}(I_{R_{q_{k'}}}) + \sum_{j=k'+1}^{l} \frac{z^*_{q_j} - z^*_{q_{j+1}}}{\alpha} \cdot \widehat{\pi}(I_{R_{q_j}}) \\
&= \widehat{\pi}(I_{R_{q_{k'}}}) + \sum_{j=k'+1}^{l} (\widehat{\pi}(I_{R_{q_j}}) - \widehat{\pi}(I_{R_{q_{j-1}}})) \cdot \frac{z^*_{q_j}}{\alpha} \\
&= \sum_{j=1}^{k'} (\pi(\mathcal{O}_{q_j}) - \pi(\mathcal{O}_{q_{j-1}})) + \sum_{j=k'+1}^{l} (\pi(\mathcal{O}_{q_j}) - \pi(\mathcal{O}_{q_{j-1}})) \cdot \frac{z^*_{q_j}}{\alpha} \\
&\leqslant \frac{1}{\alpha} \cdot \sum_{j=1}^{l} z^*_{q_j} \cdot (\pi(\mathcal{O}_{q_j}) - \pi(\mathcal{O}_{q_{j-1}})) = \frac{1}{\alpha} \cdot \widehat{\pi}(z^*),
\end{aligned}$$

where the inequality follows from the following:

$$\frac{z^*_{q_j}}{\alpha} \geqslant 1,$$

which holds because $z^*_{q_j} \geqslant z^*_{q_{k'}}$ for any $j \leq k'$, and $z^*_{q_{k'}} \geqslant \alpha$.

This completes a proof of the lemma. □

Theorem 1.

$$\mathbb{E}(W_{\max}(\mathcal{A}) + \pi(\mathcal{R})) \leqslant \frac{e}{e-1} \cdot OPT$$

where OPT is the optimal value of the $PD_m \mid w_j \mid W_{\max}(\mathcal{A}) + \pi(\mathcal{R})$ problem in which the penalty function $\pi(\cdot)$ is a submodular function.

Proof. Given an instance $(\mathcal{M}, \mathcal{O}, w, p, \pi)$ of the $P \mid w_j \mid W_{\max}(\mathcal{A}) + \pi(\mathcal{R})$ problem and two constants $\alpha \in (0, 1)$ and β uniformly at random from the interval $(0, \alpha]$, we suppose that $(\mathcal{A}', \mathcal{R}')$ is a solution produced by the randomized rounding algorithm and that $(\mathcal{A}^*, \mathcal{R}^*)$ is an optimal solution of the $P \mid w_j \mid W_{\max}(\mathcal{A}) + \pi(\mathcal{R})$ problem for the instance $(\mathcal{M}, \mathcal{O}, w, p, \pi)$. The optimal value and the output value are $W_{\max}(\mathcal{A}^*) + \pi(\mathcal{R}^*)$ and $W_{\max}(\mathcal{A}') + \pi(\mathcal{R}')$, respectively, and we denote them by OPT and OUT. We assume that $(\boldsymbol{x}^*, \boldsymbol{z}^*)$ is an optimal solution of the relaxation of convex program (4), and its objective value is $OPT^* = W_{\max}(\boldsymbol{x}^*)) + \widehat{\pi}(\boldsymbol{z}^*))$, $W_{\max}(\boldsymbol{x}^*)) = \max\limits_{i \in [m], j \in [n]} \left(w_j \sum_{k=1}^{j} p_{ik} \cdot x^*_k \right)$. Now, we construct a feasible solution $(\boldsymbol{x}, \boldsymbol{z})$ to the relaxation of convex program (4) via the following strategy: if $O_j \in \mathcal{A}^*$, then set $x_j = 1$ and $z_j = 0$. Since $(\boldsymbol{x}^*, \boldsymbol{z}^*)$ is the optimal solution of relaxation of convex program (4), we have that $OPT = W_{\max}(\boldsymbol{x}) + \widehat{\pi}(\boldsymbol{z}) \geqslant OPT^* = W_{\max}(\boldsymbol{x}^*) + \widehat{\pi}(\boldsymbol{z}^*)$.

Based on Lemmas 2,3 and 4, we have

$$\begin{aligned}
&\mathbb{E}(W_{\max}(\mathcal{A}') + \pi(\mathcal{R}')) \\
\leqslant &\frac{1}{\alpha} \cdot \ln\left(\frac{1}{1-\alpha}\right) \cdot W_{\max}(\boldsymbol{x}^*) + \frac{1}{\alpha} \cdot \widehat{\pi}(\boldsymbol{z}^*) \\
\leqslant &\max\left\{\frac{1}{\alpha} \cdot \ln\left(\frac{1}{1-\alpha}\right), \frac{1}{\alpha}\right\} \cdot (W_{\max}(\boldsymbol{x}^*) + \widehat{\pi}(\boldsymbol{z}^*)) \\
\leqslant &f(\alpha) \cdot OPT^* \\
\leqslant &f(\alpha) \cdot OPT,
\end{aligned}$$

where $f(\alpha) = \max\{\frac{1}{\alpha}, \frac{1}{\alpha}\ln(\frac{1}{1-\alpha})\}$. Let $\alpha = \frac{e-1}{e}$. Then, we have $f(\alpha) = \frac{e}{e-1}$. This completes a proof of the theorem. □

Now, we present a deterministic polynomial-time approximation algorithm based on the randomized rounding algorithm.

Algorithm 1 : $\mathcal{A}_{Lr}$

Input: An instance $(\mathcal{M}, \mathcal{O}, w, p, \pi)$ of the $PD_m \mid w_j \mid W_{\max}(\mathcal{A}) + \pi(\mathcal{R})$ problem.

Output: A schedule $(\mathcal{A}, \mathcal{R})$ and its associated cost $W_{\max}(\mathcal{A}) + \pi(\mathcal{R})$, where $\mathcal{R}$ denotes the set of rejected orders and $\mathcal{A}$ denotes the set of accepted orders.

1: Find an optimal solution $(\boldsymbol{x}^*, \boldsymbol{z}^*)$ of the relaxation of convex program (4).

2: Sort all elements of the vector $\boldsymbol{x}^*$ in non-decreasing order, and for convenience, assume that $x_1^* \geqslant x_2^* \geqslant \cdots \geqslant x_n^*$. Set $\mathcal{A}_0 := \emptyset$, $\mathcal{R}_0 := \mathcal{O}$, and $OUT_0 := \pi(\mathcal{O})$.

3: Set $\mathcal{A}' := \emptyset$, $\mathcal{R}' := \emptyset$, $OUT := 0$.

4: **for** $j = 1, 2, \cdots, n$ **do**

  Set $\mathcal{A}_j := \mathcal{A}_{j-1} \cup \{O_j\}$, $\mathcal{R}_j := \mathcal{O} \setminus \mathcal{A}_j$, and process the orders in $\mathcal{A}_j$ according to the *LW* rule, $OUT_j := W_{\max}(\mathcal{A}_j) + \pi(\mathcal{R}_j)$.

5: **end for**

6: Set $j' := \arg\min_{0 \leqslant j \leqslant n} \{OUT_j\}$.

7: Let $(\mathcal{A}', \mathcal{R}') := (\mathcal{A}_{j'}, \mathcal{R}_{j'})$, $OUT := OUT_{j'}$.

Theorem 2. *Algorithm 1 is an $\frac{e}{e-1}$-approximation algorithm for the $PD_m \mid w_j \mid W_{\max}(\mathcal{A}) + \pi(\mathcal{R})$ problem in which the penalty function $\pi(\cdot)$ is a submodular function.*

Proof. Based on Theorem 1, when $\frac{1}{\alpha} = \frac{1}{\alpha}\ln\left(\frac{1}{1-\alpha}\right)$, or equivalently $\alpha = \frac{e-1}{e}$, the function $f(\alpha)$, for $\alpha \in (0, 1)$, attains its minimum value, which is $f(\frac{e-1}{e}) = \frac{1}{e-1}$.

Similar to Lemma 4, we assume that $\boldsymbol{z}^* = (z_1^*, z_2^*, \ldots, z_n^*)$, where $z_1^* \geqslant z_2^* \geqslant \cdots \geqslant z_n^*$. Furthermore, we assume that there are exactly l distinct, nonzero values in $\{z_1^*, z_2^*, \ldots, z_n^*\}$, which satisfy $z_{q_1}^* > z_{q_2}^* > \cdots > z_{q_l}^*$, and that $z_{q_{l+1}}^* = 0$.

Note that for each $j = 1, 2, \ldots, n$, there exists an index $j' \in \{1, 2, \ldots, l\}$ such that $z_j^* = z_{q_{j'}}^*$. For convenience, we define $q_{j'} = \max\{k \mid z_k^* = z_j^*\}$, which represents the largest index k for which $z_k^* = z_j^*$.

Note that when $\beta \in (z_{q_{j+1}}^*, z_{q_j}^*)$ and $\beta \leq \alpha$, the randomized rounding algorithm outputs the same solution $(\mathcal{A}, \mathcal{R})$, where

$$\mathcal{R} = \{O_1, O_2, \ldots, O_{q_j}\} \text{ and } \mathcal{A} = \{O_{q_{j+1}}, O_{q_{j+2}}, \ldots, O_n\}.$$

Thus, there are $l+1$ different solutions with expected objective values of no more than $\frac{e}{e-1} \cdot OPT$.

Moreover, Algorithm 1 searches for the solution with the minimum objective value among $n+1$ solutions; these $n+1$ solutions contain the $l+1$ different solutions mentioned above. Thus, by choosing the solution with the minimum objective value, we obtain the desired solution that is no more than $\frac{e}{e-1} \cdot OPT$ in polynomial time.

This completes a proof of the theorem. □

4 Conclusions and Further Research

In this paper, we consider the problem of parallel-machine customer order scheduling with weighted and submodular rejection penalties. We present a deterministic polynomial-time $\frac{e}{e-1}$-approximation algorithm based on the randomized (α, β)-rounding algorithm.

A challenging task for us is to design some better approximation algorithms with smaller performance ratios and lower time complexity for this problem. Furthermore, it is often impossible to obtain the information of all customer orders in advance. Scheduling problems with the online setting are studied [21,23]. Studying this problem with the online setting is much more interesting.

Acknowledgments. This work was supported in part by the Yunnan Fundamental Research Projects [Nos. 202501AT070360 and 202401AU070027], the Open Project Program of Yunnan Key Laboratory of Intelligent Systems and Computing [No. ISC24Y01], and the Project of Yunnan Provincial Department of Education Science Research Fund (No. 2024J0573).

Disclosure of Interests. The authors declare that they have no conflicts of interest.

References

1. Ahmadi, R., Bagchi, U.: Scheduling of Multi-job Customer Orders in Multi-machine Environments. ORSA/TIMS, Philadelphia (1990)
2. Ahmadi, R., Bagchi, U.: Coordinated scheduling of customer orders. Working paper, John E. Anderson Graduate School of Management, University of California, Los Angeles (1993)
3. Ahmadi, R., Bagchi, U., Roemer, T.A.: Coordinated scheduling of customer orders for quick response. Naval Res. Logistics (NRL) **52**(6), 493–512 (2005)
4. Chen, R.-X., Li, S.-S.: Minimizing maximum delivery completion time for order scheduling with rejection. J. Comb. Optim. **40**(4), 1044–1064 (2020). https://doi.org/10.1007/s10878-020-00649-2
5. Dai, B., Li, W.: Vector scheduling with rejection on two machines. Int. J. Comput. Math. **97**(12), 2507–2515 (2020)
6. Du, D., Lu, R., Xu, D.: A primal-dual approximation algorithm for the facility location problem with submodular penalties. Algorithmica **63**(1), 191–200 (2012)
7. Fujishige, S.: Submodular functions and optimization. Elsevier, Amsterdam (2005)
8. Garey, M.R., Johnson, D.S.: Computers and intractability: a guide to NP-completeness. A Guide to NP-Completeness, New York, Computers and Intractability (1979)

9. Geunes, J., Levi, R., Romeijn, H.E., Shmoys, D.B.: Approximation algorithms for supply chain planning and logistics problems with market choice. Math. Program. **130**(1), 85–106 (2011)
10. Hou, B., Zheng, H., Liu, W., Wu, W., Du, D.Z., Gao, S.: Improved approximation algorithm for the parallel-machine customer order scheduling with delivery time and submodular rejection penalties. Optimization Lett. **18**(7), 1747–1754 (2024)
11. Leung, J.Y.T., Li, H., Pinedo, M.: Order scheduling in an environment with dedicated resources in parallel. J. Sched. **8**(5), 355–386 (2005)
12. Leung, J.Y.T., Li, H., Pinedo, M.: Scheduling orders for multiple product types with due date related objectives. Eur. J. Oper. Res. **168**(2), 370–389 (2006)
13. Leung, J.Y.T., Li, H., Pinedo, M.: Scheduling orders for multiple product types to minimize total weighted completion time. Discret. Appl. Math. **155**(8), 945–970 (2007)
14. Li, M., Ran, Y., Zhang, Z.: A primal-dual algorithm for the minimum power partial cover problem. J. Comb. Optim. **44**(3), 1913–1923 (2022)
15. Li, W., Ou, J.: Machine scheduling with restricted rejection: an application to task offloading in cloud-edge collaborative computing. Eur. J. Oper. Res. **314**(3), 912–919 (2024)
16. Li, W.: A best possible online algorithm for the parallel-machine scheduling to minimize the maximum weighted completion time. Asia-Pacific J. Oper. Res. **32**(04), 1550030 (2015)
17. Li, Y., Du, D., Xiu, N., Xu, D.: Improved approximation algorithms for the facility location problems with linear/submodular penalties. Algorithmica **73**(2), 460–482 (2015)
18. Liu, X., Li, W.: Approximation algorithms for the multiprocessor scheduling with submodular penalties. Optimization Lett. **15**(6), 2165–2180 (2021). https://doi.org/10.1007/s11590-021-01724-1
19. Liu, X., Li, W.: An approximation algorithm for the-prize-collecting multicut problem in trees with submodular penalties. Math. Struct. Comput. Sci. **34**(3), 193–210 (2024)
20. Liu, X., Li, W., Dai, H.: Approximation algorithms for the minimum power cover problem with submodular/linear penalties. Theoret. Comput. Sci. **923**, 256–270 (2022)
21. Liu, X., Xiao, M., Li, W., Zhu, Y., Ma, L.: Algorithms for single machine scheduling problem with release dates and submodular penalties. J. Comb. Optim. **45**(4), 105 (2023)
22. Lovász, L.: Submodular functions and convexity, pp. 235–257. Springer Berlin Heidelberg, Berlin, Heidelberg (1983)
23. Lu, L., Ng, C.T., Zhang, L.: Optimal algorithms for single-machine scheduling with rejection to minimize the makespan. Int. J. Prod. Econ. **130**(2), 153–158 (2011)
24. Lu, L., Zhang, L., Ou, J.: Single machine scheduling with rejection to minimize the weighted makespan. In: Algorithmic Aspects in Information and Management: 15th International Conference, AAIM 2021, Virtual Event, December 20–22, 2021, Proceedings 15, pp. 96–110. Springer (2021)
25. Wang, G., Cheng, T.E.: Customer order scheduling to minimize total weighted completion time. Omega **35**(5), 623–626 (2007)
26. Wang, W., Liu, X.: A combinatorial 2-approximation algorithm for the parallel-machine scheduling with release times and submodular penalties. Mathematics **10**(1) (2022)
27. Wang, W., Zhangm, X., Liu, X.: Parallel-machine customer order scheduling with weighted completion time and submodular rejection penalties. University of Science and Technology (Natural Science Edition), Accepted, Joumnal of Huazhong (2025)
28. Xu, D., Wang, F., Du, D., Wu, C.: Approximation algorithms for submodular vertex cover problems with linear/submodular penalties using primal-dual technique. Theoret. Comput. Sci. **630**, 117–125 (2016)
29. Xu, D., Wang, F., Du, D., Wu, C.: Approximation algorithms for submodular vertex cover problems with linear/submodular penalties using primal-dual technique. Theoret. Comput. Sci. **630**, 117–125 (2016)

30. Yang, J.: The complexity of customer order scheduling problems on parallel machines. Comput. Oper. Res. **32**(7), 1921–1939 (2005)
31. Yang, R., Xu, D., Guo, L., Zhang, D.: Regularized two-stage submodular maximization under streaming. Sci. China Inf. Sci. **65**(4), 1–10 (2022). https://doi.org/10.1007/s11432-020-3420-9
32. Zheng, H.Y., Gao, S.G., Liu, W., Hou, B.: An approximation algorithm for the parallel-machine customer order scheduling with delivery time and submodular rejection penalties. J. Operations Res. Soc. China **12**(2), 495–504 (2024)

Parallelizing Scheduling Algorithms for Resource Allocation Under V-RAN

Qinqin Gong[1], Xiankun Yu[1], Donglei Du[2], Dachuan Xu[1], and Ruiqi Yang[1](✉)

[1] Institute of Operations Research and Information Engineering, Beijing University of Technology, Beijing 100124, People's Republic of China
gongqinqin@emails.bjut.edu.cn, {xudc,yangruiqi}@bjut.edu.cn

[2] Faculty of Business Administration, University of New Brunswick, Fredericton, NB E3B 5A3, Canada
ddu@unb.ca

Abstract. This study investigates data transmission in a virtualized radio access network (RAN) architecture, where a Central Unit (CU) is interconnected with Remote Units (RUs) via a mid-haul Passive Optical Network (PON), and user equipments (UEs) are wirelessly served by the RUs over the air interface. In this architecture, data transmission follows a two-stage process. Stage 1: The Central Unit (CU) buffers incoming data until the scheduler triggers transmission to the Remote Units (RUs) via the mid-haul Passive Optical Network (PON), which is subject to capacity constraints. Stage 2: Each RU then delivers the data to user equipments (UEs) over the air interface, with wireless transmission rates that exhibit time-varying characteristics. The objective is to maximize the aggregate utility of all users, where utility is modeled as a function of long-term data rates. This problem was first investigated by (Sinha et al. 2019), who proposed a linear programming (LP)-based 1/2-approximation algorithm. Subsequently, (Sinha et al. 2019) demonstrated that the problem can be reduced to a submodular maximization problem with partition matroid constraints, and accordingly developed a greedy-based 1/2-approximation algorithm. In this study, we focus on parallel algorithms design for this problem and propose an improved parallel algorithm with an approximation ratio approaching $1/(1+\alpha)$, where α denotes the curvature parameter of the submodular utility function.

Keywords: RAN architecture · Resource allocation · Submodular maximization · Parallel algorithms

1 Introduction

We address the optimal scheduling problem arising in the virtualized Radio Access Network (RAN) architecture for communication networks, which aligns with the principle of "long wires and short wireless". In this architecture, a

Q. Gong and X. Yu—Equal contribution.

M. Li et al. (Eds.): TAMC 2025, LNCS 16084, pp. 192–204, 2026.
https://doi.org/10.1007/978-981-95-4839-2_15

Central Unit (CU) is connected to multiple Remote Units (RUs) via a Passive Optical Network (PON), while user equipments (UEs) receive data over the air from the RUs within the same time slot. Additionally, the PON mid-haul is partitioned into slices for individual RUs, each subject to capacity constraints, and the wireless interface rates exhibit time-varying characteristics.

The objective is to maximize the aggregate utility of all UEs, where utility is encoded by a strictly concave smooth utility function defined over long-term data rates. This utility formulation originates from [16], which investigated throughput allocation maximization in saturated systems. The authors proposed a gradient scheduling algorithm that achieves asymptotic optimality relative to the global optimum. Building on this asymptotic optimality condition, Sinha et al. [14] studied an approximate instantaneous Single Shot problem and developed a 1/2-approximation algorithm via linear programming rounding, specifically for the maximization Single Shot problem constrained to a special setting. Furthermore, they demonstrated that the optimized utility is submodular and reformulated the problem as a submodular function maximization problem subject to partition matroid system constraints. Leveraging the greedy method proposed in [9], one can readily derive a 1/2-approximation; however, this approach becomes computationally expensive as the input size increases.

We observe that during the implementation of the centralized greedy algorithm, each iteration always selects the assignment with the maximum marginal gain. Since the utility function is defined for the optimal assignment problem over assignments, an additional call to an optimal assignment algorithm is required for computation. As a result, the execution efficiency of the greedy algorithm is rather low. This observation motivates us to design parallelized acceleration algorithms by introducing parallelizable operations in each iteration to enhance the algorithm's efficiency. We reformulate the data transmission scheduling problem under the v-RAN architecture from a vectorized perspective. We propose a parallelized algorithm that achieves a $1/(1+\alpha)$-approximation with high probability. The algorithm has an adaptive complexity of $O(\epsilon^{-2}\log n \log(\epsilon^{-1}k))$, requires $O(\epsilon^{-2}nk\log n\log(\epsilon^{-1}k))$ queries of the function value, and $O(\epsilon^{-2}\sqrt{n}\log n\log(\epsilon^{-1}k))$ steps of independence queries.

1.1 Related Work

Resource scheduling in Radio Access Network (RAN) architectures has been extensively studied in wireless communication systems. Under orthogonal frequency division multiple access (OFDMA) technology, the minimum wireless resource unit assignable to user equipments (UEs) for data transmission is termed a Resource Block (RB). For resource allocation in virtualized RAN (v-RAN) with multiple sources and UEs, the objective is to determine the scheduling of RBs to UEs in each time slot. Most prior research has focused on throughput maximization, which quantifies the network's data transmission capacity. Typically, the utility associated with throughput is modeled by a strictly concave smooth function, reflecting the intuitive phenomenon that throughput increments diminish with additional RB assignments. This observation motivates our investigation into the utility's submodularity, a property characterized by diminishing marginal returns. In fact, submodular utility functions have been utilized in the maximization of signal-to-interference-plus-noise ratio (SINR) within heterogeneous network (HetNet) architectures [3,17], downlink radio resource allocation with carrier aggregation in LTE-Advanced networks [11], and downlink radio resource allocation with fronthaul compression [18], respectively. Analogous applications of submodular-based resource allocation can be found in prior works such as [12,13,15], among others. These studies collectively demonstrate that submodularity plays a pivotal role in resource scheduling within radio access network (RAN) architectures for wireless communication systems. Given the extensive research on centralized algorithms, we next provide a brief overview of parallel computation techniques for submodular maximization.

The seminal work by Balkanski and Singer [2] introduced a framework for parallel computation in submodular maximization. Contrasting with traditional centralized algorithms, they defined adaptive complexity as the minimum number of sequential rounds required to achieve a constant approximation guarantee, where polynomially many queries can be executed in parallel during each round. For any given $\epsilon > 0$, their framework yields a high-probability $(1/3 - \epsilon)$-approximation for submodular maximization under cardinality constraints, with an adaptive complexity of $O(\log n)$, where n denotes the input size. Notably, they also established a negative result: No algorithm can achieve an approximation better than $O(1/\log n)$ with fewer than $O(\log n/\log\log n)$ rounds of adaptivity for cardinality-constrained submodular maximization. Building on [2], numerous parallel algorithms have been developed for submodular maximization under more complex constraints. For example, Balkanski et al. [1] addressed matroid system constraints, providing a high-probability $(1-e^{-1}-\epsilon)$-approximation with an adaptive complexity of $O(\epsilon^{-3}\log n\log(\epsilon^{-3}k))$, where k is the matroid rank. Independently, [4] studied parallel submodular maximization under matroid constraints, establishing an $O(\epsilon^{-3}\log^2 n)$ adaptive complexity to achieve a $(1-e^{-1}-\epsilon)$-approximation for monotone functions in expectation and an $(e^{-1}-\epsilon)$-approximation for non-monotone functions. Further advancements in parallel submodular maximization can be found in works such as [5,6,8], among others.

1.2 Technical Overview

For the data transmission task in communication networks under the v-RAN architecture, the objective is to ensure network stability while maximizing throughput. Assuming the utility function $U(\cdot)$ is strictly concave and smooth, the long-term maximization problem can be transformed into a Single-Slot problem, where the goal is to determine the instantaneous variables $\mathbf{x}(t), \mathbf{y}(t)$ at time slot t. Specifically, the per-slot problem is formulated as:

$$\max_{\mathbf{x}(t),\mathbf{y}(t)} \dot{U}\left(R_{ij}(t)\right)\sum_k y_{ijk}(t) \tag{1}$$

where $R_{ij}(t)$ denotes an exponentially smoothed long-term transmittable data rate. While the objective function has been proven submodular when $U(\cdot)$ is a logarithmic function (as discussed in [14]), we emphasize that its structural properties may vary significantly with different choices of utility functions. Conforti and Cornu'ejols [7] introduced the concept of total curvature α for nondecreasing submodular functions, which quantifies the degree to which a submodular function deviates from modularity. We restate the submodular and it's curvature as follows. Assume $f : \{0,1\}^n \rightarrow R_{\geq 0}$ is submodular if $f(\mathbf{x} + \ell \mathbf{1}_i) - f(\mathbf{x}) \geq f(\mathbf{y} + \ell \mathbf{1}_i) - f(\mathbf{y})$ for any two vectors $\mathbf{x} \preceq \mathbf{y}$, $\mathbf{1}_i$ represents the indicator vector of coordinate i and $\ell \geq 0$. Submodular f has curvature α defined as $\alpha = 1 - \min_{i \in [n]} \frac{f(\mathbf{1}) - f(\mathbf{1} \backslash \{\mathbf{1}_i\})}{f(\mathbf{1}_i)}$. Leveraging this total curvature, we adapt a generalized parallel algorithm for submodular function maximization to this resource allocation problem and establish performance guarantees for the generalized algorithm.

The standard approach for deriving a 1/2-approximation guarantee in submodular function maximization with matroid constraints relies on the greedy algorithm proposed in [9]. This procedure can be reformulated in vector form, initiating with the $\mathbf{0}$ assignment vector and iteratively adding a single assignment vector that yields the maximum marginal utility. The algorithm terminates when no such single assignment vector with positive marginal utility exists. However, this greedy approach incurs high computational cost for large input sizes n, as it requires $O(n)$ evaluations of the optimal rate allocation at each update step.

To tackle this challenge, we explore parallel computation for submodular maximization, where performance is quantified by adaptive complexity. Specifically, leveraging the availability of an independence oracle for assignment vectors, we first generate random feasible assignment vector sequences in parallel during each round. For each inner iteration, we identify a set of assignment vectors that yields significant marginal gains while preserving independence. We then select the inner iteration with the highest average gain and update the solution assignment vector using the optimal set from this iteration. Through theoretical analysis, we demonstrate that by appropriately choosing the number of rounds and parallel machines, the output assignment vector achieves a strong performance guarantee with high probability.

The remainder of this paper is organized as follows. Section 2 introduces the system model and provides the problem formulation. Section 3 describes a generalized parallel algorithm for submodular maximization with matroid system constraints, and theoretical analysis is presented in Sect. 4. Section 5 concludes our work.

2 System Model and Problem Formulation

We consider the data transmission task in a vRAN architecture, where a Central Unit (CU) is connected to Remote Units (RUs) via a split-processing Passive Optical Network (PON), and user equipments (UEs) receive data over the air on Resource Blocks (RBs) from the RUs. The PON is partitioned into slices for individual RUs, with each slice having a limited capacity. Specifically, for RU$_i$ (where i indexes the RUs), the capacity is denoted by C_i, and the total mid-haul PON capacity is denoted by C. Let UE$_j$ denote the j-th UE associated with RU$_i$, and RB$_k$ denote the k-th RB. For any time slot t, the vector $\mathbf{x}(t) = (x_{ik}^j(t))_{ijk}$ represents the RB assignment, and the vector $\mathbf{y}(t) = (y_{ik}^j(t))_{ijk}$ represents the rate allocation for an assignment vector. Specifically, for any i, j, k:

$$x_{ik}^j(t) = \begin{cases} 1, & \text{if } \mathrm{RB}_k \text{ is assigned to } \mathrm{UE}_j \text{ via } \mathrm{RU}_i \text{ at time slot } t, \\ 0, & \text{otherwise,} \end{cases} \tag{2}$$

and $y_{ik}^j(t) \geq 0$ denotes the rate allocated by RU$_i$ to UE$_j$ on RB$_k$ at time slot t. Let $\gamma_{ik}^j(t)$ be the maximum air-interface rate at which RU$_i$ can transmit data on RB$_k$ to UE$_j$ at time slot t. Inspired by the work of [14], which transformed the resource allocation problem into a Single-Shot problem, our objective is to solve this transformed maximization problem. Specifically, we consider the following Single Shot problem:

$$\max_{\mathbf{x}(t),\mathbf{y}(t)} \sum_{i,j} \dot{U}(R_{ij}(t)) \sum_{k} y_{ik}^j(t). \tag{3}$$

Following [14,16], we define $R_{ij}(t)$ as an exponentially smoothed long-term service rate, updated as:

$$R_{ij}(t+1) = (1-\beta)R_{ij}(t) + \beta \cdot \sum_{k} y_{ik}^j(t), \quad R_{ij}(1) = 0, \tag{4}$$

where $\beta > 0$ is a small smoothing constant. For ease of presentation, the symbol t for all variables can be dropped in this context. We study the simplified Single Shot problem with limit PON and air-interface rate constraints as follows.

$$\max_{\mathbf{x},\mathbf{y}} f(\mathbf{x},\mathbf{y}) = \sum_{ij} \dot{U}(R_{ij}) \sum_{k} y_{ik}^j$$

$$\text{s.t.} \quad y_{ik}^j \leq \gamma_{ik}^j \cdot x_{ik}^j, \qquad \forall i, j, k \tag{5}$$

$$\sum_{j} x_{ik}^j \leq 1, \qquad \forall i, k \tag{6}$$

$$\sum_{j,k} y_{ik}^j \leq C_i, \qquad \forall i \tag{7}$$

$$\sum_{i,j,k} y_{ik}^j \leq C \tag{8}$$

$$x_{ik}^j \in \{0,1\}, y_{ik}^j \geq 0, \qquad \forall i, j, k$$

The inequality (5) reveals that one can allocate date rate y_{ik}^j only if the RB$_k$ has assigned to UE$_j$ according to RU$_i$. Further, the allocated data rate in this assignment at most γ_{ik}^j. The inequality (6) states that the RB$_k$ is assigned to at most one UE for any fixed i, k. The inequalities (7) and (8) denote the PON slice capacity and the total PON capacity constraints, respectively.

By introducing a matrix representation for assignments, we offer a novel perspective on the aforementioned maximization problem. Let $A = (i, j, k)_{ijk}$ denote a $3 \times n$ matrix encoding all assignments, where A_{ik} is the submatrix fixed to indices i, k. Define $X = (\mathbf{1}_{ik}^j)$ as the indicator matrix corresponding to A's indices, with X_{ik} representing sequences of indicator vectors for A_{ik}. Each column of matrix X is an n-dimensional vector with exactly one component as 1 and the rest as 0. For any $\mathbf{x}_1, \mathbf{x}_2 \in \{0,1\}^n$, let $\mathbf{x}_1 \vee \mathbf{x}_2$ denote the vector whose i-th coordinate is $(\mathbf{x}_1 \vee \mathbf{x}_2)^i = \max\{x_1^i, x_2^i\}$, and $\mathbf{x}_1 \wedge \mathbf{x}_2$ denote the vector with i-th coordinate $(\mathbf{x}_1 \wedge \mathbf{x}_2)^i = \min\{x_1^i, x_2^i\}$. Additionally, $\mathbf{x}_1 \setminus \mathbf{x}_2$ is defined as the vector with i-th coordinate $(\mathbf{x}_1 \setminus \mathbf{x}_2)^i = \max_+(x_1^i - x_2^i)$, where $\max_+(a) = \max\{0, a\}$. We use $\mathbf{x}_1 \preceq \mathbf{x}_2$ to denote $x_1^i \leq x_2^i$ for all coordinates i. A vector $\mathbf{x} = (x_{ik}^j)$ is a feasible assignment vector if $\sum_j x_{ik}^j \leq 1$ for all fixed i, k. Let $\mathcal{I} = \{\mathbf{x} : \sum_j x_{ik}^j \leq 1, \forall i, k\}$ be the set of all feasible assignment vectors. From inequality (6), we have $\|\mathbf{x} \wedge X_{ik}\|_1 \leq 1$ for any $\mathbf{x} \in \mathcal{I}$, which induces a partition matroid system $\mathcal{M} = (X, \mathcal{I})$, where $\mathbf{x} \wedge X_{ik} = \vee_{\mathbf{x}' \in X_{ik}} \mathbf{x} \wedge \mathbf{x}'$. Let $\mathbf{y}^*(\mathbf{x})$ denote the optimal data rate allocation for $\mathbf{x}$. Following [14], a polynomial-time algorithm for computing $\mathbf{y}^*(\mathbf{x})$ for $\mathbf{x} \in \mathcal{I}$ can be adapted, summarized in Algorithm 1. This reduces the problem to finding a feasible assignment vector that maximizes utility. Since the objective utility is proven submodular [14], the Single Shot problem becomes a submodular function maximization problem with a partition matroid constraint.Let $\mathbf{y}^*(\mathbf{x}) = A^*(\mathbf{x})$. For clarity, define $f^*(\mathbf{x}) = f(\mathbf{x}, \mathbf{y}^*(\mathbf{x}))$ and the marginal gain $f_{\mathbf{x}'}^*(\mathbf{x}) = f^*(\mathbf{x}' \vee \mathbf{x}) - f^*(\mathbf{x}')$ with respect to $\mathbf{x}'$. The classic greedy method from [9] initializes with $\mathbf{x} = \mathbf{0}$ and iteratively selects an augmentation vector $\mathbf{x}' \in X$ that maximizes $f^*(\mathbf{x} \vee \mathbf{x}')$ under $\mathbf{x} \vee \mathbf{x}' \in \mathcal{I}$, terminating when no such vector exists. While this method provides a 0.5-approximation guarantee, it incurs $O(n)$ invocations of A^* per augmentation to compute $\mathbf{y}^*(\mathbf{x}')$, making it computationally infeasible for large n.

In this setting, an independence oracle for the partition matroid system can be readily constructed. To reduce the number of invocations in the greedy procedure, we propose an accelerated parallel algorithm for submodular maximization in this resource allocation problem. By leveraging the concept of bounded curvature, we analyze the worst-case approximation guarantee of the proposed algorithm.

3 Accelerated Adaptive Sequence with Independence Oracle

In this section, we employ an extended parallel procedure for the Single-Shot problem using an independence oracle. An independence oracle can be efficiently constructed by verifying if $\sum_k x_{ik}^j \leq 1$ for all pairs (i, k) in $O(n)$ time for any given assignment vector $\mathbf{x} \in \{0,1\}^n$. Leveraging this oracle, we adopt the strategy proposed in [10] to generate a random sequence of assignment

Algorithm 1 A^* : Optimum Data Rate Allocation [14]

Input: An assignment vector $\mathbf{X} \in \mathcal{I}$

Set $\mathbf{y} \leftarrow \mathbf{0}$

Order the (ijk) triples of $\mathbf{X}$ in non-increasing order according to the value of $\dot{U}(R_{ij})$

Go through the each triple sequentially in the above order. When consider the triple (ijk), set

$$y^j_{ik} \leftarrow \min\left\{\gamma^j_{ik}x^j_{ik}, C - \sum_{i'j'k'} y_{i'j'k'}, C_i - \sum_{j'k'} y_{ij'k'}\right\}$$

Set $\mathbf{y}^*(\mathbf{X}) \leftarrow \mathbf{y}$

Algorithm 2 Parallel Random Sequence for Matroid System under Independence Oracle

Input: Matroid system $M = (X, \mathcal{I})$

$c \leftarrow 0, U \leftarrow X$

while $U \neq \mathbf{0}$ **do**

  sequence $\overrightarrow{b}_1, ..., \overrightarrow{b}_{|U|} \leftarrow$ random permutation of the column vectors of U

  $\widetilde{i}^* \leftarrow \max\{i : \mathbf{x}_1 \vee \cdots \vee \mathbf{x}_c \vee \mathbf{b}_1 \vee \cdots \vee \mathbf{b}_i \in \mathcal{I}\}$

  $\mathbf{x}_{c+1}, ..., \mathbf{x}_{c+\widetilde{i}^*} \leftarrow \mathbf{b}_1, ..., \mathbf{b}_{\widetilde{i}^*}$

  $c \leftarrow c + \widetilde{i}^*$

  $U = \{\uplus\mathbf{x} : \mathbf{x}_1 \vee \cdots \vee \mathbf{x}_c \vee \mathbf{x} \in \mathcal{I}\}$ ▷ symbol $\uplus$ denotes concatenation of the assignment vectors incorporating $\mathbf{x}_1 \vee \cdots \vee \mathbf{x}_c$ are still feasible.

end while

return vectors sequence $\mathbf{x}_1, \ldots, \mathbf{x}_c$

vectors, which constructs a base of the matroid system $(X, \mathcal{I})$ in $O(\sqrt{n})$ time. The pseudocode is outlined in Algorithm 2.

We initialize a candidate matrix $U = X$ and a counter $c = 0$. In each while-loop iteration, a random permutation of U's column vectors $\mathbf{b}_1, \ldots, \mathbf{b}_{|U|}$ is generated. Let $\mathbf{x}_1, \ldots, \mathbf{x}_c$ be the first c vectors obtained from prior iterations. Define $\widetilde{i}^*$ as the maximum index such that $\mathbf{x}_1 \vee \cdots \vee \mathbf{x}_c \vee \mathbf{b}_1 \vee \cdots \vee \mathbf{b}_{\widetilde{i}^*} \in \mathcal{I}$, ensuring independence when annexing vectors from the permutation. The sequence $\mathbf{x}_{c+1}, \ldots, \mathbf{x}_{c+\widetilde{i}^*}$ is then updated with $\mathbf{b}_1, \ldots, \mathbf{b}_{\widetilde{i}^*}$, and the candidate matrix is reassembled as $U = \{\uplus\mathbf{x} : \mathbf{x}_1 \vee \cdots \vee \mathbf{x}_c \vee \mathbf{x} \in \mathcal{I}\}$. The procedure outputs the vector sequence $\mathbf{x}_1, \ldots, \mathbf{x}_c$. The main parallel procedure utilizes the above random vector sequence construction as a subroutine in each round to generate feasible assignment vectors. The algorithm broadly follows the adaptive sequence++ framework introduced in [1], adapted here for finding feasible assignment vectors (summarized in Algorithm 3).

Parameters Δ (number of rounds) and ρ (number of parallel machines) govern the algorithm's structure. Let $\mathbf{X}$ denote the assignment vector returned at an iteration, and define $\mathcal{M}(\mathbf{X}, \mathcal{I}) = \{\mathbf{x}' : \mathbf{X} \vee \mathbf{x}' \in \mathcal{I}\}$ as the matroid where $\mathbf{x}'$

Algorithm 3 Accelerated Adaptive Sequence

Input: Instance $(f, M, \Delta, \rho, \epsilon)$
$\mathbf{X} \leftarrow \mathbf{0}, m \leftarrow \max_{\mathbf{x}\in X} f^*(\mathbf{x})$
for $r = 1 : \Delta$ **do**
  $U \leftarrow X$
  while $U \neq \mathbf{0}$ **do**
    for $j = 1 : \rho$ **do**
      vector sequence $\mathbf{x}_1, \mathbf{x}_2, \cdots, \mathbf{x}_{\text{Rank}(\mathbf{X},\mathcal{I})}$ % Random permutation returned by Algorithm 2.
      $U_i \leftarrow \{\mathbf{x} : \mathbf{X} \vee \mathbf{x}_1 \vee \mathbf{x}_2 \vee \cdots \vee \mathbf{x} \in \mathcal{I}$ and $f^*_{\mathbf{X}\vee\mathbf{x}_1\vee\mathbf{x}_2\vee\cdots\vee\mathbf{x}_{i-1}}(\mathbf{x}) \geq m\}$
      $i^* \leftarrow \min\{i : \frac{|U_i|}{|U|} \leq \frac{1}{1+\epsilon}\}$
      $\mathbf{x}^j \leftarrow \mathbf{X} \vee \mathbf{x}_1 \vee \mathbf{x}_2 \vee \cdots \vee \mathbf{x}_{i^*}$
      $U^j \leftarrow U_{i^*}$
      $w_j \leftarrow \frac{1}{i^*}\sum_{\ell=1}^{i^*} f^*_{\mathbf{X}\vee\mathbf{x}_1\vee\cdots\mathbf{x}_{\ell-1}}(\mathbf{x}_\ell)$
    end for
    $j^* \leftarrow \arg\max_{j\in[\rho]} w_j$
    $\mathbf{X} \leftarrow \mathbf{X}^{j^*}$
    $U \leftarrow U^{j^*}$
  end while
  $m \leftarrow \frac{m}{1+\epsilon}$
  $r \leftarrow r + 1$
end for
return $\mathbf{X} \leftarrow \arg\max_{\mathbf{X}} f^*(\mathbf{X})$

is feasible when combined with $\mathbf{x}$. Let $\mathbf{x}_1, \ldots, \mathbf{x}_{|\text{Rank}(\mathcal{M}(\mathbf{X},\mathcal{I}))|}$ be the random vector sequence from Algorithm 2 in each inner iteration, with $\text{Rank}(\mathcal{M}(\mathbf{X}, \mathcal{I}))$ denoting the matroid's rank. We initialize $m = \max_{\mathbf{x}} f^*(\mathbf{x})$ as the threshold for round r. An assignment vector $\mathbf{x}'$ is termed *good* for sequence $\mathbf{x}_1, \ldots, \mathbf{x}_\ell$ and threshold m if $\mathbf{X} \vee \mathbf{x}_1 \vee \cdots \vee \mathbf{x}_i \vee \mathbf{x}' \in \mathcal{I}$ and $f^*_{\mathbf{X}\vee\mathbf{x}_1\vee\cdots\vee\mathbf{x}_i}(\mathbf{x}') \geq m$ for all $i \in [\ell]$. Let $U_i = \{\mathbf{x}' : \mathbf{X}\vee\mathbf{x}_1\vee\cdots\vee\mathbf{x}_i\vee\mathbf{x}' \in \mathcal{I}$ and $f^*_{\mathbf{X}\vee\mathbf{x}_1\vee\cdots\vee\mathbf{x}_i}(\mathbf{x}') \geq m\}$ denote the set of good vectors for index i. In each inner iteration, we identify $i^* = \min_i \frac{|U_i|}{|U|}$ as the first index with a sufficiently large size decrement relative to U. We update $\mathbf{X}^j = \mathbf{X} \vee \mathbf{x}_1 \vee \cdots \vee \mathbf{x}_{i^*}$ and $U^j = U_{i^*}$. The inner iteration with the maximum average marginal gain $w_j = \frac{1}{i^*}\sum_{\ell=1}^{i^*} f^*_{\mathbf{X}\vee\mathbf{x}_1\vee\cdots\vee\mathbf{x}_{\ell-1}}(\mathbf{x}_\ell)$ is selected as j^*, and $\mathbf{X}$ and U are updated for the next round.

4 Theoretical Analysis

The inner iteration computation process can be executed in parallel with zero adaptivity. The core challenge lies in determining the parameters Δ and ρ, as the main algorithm's adaptivity is bounded by $O(n\Delta\rho)$. Before detailing the main determination process, we present two technical lemmas as follows. By the definition of threshold m, it can be inferred that m serves as an approximate upper bound on the maximum marginal gain of the assignment vectors.

Lemma 1. *During any round of Algorithm 3, for any inner iteration, the following inequality holds:*

$$\max_{\mathbf{x}':\mathbf{X}\vee\mathbf{x}'\in\mathcal{I}} f^*_{\mathbf{X}}(\mathbf{x}') \leq (1+\epsilon)m, \tag{9}$$

where $\mathbf{X}$ *denotes the current assignment vector and* $\mathbf{x}'$ *represents a candidate assignment vector.*

Proof. The claim holds trivially in the initial round, where $m = \max_{\mathbf{x}\in X} f^*(\mathbf{x})$, $\mathbf{X} = \mathbf{0}$, and $U = X$. Suppose the claim is valid at the beginning of an arbitrary iteration, and let $\mathbf{X}$ be updated to $\mathbf{X}\vee\mathbf{x}_1\vee\cdots\vee\mathbf{x}_{i^*}$ during some inner iteration. To prove the lemma, consider the following sequence of inequalities:

$$\begin{aligned} m &> \frac{1}{1+\epsilon}\cdot\max_{\mathbf{x}:\mathbf{X}\vee\mathbf{x}\in\mathcal{I}} f^*_{\mathbf{X}}(\mathbf{x}) \\ &\geq \frac{1}{1+\epsilon}\cdot\max_{\mathbf{x}:\mathbf{X}\vee\mathbf{x}_1\vee\cdots\vee\mathbf{X}_{i^*}\vee\mathbf{x}\in\mathcal{I}} f^*_{\mathbf{X}}(\mathbf{x}) \\ &\geq \frac{1}{1+\epsilon}\cdot\max_{\mathbf{x}:\mathbf{X}\vee\mathbf{x}_1\vee\cdots\vee\mathbf{x}_{i^*}\vee\mathbf{x}\in\mathcal{I}} f^*_{\mathbf{X}\vee\mathbf{x}_1\vee\cdots\vee\mathbf{x}_{i^*}}(\mathbf{x}). \end{aligned}$$

The second inequality follows from the set inclusion $\{\mathbf{x}' : \mathbf{X}\vee\mathbf{x}' \in \mathcal{I}\} \supseteq \{\mathbf{x}' : \mathbf{X}\vee\mathbf{x}_1\vee\cdots\vee\mathbf{x}_{i^*}\vee\mathbf{x}' \in \mathcal{I}\}$, and the third inequality is justified by the submodularity of f^*. Now, consider the scenario where the threshold value m is updated to $m' = \frac{m}{1+\epsilon}$ in an outer round. At this iteration, $U = \{\mathbf{0}\}$ with respect to the current assignment vector $\mathbf{X}$. Consider a vector $\mathbf{x}'$ that was discarded from U in a previous iteration, where the current assignment vector satisfies $\mathbf{X}'\vee\mathbf{x}_1\vee\cdots\vee\mathbf{x}_{i^*} \preceq \mathbf{x}$. We have either

$$\mathbf{X}'\vee\mathbf{x}_1\vee\cdots\vee\mathbf{x}_{i^*}\vee\mathbf{x}' \notin \mathcal{I}, \text{ or } f^*_{\mathbf{X}'\vee\mathbf{x}_1\vee\cdots\vee\mathbf{x}_{i^*}}(\mathbf{x}') < m,$$

since the assignment vector $\mathbf{x}'$ has been discarded. If $\mathbf{X}'\vee\mathbf{x}_1\vee\cdots\vee\mathbf{x}_{i^*}\vee\mathbf{x}' \notin \mathcal{I}$, following by $\mathbf{X}'\vee\mathbf{x}_1\vee\cdots\vee\mathbf{x}_{i^*} \preceq \mathbf{x}$, we acquire $\mathbf{X}\vee\mathbf{x}' \notin \mathcal{I}$, which violates the assumption of claim. We consider the case of $f^*_{\mathbf{X}'\vee\mathbf{x}_1\vee\cdots\vee\mathbf{x}_{i^*}}(\mathbf{x}') < m$. By submodularity, we yield

$$f^*_{\mathbf{X}}(\mathbf{x}') \leq f^*_{\mathbf{X}'\vee\mathbf{x}_1\vee\cdots\vee\mathbf{x}_{i^*}}(\mathbf{x}') < m = (1+\epsilon)m'.$$

Thus, the proof of this claim is complete.

The following lemma asserts that each feasible assignment vector $\mathbf{x}'$ added to $\mathbf{X}^{t-1}$ during round r possesses an approximately optimal marginal gain with respect to $\mathbf{X}^{t-1}$ in expectation. The proofs are deferred to Appendix.

Lemma 2. *During a round of Algorithm 3, let* $\mathbf{x}_1,\ldots,\mathbf{x}_{Rank(\mathcal{M}(\mathbf{X},\mathcal{I}))}$ *be the random sequence returned in an inner iteration. For any* $i \leq i^*$*, define* $U_i^{\mathcal{M}} = \{\mathbf{x}' : \mathbf{X}\vee\mathbf{x}_1\vee\cdots\vee\mathbf{x}_i\vee\mathbf{x}' \in \mathcal{I}\}$*. Then we have*

$$\begin{aligned} &\mathbb{E}_{\mathbf{x}_i\sim U(X_{i-1}^{\mathcal{M}})}[f^*_{\mathbf{X}\vee\mathbf{x}_1\vee\mathbf{x}_2\vee\cdots\vee\mathbf{x}_{i-1}}(\mathbf{x}_i)] \\ &\geq \frac{1}{(1+\epsilon)^2}\cdot\max_{\mathbf{x}:\mathbf{X}\vee\mathbf{x}_1\vee\mathbf{x}_2\vee\cdots\vee\mathbf{x}_{i-1}\vee\mathbf{x}\in\mathcal{I}} f^*_{\mathbf{X}\vee\mathbf{x}_1\vee\mathbf{x}_2\vee\cdots\vee\mathbf{x}_{i-1}}(\mathbf{x}). \end{aligned}$$

Proof. Since the assignment vector $\mathbf{x}_i$ is uniformly selected from $U_{i-1}^{\mathcal{M}}$—i.e., $\mathbf{x}_i \sim \mathrm{U}(U_{i-1}^{\mathcal{M}})$—it follows that:

$$\begin{aligned}
&\mathbb{E}_{\mathbf{x}_i \sim \mathrm{U}(X_{i-1}^{\mathcal{M}})}[f_{\mathbf{X}\vee\mathbf{x}_1\vee\mathbf{x}_2\vee\cdots\vee\mathbf{x}_{i-1}}(\mathbf{x}_i)] \\
&\geq \mathbb{P}[f_{\mathbf{X}\vee\mathbf{x}_1\vee\mathbf{x}_2\vee\cdots\vee\mathbf{x}_{i-1}}(\mathbf{x}_i) \geq m] \cdot m = \frac{|U_{i-1}|}{|U_{i-1}^{\mathcal{M}}|} \cdot m \\
&\geq \frac{1}{(1+\epsilon)^2} \cdot \max_{\mathbf{x}:\mathbf{X}\vee\mathbf{x}_1\vee\mathbf{x}_2\vee\cdots\vee\mathbf{x}_{i-1}\vee\mathbf{x}\in\mathcal{I}} f_{\mathbf{X}\vee\mathbf{x}_1\vee\mathbf{x}_2\vee\cdots\vee\mathbf{x}_{i-1}}(\mathbf{x}),
\end{aligned}$$

where the first inequality follows by the definition of the expectation and the second inequality is derived by combining the definition of i^* and Lemma 1.

Let $\mathbf{X} = \mathbf{x}_1 \vee \mathbf{x}_2 \vee \cdots \vee \mathbf{x}_{k'}$ ($k' \leq k$) be the assignment vector returned by Algorithm 3, where k denotes the rank of the partition matroid and the vectors are added in order. Define $\mathbf{X}^\ell = \mathbf{x}_1 \vee \mathbf{x}_2 \vee \cdots \vee \mathbf{x}_\ell$ as the ℓ-th vector formed by concatenating the first ℓ assignment vectors. The following technical lemma is derived analogously to the work of [9].

Lemma 3. *Let* $\mathbf{X}$ *be the ordered assignment vector generated by Algorithm 3, and let* $\mathbf{Y} \in \{0,1\}^n$ *be an arbitrary feasible assignment vector. Then we obtain*

$$f^*(\mathbf{Y}) \leq \alpha \cdot f^*(\mathbf{X}) + \sum_{\mathbf{x}_\ell \in \mathbf{X}\wedge\mathbf{Y}} f^*_{\mathbf{X}^{\ell-1}}(\mathbf{x}_\ell) + \sum_{\mathbf{x}\in\mathbf{Y}\setminus\mathbf{X}} f^*_{\mathbf{X}}(\mathbf{x}).$$

Proof. By submodularity, we have

$$f^*(\mathbf{X}\vee\mathbf{Y}) \leq f^*(\mathbf{X}) + \sum_{\mathbf{x}\in\mathbf{Y}\setminus\mathbf{X}} f^*(\mathbf{X}\vee\mathbf{x}) - f^*(\mathbf{X}).$$

By the definition of curvature α, we yield

$$\begin{aligned}
f^*(\mathbf{X}\vee\mathbf{Y}) &= f^*(\mathbf{Y}) + \sum_{\mathbf{x}_\ell\in\mathbf{X}\setminus\mathbf{Y}} f^*(\mathbf{Y}\vee\mathbf{X}^{\ell-1}\vee\mathbf{x}_\ell) - f^*(\mathbf{Y}\vee\mathbf{X}^{\ell-1}) \\
&\geq f^*(\mathbf{Y}) + (1-\alpha) \sum_{\mathbf{x}_\ell\in\mathbf{X}\setminus\mathbf{Y}} f^*(\mathbf{X}^{\ell-1}\vee\mathbf{x}_\ell) - f^*(\mathbf{X}^{\ell-1})
\end{aligned}$$

Combining the above two inequalities, we conclude

$$\begin{aligned}
f^*(\mathbf{Y}) \leq & \sum_{\mathbf{x}_\ell\in\mathbf{X}\wedge\mathbf{Y}} f^*(\mathbf{X}^{\ell-1}\vee\mathbf{x}_\ell) - f^*(\mathbf{X}^{\ell-1}) + \sum_{\mathbf{x}\in\mathbf{Y}\setminus\mathbf{X}} f^*(\mathbf{X}\vee\mathbf{x}) - f^*(\mathbf{X}) \\
&+ \alpha \cdot \sum_{\mathbf{x}_\ell\in\mathbf{X}\setminus\mathbf{Y}} f^*(\mathbf{X}^{\ell-1}\vee\mathbf{x}_\ell) - f^*(\mathbf{X}^{\ell-1}) \\
&\leq \sum_{\mathbf{x}_\ell\in\mathbf{X}\wedge\mathbf{Y}} f^*_{\mathbf{X}^{\ell-1}}(\mathbf{x}_\ell) + \sum_{\mathbf{x}\in\mathbf{Y}\setminus\mathbf{X}} f^*_{\mathbf{X}}(\mathbf{x}) + \alpha \cdot f^*(\mathbf{X}).
\end{aligned}$$

In the initial round, let $m = \max_{\mathbf{x}} f^*(\mathbf{x}) \leq \text{OPT}$, where OPT denotes the utility of the optimal assignment vector $\mathbf{x}^*$, i.e., $\text{OPT} = f^*(\mathbf{x}^*)$. We set $\Delta = O(\epsilon^{-1}\log(\epsilon^{-1}k))$, ensuring that after Δ rounds, the threshold value satisfies $m \leq \frac{1}{(1+\epsilon)^\Delta}\cdot\text{OPT} \leq O\left(\frac{\epsilon}{k}\right)\cdot\text{OPT}$. To enable pairwise comparisons between $\mathbf{x}$ and $\mathbf{x}^*$, we append dummy assignment vectors $\mathbf{x}_{k'+1}, \ldots, \mathbf{x}_k$ such that $f^*_{\mathbf{x}^\ell}(\mathbf{x}) \geq m_\Delta$ for any assignment vector $\mathbf{x}$, where m_Δ is the threshold after Δ rounds. By the matroid system's properties, the optimal assignment vector can be decomposed as $\mathbf{X}^* = \mathbf{x}_1^* \vee \mathbf{x}_2^* \vee \cdots \vee \mathbf{x}_k^*$, with $\mathbf{X}^{*\ell} = \mathbf{x}_1^* \vee \cdots \vee \mathbf{x}_\ell^*$ satisfying $\mathbf{X}^{\ell-1} \vee \mathbf{x}_\ell^* \in \mathcal{I}$. The main result is formalized in the following theorem.

Theorem 1. *Given a constant $\epsilon > 0$, with probability $1 - o(1)$, Algorithm 3 returns a feasible assignment vector $\mathbf{X}$ with the following properties.*

- *Adaptive complexity: $O\left(\epsilon^{-2}\log n \log(\epsilon^{-1}k)\right)$;*
- *Function value queries: $O\left(\epsilon^{-2}nk\log n \log(\epsilon^{-1}k)\right)$;*
- *Independence queries: $O\left(\epsilon^{-2}\sqrt{n}\log n \log(\epsilon^{-1}k)\right)$.*

Moreover, the assignment satisfies

$$f^*(\mathbf{X}^*) \leq \left(\alpha + \frac{(1-\epsilon)^2}{1+\epsilon}\right)\cdot f^*(\mathbf{X}), \tag{10}$$

where $\mathbf{X}^*$ denotes the optimal assignment vector.

Proof. We argue that the following inequality

$$\sum_{i\in[i^*]} f_{\mathbf{X}^{\ell-1}}(\mathbf{x}_i) \geq \frac{i^* m}{(1+\epsilon)^2},$$

for each inner-for loop. To prove the above claim, we notice

$$\mathbb{P}_{\mathbf{x}_i}[f^*_{\mathbf{X}\vee\mathbf{x}_1\cdots\mathbf{x}_{i-1}}(\mathbf{x}_i) \geq m] \geq \frac{1}{1+\epsilon}$$

for any $\mathbf{x}_i \sim \mathrm{U}(X^M_{i-1})$. We use binary variable $A_i \in \{0,1\}$ to denote if

$$f^*_{\mathbf{X}\vee\mathbf{x}_1\cdots\mathbf{x}_{i-1}}(\mathbf{x}_i) \geq m$$

for index $i \leq i^*$ and denote $Y = \sum_{i=1}^{i^*} A_i$ as the amount of indexes $i \leq i^*$ subject to $f^*_{\mathbf{X}\vee\mathbf{x}_1\cdots\mathbf{x}_{i-1}}(\mathbf{x}_i) \geq m$. Then we have $\mu = \mathbb{E}[Y] \geq \frac{i^*}{1+\epsilon}$. By the multiplicative Chernoff bound, we have

$$\mathbb{P}\left[Y \leq (1-\epsilon)\cdot\frac{i^*}{1+\epsilon}\right] \leq e^{-\frac{\epsilon^2}{2}\cdot\frac{i^*}{1+\epsilon}} \leq e^{-\frac{\epsilon^2}{2(1+\epsilon)}}.$$

Further, by setting variable $Z \leq \rho$ as the number of inner-iterations subject to $Y \geq (1-\epsilon)\cdot\frac{i^*}{1+\epsilon}$, we have $\nu = \mathbb{E}[Z] \geq (1 - e^{-\frac{\epsilon^2}{2(1+\epsilon)}})\rho$. We again use the multiplicative Chernoff bound and derive

$$\mathbb{P}\left[Z \leq (1-\delta)(1-e^{-\frac{\epsilon^2}{2(1+\epsilon)}})\rho\right] \leq e^{-\frac{\delta^2}{2}\cdot\left(1-e^{-\frac{\epsilon^2}{2(1+\epsilon)}}\right)\rho},$$

for any $\delta > 0$. By setting $\rho = O\left(\frac{\delta^{-2}}{1-e^{-\epsilon^2}}\log(\frac{\Delta \log n}{\delta\epsilon})\right)$, we conclude that, with probability $1-O\left(\frac{\delta\epsilon}{\Delta \log n}\right)$, at least one inner-iteration subject to $Y \geq (1-\epsilon)\cdot\frac{i^*}{1+\epsilon}$. Then by Lemma 1, we yield

$$\sum_{i\in[i^*]} f_{\mathbf{X}^{\ell-1}}(\mathbf{x}_\ell) \geq Y \cdot m \geq \frac{1-\epsilon}{(1+\epsilon)^2}\cdot \max_{\mathbf{x}:\mathbf{X}^{\ell-1}\vee\mathbf{x}\in\mathcal{I}} f_{\mathbf{X}^{\ell-1}}(\mathbf{x}).$$

Equipping with Lemma 3 by setting $\mathbf{X}^* = \mathbf{Y}$, with probability $1-\delta$, we derive the following inequalities

$$\begin{aligned} f^*(\mathbf{X}^*) - \alpha \cdot f^*(\mathbf{X}) &\leq \mathbb{E}\left[\sum_{\mathbf{x}_\ell\in\mathbf{X}^*\wedge\mathbf{X}} f^*_{\mathbf{X}^{\ell-1}}(\mathbf{x}_\ell)\right] + \mathbb{E}\left[\sum_{\mathbf{x}\in\mathbf{X}^*\setminus\mathbf{X}} f^*_{\mathbf{X}}(\mathbf{x})\right] \\ &\leq \mathbb{E}\left[\sum_{\mathbf{x}^*_\ell\in\mathbf{X}^*} f^*_{\mathbf{X}^{\ell-1}}(\mathbf{x}^*_\ell)\right] \leq \max_{\mathbf{x}:\mathbf{X}^{\ell-1}\vee\mathbf{x}\in\mathcal{I}} f^*_{\mathbf{X}^{\ell-1}}(\mathbf{x}) \\ &\leq \frac{(1+\epsilon)^2}{1-\epsilon}\cdot\mathbb{E}\left[\sum_{i\in k} f^*_{\mathbf{X}^{\ell-1}}(\mathbf{x}_\ell)\right] = \frac{(1+\epsilon)^2}{1-\epsilon}\cdot f^*(\mathbf{X}). \end{aligned}$$

The second inequality holds because $\mathbf{X}^{\ell-1} \vee \mathbf{x}^*_\ell \in \mathcal{I}$ and the last inequality follows from Lemma 2. By appropriately choosing the parameters Δ and ρ, the number of independence queries can be bounded by $O\left(\epsilon^{-2}\sqrt{n}\log n \log(\epsilon^{-1}k)\right)$. The query complexity for function values is at most $O\left(\epsilon^{-2}nk\log n\log(\epsilon^{-1}k)\right)$, and the average query complexity for function values is $O\left(\epsilon^{-2}n\log n\log(\epsilon^{-1}k)\right)$.

5 Conclusion

This paper addresses resource allocation in v-RAN by transforming it into a submodular maximization problem with partition matroid constraints. We propose an accelerated parallel algorithm using random permutations and an independence oracle, achieving a parameterized approximation ratio with high probability. The algorithm reduces the computational cost of traditional greedy methods. This work enhances scalability for large networks, though future research may explore dynamic channels or non-monotone utilities.

Acknowledgements. This work was partially supported by Beijing Natural Science Foundation (Nos. Z220004, IS24001, QY24081).

References

1. Balkanski, E., Rubinstein, A., Singer, Y.: An optimal approximation for submodular maximization under a matroid constraint in the adaptive complexity model. In: Proceedings of STOC, pp. 66–77 (2019)

2. Balkanski, E., Singer, Y.: The adaptive complexity of maximizing a submodular function. In: Proceedings of STOC, pp. 1138–1151 (2018)
3. Bikram Kumar, B., Sharma, L., Wu, S.L.: Online distributed UE association for heterogeneous radio access network. Sensors **19**(6), 1412 (2019)
4. Chekuri, C., Quanrud, K.: Parallelizing greedy for submodular set function maximization in matroids and beyond. In: Proceedings of STOC, pp. 78–89 (2019)
5. Chekuri, C., Quanrud, K.: Submodular function maximization in parallel via the multilinear relaxation. In: Proceedings of SODA, pp. 303–322 (2019)
6. Chen L, Feldman M, Karbasi A. Unconstrained submodular maximization with constant adaptive complexity. In: Proceedings of STOC, pp. 102–113 (2019)
7. Conforti, M., Cornuéjols, G.: Submodular set functions, matroids and the greedy algorithm: tight worst-case bounds and some generalizations of the Rado-Edmonds theorem. Discret. Appl. Math. **7**(3), 251–274 (1984)
8. Ene, A., Nguyễ, H.L., Vladu, A.: Submodular maximization with matroid and packing constraints in parallel. In: Proceedings of STOC, pp. 90–101 (2019)
9. Fisher, M.L., NemhaUE, G.L., Wolsey, L.A.: An analysis of approximations for maximizing submodular set functions-II. Polyhedral Combinatorics, 73–87 (2019)
10. Karp, R.M., Upfal, E., Wigderson, A.: The complexity of parallel search. J. Comput. Syst. Sci. **36**(2), 225–253 (1988)
11. Liao, H.S., Chen, P.Y., Chen, W.T.: An efficient downlink radio resource allocation with carrier aggregation in LTE-advanced networks. IEEE Trans. Mob. Comput. **13**(10), 2229–2239 (2014)
12. Liu, J.S.: Joint downlink resource allocation in LTE-Advanced heterogeneous networks. Comput. Netw. **146**, 85–103 (2018)
13. Roy, A., Chaporkar, P., Karandikar, A.: Optimal radio access technology selection algorithm for LTE-WIFI network. IEEE Trans. Veh. Technol. **67**(7), 6446–6460 (2018)
14. Sinha, A., Andrews, M., Ananth, P.: Scheduling algorithms for 5G networks with mid-haul capacity constraints. In: Proceedings of of WiOPT, pp. 1–8 (2019)
15. Stephen, R.G., Zhang, R.: Joint millimeter-wave fronthaul and OFDMA resource allocation in ultra-dense CRAN. IEEE Trans. Commun. **65**(3), 1411–1423 (2017)
16. Stolyar, A.L.: On the asymptotic optimality of the gradient scheduling algorithm for multiUE throughput allocation. Oper. Res. **53**(1), 12–25 (2005)
17. Su, G., Chen, B., Lin, X., et al.: A submodular optimization framework for outage-aware cell association in heterogeneous cellular networks. Math. Probl. Eng. **1**, 4567625 (2016)
18. Xia, W., Quek, T.Q.S., Zhang, J., et al.: Programmable hierarchical C-RAN: from task scheduling to resource allocation. IEEE Trans. Wireless Commun. **18**(3), 2003–2016 (2019)

Computational Geometry and Graph Theory

Constructive Upper Bounds on the Rectangle Eliminating Number in Grid Graphs

Enguo Niu[1,2] and Zhongzheng Tang[1,2](✉)

[1] School of Mathematical Sciences, Beijing University of Posts and Telecommunications, Beijing 100876, China
niuenguo@bupt.edu.cn

[2] Key Laboratory of Mathematics and Information Networks (Beijing University of Posts and Telecommunications), Ministry of Education, Beijing 100876, China
tangzhongzheng@bupt.edu.cn

Abstract. The hitting set is a fundamental concept in graph theory and combinatorial optimization. In this paper, we study a special class of the hitting set problem in grid graphs. Given an $n \times m$ grid graph $G = (V, E)$, a rectangle eliminating set is an edge subset $A \subseteq E$ such that $G - A$ contains no rectangular outlines. The minimum cardinality of the rectangle eliminating set is the rectangle eliminating number, denoted by $S(n, m)$. We derive several upper bounds for $S(n, m)$ by periodic sampling, binary decomposition and hybrid constructions. In particular, for square grids, we show that $S(n, n) \leq \frac{2n^2}{3} + \frac{40}{3}\sqrt{2} \cdot \sqrt{n} + 3$ when $n \geq 32$.

Keywords: Hitting Set · Rectangle Eliminating Number · Constructive Upper Bounds

1 Introduction

The Hitting Set problem is one of the 21 classic NP-complete problems introduced by Richard Karp in 1972 [7]. This problem has stimulated related theoretical research [1,8,9] and has extensive practical applications [10,11]. The hitting set problem refers to: given a collection $\mathcal{C}$ of subsets of a finite set S, a *hitting set* for $\mathcal{C}$ is a subset $S' \subseteq S$ such that S' contains at least one element from each subset in $\mathcal{C}$, i.e., for all $T \in \mathcal{C}$, $S' \cap T \neq \varnothing$. This paper focuses on a specialized variant of the hitting set problem that was first introduced by Paul A. Dreyer and Therese C. Biedl in 1997 [5]. Let's start with an example from the New York Times:

60 matchsticks are used to form a 5×5 square grid. In this grid graph are the outlines of 1×1 squares, 2×2 squares, and so on. The aim is to remove

Supported by National Natural Science Foundation of China under Grant No. 12101069.

M. Li et al. (Eds.): TAMC 2025, LNCS 16084, pp. 207–220, 2026.
https://doi.org/10.1007/978-981-95-4839-2_16

the minimum number of matches so that the resulting graph contains no square outlines.

We now consider more general cases. Let's define a *grid graph* in graph theory. An $n \times m$ grid graph $G = (V, E)$ consists of vertices arranged in a two-dimensional lattice with $n + 1$ rows and $m + 1$ columns, where edges connect pairs of horizontally or vertically adjacent vertices. An example of a 4×6 grid graph is illustrated in Fig. 1. Formally, the vertex and edge sets are given by:

$$V = \{(i, j) \mid 0 \leq i \leq n, 0 \leq j \leq m\}, \quad E = \{\{(i, j), (i', j')\} \mid |i - i'| + |j - j'| = 1\}.$$

Notice that the vertex number $|V| = nm + n + m + 1$ and the edge number $|E| = 2nm + n + m$.

Given an $n \times m$ grid graph $G = (V, E)$, a *rectangle eliminating set* is an edge subset $A \subseteq E$ such that $G - A = (V, E - A)$ contains no rectangular outlines. Let $S(n, m)$ be the minimum cardinality of the rectangle eliminating set, referred to as the *rectangle eliminating number.*

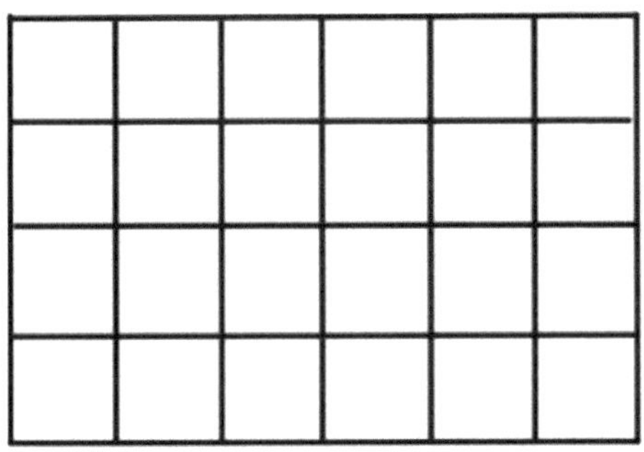

Fig. 1. A 4×6 Grid Graph.

The following explains that the rectangle eliminating set in grid graph $G = (V, E)$ can be viewed as a hitting set. Define the family of subsets $\mathcal{C} \subseteq 2^E$ as follows: each set in $\mathcal{C}$ consists of the edges that form a rectangular outline in G.

The rectangle eliminating set $A \subseteq E$ interacts with every subset in $\mathcal{C}$, i.e., A and any rectangle in the grid graph share at least one edge. Thus, A is a hitting set of $\mathcal{C}$.

The rectangle eliminating problem in grid graphs is also closely related to the *tromino tiling problem*, as deleting the shared edges of two adjacent squares can be interpreted as joining them into a single unit. Figure 2a shows a *tromino*, which is the smallest structure that does not contain any rectangles. As we will demonstrate later, tiling a grid graph with trominoes while avoiding rectangular outlines leads to a good rectangle eliminating set.

1.1 Related Works

Paul A. Dreyer and Therese C. Biedl proposed a classical construction in 1997 to solve the rectangle eliminating problem in $n \times m$ grid graphs [5]. This construc-

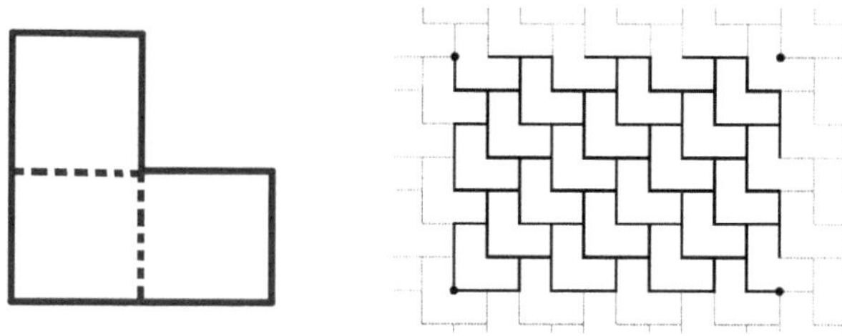

(a) Tromino (b) Infinite Tiling

Fig. 2. A Tromino and one of its Infinite Tiling Patterns.

tion produces a tiling of the plane using trominoes (see Fig. 2b). The authors then tailored an $n \times m$ subgrid graph from this infinite tiling to form an approximate solution. An upper bound on the rectangle eliminating number was provided: $\frac{2nm}{3} + \frac{4(m+n)}{3} + 2$.

We now present some research on tromino tiling problems that are closely related to the rectangle eliminating problem.

In 1954, S. W. Golomb [6] discussed how to tile grid graphs with polyominoes. The author proved the following result:

Proposition 1. *$2^k \times 2^k$ grids can be tiled with a 1×1 square and a number of trominoes, no matter where the 1×1 square is placed on the grid.*

In 1986, I-Ping Chu and Richard Johnsonbaugh [4] established a necessary and sufficient condition for using trominoes to tile a *deficient $n \times n$ grid graph*(i.e., an $n \times n$ grid missing one square).

Proposition 2. *If $n \neq 5$, then a deficient $n \times n$ board can be tiled with trominoes if and only if $3 \nmid n$.*

J. Marshall Ash and Solomon W. Golomb studied how to tile *dog-eared rectangles* using trominoes in 2004 [2], where dog-eared rectangle is an $n \times m$ rectangle with the upper right 1×1 square removed.

In 2018, Befumo, Arthur and Lenchner, Jonathan extended the problem to higher dimensions [3]. A *L-tilable board* is a board that can be tiled with trominoes in a high dimension. For example, $3 \times 2 \times 1$ board is L-tilable because it can be tiled with 2 trominoes in three-dimensional space. A $(K_1 \times \cdots \times K_N) - 1$ board is a $K_1 \times \cdots \times K_N$ dimensional grid graph with a $1 \times \cdots \times 1$ cube removed, where K_i denotes the length of the i-th dimension of the grid graph.

Proposition 3. *A $(K_1 \times \cdots \times K_N) - 1$ board, where $K_1 \times \cdots \times K_N \equiv 1 \pmod 3$, for $N \geq 3$ and some three of the $K_i > 1$, is always L-tilable.*

1.2 Main Results

Periodic Sampling Construction. Considering an $n \times m$ grid graph, the idea of periodic sampling construction is inspired by the classical construction in [5]. By introducing a new infinite tiling pattern using trominoes, we derive an improved upper bound:

$$S(n,m) \le \min\{f(n,m), f(m,n)\},$$

where $f : \mathbb{Z}^+ \times \mathbb{Z}^+ \to \mathbb{R}^+$ is defined as follows:

$$f(x,y) = \begin{cases} \frac{2xy}{3} + \frac{x}{3}, & x \equiv 0 \pmod 3 \\ \frac{2xy}{3} + \frac{x+y}{3} - \frac{1}{3}, & x \equiv 1 \pmod 3 \\ \frac{2xy}{3} + \frac{y}{6} + \frac{x}{3} - \frac{1}{6}, & x \equiv 2 \pmod 3 \text{ and } y \text{ is odd} \\ \frac{2xy}{3} + \frac{y}{6} + \frac{x}{3} - \frac{2}{3}, & x \equiv 2 \pmod 3 \text{ and } y \text{ is even.} \end{cases}$$

Binary Decomposition Construction. Define a function $l : \mathbb{N}^+ \to \mathbb{N}^+$ as follows: For every $n \in \mathbb{N}^+$, $l(n)$ represents the number of 1*s* in the binary expression of n. We systematically breakdown the $n \times n$ grid graph layer by layer based on the binary representation of n, with each layer being tiled by multiple solutions of the $2^i \times 2^i$ grid graphs ($1 \le i \le k$) as outlined in Golomb's paper [6]. The final upper bound obtained is:

$$S(n,n) \le \frac{2}{3}n^2 + \frac{2}{3} \sum_{1 \le j < i \le l(n)} 2^{k_j - k_i} - \frac{2l(n)}{3} + 1,$$

where $n = 2^{k_1} + 2^{k_2} + \cdots + 2^{k_l}$ with $k_1 > k_2 > \cdots > k_l \ge 1$.

Hybrid Construction. In an $n \times n$ square grid graph where $n \ge 32$, we divide the graph into two sections: one section is tiled using the binary decomposition construction, while the other section is tiled using the periodic sampling construction. Subsequently, we establish an improved upper bound by integrating the two preceding constructions:

$$S(n,n) \le \frac{2}{3}n^2 + \frac{40}{3}\sqrt{2} \cdot \sqrt{n} + 3.$$

2 Preliminaries

In this section, we give a lower bound on $S(n,m)$ and show that $S(n,n) = \frac{2}{3}n^2 + \frac{1}{3}$ where $n = 2^k$ for all $k \in \mathbb{N}^+$.

Lemma 1. *For an $n \times m$ grid graph $G = (V,E)$, let $A^* \subset E$ be an arbitrary optimal rectangle eliminating set. Then $G - A^*$ is connected.*

Proof. Denote the graph obtained after edge deletion as $G' = G - A^*$. Suppose that G' has $p \ge 2$ connected components: $V_1, V_2, \ldots, V_p$. Assume that every edge in G has a length of 1. For any vertices $u, v \in V$, denote the shortest distance between u and v by $d(u,v)$. Define the distance $d(V_i,V_j) = \min_{v_i \in V_i, v_j \in V_j}\{d(v_i,v_j)\}$ between the connected components. Clearly, $\min_{1 \le i < j \le p} d(V_i,V_j) = 1$.

Without loss of generality, let $d(V_1,V_2) = d(v_1,v_2) = 1$ where $v_1 \in V_1$ and $v_2 \in V_2$. The resulting graph is then designated as G'' after we add the edge (v_1,v_2) to G'. If G'' contains a rectangle by adding edge (v_1,v_2), then there must exist another path from v_1 to v_2, which contradicts the fact that V_1 and V_2 are not connected in G'. Therefore, $A^* \setminus \{(v_1,v_2)\}$ is a rectangle eliminating set but $|A^* \setminus \{(v_1,v_2)\}| < |A^*|$ which derives a contradiction. Thus, G' is connected. □

It is shown in Lemma 1 that the grid graph remains connected after deleting an optimal rectangle eliminating set. Let F represent the set of faces, and define A^* as an optimal rectangle eliminating set. Then the number of remaining edges can be expressed as $|E| - |A^*| = 2nm + m + n - |A^*|$. It follows from Euler's theorem for planar graphs: $(nm+n+m+1) - (2nm+m+n-|A^*|) + |F| = 2$. Then we derive the relationship between $|F|$ and $|A^*|$:

$$|F| + |A^*| = nm + 1. \tag{1}$$

We observe that maximizing the number of faces in the resulting planar graph is equivalent to minimizing the rectangle eliminating number. This graph attains the maximum number of faces because each tromino comprises exactly three squares. The following provides a lower bound for $S(n,m)$.

Lemma 2.

$$S(n,m) \ge nm - \lfloor \frac{nm}{3} \rfloor.$$

Proof. The maximum number of trominoes that can tile an $n \times m$ grid graph is given by $\lfloor \frac{nm}{3} \rfloor$. When including the outer unbounded face, the total number of faces in the tiling satisfies:

$$|F| \le |\frac{nm}{3}| + 1.$$

By (1), we establish the lower bound for $|A^*|$:

$$|A^*| \ge nm - \lfloor \frac{nm}{3} \rfloor.$$

□

For square grid graph with $n = 2^k$, an upper bound on $S(n,n)$ can be derived essentially from S. W. Golomb's construction (Fig. 3).

Lemma 3. *[6]*

$$S(n,n) \le \frac{2}{3}n^2 + \frac{1}{3},$$

where $n = 2^k$ for all $k \in \mathbb{N}^+$.

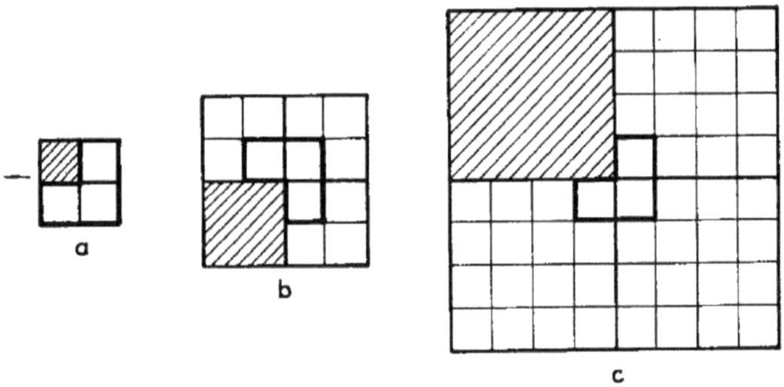

Fig. 3. Diagram for Inductive Proof.

Proof. One tromino and one 1×1 square (which is assumed, without losing generality, to be in the upper left corner) can be used to tile a 2×2 grid. Assume that a $2^k \times 2^k$ grid can be tiled without creating any rectangular outlines using a number of trominoes and one 1×1 square. Suppose that the 1×1 square is placed in the upper left corner.

Now, consider a $2^{k+1} \times 2^{k+1}$ grid. Divide this grid into four equal quadrants. Assume the 1×1 square is located in the upper left corner of the second quadrant.

Consider three unit squares positioned as follows: one in the lower left corner of the first quadrant, one in the third quadrant's upper right corner, and one in the fourth quadrant's upper left corner. A tromino is formed by the union of these three squares.

The remaining areas of each quadrant can now be completely tiled with trominoes using the $2^k \times 2^k$ tiling procedure by induction. Since the area where the four quadrants meet is fully enclosed, the tiling process will not produce any rectangular outlines.

Lastly, a rectangle eliminating set for the $2^{k+1} \times 2^{k+1}$ grid is produced by deleting one edge from the boundary around the upper left 1×1 square. The upper bound for the rectangle eliminating number is $1 + \frac{2(n^2-1)}{3} = \frac{2}{3}n^2 + \frac{1}{3}$ where $n = 2^k$. □

The exact value of $S(n, n)$ is in fact the upper bound established by Lemma 3.

Lemma 4.

$$S(n,n) = \frac{2}{3}n^2 + \frac{1}{3}, \tag{2}$$

where $n = 2^k$ *for all* $k \in \mathbb{N}^+$.

Proof. Since $2^{2k} = 4^k \equiv 1 \pmod 3$:

$$S(n,n) \geq 2^{2k} - \lfloor \frac{2^{2k}}{3} \rfloor = n^2 - (\frac{n^2}{3} - \frac{1}{3}) = \frac{2}{3}n^2 + \frac{1}{3}.$$

The upper bound of $S(n,n)$ is given by Lemma 3, so we conclude that the exact value of $S(n,n)$ is:

$$S(n,n) = \frac{2}{3}n^2 + \frac{1}{3}.$$

□

3 Periodic Sampling Construction

As illustrated in Fig. 4a, we tile the trominoes for an $n \times m$ grid graph.

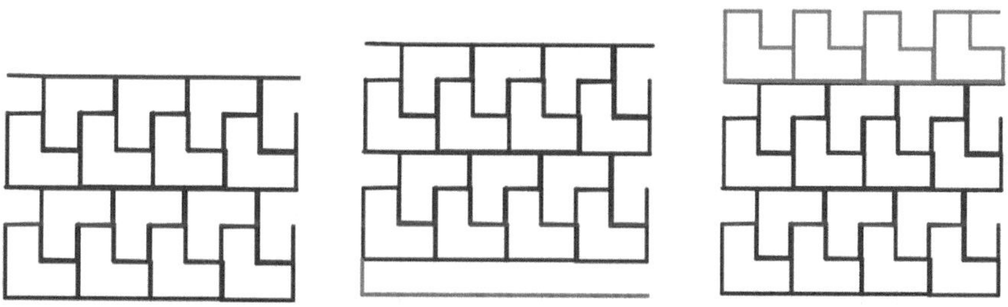

(a) $n \equiv 0 \pmod 3$ (b) $n \equiv 1 \pmod 3$ (c) $n \equiv 2 \pmod 3$

Fig. 4. Periodic Tiling.

Theorem 1. *The periodic sampling construction gives an upper bound for the rectangle eliminating number:*

$$S(n,m) \leq \min\{f(n,m), f(m,n)\},$$

where $f : \mathbb{Z}^+ \times \mathbb{Z}^+ \to \mathbb{R}^+$ is defined as follows:

$$f(x,y) = \begin{cases} \frac{2xy}{3} + \frac{x}{3}, & x \equiv 0 \pmod 3 \\ \frac{2xy}{3} + \frac{x+y}{3} - \frac{1}{3}, & x \equiv 1 \pmod 3 \\ \frac{2xy}{3} + \frac{y}{6} + \frac{x}{3} - \frac{1}{6}, & x \equiv 2 \pmod 3 \text{ and } y \text{ is odd} \\ \frac{2xy}{3} + \frac{y}{6} + \frac{x}{3} - \frac{2}{3}, & x \equiv 2 \pmod 3 \text{ and } y \text{ is even.} \end{cases} \tag{3}$$

Proof. The proof is divided into 3 cases of n.

The case of $n \equiv 0 \pmod 3$ is considered first. According to Fig. 4a, there are precisely $m-1$ trominoes and 2 deficient trominoes for every three rows. Therefore, the cardinality of the rectangle eliminating set is as follows:

$$S_1 = \frac{n}{3}[2(m-1)+3] = \frac{2nm}{3} + \frac{n}{3}.$$

Consider the case where $n \equiv 1 \pmod 3$. An $(n-1)\times m$ grid can be extended to an $n \times m$ grid by adding one additional row (see Fig. 4b). The cardinality of the rectangle eliminating set is:

$$S_2 = \frac{2(n-1)m}{3} + \frac{n-1}{3} + m = \frac{2nm}{3} + \frac{n+m}{3} - \frac{1}{3}.$$

Consider the case where $n \equiv 2 \pmod 3$. An $n \times m$ grid can be obtained from an $(n-2) \times m$ grid by adding 2 rows. (see Fig. 4c). When m is even, the cardinality of the rectangle eliminating set is:

$$S_3 = \frac{2\,m(n-2)}{3} + \frac{n-2}{3} + \frac{m}{2} \times 3 = \frac{2nm}{3} + \frac{m}{6} + \frac{n}{3} - \frac{2}{3}.$$

When m is odd, the cardinality of the rectangle eliminating set is written as:

$$S_3 = \frac{2\,m(n-2)}{3} + \frac{n-2}{3} + \frac{m-1}{2} \times 3 + 2 = \frac{2nm}{3} + \frac{m}{6} + \frac{n}{3} - \frac{1}{6}.$$

We divide the values of m into three cases in a similar way. An upper bound for the rectangle eliminating number is obtained by choosing the minimum of the upper bounds in each situation:

$$S(n,m) \le \min\{f(n,m), f(m,n)\}.$$

□

4 Binary Decomposition Construction

For every $n \in \mathbb{N}^+$, recall that $l(n)$ represents the number of 1s in the binary expression of n.

4.1 Construction Method

For $2^k \times 2^k$ grid graphs, we utilize S. W. Golomb's construction. In this construction, the optimal cardinality of the rectangle eliminating set is $(2n^2+1)/3$ where $n = 2^k$.

It can be expressed as follows for an even number n:

$$n = a_k 2^k + a_{k-1} 2^{k-1} + \cdots + a_1 2 + a_0,$$

where $a_k = 1$ and $a_0 = 0$. From now on we will simply say:

$$n = (a_k, a_{k-1}, a_{k-2}, \ldots, a_1, a_0).$$

It is evident that $2^k \le n < 2^{k+1}$ for $n = (a_k, a_{k-1}, a_{k-2}, \ldots, a_1, a_0)$. We assume the situation $k \ge 5$ in this study.

The $n \times n$ grid graph is decomposed layer by layer using the binary expression for n. Multiple solutions of the $2^i \times 2^i$ grid graphs $(1 \le i \le k)$ are used to tile

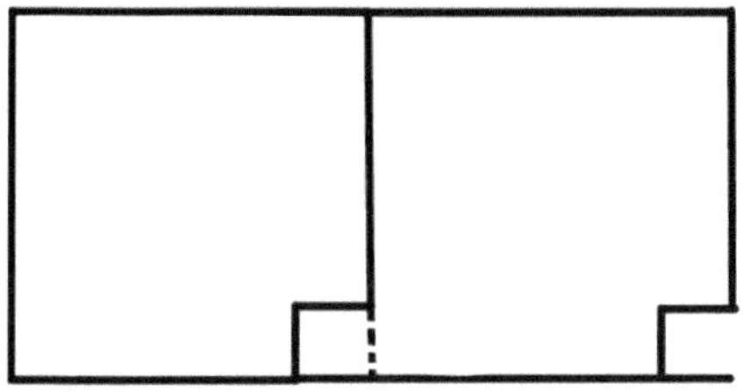

Fig. 5. Diagram of the Composition of Two Solutions of $2^i \times 2^i$.

each layer. During the tiling process, we remove the dotted lines (see Fig. 5) and assign this removed edge to the $2^i \times 2^i$ on the left. Then the size of the rectangle eliminating set for each $2^i \times 2^i$ still is: $s_i = \frac{2}{3}(2^i) + \frac{1}{3}$.

For example: $6 = 2^2 + 2$, we can decompose 6×6 graph as shown in Fig. 6a.

Theorem 2. *The binary decomposition construction gives an upper bound for the rectangle eliminating number:*

$$S(n,n) \le \frac{2}{3}n^2 + \frac{2}{3}\sum_{1 \le j < i \le l(n)} 2^{k_j - k_i} - \frac{2l(n)}{3} + 1,$$

where $n = 2^{k_1} + 2^{k_2} + \cdots + 2^{k_l}$ *with* $k_1 > k_2 > \cdots > k_l \ge 1$.

Proof. Let the set $U = \{i \mid a_i = 1\}$ and $|U| = l(n)$. Denote the elements of the set U by a_{k_j}, $j = 1, 2, \ldots, l(n)$ and $a_{k_j} > a_{k_{j+1}}$. Obviously $a_{k_1} = a_k$. We decompose the $n \times n$ gird graph into $l(n)$ layers, and let the ith layer consist of optimal solutions of $2^{k_i} \times 2^{k_i}$. Suppose that the number of optimal solutions of $2^{k_i} \times 2^{k_i}$ is n_i, then we notice that n_i satisfies:

$$\begin{aligned} n_1 &= 1 \\ n_2 &= 2(2^{k_1})2^{-k_2} + 1 \\ &\ldots \\ n_i &= 2(\sum_{j=1}^{i-1} 2^{k_j})2^{-k_i} + 1. \end{aligned}$$

Three trominoes in the upper left corner combine to generate an extra tromino for the ith layer ($i > 1$). The shaded area in Fig. 6b illustrates this. The redesigned tromino configuration reduces the required edge removals from three to two. As a result, the rectangle eliminating set for each layer decreases by one element. Therefore, $S(n,n)$'s upper bound is:

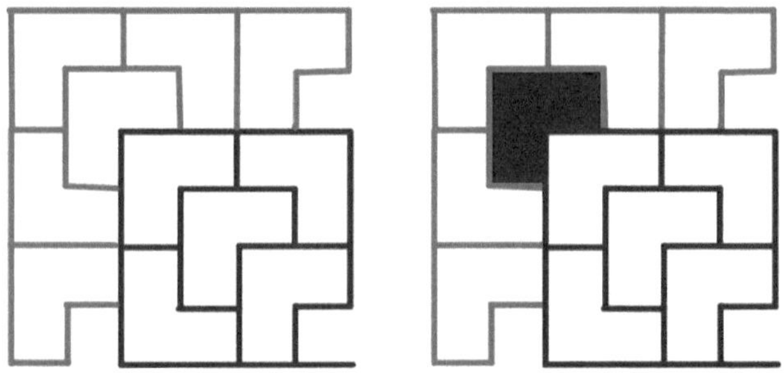

(a) A 6 × 6 Graph (b) Shaded Tromino

Fig. 6. An Example of Binary Decomposition Construction.

$$
\begin{aligned}
&S(n,n)\\
&\leq \sum_{i=1}^{l(n)} s_i n_i - (l(n)-1)\\
&= \sum_{i=1}^{l(n)} (\frac{2}{3}(2^{2k_i}) + \frac{1}{3}) \times (2(\sum_{j=1}^{i-1} 2^{k_j})2^{-k_i} + 1) - (l(n)-1)\\
&= \frac{4}{3} \sum_{1\leq j<i\leq l(n)} 2^{k_j+k_i} + \frac{2}{3}\sum_{i=1}^{l} 2^{2k_i} + \frac{2}{3} \sum_{1\leq j<i\leq l(n)} 2^{k_j-k_i} + \frac{l(n)}{3} - (l(n)-1)\\
&= \frac{4}{3}\cdot\frac{1}{2}\left(n^2 - \sum_{i=1}^{l} 2^{2k_i}\right) + \frac{2}{3}\sum_{i=1}^{l} 2^{2k_i} + \frac{2}{3} \sum_{1\leq j<i\leq l(n)} 2^{k_j-k_i} + \frac{l(n)}{3} - (l(n)-1)\\
&= \frac{2}{3}n^2 + \frac{2}{3} \sum_{1\leq j<i\leq l(n)} 2^{k_j-k_i} - \frac{2l(n)}{3} + 1,
\end{aligned}
$$

which completes the proof. □

4.2 Special Cases

Now we consider the cases where n takes on special values.

Lemma 5. *Suppose that* $n = (1,1,\ldots,1,0,\ldots,0,0)$, *where the number of* 0*s is* $k+1-l(n)$. *Then the upper bound can be written as:*

$$
\begin{aligned}
S(n,n) &\leq \frac{2}{3}n^2 + \frac{2}{3} \sum_{1\leq j<i\leq l(n)} 2^{k_j-k_i} - \frac{2l(n)}{3} + 1\\
&= \frac{2}{3}n^2 + \frac{1}{3\cdot 2^{k-l(n)-1}}n - 2l(n) + 1.
\end{aligned}
$$

Proof. It is clear to note that:

$$\begin{aligned} n &= 2^k + 2^{k-1} + \cdots + 2^{k-l(n)+1} \\ &= 2^{k+1} - 2^{k-l(n)+1} \\ &= 2^{k+1-l(n)}(2^{l(n)} - 1). \end{aligned} \tag{4}$$

For any $1 \leq j < i \leq l(n)$, let $p = k_j - k_i$. For each p (with values in the range $1 \leq p \leq l(n) - 1$), the number of (i, j) pairs satisfying $k_j - k_i = p$ is $l(n) - p$. Consequently, it follows that:

$$\begin{aligned} \sum_{1 \leq j < i \leq l(n)} 2^{k_j - k_i} &= \sum_{p=1}^{l(n)-1} (l(n) - p) \cdot 2^p \\ &= l(n) \sum_{p=1}^{l(n)-1} 2^p - \sum_{p=1}^{l(n)-1} p \cdot 2^p \\ &= l(n)(2^{l(n)} - 2) - \left[2 + (l(n) - 2) \cdot 2^{l(n)}\right] \\ &= 2^{l(n)+1} - 2l(n) - 2. \end{aligned}$$

As shown in (4), it follows that:

$$2^{l(n)} = \frac{n}{2^{k+1-l(n)}} + 1.$$

The following is a simplified version of the upper bound:

$$S(n, n) \leq \frac{2}{3}n^2 + \frac{1}{3 \cdot 2^{k-l(n)-1}}n - 2l(n) + 1.$$

□

Corollary 1. *When* $n = (1, 1, 0, \ldots, 0, 0)$*, the solution obtained by binary decomposition construction is optimal.*

Proof. We have $l(n) = 2$ for $n = (1, 1, 0, \ldots, 0, 0)$, and the following equation is valid:

$$\frac{2}{3} \sum_{1 \leq j < i \leq l(n)} 2^{k_j - k_i} = \frac{4}{3}.$$

The size of the rectangle eliminating set is therefore $\frac{2}{3}n^2 + 1$. We compute and obtain the following expression for n:

$$n = 2^k + 2^{k-1} = 2 \cdot 2^{k-1} + 2^{k-1} = 3 \cdot 2^{k-1}.$$

Substitute $n = 3 \cdot 2^{k-1}$ for the lower bound in Lemma 2:

$$\begin{aligned} n^2 - \lfloor \frac{n^2}{3} \rfloor &= 6 \cdot 2^{2k-2} = 3 \cdot 2^{k-1} \times 2 \cdot 2^{k-1} \\ &= \frac{2}{3}(3 \cdot 2^{k-1})^2 = \frac{2}{3}n^2. \end{aligned}$$

When the size of the rectangle eliminating set reaches the lower bound $\frac{2}{3}n^2$, the grid graph is perfectly tiled with trominoes. Then we still need to eliminate the outermost $n \times n$ rectangle and therefore the optimal value is $\frac{2}{3}n^2 + 1$. □

5 Hybrid Construction

The $n \times n$ grid graph is divided into 3 regions via the hybrid construction (see Fig. 7). Binary decomposition construction is used to solve $(n-c) \times (n-c)$, and periodic sampling construction is used to solve the remaining components. The size of the rectangle eliminating set of these three regions are denoted by $r(n-c)$, $r(c, n)$ and $r(n-c, c)$.

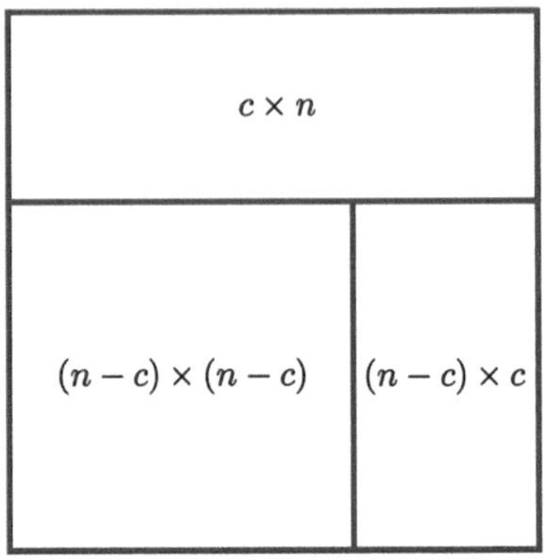

Fig. 7. Hybrid Construction.

Theorem 3. *The hybrid construction gives an upper bound for the rectangle eliminating number:*

$$S(n, n) \leq \frac{2}{3}n^2 + \frac{40}{3}\sqrt{2} \cdot \sqrt{n} + 3. \tag{5}$$

Proof. Let $n = (a_k, a_{k-1}, \ldots, a_1, a_0)$ and $c = (a_{p-1}, a_{p-2}, \ldots, a_1, a_0)$, where $a_k = 1$ and $p = \lfloor \frac{k+1}{2} \rfloor$. Then $n - c = (a_k, a_{k-1}, \ldots, a_{p+1}, a_p, 0, \ldots, 0)$. Let $n' = (1, 1, \ldots, 1, a_{p-1}, a_{p-2}, \ldots, a_1, a_0)$. The proof is divided into 3 cases of n.

Consider the case where $c \equiv 0 \pmod 3$:

$$\begin{aligned} S(n, n) &\leq r(n-c) + r(c, n) + r(n-c, c) \\ &= \frac{2}{3}(n-c)^2 + \frac{2}{3} \sum_{1 \leq j < i \leq l(n-c)} 2^{k_j - k_i} - \frac{2l(n-c)}{3} + 1 \\ &+ \frac{2}{3}cn + \frac{1}{3}c \\ &+ \frac{2}{3}c(n-c) + \frac{1}{3}c \\ &= \frac{2}{3}n^2 + \frac{2}{3} \sum_{1 \leq j < i \leq l(n-c)} 2^{k_j - k_i} - \frac{2l(n-c)}{3} + 1 + \frac{2}{3}c. \end{aligned}$$

It is easy to notice that:

$$\frac{2}{3}\sum_{1\le j<i\le l(n-c)} 2^{k_j-k_i} \le \frac{2}{3}\sum_{1\le j<i\le l(n'-c)} 2^{k_j-k_i} = \frac{2}{3}\cdot\frac{1}{2^{p-1}}\cdot(n'-c).$$

Since $n'-c=(1,1,\ldots,1,0,\ldots,0)\ge c\cdot 2^p=(a_{p-1},a_{p-2},\ldots,a_1,a_0,0,\ldots,0)$ and $2\cdot n\ge 2^{k+1}>n'$:

$$c\cdot 2^p\le n'-c,\quad n'<2\cdot n.$$

Then we derive the upper bound for $S(n,n)$:

$$\begin{aligned}S(n,n)&\le\frac{2}{3}n^2+\frac{2}{3}\cdot\frac{1}{2^{p-1}}\cdot(n'-c)-\frac{2l(n'-c)}{3}+1+\frac{2}{3}c\\&\le\frac{2}{3}n^2+\frac{2}{2^p+1}\cdot n'-\frac{2l(n'-c)}{3}+1\\&\le\frac{2}{3}n^2+\frac{4}{2^p+1}\cdot n-\frac{2l(n'-c)}{3}+1.\end{aligned}$$

According to the definition of p:

$$2p=2\cdot\lfloor\frac{k+1}{2}\rfloor\ge 2\cdot\frac{k}{2}=k.$$

Therefore:

$$\begin{aligned}S(n,n)&\le\frac{2}{3}n^2+\frac{4n}{\sqrt{2^k}}+\frac{1}{3}\\&\le\frac{2}{3}n^2+\frac{4\sqrt{2}n}{\sqrt{n}}+\frac{1}{3}\\&=\frac{2}{3}n^2+4\sqrt{2}\cdot\sqrt{n}+\frac{1}{3}.\end{aligned}$$

Consider the case where $c\not\equiv 0 \pmod 3$. For any $i\in N^+$, $2^i\equiv 1 \pmod 3$ or $2^i\equiv 2 \pmod 3$. Furthermore, $2^i+2^{i+1}\equiv 0 \pmod 3$. So, either $c+2^p\equiv 0 \pmod 3$ or $c+2^{p+1}\equiv 0 \pmod 3$.

Let $c'=c+2^p$ or $c'=c+2^{p+1}$ such that $c'\equiv 0 \pmod 3$. Through the same calculation, we obtain the new upper bounds. The difference from the established case appears in the following inequalities:

$$c'\cdot 2^{p-2}\le n'-c',\quad\text{or}\quad c'\cdot 2^{p-3}\le n'-c'.$$

When $c'=c+2^p$, we have the upper bound:

$$\begin{aligned}S(n,n)&\le\frac{2}{3}n^2+\frac{2}{3}\cdot\frac{1}{2^p}\cdot(n'-c')-\frac{2l(n'-c')}{3}+1+\frac{2}{3}c'\\&\le\frac{2}{3}n^2+\frac{20}{3\cdot 2^p}\cdot n-\frac{2l(n'-c')}{3}+1\\&\le\frac{2}{3}n^2+\frac{20}{3}\sqrt{2}\cdot\sqrt{n}+\frac{1}{3}.\end{aligned}$$

When $c' = c + 2^{p+1}$, we have the upper bound:

$$\begin{aligned} S(n,n) &\le \frac{2}{3}n^2 + \frac{2}{3} \cdot \frac{1}{2^{p-1}} \cdot (n' - c' + 2^{p+1}) - \frac{2l(n'-c')}{3} + 1 + \frac{2}{3}c' \\ &\le \frac{2}{3}n^2 + \frac{40}{3 \cdot 2^p} \cdot n - \frac{2l(n'-c')}{3} + \frac{11}{3} \\ &\le \frac{2}{3}n^2 + \frac{40}{3}\sqrt{2} \cdot \sqrt{n} + 3. \end{aligned}$$

Based on the above analysis, the conclusion holds. □

References

1. Agrawal, M., Saha, C., Saptharishi, R., Saxena, N.: Jacobian hits circuits: hitting-sets, lower bounds for depth-*d* occur-*k* formulas and depth-3 transcendence degree-*k* circuits. In: Proceedings of the forty-fourth annual ACM symposium on Theory of computing, pp. 599–614 (2012)
2. Ash, J.M., Golomb, S.W.: Tiling deficient rectangles with trominoes. Math. Mag. **77**(1), 46–55 (2004)
3. Befumo, A., Lenchner, J.: Tiling one-deficient rectangular solids with trominoes in three and higher dimensions. Math. Mag. **91**(1), 62–69 (2018)
4. Chu, I.P., Johnsonbaugh, R.: Tiling deficient boards with trominoes. Math. Mag. **59**(1), 34–40 (1986)
5. Dreyer, P.A., Jr., Biedl, T.C.: Rectangle breaking in grids. Congr. Numer. **128**, 19–32 (1997)
6. Golomb, S.W.: Checker boards and polyominoes. Am. Math. Mon. **61**(10), 675–682 (1954)
7. Karp, R.M.: Reducibility among combinatorial problems. In: 50 Years of Integer Programming 1958-2008: from the Early Years to the State-of-the-Art, pp. 219–241. Springer (2009)
8. McGuire, G., Tugemann, B., Civario, G.: There is no 16-clue sudoku: solving the sudoku minimum number of clues problem via hitting set enumeration. Exp. Math. **23**(2), 190–217 (2014)
9. Moreno-Centeno, E., Karp, R.M.: The implicit hitting set approach to solve combinatorial optimization problems with an application to multigenome alignment. Oper. Res. **61**(2), 453–468 (2013)
10. Wang, Q., Jin, T., Mohamed, M.A.: An innovative minimum hitting set algorithm for model-based fault diagnosis in power distribution network. IEEE Access **7**, 30683–30692 (2019)
11. Wang, Q., Jin, T., Wang, M.: A hierarchical minimum hitting set calculation method for multiple multiphase faults in power distribution networks. IEEE Trans. Industr. Electron. **68**(1), 4–14 (2020)

Some Combinatorial Algorithms on the Eliminating Edge Feedback Number of Hypergraphs

Zhongzheng Tang[1,2], Haoyang Zou[3], and Zhuo Diao[3(✉)]

[1] School of Mathematical Sciences, Beijing University of Posts and Telecommunications, Beijing 100876, China
tangzhongzheng@amss.ac.cn

[2] Key Laboratory of Mathematics and Information Networks (Beijing University of Posts and Telecommunications), Ministry of Education, Beijing 100876, China

[3] School of Statistics and Mathematics, Central University of Finance and Economics, Beijing 100081, China
diaozhuo@amss.ac.cn

Abstract. Let $H = (V, E)$ be a hypergraph with n vertices and m edges. $A \subseteq E$ is an eliminating edge feedback set of H if $H /\!/ A$ has no cycle and $D'_c(H)$ denote the minimum cardinality of an eliminating edge feedback set of H. In this paper, we prove (i) for any hypergraph H, $D'_c(H) \leq n/2$. (ii) for any hypergraph H with maximum degree $\Delta \leq 2$ and girth $k \geq 4$, $D'_c(H) \leq \frac{n}{4}$. (iii) for any hypergraph H with maximum degree $\Delta \leq 3$ and girth $k \geq 3$, $D'_c(H) \leq \frac{n}{3}$. Based on the proofs, some combinatorial algorithms on the eliminating edge feedback number are designed.

Keywords: hypergraphs · eliminating edge · feedback set · combinatorial algorithms

1 Introduction

A hypergraph is a generalization of a graph in which an edge can join any number of vertices. The loop is an edge which contains only one vertex. The multiple edges are edges which contain same vertices. A multiple hypergraph is a hypergraph allowing loops and multiple edges. Let $H(V, E)$ be a multiple hypergraph with vertex set V and edge set E. As for a graph, the order of H, denoted by n, is the number of vertices. The number of edges is denoted by m. The studies of this paper are applicable to multiple hypergraphs.

For each vertex $v \in V$, the **degree** $d(v)$ is the number of edges in E that contains v. The maximum degree is denoted by Δ. We say v is an isolated vertex

Supported by National Natural Science Foundation of China under Grant No.11901605, No.12101069, the Fundamental Research Funds for the Central Universities, the disciplinary funding of Central University of Finance and Economics.

M. Li et al. (Eds.): TAMC 2025, LNCS 16084, pp. 221–234, 2026.
https://doi.org/10.1007/978-981-95-4839-2_17

of H if $d(v) = 0$. Hypergraph H is **k-regular** if each vertex's degree is k. For each edge $e \in E$, $|e|$ is the number of vertices in e. Hypergraph H is **k-uniform** if each edge has k vertices. The **rank** of H is the maximum number of vertices contained in any edge of H, denoted by R. Hypergraph H is **linear** if every two distinct edges share at most one common vertex.

Let $k \geq 2$ be an integer. A cycle of length k, denoted as k-cycle, is a vertex-edge sequence $C = v_1 e_1 v_2 e_2 \cdots v_k e_k v_1$ with: (1)$\{e_1, e_2, \ldots, e_k\}$ are distinct edges of H. (2)$\{v_1, v_2, \ldots, v_k\}$ are distinct vertices of H. (3)$\{v_i, v_{i+1}\} \subseteq e_i$ for each $i \in [k]$, here $v_{k+1} = v_1$. The **girth** of H is the length of the minimum cycle in H. For any vertex set $S \subseteq V$, we write $H \setminus S$ for the subhypergraph of H obtained from H by deleting all vertices in S and all edges incident with some vertices in S. For any edge set $A \subseteq E$, we write $H \setminus A$ for the subhypergraph of H obtained from H by deleting all edges in A and keeping vertices. If S is a singleton set $\{s\}$, we write $H \setminus s$ instead of $H \setminus \{s\}$. For any vertex set $S \subseteq V$, the **induced subhypergraph** by S is $H_S = (S, E_S)$ where the edge set $E_S = \{e \cap S \mid e \in E\} \setminus \{\varnothing\}$.

1.1 Related Works

Let $G = (V, E)$ be a simple graph with vertex set V and edge set E. $S \subseteq V$ is a feedback vertex set of G if $G \setminus S$ has no cycle and the feedback vertex number $\tau_c(G)$ denotes the minimum cardinality of a feedback vertex set of G. Feedback vertex sets have been extensively studied [1–9]. A direction in the study of feedback vertex sets is to find an upper bound on the minimum size of a feedback vertex set.

An old conjecture due to Albertson and Berman is challenging:

Conjecture 1. (Albertson and Berman [10]) Every planar graph G of order n admits an induced forest of order at least $n/2$, that is $\tau_c(G) \leq n/2$.

Albertson and Berman's conjecture remains open. The best known result is that every planar graph G on n vertices has a feedback vertex set with size at most $3n/5$, due to Borodin's acyclic 5-coloring theorem [11].

Let the girth of G be the minimum cycle length of G. Define k-girth graph as a graph with the girth at least k. The minimum size of feedback vertex sets is related to the girth of a graph. There are many literatures on this issue [12–15]:

Theorem 1. *(Alon et al. [12]) Every triangle-free graph of order n and size m admits a feedback vertex set at most $m/4$.*

For a planar k-girth graph G with size m, it is easy to prove G admits a feedback vertex set of size at most $2m/k$. Furthermore, Dross et al. [13] made the following conjecture:

Conjecture 2. (Dross et al. [13]) Let G be a planar graph of size m and girth k. There exists a feedback vertex set S of G of size at most m/k.

In the same paper, Dross et al. [13] proved the following theorem, which is a first non-trivial step toward conjecture:

Theorem 2. *(Dross et al. [13]) Let G be a planar graph of size m and girth k. There exists a feedback vertex set S of G of size at most $\frac{4m}{3k}$.*

Tom Kelly and Chun-Hung Liu [14] proved the following theorem and corollary, which solved Conjecture 2 for the case $k = 5$:

Theorem 3. *(Kelly and Liu [14]) If G is a connected planar graph of girth at least five on n vertices and m edges, then $\tau_c(G) \leq (2m - n + 2)/7$.*

Corollary 1. *(Kelly and Liu [14]) If G is a connected planar graph of girth at least five on n vertices and m edges, then $\tau_c(G) \leq m/5$.*

1.2 Our Results

In this paper, along a different line, we define a variation of feedback number called eliminating edge feedback number. We prove some sharp upper bounds for the eliminating edge feedback number. The results are stated as follows:

- H is a hypergraph with n vertices, the eliminating edge feedback number $D'_c(H) \leq \frac{n}{2}$.
- H is a hypergraph with maximum degree $\Delta \leq 2$ and girth $k \geq 4$, the eliminating edge feedback number $D'_c(H) \leq \frac{n}{4}$. A polynomial time algorithm is designed to compute an eliminating edge feedback set with cardinalities at most $\frac{n}{4}$.
- H is a hypergraph with maximum degree $\Delta \leq 3$ and girth $k \geq 3$, the eliminating edge feedback number $D'_c(H) \leq \frac{n}{3}$. A polynomial time algorithm is designed to compute an eliminating edge feedback set with cardinalities at most $\frac{n}{3}$.

2 The Eliminating Feedback Number

In this section, for a hypergraph with maximum degree Δ and girth k, some sharp upper bounds of the eliminating edge feedback number are proved. The contents are organized as follows:

- In Definition 1, 2 and 3, the eliminating edge feedback number is introduced.
- In Theorem 5, for every hypergraph, an upper bound of the eliminating edge feedback number is proved.
- In Theorem 6 and Corollary 2, for every hypergraph with maximum degree $\Delta \leq 2$ and girth $k \geq 4$, an upper bound of the eliminating edge feedback number is proved.

- In Theorem 7 and Corollary 3, for every hypergraph with maximum degree $\Delta \leq 3$ and girth $k \geq 3$, an upper bound of the eliminating edge feedback number is proved.

Our main results are summarized in Theorem 4:

Theorem 4. *Let H be a hypergraph on vertices with maximum degree Δ and girth k, $D'_c(H)$ is the eliminating edge feedback number of H. The next conclusions hold:*

- $D'_c(H) \leq n/2$.
- *If maximum degree $\Delta \leq 2$ and girth $k \geq 4$, then $D'_c(H) \leq n/4$.*
- *If maximum degree $\Delta \leq 3$ and girth $k \geq 3$, then $D'_c(H) \leq n/3$.*

For convenience, we introduce two useful operations: eliminating vertices and eliminating edges.

Definition 1. *In hypergraph $H(V, E)$, given a vertex set $S \subseteq V$, let $H /\!/ S$ be the sub-hypergraph induced by the vertex set $V \setminus S$. We write $H /\!/ S$ for eliminating vertex set S. An instance is shown in Fig. 1(b).*

Definition 2. *In hypergraph $H(V, E)$, given an edge set $A \subseteq E$, let $H /\!/ A$ be the sub-hypergraph induced by the vertex set $V \setminus (\cup_{e \in A} e)$. We write $H /\!/ A$ for eliminating edge set A. An instance is shown in Fig. 1(c).*

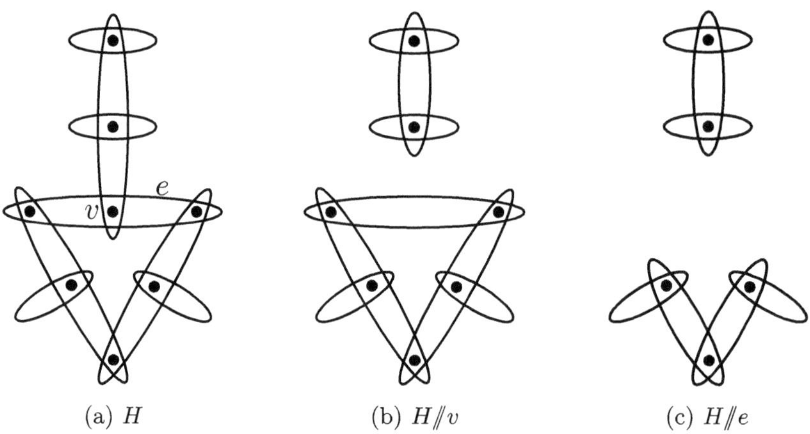

Fig. 1. The operations of vertex eliminating and edge eliminating

Definition 3. *Given a hypergraph $H(V, E)$, a set of edges $A \subseteq E$ is called an eliminating edge feedback set if $H /\!/ A$ has no cycle. The minimum cardinality of the eliminating edge feedback set is called eliminating edge feedback number, denoted as $D'_c(H)$. An instance is shown in Fig. 2.*

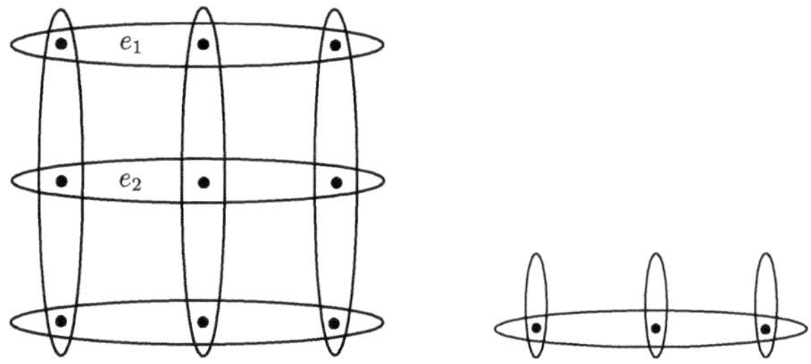

(a) H (b) $H/\!/A$

Fig. 2. The minimum eliminating edge feedback set $A = \{e_1, e_2\}$ of H

Remark 1. The operation of vertex eliminating docs not change the degree of other vertices in hypergraphs. Specifically, when a vertex eliminating operation is performed on a k-regular hypergraph, the resulting hypergraph is still k-regular.

We first establish a simple upper bound for the eliminating edge feedback number to warm-up.

Theorem 5. *Let H be a hypergraph with n vertices. Then $D'_c(H) \leq n/2$.*

Proof. We will prove $D'_c(H) \leq n/2$. Suppose the theorem fails. Let us take out a counterexample $H = (V, E)$ with minimum number of vertices, thus $D'_c(H) > n/2$. We will get a contradiction through the next Claim 1.

Claim 1. $\forall e \in H, |e| \leq 1$.

If there exists $e \in H$ with $|e| \geq 2$, because $H = (V, E)$ is a counterexample with minimum number of vertices, thus we have

$$D'_c(H/\!/e) \leq (n - |e|)/2 \leq (n - 2)/2$$

Considering a minimum eliminating edge feedback set A of $H' = H/\!/e$, we have $A \subseteq E \backslash e$ and $|A| \leq (n-2)/2$. Thus $A \cup e$ is an eliminating edge feedback set for H and $|A \cup e| = |A| + 1 \leq n/2$, which is in contradiction with $D'_c(H) > n/2$.

The Claim 1 tells us $\forall e \in H, |e| \leq 1$, which means H has no cycle. Thus $D'_c(H) = 0$ leads to a contradiction with $D'_c(H) > n/2$ immediately. □

We then establish the upper bound on the eliminating edge feedback number for n-vertex hypergraphs with maximum degree $\Delta \leq 2$ and girth $k \geq 4$. Theorem 6 first addresses the 2-regular case, while Corollary 2 subsequently extends this result to general hypergraphs with $\Delta \leq 2$.

Theorem 6. *Let H be a 2-regular hypergraph with n vertices. If the girth of H is at least 4, $D'_c(H) \leq n/4$.*

Proof. Suppose the theorem fails. Let us take out a counterexample $H = (V, E)$ with minimum number of vertices, thus $D'_c(H) > n/4$. We will break the proof into a series of claims.

Claim 2. H is connected.

Suppose that H is not connected. Denote the components of H as $\{H_i \mid i \in [k]\}$ for $k \geq 2$ and n_i is the vertex number of H_i for $i \in [k]$. Because every component has at least one vertex and H is a counterexample with minimum number of vertices, thus we have

$$D'_c(H_i) \leq n_i/4,\ i \in [k] \ \Rightarrow\ D'_c(H) = \sum_{i\in[k]} D'_c(H_i) \leq \sum_{i\in[k]} n_i/4 = n/4.$$

This is a contradiction with $D'_c(H) > n/4$.

Claim 3. Every vertex is contained in some cycles of H.

If there exists $v \in V$ which doesn't belong to any cycle of H, then $D'_c(H) = D'_c(H/\!\!/v)$. Because H is a counterexample with minimum number of vertices, we have $D'_c(H/\!\!/v) \leq (n-1)/4$. Thus we have

$$D'_c(H) = D'_c(H/\!\!/v) \leq (n-1)/4.$$

This is a contradiction with $D'_c(H) > n/4$.

Claim 4. $\forall e \in H,\ |e| \leq 3$.

If there exists $e \in H$ with $|e| \geq 4$ in H, we have $D'_c(H/\!\!/e) \leq (n-|e|)/4 \leq (n-4)/4$ because H is a counterexample with minimum number of vertices. Considering a minimum eliminating edge feedback set A of $H' = H/\!\!/e$, we have $A \subseteq E \setminus e$ and $|A| \leq (n-4)/4$. Thus $A \cup \{e\}$ is an eliminating edge feedback set for H and $|A \cup \{e\}| = |A| + 1 \leq n/4$. Thus we have

$$D'_c(H) \leq D'_c(H/\!\!/e) + 1 \leq |A| + 1 \leq (n-4)/4 + 1 = n/4.$$

This is a contradiction with $D'_c(H) > n/4$.

Claim 5. $\forall e \in H,\ |e| \geq 2$.

Take a vertex $v \in e$. By Claim 3, v is contained in some cycle C, thus there are two adjacent edges of C connected to v. Due to 2-regularity of H, we know that e is one of two adjacent edges incident to v. Thus $|e| \geq 2$.

Claim 6. There are no two adjacent edges e_1, e_2 such that $|e_1| = 2, |e_2| = 3$.

Assume e_1, e_2 are two adjacent edges and $|e_1| = 2, |e_2| = 3$. H is linear, so let us denote $e_1 = \{v_1, v_2\}$ and $e_2 = \{v_2, v_3, v_4\}$. We have $D'_c(H/\!\!/\{v_1, v_2, v_3, v_4\}) \leq (n-4)/4$ because H is a counterexample with minimum number of vertices. Considering a minimum eliminating edge feedback set A of $H' = H/\!\!/\{v_1, v_2, v_3, v_4\}$, we have $A \subseteq E \setminus \{e_1, e_2\}$ and $|A| \leq (n-4)/4$. Thus $A \cup \{e_2\}$ is an eliminating edge feedback set for H and $|A \cup \{e_2\}| = |A| + 1 \leq n/4$. Thus we have

$$D'_c(H) \leq D'_c(H/\!\!/\{v_1, v_2, v_3, v_4\}) + 1 \leq (n-4)/4 + 1 = n/4.$$

This is a contradiction with $D'_c(H) > n/4$. A figure is shown in Fig. 3(a).

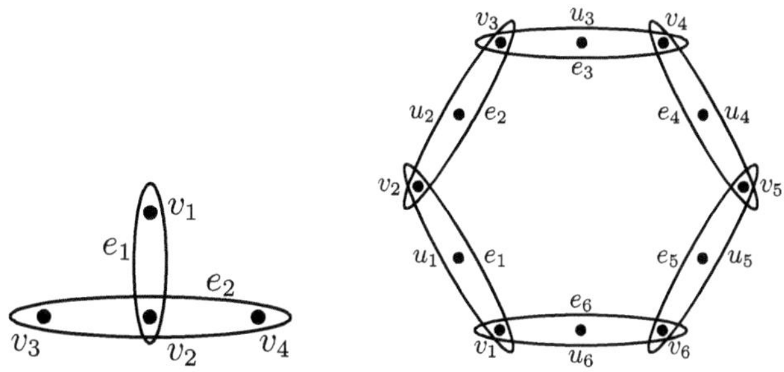

(a) e_1 and e_2 (b) An even cycle C

Fig. 3. Schematic diagrams in Claims 6 and 8

Claim 7. H is 3-uniform.

By Claims 4 and 5, any edge of H contains 2 or 3 vertices. H is connected. By Claim 6, H is 2-uniform or 3-uniform. If H is 2-uniform and 2-regular, H is a simple cycle C in graph. The girth of H is at least 4, so the cycle C has at least 4 vertices. Thus $D_c'(H) = 1 \leq n/4$. This is a contradiction with $D_c'(H) > n/4$. So we know H is 3-uniform.

Let $C = v_1 e_1 v_2 e_2 \dots v_k e_k v_1$ be a shortest cycle in H, here $k \geq 4$. Due to 3-uniform, for $1 \leq i \leq k$, e_i has a vertex u_i that is distinct from v_i and v_{i+1}. C is the shortest cycle, thus $u_i, 1 \leq i \leq k$ are distinct vertices. Denote $V(C) = \{v_i \mid 1 \leq i \leq k\} \cup \{u_i \mid 1 \leq i \leq k\}$, $E(C) = \{e_i \mid 1 \leq i \leq k\}$ for convenience.

Claim 8. C is not an even cycle.

If C is an even cycle, we can assume $k = 2t$ with $t \geq 2$. Then $D_c'(H /\!\!/ V(C)) \leq (n-4t)/4$ because H is a counterexample with minimum number of vertices. Considering a minimum eliminating edge feedback set A of $H' = H /\!\!/ V(C)$, we have $A \subseteq E$ and $|A| \leq (n-4t)/4$. Due to 2-regularity of H, we have $A \cup \{e_{2i} \mid 1 \leq i \leq t\}$ is an eliminating edge feedback set for H and $|A \cup \{e_{2i} \mid 1 \leq i \leq t\}| \leq |A| + t \leq n/4$, which is a contradiction with $D_c'(G) > n/4$. A figure is shown in Fig. 3(b).

Claim 9. C is not an odd cycle.

If C is an odd cycle, we can assume $k = 2t+1$ with $t \geq 2$.

- Consider the hypergraph $H' = H /\!\!/ (V(C) \setminus \{v_1, u_1, u_{2t+1}\})$. Then $D_c'(H') \leq [n-(4t-1)]/4$ because H is a counterexample with minimum number of vertices. A is a minimum eliminating edge feedback set of H'. Due to 2-regularity of H, we have $A \cup \{e_{2i} \mid 1 \leq i \leq t\}$ is an eliminating edge feedback set for H so we have $D_c'(H) \leq D_c'(H') + t$. A figure is shown in Fig. 4(a)(b).
- Consider the hypergraph $H' /\!\!/ \{v_1, u_{2t+1}\}$. H'' is the hypergraph by adding a new vertex $\tilde{v_1}$ to connect e_1 and e_{2t+2} in $H' /\!\!/ \{v_1, u_{2t+1}\}$. On the one side, H'' is 2-regular and the girth of H'' is at least 4. Otherwise, if H'' has a

triangle, H' has a 4-cycle, which is a contradiction with the odd length of the minimum cycle in H. On the other side, H'' has $(n-4t)/4$ vertices and any eliminating edge feedback set of H'' is also an eliminating edge feedback set of H'. Thus $D'_c(H') \leq D'_c(H'') \leq (n-4t)/4$. A figure is shown in Fig. 4(b)(c).

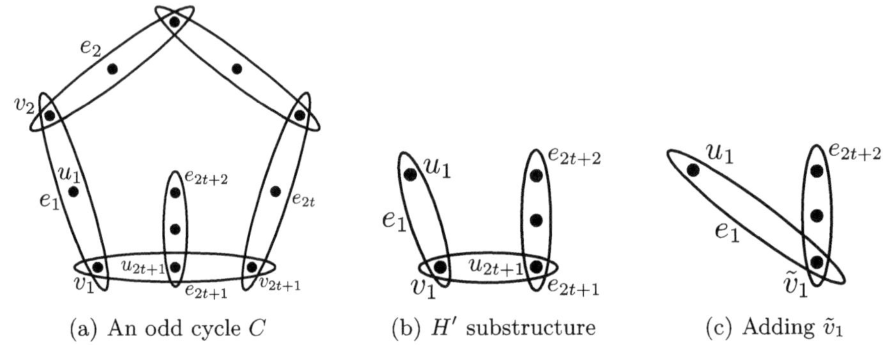

Fig. 4. Schematic diagrams in Claim 9

So we have $D'_c(H) \leq D'_c(H')+t \leq D'_c(H'')+t \leq (n-4t)/4+t = \frac{n}{4}$. This is a contradiction with $D'_c(H) > n/4$. Obviously, it is impossible that both Claim 8 and Claim 9 hold. Thus there is always a contradiction. We finish the proof. □

For a hypergraph H with maximum degree $\Delta \leq 2$ and girth $k \geq 4$, we can construct a 2-regular hypergraph H' by adding some loops to the vertices and these operations keep the eliminating edge feedback number. According to Theorem 6, the next corollary is instant:

Corollary 2. *Let H be a hypergraph with maximum degree $\Delta \leq 2$ and girth $k \geq 4$. Then $D'_c(H) \leq n/4$.*

Following the proof of Theorem 6, we design a combinatorial algorithm for finding an eliminating edge feedback set of cardinality at most $n/4$ for n-vertex hypergraphs with maximum degree $\Delta \leq 2$ and girth $k \geq 4$. The corresponding algorithm (Algorithm 1) is presented in the appendix.

Next, we establish the upper bound on the eliminating edge feedback number for n-vertex hypergraphs with maximum degree $\Delta \leq 3$ and girth $k \geq 3$. Theorem 7 first addresses the 3-regular case, while Corollary 3 subsequently extends this result to general hypergraphs with $\Delta \leq 3$.

Theorem 7. *Let H be a 3-regular hypergraph with n vertices. If the girth of H is at least 3, $D'_c(H) \leq n/3$.*

Proof. Suppose the theorem fails. Let us take out a counterexample $H = (V, E)$ with minimum number of vertices, thus $D'_c(H) > n/3$. Obviously $n \geq 3$. We will break the proof into a series of claims.

Claim 10. Every vertex is contained in some cycles of H.

If there exists $v \in V$ which doesn't belong to any cycle of H, then $D'_c(H) = D'_c(H/\!\!/v)$. Because H is a counterexample with minimum number of vertices, we have $D'_c(H/\!\!/v) \leq (n-1)/3$. Thus we have

$$D'_c(H) = D'_c(H/\!\!/v) \leq (n-1)/3.$$

This is a contradiction with $D'_c(H) > n/3$.

Claim 11. $\forall e \in H$, $|e| \leq 2$.

If there exists $e \in H$ with $|e| \geq 3$ in H, we have $D_c(H/\!\!/e) \leq (n-|e|)/3 \leq (n-3)/3$ because H is a counterexample with minimum number of vertices. Considering a minimum eliminating edge feedback set A of $H' = H/\!\!/e$, we have $A \subseteq E \setminus e$ and $|A| \leq (n-3)/3$. Thus $A \cup \{e\}$ is an eliminating edge feedback set for H and $|A \cup \{e\}| = |A| + 1 \leq n/3$. Thus we have

$$D'_c(H) \leq D'_c(H/\!\!/e) + 1 \leq |A| + 1 \leq (n-3)/3 + 1 = n/3.$$

This is a contradiction with $D'_c(H) > n/3$.

According to Claim 11, the hypergraph H is a graph, denoted as G.

Claim 12. G is triangle free.

If there exists $C = v_1e_1v_2e_2v_3e_3v_1$ in G, then $D'_c(G/\!\!/\{v_1, v_2, v_3\}) \leq (n-3)/3$ because G is a counterexample with minimum number of vertices. Considering a minimum eliminating edge feedback set A of $G' = G/\!\!/\{v_1, v_2, v_3\}$, we have $A \subseteq E$ and $|A| \leq (n-3)/3$. Due to 3-regularity of G, we have $A \cup \{e_1\}$ is an eliminating edge feedback set for G and $|A \cup \{e_1\}| = |A| + 1 \leq n/3$, which is a contradiction with $D'_c(G) > n/3$. A figure is shown in Fig. 5(a).

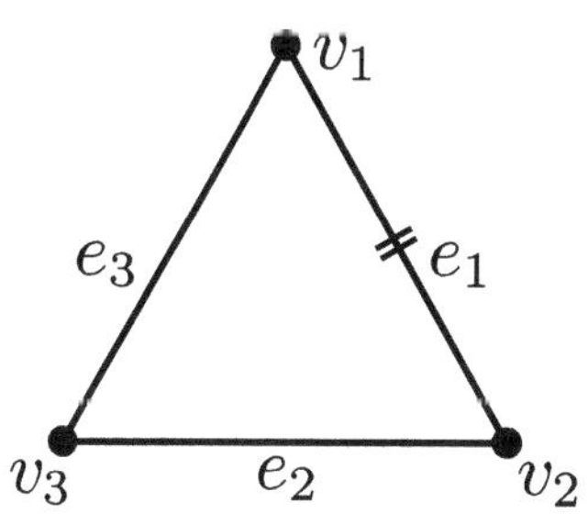

(a) Eliminating triangles

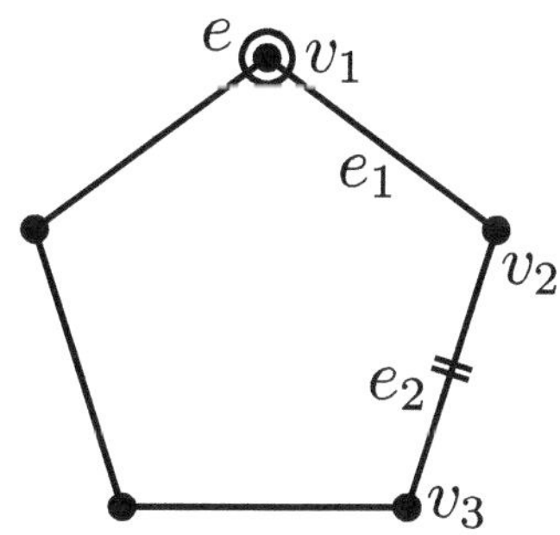

(b) Eliminating loops

Fig. 5. Schematic diagrams in Claims 12 and 13

Claim 13. G is loop-free.

If there is a loop $e \in E$ with $e = \{v_1\}$ in G. Due to Claim 10 and Claim 12, we know there is a cycle $C = v_1 e_1 v_2 e_2 v_3 e_3 v_4 \cdots e_k v_1$ with $k \geq 4$ in G. Then $D'_c(G /\!\!/ \{v_1, v_2, v_3\}) \leq (n-3)/3$ because G is a counterexample with minimum number of vertices. Considering a minimum eliminating edge feedback set A of $G' = G /\!\!/ \{v_1, v_2, v_3\}$, we have $A \subseteq E$ and $|A| \leq (n-3)/3$. Due to 3-regularity of G, we have $A \cup \{e_2\}$ is an eliminating edge feedback set for G and $|A \cup \{e_2\}| = |A| + 1 \leq n/3$, which is a contradiction with $D'_c(G) > n/3$. A figure is shown in Fig. 5(b).

G is loop-free and linear, thus G is a simple graph.

Claim 14. There is no cycle with length 4 in G.

Suppose on the contrary, there is a cycle $C = v_1 e_1 v_2 e_2 v_3 e_3 v_4 e_4 v_1$ in G. Due to 3-regularity, for $1 \leq i \leq 4$, v_i has a neighbor u_i. Because G is triangle free, we have $u_1 \neq u_2, u_2 \neq u_3$.

If $u_1 = u_3$, we have $D'_c(G /\!\!/ \{v_1, v_2, v_3, v_4, u_1, u_2\}) \leq (n-6)/3$ because G is a counterexample with minimum number of vertices. Considering a minimum eliminating edge feedback set A of $G' = G /\!\!/ \{v_1, v_2, v_3, v_4, u_1, u_2\}$, we have $A \subseteq E$ and $|A| \leq (n-6)/3$. Due to 3-regularity of G, we have $A \cup \{e_3, e_5\}$ is an eliminating edge feedback set for G and $|A \cup \{e_3, e_5\}| = |A| + 2 \leq n/3$, which is a contradiction with $D'_c(G) > n/3$. See Fig. 6(a) for an illustration.

If $u_1 \neq u_3$, we have $D'_c(G /\!\!/ \{v_1, v_2, v_3, v_4, u_1, u_3\}) \leq (n-6)/3$ because G is a counterexample with minimum number of vertices. Considering a minimum eliminating edge feedback set A of $G' = G /\!\!/ \{v_1, v_2, v_3, v_4, u_1, u_3\}$, we have $A \subseteq E$ and $|A| \leq (n-6)/3$. Due to 3-regularity of G, we have $A \cup \{e_5, e_6\}$ is an eliminating edge feedback set for G and $|A \cup \{e_5, e_6\}| = |A| + 2 \leq n/3$, which is a contradiction with $D'_c(G) > n/3$. See Fig. 6(b) for an illustration.

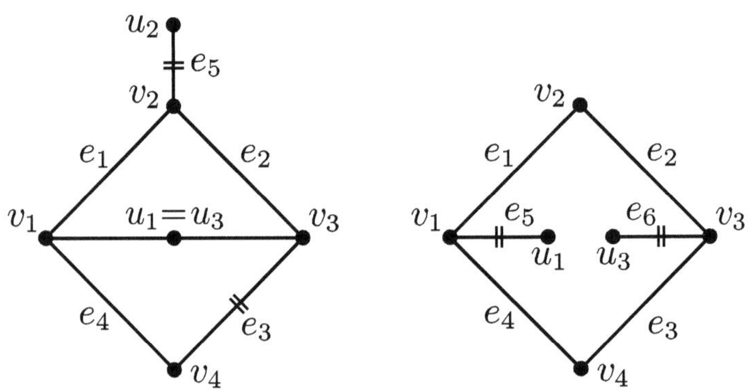

(a) Case: $u_1 = u_3$ (b) Case: $u_1 \neq u_3$

Fig. 6. Schematic diagrams in Claim 14

Let $C = v_1e_1v_2e_2\dots v_ke_kv_1$ be a shortest cycle in G, here $k \geq 5$. Due to 3-regularity, for $1 \leq i \leq k$, v_i has a neighbor u_i. C is the shortest cycle, thus u_i are distinct vertices for $\leq i \leq k$. Denote $V(C) = \{v_i \mid 1 \leq i \leq k\}$ and $E(C) = \{e_i \mid 1 \leq i \leq k\}$ for convenience.

Claim 15. $k \not\equiv 0 \pmod 3$

If $k \equiv 0 \pmod 3$, we can assume $k = 3t$ with $t \geq 2$. Then $D'_c(G/\!\!/V(C)) \leq (n - 3t)/3$ because G is a counterexample with minimum number of vertices. Considering a minimum eliminating edge feedback set A of $G' = G/\!\!/V(C)$, we have $A \subseteq E$ and $|A| \leq (n-3t)/3$. Due to 3-regularity of G, we have $A \cup \{e_{3i} \mid 1 \leq i \leq t\}$ is an eliminating edge feedback set for G and $|A \cup \{e_{3i} \mid 1 \leq i \leq t\}| \leq |A| + t \leq n/3$, which is a contradiction with $D'_c(G) > n/3$. See Fig. 7(a) for an illustration.

Claim 16. $k \not\equiv 1 \pmod 3$

If $k \equiv 1 \pmod 3$, assume $k = 3t + 1$ with $t \geq 2$. Then $D'_c(G/\!\!/(V(C) \cup \{u_2, u_4\})) \leq (n-3t-3)/3$ because G is a counterexample with minimum number of vertices. Considering a minimum eliminating edge feedback set A of $G' = G/\!\!/(V(C) \cup \{u_2, u_4\})$, we have $A \subseteq E$ and $|A| \leq (n - 3t - 3)/3$. Due to 3-regularity of G, we have $A \cup \{e'_2, e'_4\} \cup \{e_{3i} \mid 2 \leq i \leq t\}$ is an eliminating edge feedback set for G and $|A \cup \{e'_2, e'_4\} \cup \{e_{3i} \mid 2 \leq i \leq t\}| = |A| + t + 1 \leq n/3$, which is a contradiction with $D'_c(G) > n/3$. See Fig. 7(b) for an illustration.

Claim 17. $k \not\equiv 2 \pmod 3$

If $k \equiv 2 \pmod 3$, assume $k = 3t + 2$ with $t \geq 1$. Then $D'_c(G/\!\!/(V(C) \cup \{u_1\})) \leq (n - 3t - 3)/3$ because G is a counterexample with minimum number of vertices. Considering a minimum eliminating edge feedback set A of $G' = G/\!\!/(V(C) \cup \{u_1\})$, we have $A \subseteq E$ and $|A| \leq (n - 3t - 3)/3$. Due to 3-regularity of G, we have $A \cup \{e'_1\} \cup \{e_{3i} \mid 1 \leq i \leq t\}$ is an eliminating edge feedback set for G and $|A \cup \{e'_1\} \cup \{e_{3i} \mid 1 \leq i \leq t\}| = |A| + t + 1 \leq n/3$, which is a contradiction with $D'_c(G) > n/3$. See Fig. 7(c) for an illustration.

Obviously, it is impossible that Claim 15, Claim 16 and Claim 17 all holds. Thus there is always a contradiction. We finish the proof. □

For a hypergraph H with maximum degree $\Delta \leq 3$ and girth $k \geq 3$, we can construct a 3-regular hypergraph H' by adding some loops to the vertices and these operations keep the eliminating edge feedback number. According to Theorem 7, the next corollary is instant:

Corollary 3. *Let H be a hypergraph with maximum degree $\Delta \leq 3$ and girth $k \geq 3$. Then $D'_c(H) \leq n/3$.*

Based on the proof of Theorem 7, we design a combinatorial algorithm for finding an eliminating edge feedback set of cardinality at most $n/3$ for n-vertex hypergraphs with maximum degree $\Delta \leq 3$ and girth $k \geq 3$. The corresponding algorithm (Algorithm 2) is presented in the appendix.

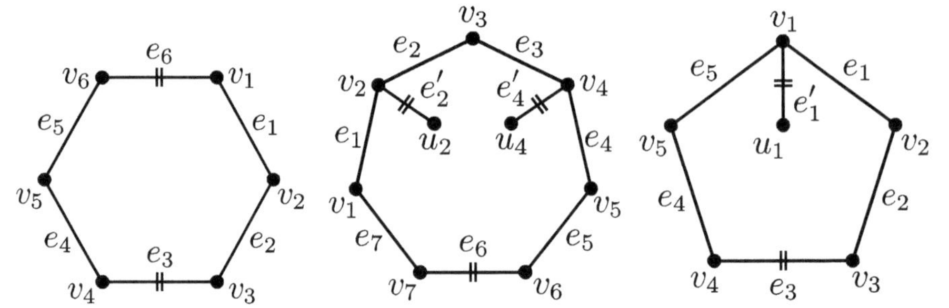

(a) Case: $k \equiv 0 \pmod 3$ (b) Case: $k \equiv 1 \pmod 3$ (c) Case: $k \equiv 2 \pmod 3$

Fig. 7. Schematic diagrams in Claims 15, 16 and 17

Appendix: Two Combinatorial Algorithms for Eliminating Edge Feedback Set

Algorithm 1. Eliminating edge set of hypergraphs with $\Delta \leq 2$ and girth $k \geq 4$

Input: A hypergraph H with maximum degree $\Delta \leq 2$ and girth $k \geq 4$.
Output: An eliminating edge feedback set with size at most $n/4$.

if H contains no cycles **then**
    return $\varnothing$
if H is a cycle and e is any edge in H **then**
    return $\{e\}$
if H has l connected components $H_1, \ldots, H_l$ **then**
    return $\cup_{i=1}^{l} \text{ALG1}(H_i)$
if there exists a vertex v that is not contained in some cycles of H **then**
    return $\text{ALG1}(H /\!\!/ \{v\})$
if there is an edge e with $|e| \geq 4$ **then**
    return $\text{ALG1}(H /\!\!/ \{e\}) \cup \{e\}$
if there are two adjacent edges e_1, e_2 such that $|e_1| = 2$, $|e_2| = 3$ **then**
    return $\text{ALG1}(H /\!\!/ \{e_2\}) \cup \{e_2\}$
Let $C = v_1 e_1 v_2 e_2 \cdots v_k e_k v_1$ be a shortest cycle in H and the vertex set $V(C) = \{v_i \mid 1 \leq i \leq k\} \cup \{u_i \mid 1 \leq i \leq k\}$. For each $i \in [k]$, $e_i = \{v_i, u_i, v_{i+1}\}$.
Set $A = \{e_{2i} \mid 1 \leq i \leq \lfloor \frac{k}{2} \rfloor\}$.
if $k = 2t$ **then**
    return $\text{ALG1}(H /\!\!/ V(C)) \cup A$
if $k = 2t + 1$ **then**
    H'' is the hypergraph by adding a new vertex $\tilde{v}_1$ to connect e_1 and e_{2t+2} in $H /\!\!/ (V(C) \setminus \{u_1\})$ as shown in Fig. 4
    return $\text{ALG1}(H'') \cup A$

Algorithm 2. Eliminating edge set of hypergraphs with $\Delta \leq 3$ and girth $k \geq 3$

Input: A hypergraph H with maximum degree $\Delta \leq 3$ and girth $k \geq 3$.

Output: An eliminating edge feedback set with size at most $n/3$.

if H contains no cycles **then**
  return $\varnothing$
if H is a cycle and e is any edge in H **then**
  return $\{e\}$
if there exists a vertex v that is not contained in some cycles of H **then**
  return ALG2($H /\!\!/ \{v\}$)
if there is an edge e with $|e| \geq 3$ **then**
  return ALG2($H /\!\!/ \{e\}$) $\cup \{e\}$
if H contains a triangle $C = v_1e_1v_2e_2v_3e_3v_1$ **then**
  return ALG2($H /\!\!/ \{e_1\}$) $\cup \{e_1\}$
if there exists a 2-degree vertex v_1 or a loop $e = \{v_1\}$ **then**
  Let $C = v_1e_1v_2e_2 \cdots v_ke_kv_1$ with $k \geq 4$ be a cycle containing v_1 in H.
  return ALG2($H /\!\!/ \{e_2\}$) $\cup \{e_2\}$
Let $C = v_1e_1v_2e_2 \cdots v_ke_kv_1$ be a shortest cycle in H and the vertex set $V(C) = \{v_i \mid 1 \leq i \leq k\}$. For each $i \in [k]$, let $u_i \notin V(C)$ be a neighbor of v_i. Set $A = \{e_{3i} \mid 1 \leq i \leq \lfloor \frac{k}{3} \rfloor\}$.
if $k = 4$ and $u_1 = u_3$ **then**
  return ALG2($H /\!\!/ \{e_3, e_5\}$) $\cup \{e_3, e_5\}$ as shown in Fig. 6(a)
else if $k = 4$ and $u_1 \neq u_3$ **then**
  return ALG2($H /\!\!/ \{e_5, e_6\}$) $\cup \{e_5, e_6\}$ as shown in Fig. 6(b)
else if $k \equiv 0 \pmod 3$ **then**
  return ALG2($H /\!\!/ V(C)$) $\cup A$ as illustration in Fig. 7(a)
else if $k \equiv 1 \pmod 3$ **then**
  return ALG2($H /\!\!/ (V(C) \cup \{u_2, u_4\})$) $\cup (A \setminus \{e_3\} \cup \{e'_2, e'_4\})$ as illustration in Fig. 7(b)
else
  return ALG2($H /\!\!/ (V(C) \cup \{u_1\})$) $\cup (A \cup \{e'_1\})$ as illustration in Fig. 7(c)

References

1. Bafna, V., Berman, P., Fujito, T.: A 2-approximation algorithm for the undirected feedback vertex set problem. SIAM J. Discret. Math. **12**(3), 289–297 (1999)
2. Erdös, P., Pósa, L.: On independent circuits contained in a graph. Canadian J. Math. **17**, 347–352 (1965)
3. Even, G., Naor, J.S., Schieber, B., Sudan, M.: Approximating minimum feedback sets and multi-cuts in directed graphs. In: Balas, E., Clausen, J. (eds.) IPCO 1995. LNCS, vol. 920, pp. 14–28. Springer, Heidelberg (1995). https://doi.org/10.1007/3-540-59408-6_38
4. Garey, M.R., Johnson, D.S.: Computers and Intractability: A Guide to the Theory of NP-Completeness. W. H. Freeman (1979)
5. Reed, B.A., Robertson, N., Seymour, P.D., Thomas, R.: Packing directed circuits. Combinatorica **16**(4), 535–554 (1996)
6. Ren, H., Yang, C., Zhao, T.: A new formula for the decycling number of regular graphs. Discret. Math. **340**(12), 3020–3031 (2017)
7. Shi, L., Hongyu, X.: Large induced forests in graphs. J. Graph Theory **85**(4), 759–779 (2017)
8. Wang, T., Baoyindureng, W.: The size of graphs with given feedback vertex number. Discret. Appl. Math. **314**, 213–222 (2022)
9. Wang, T., Baoyindureng, W.: Maximum induced forests of product graphs. Bull. Malaysian Math. Sci. Soc. **46**(1), 7 (2023)
10. Albertson, M.O., Berman, D.M.: A conjecture on planar graphs. Graph Theory Related Topics **357** (1979)
11. Borodin, O.V.: A proof of grünbaum's conjecture on the acyclic 5-colorability of planar graphs. In: Doklady Akademii Nauk, vol. 231, pp. 18–20. Russian Academy of Sciences (1976)
12. Alon, N., Mubayi, D., Thomas, R.: Large induced forests in sparse graphs. J. Graph Theory **38**(3), 113–123 (2001)
13. Dross, F., Montassier, M., Pinlou, A.: A lower bound on the order of the largest induced forest in planar graphs with high girth. Discret. Appl. Math. **214**, 99–107 (2016)
14. Kelly, T., Liu, C.-H.: Minimum size of feedback vertex sets of planar graphs of girth at least five. Eur. J. Comb. **61**, 138–150 (2017)
15. Kelly, T., Liu, C.-H.: Size of the largest induced forest in subcubic graphs of girth at least four and five. J. Graph Theory **89**(4), 457–478 (2018)

A Space-Partition Based Approach to the 2-Center Problem in Three and Higher Dimensions

Xuehou Tan[1,2(✉)] and Rong Chen[1]

[1] Dalian Maritime University, Linghai Road 1, Dalian, China
th7625@tokai.ac.jp

[2] Tokai University, 4-1-1 Kitakaname, Hiratsuka 259-1292, Japan

Abstract. For a set S of n points in three dimensions, the 2-center problem asks for two congruent balls of the minimum radius, whose union covers all points of S. We first present an $O(n^2 \log n)$ time and $O(n)$ space algorithm for the set S of points in convex position. Our algorithm is characterized by finding a (small) set of planes such that it contains an optimal plane that partitions S into two disjoint subsets; each can be covered by a ball of the minimum radius. The convexity of S allows to efficiently enumerate all possible situations needed in computing the 2-center of S. It is further extended to solve the 2-center problem for a set of arbitrarily given points in $O(n^2 \log n)$ time and $O(n)$ space. This significantly improves upon two previously known results; one is a deterministic algorithm with $O(n^{3+\epsilon})$ running time for any $\epsilon > 0$ [Agarwal et al., SIAM J. Computing 29 (2000) 912-953], and the other is a randomized algorithm with roughly $O(n^2 \log^4 n \log \log n)$ expected time and $O(n^2)$ space [Agarwal et al., Comput. Geometry 46 (2013) 734-746]. Moreover, our algorithm can be easily generalized to the 2-center problem in fixed d-dimension, $d \geq 4$. For small values of d, our result also gives an improvement upon the previously known $n^{O(1)}$ time bound [Agarwal and Procopiuc, Algorithmica 33 (2002) 201-226].

Keywords: Computational geometry · 2-center problems · The smallest enclosing ball problem · Space partitions

1 Introduction

Let S be a set of n points in three dimensions. Assume that all points of S are in general position, i.e., no five points are on (the surface of) a same ball. The *p-center problem* asks for p congruent balls of the minimum radius such that their union covers S. If p is part of input, then the problem is NP-complete [12] even in the plane. This intractability result has led to extensive work on developing efficient algorithms for small values of p. The one-center problem in fixed d-dimension, known as the *smallest enclosing ball* problem, can be solved in deterministic linear time [6, 8, 10].

M. Li et al. (Eds.): TAMC 2025, LNCS 16084, pp. 235–247, 2026.
https://doi.org/10.1007/978-981-95-4839-2_18

It has been known that the 2-center problem in d-dimension is NP-complete, if d is not fixed [11]. In the past three decades, the planar 2-center problem has been investigated extensively. After several near-quadratic time algorithms were given, Sharir [14] was the first to present a near-linear algorithm with running time $O(n \log^9 n)$, using the powerful (but complicated) parametric search paradigm. Later, Eppstein [9] presented a randomized algorithm with expected $O(n \log^2 n)$ time, and Chan [5] proposed an $O(n(\log n \log\log n)^2)$ time deterministic algorithm. Recently, Wang [16] (as well as Choi and Ahn [7]) has given an $O(n \log^2 n)$ time solution, and Tan [15] has presented an $O(n \log n)$ time algorithm, without using parametric search.

For the 2-center problem in three dimensions, Agarwal, Efrat and Sharir [2] were the first to give an algorithm with $O(n^{3+\epsilon})$ running time, for any $\epsilon > 0$. It requires a rather complicated technique for dynamically maintaining upper and lower envelopes of bivariate functions. Later, Agarwal, Avraham and Sharir [1] presented a randomized algorithm with $O((1 - R^*/R')^{-3} n^2 \log^4 n \log\log n)$ expected time and $O(n^2)$ space, where R' and R^* are the optimum solutions (radii) to the one-center and two-center problems for S respectively. (It is clear that $R^* < R'$.) Note that the performance of Agarwal et al.'s randomized algorithm also depends on the value of R^*/R'. They mainly extended Sharir's planar 2-center algorithm [14] and its refinements [5,9] to three dimensions, with some new geometric insights and more involved data structures. Whether or not a near-quadratic time deterministic algorithm for the 2-center problem in three dimensions can be developed is left as an open problem [1].

Apart from theoretic interests, the motivation of this research also comes practical applications. For instance, the three-dimensional p-center problem has an important application in modern wireless communication systems. It is a natural requirement that some transmissions should be set up so that mobile phones in a city (or some area) can be serviced. The spatial region covered by a transmission is represented by a closed ball with radius r. The solution to the p-center problems helps give an optimal placement of transmissions.

1.1 Geometric Properties of 2-Centers of Point Sets in Space

For a point p in space, denote the coordinates of p by $p.x$, $p.y$ and $p.z$, respectively. For two points p and q, denote by pq the line segment connecting p and q, and $|pq|$ the length of segment pq. The *bisector* of p and q, denoted by $N_{p,q}$, is the perpendicular plane through the midpoint of pq. A simply property of the smallest enclosing ball of a set of points is the following.

Lemma 1. *Suppose that A is the smallest ball enclosing points p_1, p_2, p_3 and p_4. Then, p_1, p_2, p_3 and p_4 can't be contained in any semi-ball of A, except for that the radius of the smallest circle enclosing two or three points of them is equal to that of A.*

Proof. Simple and omitted in this extended abstract. □

Denote by (B_1^*, B_2^*) a pair of optimum balls of the minimum radius such that $S \subseteq B_1^* \cup B_2^*$. Let b_1 and b_2 be the centers of balls B_1^* and B_2^*, respectively. Since the radius of B_1^* and B_2^* is strictly smaller than that of the smallest ball D enclosing S, one can consider to obtain ball B_1^* (resp., B_2^*) by shrinking a ball from D appropriately, say, by moving the center of D into b_1 (resp. b_2) and dropping off the points not covered by B_1^* (resp., B_2^*). By noticing the fact that $S \subseteq B_1^* \cup B_2^*$, if B_1^* and B_2^* does not intersect, then N_{b_1,b_2} separates B_1^* from B_2^*. Otherwise, N_{b_1,b_2} contains the intersection (a circle) between the surfaces of B_1^* and B_2^*. See Fig. 1.

Observation 1. *For any pair of optimum balls B_1^* and B_2^* such that $S \subseteq B_1^* \cup B_2^*$, plane N_{b_1,b_2} partitions S into two disjoint subsets S_1 and S_2 such that S_1 and S_2 are covered by B_1^* and B_2^*, respectively.*

Moreover, the points on the surface of ball B_1^* (resp., B_2^*) do not belong to $S_1 \cap S_2$. See also Fig. 1.

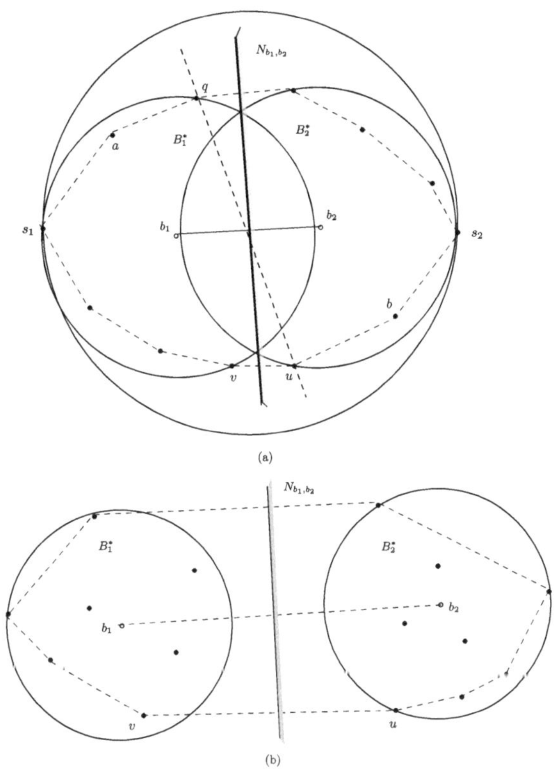

Fig. 1. Plane N_{b_1,b_2} separates S into two disjoint subsets.

Lemma 2. *For any pair of optimum balls B_1^* and B_2^* such that $S \subseteq B_1^* \cup B_2^*$, plane N_{b_1,b_2} partitions S into two disjoint subsets S_1 and S_2 such that S_1 (resp., S_2) is covered by B_1^* (resp., B_2^*), and the points on surface of ball B_1^* (resp., B_2^*) do not belong to $S_1 \cap S_2$.*

Proof. The former part of the lemma is same as Observation 1.

We show below the latter part of the lemma. Assume that a point x on the ball, say, B_1^*, belongs to $S_1 \cap S_2$. Since point x is covered by B_2^*, it needn't be covered by B_1^* at all. Let B_1' be the smallest ball enclosing all the points covered by B_1^*, excluding x. Then, the radius of B_1' is no more than that of B_1^*, and $S \subseteq B_1' \cup B_2^*$. A repeated application of the above operation implies that either the lemma holds for a pair of balls (e.g.,, B_1' and B_2^*), or a contradiction with the radius-optimality of B_1^* and B_2^* occurs. □

Without loss of generality, assume that both b_1 and b_2 are contained in the xy-plane. (So, N_{b_1,b_2} is perpendicular to the xy-plane.) Although it is difficult to compute plane N_{b_1,B_2} exactly, some (many) planes playing the same role as N_{b_1,b_2} can be found. Let M_{b_1,b_2} be the plane, which is perpendicular to N_{b_1,b_2} and passes both b_1 and b_2. Denote by a (resp., b) a point that is to the left (resp., right) of N_{b_1,b_2} and vertically above (resp., below) M_{b_1,b_2}. See Fig. 1. Also, let c (resp., d) be a point that is to the right (resp., left) of N_{b_1,b_2} and vertically above (resp., below) M_{b_1,b_2}. From Lemmas 1 and 2, and dimensionality of the point set S, a point pair (a, b) or (c, d) always exists; otherwise, all points of S are on a plane, which is thus degenerated to the planar 2-center problem [15].

Lemma 3. *There exists a plane H such that it is perpendicular to the xy-plane, contains two points $a, b \in S$ (resp., $c, d \in S$), and can be rotated into plane N_{b_1,b_2} clockwise (resp., counterclockwise) around the intersection of H and N_{b_1,b_2}, without touching any other point.*

Proof. From the definition of points a and b (resp., c and d), any plane containing a and b (resp., c and d) intersects N_{b_1,b_2}. Denote by H such a plane that it is perpendicular to the xy-plane and contains two points $a, b \in S$ (resp., $c, d \in S$). The existence of plane H follows from the above discussion.

Consider a rotation of plane H into N_{b_1,b_2} clockwise (resp., counterclockwise), around their intersection. Assume that some points are encountered in the rotation. (Otherwise, the lemma is trivially true.) By replacing a or b (resp., c or d) with the corresponding, encountered point repeatedly, so as to define or specifiy a new plane H, the lemma is true. □

Following from Lemma 3, plane H can play the same role of N_{b_1,b_2}, i.e., it can be used to partition S into subsets S_1 and S_2 such that S_1 and S_2 are covered by B_1^* and B_2^* respectively. The formal definition of planes H as well as the algorithm for finding a plane H will be given in Sect. 2.2.

1.2 Our Result

We present a new, space-partition based approach to the 2-center problem for a set S of points in three (and higher) dimensions. Denote by R^* the optimum radius of the two-center balls of S, and (B_1^*, B_2^*) a pair of optimum balls of radius R^* such that the union of B_1^* and B_2^* covers S.[1] Denote by $H(p, q, w)$ the plane passing through points p, q and w.

We will describe our algorithm for the 2-center problem for a set S of three-dimensional points in convex position and then generalize it to the set of arbitrarily given points. A set of points is said to be in *convex position* if it is the set of vertices of a convex polyhedron P. Our algorithm first makes use of the smallest ball D enclosing S, whose radius is R'. Since all points of S are in general position, at most four points s_1, s_2, s_3 and s_4 (two of them may not exist) are on the surface of D. From a simple fact that $R^* < R'$ as well as the pigeonhole principle, an optimum ball B_i^* $(i = 1, 2)$ covers at least one and at most three of the points on the surface of D. By considering all possible combinations of the points, which are on the surface of D and covered by B_1^* and B_2^* respectively, the 2-center problem can be reduced to at most ten subproblems; the smallest among the solutions of those subproblems is thus R^*.

To solve a considered subproblem, say, $\{s_1\}$ is required to be covered by an optimal ball and $\{s_1, s_3, s_4\}$ by the other, the concept of separating planes is employed. For an edge uv of polyhedron P, $u \neq s_i$ and $v \neq s_i$ for all $1 \leq i \leq 4$, it together with s_1 (resp., one of s_2, s_3 and s_4) can be used to define an *atomic* separating plane that separates $\{s_1\}$ from $\{s_2, s_3, s_4\}$. This atomic separating plane is then swept through the polyhedron P around the lines containing uv, until one of s_2, s_3, and s_4 (resp., s_1) is encountered.

Consider how the procedure of sweeping the plane, say, $H(u, v, s_1)$ is performed. Whenever a new vertex p of polyhedron P is encountered, the plane $H(u, v, p)$ clearly separates $\{s_1\}$ from $\{s_2, s_3, s_4\}$. This sweep then gives a sequence of two disjoint subsets, whose union is S, such that $\{s_1\}$ and $\{s_2, s_3, s_4\}$ are always separated. From the convexity of S, a locally optimal solution that is related to (uv, s_1) exists, and a binary search on the sequence of two disjoint subsets can be performed in $O(n \log n)$ time and $O(n)$ space, so as to compute that locally optimal solution. As will be seen, points v and p are required to be covered together with set $\{s_1\}$, while point u is covered together with set $\{s_2, s_3, s_4\}$. Hence, a total of $O(n)$ optimal solutions to that 2-center subproblem can be introduced or defined. The smallest among those locally optimal solutions then gives the solution to that subproblem. Since all other situations can be dealt with analogously (see the proof of Theorem 1), an $O(n^2 \log n)$ time and $O(n)$ space solution for the 2-center problem for S can be obtained.

Consider now the 2-center problem for a set S of arbitrarily given points. Denote by $CH(S)$ be the convex hull of S, and K the set of points of $CH(S)$. Without loss of generality, assume that all four points s_1, s_2, s_3 and s_4, which

[1] Although the number of pairs of optimum balls B_1^* and B_2^* is infinite, the solution R^* is unique. Also, it is sufficient for us to find only one pair of balls B_1^* and B_2^*.

are on the surface of the smallest enclosing ball of K, exist. Again, the 2-center problem of S is first divided into (at most) ten subproblems, according to the combinations of s_1, s_2, s_3 and s_4.

Take as an example below the 2-center subproblem for S in which sets $\{s_1\}$ and $\{s_2, s_3, s_4\}$ are always covered by different balls. For an edge uv of $CH(S)$, denote by $r^*(uv, s_1)$ the locally optimal solution to the corresponding 2-center subproblem for K, which is obtained in the procedure of sweeping plane $H(u, v, s_1)$ around the line containing uv. (As mentioned above, two sets $\{s_1, v\}$ and $\{s_2, s_3, s_4, u\}$ are covered by different optimal balls.) We can then show that there is a locally optimal solution for the 2-center subproblem for S, denoted by $R^*(uv, s_1)$, which is based on $r^*(uv, s_1)$ in the sense that when S is limited to K, its solution is just $r^*(uv, s_1)$. For this purpose, we describe a sufficient condition, termed the Sufficient Lemma (Lemma 4), for the radius of two congruent balls covering S to be $R^*(uv, s_1)$, under the requirement that when S is limited to K, its initial (or partial) solution should be $r^*(uv, s_1)$. So, $r^*(uv, s_1)$ is embedded as a sub-condition.

The sufficient lemma can be used to develop an efficient algorithm for computing $R^*(uv, s_1)$. First, the solution $r^*(uv, s_1)$ is computed. Two optimal balls, say, D_1 and D_2 of radius $r^*(uv, s_1)$ such that $K \subseteq D_1 \cup D_1$, are thus found. Denote by S_1 (resp., S_2) the points of S, which are covered by ball D_1 (resp., D_2). Next, we show that it is possible to find a half of uncovered points such that an "clever" addition of them to either of S_1 and S_2 makes the larger radius of two smallest enclosing balls of the *newly* obtained sets S_1 and S_2 be the (locally) optimal solution for the set $S_1 \cup S_2$ (Lemma 5). The solution $R^*(uv, s_1)$ can be thus found by divide-and-conquer. Moreover, the operation of adding a half of uncovered points to either of S_1 and S_2 can be done in $O(n)$ time and space. It thus takes $O(n \log n)$ time and $O(n)$ space to compute $R^*(uv, s_1)$ (see the proof of Theorem 1). As the total number of those locally optimal solutions (like $R^*(uv, s_1)$) is $O(n)$, the 2-center solution for the subproblem in which $\{s_1\}$ and $\{s_2, s_3, s_4\}$ are covered by different balls can be computed in $O(n^2 \log n)$ time and $O(n)$ space. In conclusion, the overall 2-center solution for S can be obtained in $O(n^2 \log n)$ time and $O(n)$ space, too.

Our result not only gives a positive answer to the open problem posed by Agarwal et al. [1], but also is a radical breakthrough, as compared with two previously known results [1,2]. Moreover, our algorithm is (conceptually) rather simple, and can be easily generalized to the 2-center problem in fixed d-dimension, $d \geq 4$ (see Corollary 1). For small values of d (e.g., $d = 4$ or 5), it also gives an improvement upon the previously known $n^{O(1)}$ time bound [3].

2 The 2-Centers of Sets of Three-Dimensional Points in Convex Position

In this section, we present an $O(n^2 \log n)$ time and $O(n)$ space algorithm for the 2-center problem for a set S of three-dimensional points in convex position. Denote by P the convex polyhedron formed by all points of S. Assume that all points of S are in general position, i.e., no five points can be on (the surface of) a ball. For a ball B, denote $B(S)$ the set of the points, which are covered by B.

2.1 The First Round Reduction: Dividing the 2-Center problem Into a Constant Number of Subproblems

Without loss of generality, assume that four points s_1, s_2, s_3 and s_4 are on the surface of D, where D is the smallest ball enclosing all points of S. (Other situations will be discussed in the proof of Theorem 1.) Denote by $r^*_{s_i}$ the optimal radius of the 2-center balls of S, under the assumption that s_i is covered by an optimal ball, and three others are covered by the other, where $i \in \{1, 2, 3, 4\}$. Also, denote by $r^*_{s_j,s_k}$ the optimal radius of the 2-center balls of S, where $j, k \in \{1, 2, 3, 4\}$ and $j \neq k$, under the assumption that s_j and s_k are covered by an optimal ball, and two others by the other. The overall optimum solution R^* for S is then the minimum among those ten solutions.

In the rest of this section, we will discuss how to compute only $r^*_{s_1}$. For ease of presentation, denote $r^*_{s_1}$ by r^*. Still, denote by (B^*_1, B^*_2) a pair of optimal balls of radius r^* such that $\{s_1\}$ and $\{s_2, s_3, s_4\}$ are covered by B^*_1 and B^*_2, respectively. Note that Lemma 2 as well as other discussions in Section 1.1 also work for any pair of optimal balls B^*_1 and B^*_2 of radius r^*. For instance, the following observation is can be made by a simple application of Lemma 2 to r^*.

Observation 2. *Suppose that r^* is the solution to a 2-center subproblem described above. For any pair of optimal balls B^*_1 and B^*_2 of radius r^* such that $K \subseteq B^*_1 \cup B^*_2$, plane N_{b_1,b_2} separates S into two disjoint subsets S_1 and S_2 such that S_1 (resp., S_2) is covered by B^*_1 (resp., B^*_2), and the points on surface of ball B^*_1 (resp., B^*_2) do not belong to $S_1 \cap S_2$.*

2.2 The Algorithm

We will focus on the 2-center subproblem in which $\{s_1\}$ and $\{s_2, s_3, s_4\}$ are always covered by different balls. (All other subproblems can be dealt with analogously.) Let us first define the concept of separating planes. A plane, if it separates $\{s_1\}$ from $\{s_2, s_3, s_4\}$, is called the *separating plane* for our considered 2-center subproblem. We will show that at least one of them has the same role as plane N_{b_1,b_2} (Observation 2 as well as Lemma 3).

For an edge uv of polyhedron P, $u \neq s_1$ and $v \neq s_1$, assume that plane $H(u, v, s_1)$ separates $\{s_1\}$ from $\{s_2, s_3, s_4\}$; otherwise, it is ignored, as it cannot separate $\{s_1\}$ from $\{s_2, s_3, s_4\}$. The plane $H(u, v, s_1)$ is thus an *atomic* separating plane for the considered 2-center subproblem. Without loss of generality, assume that $s_1.x < s_i.x$ for all $2 \leq i \leq 4$.

Consider how to sweep plane $H(u, v, s_1)$ through polyhedron P around the line containing uv, until one of s_2, s_3 and s_4 is encountered. Whenever a vertex p of polyhedron P is met, the separating plane (at that time) separates S into two new subsets S_1 and S_2. In the following, we will show that points v and p are required to be covered together with $\{s_1\}$ and point u is covered with

$\{s_2, s_3, s_4\}$. It then follows from the convexity of S that two disjoint subsets (of S), which are obtained in the procedure of sweeping plane $H(u, v, s_1)$ around the line containing uv, are changed monotonically; one subset increases in its size and the other decreases. Thus, binary search can be used to find a locally optimal solution that is related to (uv, s_1).

Let us now show that the above method can find a plane that plays the same role as N_{b_1,b_2} for the 2-center subproblem, in which $\{s_1\}$ and $\{s_2, s_3, s_4\}$ are respectively covered by two optimal balls. Again, consider the procedure of sweeping $H(u, v, s_1)$ through polyhedron P. Without loss of generality, assume that edge uv is vertically below plane M_{b_1,b_2}, and plane N_{b_1,b_2} intersects edge uv such that u and v are to right and left of N_{b_1,b_2}, respectively. (So, $v.x < u.x$.) Moreover, assume that the distance of u to plane N_{b_1,b_2} is the minimum among those of all possible points. Assume also that the distance of v to plane N_{b_1,b_2} is the minimum among those of all possible points. (From the convexity of S, these assumptions can be made, although neither N_{b_1,b_2} nor M_{b_1,b_2} is known.) Assume that point q to the left of plane N_{b_1,b_2} and vertically above M_{b_1,b_2}, and its distance to plane N_{b_1,b_2} is the minimum among those of all possible points. See also Fig. 1.

Let $H(u, v, q)$ be the separating plane obtained when point q is met in the procedure of sweeping $H(u, v, s_1)$ from left to right. From the convexity of S and the above definition of points u and v, $H(u, v, q)$ can be further rotated into the plane, which is perpendicular to the xy-plane, around the line containing uq without encountering any point of S. Denote by $H_{u,q}$ the resulting plane. Moreover, plane $H_{u,q}$ can be rotated into N_{b_1,b_2} clockwise around their intersection, without touching any point of S. (Otherwise, a contradiction occurs with the convexity of S or the definition of points u and q.) Hence, plane $H(u, v, q)$ can play the same role as N_{b_1,b_2} for this 2-center subproblem.

For an edge uv with $v.x < u.x$, denote by $r^*(uv, s_1)$ the optimal solution to the 2-center subproblem, which is obtained in the procedure of sweeping $H(u, v, s_1)$ through polyhedron P around the line containing edge uv. From the above discussion, point v (resp., u) as well as the point (e.g., p or q) met in the sweep is covered together with set $\{s_1\}$ (resp., $\{s_2, s_3, s_4\}$). And, the solution $r^*(uv, s_1)$ can be found by a binary search on the sequence of two disjoint subsets, whose union is S and whose sizes change monotonically. Since it takes linear time and space to find the smallest enclosing ball of a point set, $r^*(uv, s_1)$ can be computed in $O(n \log n)$ time and $O(n)$ space.

Remember that an atomic separating plane, say, $H(u, v, s_2)$, has also to sweep through P from right to left, until point s_1 is encountered. (This helps find the point pair (c, d), if exists, as stated in Lemma 3.) In conclusion, at most $O(n)$ locally optimal solutions are found, and each of them can be computed in $O(n \log n)$ time and $O(n)$ space. The 2-center subproblem in which $\{s_1\}$ and $\{s_2, s_3, s_4\}$ are covered by different optimal balls can be thus solved in $O(n^2 \log n)$ time and $O(n)$ space.

Theorem 1. *Given a set S of n three-dimensional points in convex position, one can compute in $O(n^2 \log n)$ time the minimum radius R^* such that there are two balls of radius R^*, whose union covers S.*

Proof. Assume first that there are four points s_1, s_2, s_3, and s_4 on the surface of the smallest enclosing ball D of S. As discussed above, the 2-center subproblem in which one of s_1, s_2, s_3, and s_4 is covered by a ball and three others by the other can be solved in $O(n^2 \log n)$ time and $O(n)$ space.

Analogously, the same result can be obtained for the 2-center subproblem in which two points s_j and s_k, $j \neq k$ and $1 \leq j, k \leq 4$, are covered by an optimal ball and the others of s_1, s_2, s_3 and s_4 by the other. Take it as an example that set $\{s_1, s_2\}$ is always covered by a ball and set $\{s_3, s_4\}$ by the other. In this case, $H(u, v, s_1)$ (resp., $H(u, v, s_2)$, $H(u, v, s_3)$ and $H(u, v, s_4)$), if it separates $\{s_1, s_2\}$ from $\{s_3, s_4\}$, is an atomic separating plane. The plane, say, $H(u, v, s_1)$ is then swept through polyhedron P around the lines containing uv, until either s_3 or s_4 is encountered. As discussed above, a locally optimal 2-center solutions can found. Therefore, this 2-center subproblem can be solved in $O(n^2 \log n)$ time and $O(n)$ space, too.

Consider now the special situation in which only three points s_1, s_2, and s_3 are on the surface of ball D. In this case, one of s_1, s_2, and s_3 is covered by an optimum ball and two other by the other. Let C be the smallest enclosing disk of $\{s_1, s_2, s_3\}$. For an edge uv of polyhedron P, assume that both u and v are vertically below (resp., above) disk C. (The existence of uv simply follows from the convexity of P and the definition of disk C.) Also, we can define three atomic separating planes $H(u, v, s_i)$, where $1 \leq i \leq 3$. The rest discussion is the same as described above.

Finally, consider the case that only two points s_1 and s_2 are on the surface of ball D, i.e., $s_1 s_2$ is a diameter of D. Again, let C be a disk of diameter $s_1 s_2$. We can similarly define two atomic separating plane $H(u, v, s_1)$ and $H(u, v, s_2)$, which always separates $\{s_1\}$ from $\{s_2\}$. The rest discussion is also the same as described above.

The above treatment enumerates all possible situations in which locally optimal solutions need be found. As the optimum solution R^* is the minimum among the solutions of at most ten subproblems, the theorem follows. □

3 The 2-Centers of Sets of Arbitrarily Given Points

Let S be a set of n arbitrarily given points in three dimensions. Denote by $CH(S)$ the convex hull of S, and K the set of points on $CH(S)$. Let s_1, s_2, s_3 and s_4 (at most two of them may not exist) be the points on the smallest enclosing ball of K. Again, the 2-center problem of S is first divided to at most ten subproblems (according to the combinations of s_1, s_2, s_3 and s_4). To solve one of those subproblems, we further compute $O(n)$ locally optimal solutions. Also, the minimum among them gives the (optimal) solution to that 2-center subproblem.

Without loss of generality, assume that all four points s_1, s_2, s_3 and s_4 are on the surface of the smallest enclosing ball of K. Again, take as an example below the 2-center subproblem in which two sets $\{s_1\}$ and $\{s_2, s_3, s_4\}$ are always covered by different balls. For an edge uv of $CH(S)$ with $v.x < u.x$, let $r^*(uv, s_1)$ be the locally optimal solution to the considered 2-center subproblem for set K, in which $\{s_1, v\}$ is covered by an optimal ball and $\{s_2, s_3, s_4, u\}$ by the other.

Beginning with $r^*(uv, s_1)$, a locally optimal solution for the 2-center subproblem for S, denoted by $R^*(uv, s_1)$, can be found. For this purpose, we describe a sufficient condition for the radius of two congruent disks covering S to be the locally optimal solution $R^*(uv, s_1)$, in the sense that when S is limited to K, its solution is just $r^*(uv, s_1)$.[2] Next, a divide-and-conquer algorithm for computing $R^*(uv, s_1)$ is presented.

3.1 The Sufficient Lemma

Let A and A' be two (different) balls of equal radius such that $S \subseteq A \cup A'$. Since A and A' are of equal radius, any point on the *surface* of $CH(A \cup A')$ (it is the convex hull of two balls) is covered by exactly one of A and A'.

Lemma 4. *(The Sufficient Lemma). Suppose that E_1 and E_2 are the balls of equal radius such that $S \subseteq E_1 \cup E_2$, at least one of them, say, E_2 is the smallest ball enclosing points $q_1, q_2, q_3, q_4 \in S$ (q_3 and/or q_4 may not exist), and a point $p \in E_1(S)$ is on $CH(E_1 \cup E_2)$. The radius of E_1 and E_2 gives the locally optimal solution $R^*(uv, s_1)$ for S if (i) two subsets of K, which are separated by the bisector (plane) of the line segment connecting the centers of E_1 and E_2, can be covered by two balls of radius at most $r^*(uv, s_1)$, and (ii) the radius of every smallest ball enclosing $(S \setminus E_2(S) \cup \{q_i\})$, $i = 1, 2, 3$ or 4, is at least that of E_2.*[3]

Proof. Two instances of the situation described in the lemma are shown in Fig. 2. (The condition (i) is implicitly shown in Fig. 2.) From the lemma assumption, the radius of E_1 and E_2 is equal to that of E_2. Let K_1 and K_2 be the sets of vertices of K, which are separated by the bisector (plane) of the line segment connecting the centers of E_1 and E_2. Then, $K_1 \cup K_2 = K$ and $K_1 \cap K_2 = \phi$.

Suppose first that $S = K$. From the condition (i), K_1 and K_2 are covered by E_1 and E_2 respectively, and the larger radius between two smallest disks enclosing K_1 and K_2 is $r^*(uv, s_1)$. From the assumption made on ball E_2, the radius of E_2 is equal to $r^*(uv, s_1)$. Since $S = K$, we have that $r^*(uv, s_1)$ is identical to $R^*(uv, s_1)$ in this case. Hence, the lemma follows.

Consider now the situation in which $S \supset K$. Since point $p \in E_1(S)$ is on $CH(E_1 \cup E_2)$, ball E_1 cannot be translated to cover any of q_1, q_2, q_3 and q_4, without changing its coverage of $E_1(S)$, except that at most three points q_i ($1 \leq i \leq 4$) happen to be on the surface of E_1. From the condition (ii), one

[2] Note that Lemma 2 describes a necessary condition.

[3] A slightly simpler planar-counterpart can be found in [15]. For the planar 2-center problem, only the first round reduction (see Sect. 2.1) is needed, i.e., at most three 2-center subproblems need be wholly considered [15].

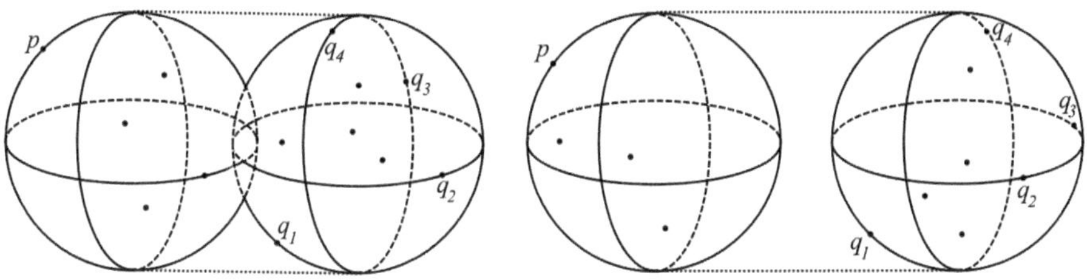

Fig. 2. An illustration of Lemma 4.

can then conclude that neither ball E_1 itself (including the exception case) nor a rotation of E_1 can produce two disks of smaller radius than that of E_2 such that S is still covered by the union of them. □

Note that the proof of Lemma 4 is independent of the definition of K, but it depends on the fact that an initial 2-center solution to the considered subproblem is embedded, when S is equal to some set $K' \supseteq K$. Beginning with $r^*(uv, s_1)$, Lemma 4 can be thus repeatedly applied, until the locally optimal solution $R^*(uv, s_1)$ for S is found (see the subsequent section).

3.2 The Algorithm

Suppose that a locally optimal solution, say, $r^*(uv, s_1)$ for set K has been obtained, where uv is an edge of polyhedron P such that $v.x < u.x$, $u \neq s_1$ and $v \neq s_1$. Denote by D_1 and D_2 two balls of radius $r^*(uv, s_1)$, whose union covers K. Assume also that D_1 and D_2 satisfy the conditions as well as assumptions stated in Lemma 4. Let c_1 and c_2 be the center of D_1 and D_2, respectively.

Suppose that at least one point of S is not covered by $D_1 \cup D_2$; otherwise, $r^*(uv, s_1)$ is trivially identical to $R^*(uv, s_1)$. Denote by L the line through c_1 and c_2. Since the uncovered points (outside of $D_1 \cup D_2$) are contained in $CH(D_1 \cup D_2)$, their orthogonal projections on L are all contained in segment c_1c_2. The set of all orthogonal projections on c_1c_2 can be partitioned into two disjoint sets SL_1 and SL_2 using the median value of those projections on c_1c_2, in $O(n)$ time [4]. (So, the difference between the sizes of sets SL_1 and SL_2 is at most one.)

Assume below that the orthogonal projections of all points of SL_1 onto c_1c_2 is closer to c_1 than that of any point of SL_2. Compute in $O(n)$ time and space the smallest enclosing ball of $D_1(S) \cup SL_1$, and denote it D'_1. Also, compute the smallest enclosing ball of $D_2(S) \cup SL_2$, and denote it D'_2. Let r_1 and r_2 be the radii of D'_1 and D'_2, respectively. Then, we have the following result.

Lemma 5. *If $r_1 \leq r_2$, then r_1 is the locally optimal solution to the 2-center subproblem for $(SL_1 \cup D_1(S)) \cup \cup D_2(S)$, under the assumption that two sets $\{s_1, v\}$ and $\{s_2, s_3, s_4, u\}$ are respectively covered by different optimal balls. (Otherwise, r_2 is the solution to the corresponding 2-center subproblem for $D_1(S) \cup (D_2(S) \cup SL_2)$.)*

Proof. Omitted in this extended abstract. □

Assume without loss of generality that $r_1 \leq r_2$. Then, the locally optimal solution r_1 to the 2-center subproblem for $(SL_1 \cup D_1(S)) \cup D_2(S)$ is found. Denote by C_2 a ball of radius r_1 such that it covers all points of $D_2(S)$ and two balls D_1', C_2 satisfy the conditions as well as assumptions in Lemma 4. (In the proof of Lemma 5, the method of finding ball C_2 is also given.) Beginning with (D_1', C_2), we can continue the procedure of adding another half of uncovered point to either of present sets $D_1'(S)$ and $C_2(S)$. By performing this operation repeatedly with Lemma 5, we can eventually find the optimal 2-center solution $R^*(uv, s_1)$ for S, under the assumption that sets $\{s_1, v\}$ and $\{s_2, s_3, s_4, u\}$ are respectively covered by different optimal balls.

The algorithm for computing other locally optimal solutions can be given analogously. We thus obtain the main result of this paper.

Theorem 2. *The 2-center problem for a set S of n points in three dimensions can be solved in $O(n^2 \log n)$ time and $O(n)$ space.*

Proof. First, compute the convex hull $CH(S)$ of set S, and then find the 2-center solution for the set K of all vertices on $CH(S)$ (Theorem 1). Let uv be an edge of $CH(S)$ with $v.x < u.x$. Denote by (D_1, D_2) be a pair of optimal balls of radius $r^*(uv, s_1)$, whose union covers all points of K, such that two sets $\{s_1, v\}$ and $\{s_2, s_3, s_4, u\}$ are covered by different optimal balls. Next, by applying Lemma 5 repeatedly, the points outside of $D_1 \cup D_2$ can be all covered in a divide-and-conquer manner. As described above, the locally optimal solution $R^*(uv, s_1)$ can be computed in $O(n \log n)$ time and $O(n)$ space. The solution for that 2-center subproblem for S is reported as the minimum of those $O(n)$ locally optimal solutions. And, R^* is the minimum among solutions of at most ten subproblems. Since the total number of locally optimal solutions (like $R^*(uv, s_1)$) is $O(n)$, the theorem follows. □

It has been conjectured that the 2-center problem in three dimensions is *3SUM-hard* [1]. Thus, our near-quadratic time algorithm might be optimal in the worst case.

Remark. The divide-and-conquer approach for arbitrary point sets is conceptually quite simple. It can be also applied to the planar 2-center problem, so as to simplify the known optimal $O(n \log n)$ time algorithm [15, Section 7.2]. Specifically, after the 2-center solution for points in convex position is obtained, a planar 2-center subproblem for the set of arbitrarily given points can be solved in $O(n \log n)$ time using the divide-and-conquer method described above, instead of the relatively complicated disk-enlarging process described in [15]. (For the 2-center problem in the plane, a total of at most three 2-center subproblems need be considered [15].)

4 Extensions

Our algorithm can analogously be adopted to the 2-center problem in fixed d-dimension, where $d \geq 4$ is a constant. Since we have assumed that all given points are in general position, the convex hull of n points in d-dimension is of complexity $\Theta(n^{\lfloor d/2 \rfloor})$ [13]. Hence, we can have the following result.

Corollary 1. *For a fixed integer $d > 0$, the 2-center problem for a set S of n points in d-dimension can be solved in $O(n^{1+\lfloor d/2 \rfloor} \log n)$ time and $O(n)$ space.*

For $d = 4$ or 5, our algorithm runs in $O(n^3 \log n)$ time. For small values of d, our result thus gives an improvement upon the previously known time bound $n^{O(1)}$ [3]. On the other hand, developing a more efficient algorithm is an interesting work.

Finally, it is an open question whether the space-partition based approach can be employed to efficiently solve other p-center problems, $p \geq 3$.

References

1. Agarwal, P.K., Avraham, R.B., Sharir, M.: The 2-center problem in three dimensions. Comput. Geometry: Theory Appl. **46**, 734–746 (2013)
2. Agarwal, P.K., Efrat, A., Sharir, M.: Vertical decomposition of shallow levels in 3-dimensional arrangements and its applications. SIAM J. Comput. **29**, 912–953 (2000)
3. Agarwal, P.K., Procopiuc, C.M.: Exact and approximation algorithms for clustering. Algorithmica **33**, 201–226 (2002)
4. Blum, M., Floyd, R.W., Pratt, V.R., Rivest, R.L., Tarjan, R.E.: Time bounds for selection. J. Comput. Syst. Sci. **7**, 448–461 (1972)
5. Chan, T.M.: More planar two-center algorithms. Comput. Geometry: Theory Appl. **13**, 189–198 (1999)
6. Chazelle, B., Matoušek, J.: On linear-time deterministic algorithms for optimization problems in fixed dimension. J. Algorithms **21**, 579–597 (1996)
7. Choi, J., Ahn, H.-K.: Efficient planar two-center algorithms. Comput. Geometry: Theory Appl. **97**(101768), 1–10 (2021)
8. Dyer, M.E.: On a multidimensional search technique and its application to the Euclidean one-center problem. SIAM J. Comput. **15**, 725–738 (1986)
9. Eppstein, D.: Faster construction of planar two-centers. In: Proceedings of 8th ACM-SIAM Symposium on Discrete Algorithms, pp. 131–138 (1997)
10. Megiddo, N.: Linear time algorithms for linear programming in R^3 and related problems. SIAM J. Comput. **12**, 759–776 (1983)
11. Megiddo, N.: On the complexity of some geometric problems in unbounded dimensions. J. Symbolic Comput. **10**, 27–334 (1990)
12. Megiddo, N., Supowit, K.: On the complexity of some common geometric location problems. SIAM J. Comput. **13**, 1182–1196 (1984)
13. Seidel, R.: Convex hull computation. In: Goodman, J.R., O'Rourke, J., Tóth, C.S. (eds.) Handbook of Discrete and Computational Geometry, pp. 687–704. CRC Press (2018)
14. Sharir, M.: A near-linear time algorithm for the planar 2-center problem. Discrete Comput. Geom. **18**, 125–134 (1997)
15. Tan, X.: An optimal and practical algorithm for the planar 2-center problem. Theory Comput. Syst. **69**, 29 (2025). https://doi.org/10.1007/s00224-025-10228-9
16. Wang, H.: On the planar two-center problem and circular hulls. Discrete Comput. Geom. **68**, 1175–1226 (2022)

Combinatorial Optimization

An LP-Rounding Based Algorithm for Hard Capacitated Uniform Facility Location Problem with Soft Penalties

Hanyin Xiao, Ruiqing Sun, Zhikang Zhang(✉), and Weidong Li

School of Mathematics and Statistics, Yunnan University, Kunming, People's Republic of China
zhikang2024@163.com

Abstract. The hard capacitated facility location problem (HCFLP) is a classical NP-hard problem in combinatorial optimization, with applications in operations research and computer science. Given a set of facilities $\mathcal{F}$ and clients $\mathcal{D}$, each facility has a capacity and open cost, and each client has splittable demand. The goal is to open a subset of facilities and assign client demands to minimize total cost. This paper studies a variant, the hard capacitated uniform facility location problem with soft penalties (HCUFLPSP), where all facilities have uniform opening cost and client demands may be partially rejected at a penalty. We propose a 5-approximation algorithm via LP-rounding, improving upon the 5.5122 ratio of Miao et al. [14].

Keywords: Hard capacitated · Facility location problem · Approximation algorithm · Soft penalties · LP-rounding

1 Introduction

The facility location problem [15], a core model in operations research and computer science, is used to locate factories, schools, warehouses, hospitals, and sensors. This paper examines the hard capacitated uniform facility location problem with soft penalties (HCUFLPSP), which is the variant of the hard capacitated facility location problem (HCFLP) and the uncapacitated facility location problem with linear penalties (UFLPLP).

In the HCFLP, each facility $i \in \mathcal{F}$ has a capacity u_i and can be opened at most once, which makes LP-based methods, such as LP rounding and primal dual techniques, more challenging. For general splittable case, An et al. [1] gave a 288-approximation without violating capacities via LP-rounding, based on multicommodity flow and matching. Kao [6] improved this to $\frac{10+\sqrt{67}}{2} \approx 9.093$. For uniform capacities, Grover et al. [4] presented an $O(\frac{1}{\varepsilon})$-approximation allowing a $1+\varepsilon$ capacity violation. For uniform cost, Levi et al. [7] designed a 5-approximation, later improved to 4 by Kao [5] using two-stage LP-rounding. The current best ratio is $\sqrt{3}+2+\varepsilon \approx 3.732+\varepsilon$ by Dabas et al. [3] via local search.

M. Li et al. (Eds.): TAMC 2025, LNCS 16084, pp. 251–261, 2026.
https://doi.org/10.1007/978-981-95-4839-2_19

In the UFLPLP, a penalty is incurred for unserved clients, representing the cost of excluding distant demands. Charikar et al. [2] gave the first 3-approximation via a primal-dual method. Xu and Xu [16] improved this to $(2+\frac{2}{e})$ using LP-rounding. The current best ratio, 1.5148, is due to Li et al. [8] using LP-rounding with non-uniform distributions. A generalized concept, submodular penalty, is widely used in multicut problem [9,10], multiprocessor scheduling problem [11], cover problem [12,13], and facility location problem [8]. What's more, Miao et al. [14] introduced HCUFLPSP, allowing partial client rejection with penalty, and gave a 5.5122-approximation via LP-rounding.

This paper studies the hard capacitated uniform facility location problem with soft penalties (HCUFLPSP), introduced by Miao et al. [14]. Section 2 formalizes the problem, Sect. 3 presents an approximation algorithm, Sect. 4 provides approximation guarantee, and Sect. 5 concludes the paper.

2 Preliminaries

2.1 Problem Statement

In the HCUFLPSP, we are given a set of facilities $\mathcal{F}$ and clients $\mathcal{D}$. Each client $j \in \mathcal{D}$ has splittable demand $d_j \in \mathbb{Z}_+$, and each facility $i \in \mathcal{F}$ has capacity $u_i \in \mathbb{Z}_+$ and can be opened at most once with uniform cost f. The unit connection cost c_{ij} between client $j \in \mathcal{D}$ and facility $i \in \mathcal{F}$ satisfies the triangle inequality. In this problem, clients may be partially unserved by paying a unit penalty p_j. The goal is to open facilities and assign demands to minimize the total cost, including open, connection, and penalty costs.

To formulate the HCUFLPSP, let x_{ij} denote the demand of client j served by facility i, y_i indicate whether facility i is opened, and z_j represent the rejected fraction of client j's demand. The problem can be modeled as the following mixed-integer program.

$$\begin{aligned}
\text{MILP: min} \quad & \sum_{i\in\mathcal{F}} f y_i + \sum_{i\in\mathcal{F}}\sum_{j\in\mathcal{D}} c_{ij}x_{ij} + \sum_{j\in\mathcal{D}} p_j d_j z_j \\
\text{s.t.} \quad & \sum_{i\in\mathcal{F}} \frac{x_{ij}}{d_j} + z_j \geq 1, \quad \forall j \in \mathcal{D}, && (1)\\
& \frac{x_{ij}}{d_j} \leq y_i, \quad \forall i \in \mathcal{F}, j \in \mathcal{D}, && (2)\\
& \sum_{j\in\mathcal{D}} x_{ij} \leq u_i y_i, \quad \forall i \in \mathcal{F}, && (3)\\
& x_{ij} \in \mathbb{R}_+ \cup \{0\}, \quad \forall i \in \mathcal{F}, j \in \mathcal{D}, && (4)\\
& y_i \in \{0,1\}, \quad \forall i \in \mathcal{F}, && (5)\\
& z_j \in [0,1], \quad \forall j \in \mathcal{D}. && (6)
\end{aligned}$$

In the MILP, constraint (1) ensures that for each client, the total served demand plus the penalized portion equals its full demand, i.e., any unserved portion must be penalized, and the remaining demand must be fully satisfied.

Constraint (2) enforces that a facility can serve demand only if it is opened. Constraint (3) guarantees that the total demand assigned to any facility does not exceed its capacity u_i.

The linear program relaxation LP(1) of the MILP can be obtained by replacing the constraints (4)–(6) with the following constraints:

$$\begin{aligned} & y_i \leq 1, \quad i \in \mathcal{F}, \\ & x_{ij}, y_i, z_j \geq 0, \quad \forall i \in \mathcal{F}, j \in \mathcal{D}. \end{aligned}$$

The dual program of LP(1) is

$$\begin{aligned} DP = \max \quad & \sum_{j \in \mathcal{D}} \alpha_j - \sum_{i \in \mathcal{F}} \xi_i \\ \text{s.t.} \quad & \frac{\alpha_j}{d_j} \leq c_{ij} + \frac{\beta_{ij}}{d_j} + \gamma_i, \quad \forall i \in \mathcal{F}, j \in \mathcal{D}, \\ & \sum_{j \in \mathcal{D}} \beta_{ij} \leq f - u_i \gamma_i + \xi_i, \quad \forall i \in \mathcal{F}, \\ & \alpha_j \leq p_j d_j, \quad \forall j \in \mathcal{D}, \\ & \alpha_j, \beta_{ij}, \gamma_i, \xi_i \geq 0, \quad \forall i \in \mathcal{F}, j \in \mathcal{D}. \end{aligned}$$

3 Algorithm

This section presents an algorithm for the HCUFLPSP. It begins by setting a threshold and identifying the penalty set. The remaining clients are then rounded in two stages. Throughout this paper, let OPT be the optimal value of program MILP. Let (x^*, y^*, z^*) and $(\alpha^*, \beta^*, \gamma^*, \xi^*)$ be the optimal solutions of program LP(1) and DP, respectively.

3.1 Penalized Client Set

In this section, we define the set of the penalized clients and determine $\hat{z}$, which corresponds to the penalized part (see Algorithm 1).

Algorithm 1: Set penalized client set

---

initialization: set $\hat{P} = \emptyset$, and $\hat{z}_j = z_j^*$ for any $j \in \mathcal{D}$.
for $j \in \mathcal{D}$ **do**
  if $z_j^* \geq \rho$ **then**
    set $\hat{z}_j = 1$, and $\hat{P} = \hat{P} \cup \{j\}$.

After setting up fully penalized clients and partially penalized clients, it is only necessary to round $D = \mathcal{D} \backslash \hat{P}$ and $\mathcal{F}$. For any allocation x, define the vicinity sets as follows: for any $A \subseteq \mathcal{F}$ and any $j \in D$, let $N_{(A,x)}(j) = \{i \in A : x_{ij} > 0\}$ be the facilities in A serving client j; for any $B \subseteq D$ and $i \in \mathcal{F}$, let $N_{(B,x)}(i) = \{j \in B : x_{ij} > 0\}$ be the clients in B served by facility i.

Let $I = \{i \in \mathcal{F} : 0 < y_i^* \le \frac{1}{2}(1-\rho)\}$ be the set of small facilities; $U = \{i \in \mathcal{F} : \frac{1}{2}(1-\rho) < y_i^* \le 1\}$ be the set of large facilities. The clients in D are divided into three categories, those that served by facilities only in I, those that served by facilities both in I and U, those that served by facilities only in U. Let $D^{(I)} = \{j \in D : x_{ij}^* = 0, \forall i \in U\}$, $D^{(\leftrightarrow)} = \{j \in D : \min(\max_{i \in I} x_{ij}^*, \max_{i \in U} x_{ij}^*) > 0\}$, and $D^{(U)} = D \backslash (D^{(I)} \cup D^{(\leftrightarrow)})$ denote the clients in three categories, respectively.

For the facilities in U, directly round up and keep the original allocation, that is, $\hat{y}_i = 1, \hat{x}_{ij} = x_{ij}^*$ for any $i \in U$ and any $j \in D$. The remaining rounding problem is for the facility set I and the client set $D^{(I)} \cup D^{(\leftrightarrow)}$.

3.2 First Stage of the Rounding Algorithm

In this stage, the algorithm iteratively forms clusters. Let $F' = I$ and $D' = D^{(I)} \cup D^{(\leftrightarrow)}$ be the facility and client sets to process. It maintains a rounding assignment $\overline{x}$ and two sets H (outlier clients) and $H' \subseteq H$ (unprocessed outliers), all initialized as $\overline{x} = 0$, $H = H' = \emptyset$.

Algorithm 2: The first stage of rounding algorithm

1 **while** $D' \cup H' \neq \emptyset$ **do**
2   **for** $j \in D'$ **do**
3   **if** $\sum_{i \in F'} \frac{x_{ij}^*}{d_j} < \frac{1}{2}(1 - z_j^*)$ **then**
4   Apply Algorithm 3 (Creating outliers).
5   set $D' = D' \backslash \{j\}$, and $x_{ij}^* = 0$ for any $i \in F'$.
6   Apply Algorithm 4 (Forming clusters).
7   **for** $j \in D' \cap D^{(I)}$ **do**
8   **if** $\sum_{i \in F'} \frac{x_{ij}^*}{d_j} < \frac{1}{2}(1 - z_j^*)$ **then**
9             $D' = D' \backslash \{j\}$, and set $x_{ij}^* = 0$ for any $i \in F'$.

Notice that, in Algorithm 2 Step 2, j must be in $D' \cap D^{(\leftrightarrow)}$, because for any $j \in D' \cap D^{(I)}$, if $\sum_{i \in F'} \frac{x_{ij}^*}{d_j} < \frac{1}{2}(1 - z_j^*)$, it will be moved in Algorithm 2 Step 9.

According to the definition of $\overline{d}_{jw}$ and red_j in Algorithm 3, we have

$$\sum_{w \in N_{(U,x^*)}(j)} \overline{d}_{jw} = \sum_{w \in N_{(U,x^*)}(j)} \frac{red_j}{\sum_{i \in U} x_{ij}^*} \cdot x_{wj}^* = red_j \cdot \frac{\sum_{w \in N_{(U,x')}(j)} x_{wj}^*}{\sum_{i \in U} x_{ij}^*}, \tag{7}$$

$$\sum_{k \in N_{(F',x^*)}(j)} x_{kjw}^* = \sum_{k \in N_{(F',x^*)}(j)} \frac{\overline{d}_{jw}}{\sum_{k \in F'} x_{kj}^*} \cdot x_{ij}^* = \overline{d}_{jw}. \tag{8}$$

Algorithm 3: Creating outliers

let $red_j = \min\{\sum_{i \in F'} \frac{x^*_{ij}}{d_j}, \sum_{i \in U} \frac{x^*_{ij}}{d_j}\}$ represent the proportion of remaining demand that needs to be reallocated for client j.
for $w \in N_{(U,x^*)}(j)$ **do**
  create client j_w at w, and set its demand to $\overline{d}_{j_w} = \frac{red_j}{\sum_{i \in U} x^*_{ij}} \cdot x^*_{wj}$.
  for $i \in N_{(F',x^*)}(j)$ **do**
    set $x^*_{ij_w} = \frac{\overline{d}_{j_w}}{\sum_{k \in F'} x^*_{kj}} \cdot x^*_{ij}$.
  set $H = H \cup \{j_w\}, H' = H' \cup \{j_w\}$, and $\frac{\alpha^*_{j_w}}{d_{j_w}} = \frac{\alpha^*_j}{d_j} + c_{wj}$.

For clarity, we introduce the following notations related to cluster formation. Let $C_{D'}$ and $C_{H'}$ denote the sets of clusters centered at non-outlier and outlier clients, respectively, with $C_{D'}, C_{H'} \subseteq \mathcal{F}$. For each $q \in C_{D'}$, let $j(q)$ be the central client and $i(q)$ the facility selected for rounding when q is formed. Define $F^*_{D'} = \{i(q) : q \in C_{D'}\}$ as the set of rounded facilities in $C_{D'}$, and let $G = \bigcup_{j \in H} B(j)$. According to Algorithm 2, the following properties hold after the first stage.

Algorithm 4: Forming clusters

initialization: $\overline{x}_{ik} = 0$ for any $i \in \mathcal{F}, j \in D$.
select client $j \in D' \cup H'$ with minimum value $\frac{\alpha^*_j}{d_j}$, i.e., $j = \arg\min_{k \in D' \cup H'} \frac{\alpha^*_k}{d_k}$.
if $j \in H'$ **then**
  let $B(j) = N_{(F',x^*)}(j)$ represent the satellite facilities. Form a cluster $(j, B(j))$ with facility $B(j)$ centered at j. The rounding decision for this cluster is postponed to the second stage of the algorithm.
  set $H' = H' \backslash \{j\}, F' = F' \backslash B(j)$.
if $j \in D'$ **then**
  select facility $i \in F'$ with maximum capacity u_i from $N_{(F',x^*)}(j)$.
  let $\delta_i = (\frac{1}{2}(1-\rho) - y^*_i) \cdot \frac{1}{\sum_{k \in N_{(F',x^*)}(j) \backslash \{i\}} y^*_i}$ be the redistribution factor for each in $N_{(F',x^*)}(j) \backslash \{i\}$.
  for $l \in N_{(F',x^*)}(j) \backslash \{i\}$ **do**
    $y^*_l = (1-\delta_i) y^*_l$.
    for $k \in N_{(D' \cup H',x^*)}(l)$ **do**
      $x^*_{lk} = (1-\delta_i) x^*_{lk}$.
      set $\overline{x}_{ik} = \delta_i \cdot x^*_{lk} + \overline{x}_{ik}$.
  for $k \in D'$ **do**
    set $\overline{x}_{ik} = x^*_{ik} + \overline{x}_{ik}$.
  set $F' = F' \backslash \{i\}$.

Property 1: $F^*_{D'}$ and G are mutually exclusive, i.e., $F^*_{D'} \cap G = \emptyset$.

Property 2: For each $j \in H$, the associated satellite set $B(j)$ forms a partition of G, i.e. $G = \bigcup_{j \in H} B(j)$ and $B(j_1) \cap B(j_2) = \emptyset$ for any $j_1 \neq j_2 \in H$.

To avoid ambiguity, $(x^{*(0)}, y^{*(0)})$ denotes the initial solution used in the rounding algorithm. For an outlier client $j \in H$ and facility $i \in \mathcal{F}$, $x_{ij}^{*(0)}$ is the allocation at the time j is created. Let $(x^{*(2)}, y^{*(2)})$ denote the solution maintained when the algorithm enters the second stage.

3.3 Second Stage of Rounding Algorithm

In the second stage, the algorithm formulates the rounding decision left for outlier clusters as a global problem. For any $j \in D \cup H$, define the scaling factor: if $j \in D$, that is, j is a non-outlier client, and $\sum_{i \in I} \frac{\overline{x}_{ij}}{d_j} + \sum_{i \in G} \frac{x_{ij}^{*(2)}}{d_j} > 0$, then

$$t'_j = \frac{d_j}{\sum_{i \in I} \overline{x}_{ij} + \sum_{i \in G} x_{ij}^{*(2)}} ((1 - z_j^*) - \sum_{i \in U} \frac{x_{ij}^{*(0)}}{d_j} - red_j).$$

$t'_j = 1$, otherwise.

For each outlier client $j \in H$, let $w(j)$ denote the facility in U where j is located, $p(j)$ denote the specific parent client that created j in $D^{(\leftrightarrow)}$. Conversely, for any $j \in D^{(\leftrightarrow)}$, let $H(j)$ denote the set of outlier clients created from j. For each $w \in U$, let $H(w)$ denote the outlier client located at w.

Algorithm 5: The second stage of rounding algorithm

---

1 construct a new instance of the hard capacitated uniform facility location problem (HCUFLP): the facility set $G = \bigcup_{j \in H} B(j)$, the client set U, each $w \in U$ has a demand d'_w, defined as $d'_w = \sum_{k \in H(w)} \sum_{i \in B(k), l \in D \cup H} t'_l x_{il}^{*(2)}$, and each facility has an open cost and a certain capacity u_i. The linear relaxation program is denoted as $LP(2)$.

2 solve the program $LP(2)$ to obtain a basic optimal solution (x'', y'').

$$\begin{aligned}
LP(2) = \min \quad & \sum_{i \in G} f y_i + \sum_{i \in G} \sum_{j \in U} c_{ij} x_{ij} \\
\text{s.t.} \quad & \sum_{i \in G} x_{ij} = d'_j, \quad \forall j \in U, && (9) \\
& \sum_{j \in U} x_{ij} \leq u_i y_i, \quad \forall i \in G, && (10) \\
& y_i \leq 1, \quad \forall i \in G, && (11) \\
& x_{ij}, y_i \geq 0, \quad \forall i \in G, j \in U.
\end{aligned}$$

3.4 Output

At the end of rounding, the integer-valued function $\hat{y}$ is constructed.

$$\hat{y}_i = \begin{cases} 1, & \text{if} \quad i \in F^*_{D'} \quad \text{or} \quad i \in U, \\ \lceil y''_i \rceil, & \text{if} \quad i \in G, \\ 0, & \text{otherwise.} \end{cases}$$

To construct the allocation function $\hat{x}$, we define two assignment g and h.

(1) The bundled assignment g of $LP(2)$

For any $w \in U$ and any $i \in G$ such that for some $k \in H(w)$, $i \in B(k)$, i.e., i belongs to a cluster centered at some $k \in H(w)$, we define the bundle assignment

$$g_{iw} = \sum_{l \in D \cup H} t'_l \cdot x^{*(2)}_{il}. \tag{12}$$

(2) The unbundled assignment h from x''

Consider the basic optimal solution (x'', y'') of $LP(2)$. Next, unbundle the assignment function x'' to the assignment function h of the original client in $D \cup H$. For each $i \in G$ and $j \in D \cup H$, define the unbundled assignment

$$h_{ij} = \sum_{w \in U} x''_{iw} \cdot \frac{1}{d'_w} \sum_{k \in H(w), i' \in B(k)} t'_j x^{*(2)}_{ij}.$$

Intuitively, the original client assignment x'' in $D \cup H$ is redistributed back in proportion in h. Therefore, for $j \in D \cup H$, we have

$$\begin{aligned} \sum_{i \in G} h_{ij} &= \sum_{i \in G, w \in U} x''_{iw} \cdot \frac{1}{d'_w} \cdot \sum_{k \in H(w), i' \in B(k)} t'_j x^{*(2)}_{ij}, \\ &= \sum_{w \in U} \sum_{k \in H(w), i' \in B(k)} t'_j x^{*(2)}_{i'j} = \sum_{i \in G} t'_j \cdot x^{*(2)}_{ij}. \end{aligned} \tag{13}$$

(3) The assignment function $\hat{x}$

Under the above conditions, the assignment $\hat{x}$ of each $j \in D$ is defined as

$$\hat{x}_{ij} = \begin{cases} x^{*(0)}_{ij}, & \text{if} \quad i \in U, \\ t'_j \overline{x}_{ij} + \sum_{k \in H(j)} \overline{x}_{ik}, & \text{if} \quad i \in F^*_{D'}, \\ h_{ij} + \sum_{k \in H(j)} h_{ik}, & \text{if} \quad i \in G, \\ 0, & \text{otherwise.} \end{cases}$$

Intuitively, the assignment of each $j \subset D$ in $\hat{x}$ consists of the original assignment to U and the rounded assignment of clients in $\{j\} \cup H(j)$ to facilities in $F^*_{D'} \cup G$.

4 Algorithm Analysis

In this section, we prove the feasibility of the algorithm to output the solution $(\hat{x}, \hat{y}, \hat{z})$, and give the approximate guarantee of the solution.

4.1 Algorithm Feasibility

(1) First stage of rounding algorithm

At any point in the first stage, let (x', y'), F', and H' be the inputs to Algorithm 3, and let (x^*, y^*) be the parameters maintained during its execution. The following constraints hold throughout the first stage, from which we derive the following lemmas.

$$\sum_{j \in D' \cup H'} x^*_{ij} \le u_i y^*_i, \quad \forall i \in F', \tag{14}$$

$$\frac{x^*_{ij}}{d_j} \le y^*_i, \quad \forall i \in F', j \in D' \cup H'. \tag{15}$$

Lemma 1. *The process of creating clients (Algorithm 3) does not invalidate (14) and (15).*

Lemma 2. *The process of cluster formation (Algorithm 4) is followed that*

(i) For any $q \in C_{D'}$, $0 < \delta_{i(q)} \le 1$.
(ii) The scaling performed by Algorithm 4 when rounding cluster q does not invalidate constraints (14) and (15).

According to Lemma 2 and Lemma 1, the following corollary holds.

Corollary 1. *(14) and (15) hold in the first phase of the rounding algorithm.*

Therefore, at any moment in the first stage, the inequality (16) always holds.

$$\sum_{i \in F'} \frac{x^*_{ij}}{d_j} \le \sum_{i \in F'} y^*_i, \quad \forall j \in D' \cup H'. \tag{16}$$

Due to the Algorithm 3 and the fact that the constraint (14) hold throughout the entire process, the following lemma can be obtained.

Lemma 3. *For any $q \in C_{D'}$, we have $\sum_{j \in D \cup H} \overline{x}_{i(q)j} \le \frac{1}{2}(1 - \rho) u_{i(q)}$.*

(2) Second stage of rounding algorithm

We now consider the second rounding stage and the clustering in $C_{H'}$. The following lemma summarizes the facility and client status.

Lemma 4. *When entering the second phase of the rounding algorithm, the following conditions hold:*

(i) For any $i \in G, \sum_{j \in D \cup H} x^{*(2)}_{ij} \le u_i y^{*(2)}_i$;
(ii) For any $j \in D^{(I)}, \sum_{i \in I} \frac{\overline{x}_{ij}}{d_j} + \sum_{i \in G} \frac{x^{*(2)}_{ij}}{d_j} \ge \frac{1}{2}(1 - z^*_j)$;
(iii) For any $j \in D^{(\leftrightarrow)}, \sum_{i \in I} \frac{\overline{x}_{ij}}{d_j} + \sum_{i \in G} \frac{x^{*(2)}_{ij}}{d_j} + \sum_{i \in U} \frac{x^{*(0)}_{ij}}{d_j} \ge \frac{1}{2}(1 - z^*_j)$;
(iv) For any $j \in H, \sum_{i \in F^*_{D'}} \overline{x}_{ij} + \sum_{i \in G} x^{*(2)}_{ij} = \sum_{i \in I} x^{*(0)}_{ij}$.

According to Lemma 4, we can immediately obtain corollary. According to the construction of g and $(\hat{x}, \hat{y}, \hat{z})$, we can obtain Lemma 5 and Lemma 6.

Corollary 2. *For any $j \in D, 0 \le t'_j \le 2$.*

Lemma 5. *$(g, 2y^{*(2)})$ is a feasible solution of $LP(2)$.*

Lemma 6. *$(\hat{x}, \hat{y}, \hat{z})$ is a feasible solution of $MILP$.*

4.2 Approximate Guarantee

In this section, we establish approximation guarantees for $(\hat{x}, \hat{y}, \hat{z})$ from Sect. 3. For $j \in H$, let $p(j)$ denote the client in D from which j was created, extended by $p(k) = k$ for $k \in D$. For any assignment x, let $x|_{(A,B)}$ denote the allocation between $A \subseteq \mathcal{F}$ and $B \subseteq D \cup H$; similarly, $y|_A$ denotes facility values for $A \subseteq \mathcal{F}$.

(1) Penalized client set $\hat{P}$

Due to the definition of $\hat{z}$, we can immediately obtain the following lemma.

Lemma 7. *The penalized cost generated by* $\hat{z}$ *is no more than* $\frac{1}{\rho}\sum_{j\in\mathcal{D}} p_j d_j z_j^*$.

(2) Clusters in $C_{H'}$

According to the design of algorithm and the complementary slackness, we can obtain the following lemma.

Lemma 8. *For any* $j \in H$ *and any* $i \in G$ *such that* $x_{ij}^{*(2)} > 0$*, we have* $c_{ij} \leq \frac{\alpha_j^*}{d_j}$.

The following lemma bounds the total cost in $\hat{x}|_{(G,D)}$ using $x''|_{(G,U)}$ and $x^{*(2)}|_{(G,D\cup H)}$, based on the definitions of $\hat{x}$, h, and Lemma 8.

Lemma 9. $\sum_{i\in G, j\in D} c_{ij}\hat{x}_{ij} \leq \sum_{i\in G, j\in U} c_{ij}x''_{ij} + \sum_{i\in G, j\in D\cup H} t'_j(x_{ip(j)} + \frac{\alpha_j^*}{d_j})x_{ij}^{*(2)}$.

According to the property of basic feasible (x'', y''), we can obtain the following lemma.

Lemma 10. $|L| \leq |U|$, *where* $L = \{i \in G : 0 < y''_i < 1\}$.

The following lemma bounds the total cost of $\hat{y}|_G$ and x'' using $y^{*(2)}|_G$, $y^{*(0)}|_G$, and $x^{*(2)}|_{(G,D\cup H)}$, based on the optimality of (x'', y'') and Lemma 5, Lemma 8, and Lemma 10.

Lemma 11. $\sum\limits_{i\in G} f\lceil y''_i\rceil + \sum\limits_{i\in G, j\in U} c_{ij}x''_{ij} \leq 2f\sum\limits_{i\in G} y_i^{*(2)} + f|U| + \sum\limits_{i\in G, j\in D\cup H} t'_j \frac{\alpha_j^*}{d_j} x_{ij}^{*(2)}$.

Combining Lemma 9, and Lemma 11, and the fact that for any $i \in U, y_i^{*(0)} \geq \frac{1}{2}(1-\rho)$, we can get an upper bound on the cost generated by $(\hat{x}|_{(G,D)}, \hat{y}|_G)$.

$$\begin{aligned}\sum_{i\in G} f\hat{y}_i + \sum_{i\in G, j\in D} c_{ij}\hat{x}_{ij} &\leq 2\sum_{i\in G} fy_i^{*(2)} + \sum_{i\in G, j\in D\cup H} t'_j(c_{ip(j)} + 2\cdot\frac{\alpha_j^*}{d_j})x_{ij}^{*(2)} \\ &+ \frac{2}{1-\rho}\sum_{i\in U} fy_i^{*(0)} + \sum_{i\in G, j\in D}(2c_{ip(j)} + 2t'_j\frac{\alpha_j^*}{d_j})x_{ij}^{*(2)}.\end{aligned} \tag{17}$$

(3) Clusters in $C_{D'}$

Considering the cost of clustering in $C_{D'}$, the following lemma bounds the cost of the rounding process for each individual cluster q in $C_{D'}$. The following lemma follows directly from the rounding process of the $C_{D'}$ cluster.

Lemma 12. *For any $q \in C_{D'}$, we have*
(i) $\hat{y}_{i(q)} \leq \frac{2}{1-\rho} y^{*(q)}_{i(q)} + \frac{2}{1-\rho}\delta_{i(q)} \sum_{k\in B(q)\setminus\{i(q)\}} y^{*(q)}_k$;

$$(ii) \sum_{j\in D} c_{i(q)j}\hat{x}_{i(q)j} \leq \sum_{i\in D} 2c_{i(q)j}x^{*(q)}_{i(q)j} + \sum_{j\in H} c_{i(q)p(j)}x^{*(q)}_{i(q)j} + \sum_{j\in D} 2t'_j \frac{\alpha^*_j}{d_j}\overline{x}_{i(q)j}$$

$$+ \sum_{j\in H} 2\frac{\alpha^*_j}{d_j}\overline{x}_{i(q)j} + \delta_{i(q)} \sum_{k\in B(q)\setminus\{i(q)\}} (\sum_{j\in D} 2c_{kj}x^{*(q)}_{kj} + \sum_{j\in H} c_{kp(j)}x^{*(q)}_{kj}).$$

Summing all clusters of $C_{D'}$ in Lemma 12 yields the following lemma.

Lemma 13. *(i)* $\sum\limits_{i\in F^*_{D'}} \hat{y}_i \leq \frac{2}{1-\rho} \sum\limits_{i\in I\setminus G} y^{*(0)}_i + \frac{2}{1-\rho} \sum\limits_{i\in G} (y^{*(0)}_i - y^{*(2)}_i)$;

$$(ii) \sum_{i\in F^*_{D'},j\in D} c_{ij}\hat{x}_{ij} \leq \sum_{i\in F_{D'*},j\in D} 2t'_j\frac{\alpha^*_j}{d_j}\overline{x}_{ij} + \sum_{i\in F^*_{D'},j\in H} 2\frac{\alpha^*_j}{d_j}\overline{x}_{ij}$$
$$+ \sum_{i\in G,j\in D} 2c_{ij}(x^{*(0)}_{ij} - \sum_{l\in H(j)} x^{*(0)}_{il} - x^{*(2)}_{ij}) + \sum_{i\in I\setminus G,j\in H} c_{ip(j)}x^{*(0)}_{ij}$$
$$+ \sum_{i\in G,j\in H} c_{ip(j)}(x^{*(0)}_{ij} - x^{*(2)}_{ij}) + \sum_{i\in I\setminus G,j\in D} 2c_{ij}(x^{*(0)}_{ij} - \sum_{l\in H(j)} x^{*(0)}_{il}). \tag{18}$$

(4) Overall guarantee
The following lemma is obtained by the complementary slackness condition.

Lemma 14. $\sum_{j\in D}(1 - z^*_j - \sum_{i\in U}\frac{x^*_{ij}}{d_j})\alpha^*_j \leq \sum_{i\in I} y^*_i + \sum_{i\in I,j\in D} c_{ij}x^*_{ij}$.

Combining all lemmas, we can obtain our main result.

Theorem 1. *There exists a polynomial-time algorithm with the approximation ratio of 5 for the HCUFLPSP.*

5 Conclusion

This paper studies the hard capacitated uniform facility location problem with soft penalties (HCUFLPSP), allowing partial service for clients. Using LP-rounding, we develop a 5-approximation algorithm, improving the 5.5122 ratio of Miao et al. [14]. To establish this guarantee, we design a two-stage rounding algorithm. Our analysis reveals loose inequalities, indicating room for further improvement through refined techniques.

Acknowledgments. The work was supported in part by Yunnan Fundamental Research Projects (Grants No. 202501AS070076, No. 202501AS070170), and Research and Innovation Program of Yunnan University (Grants No. TM-23236822).

References

1. An, H.C., Singh, M., Svensson, O.: LP-based algorithms for capacitated facility location. SIAM J. Comput. **46**(1), 272–306 (2017)
2. Charikar, M., Khuller, S., Mount, D.M., Narasimhan, G.: Algorithms for facility location problems with outliers (extended abstract). In: Proceedings of the 12th Annual ACM-SIAM Symposium on Discrete algorithms of the Conference, Washington D.C., USA (2001). https://doi.org/10.5555/365411.365555
3. Dabas, R., Garg, N., Gupta, N.: Capacitated facility location with outliers and uniform facility costs. In: Proceedings of International Conference on Integer Programming and Combinatorial Optimization of the Conference, Wrocław, Poland (2024). https://doi.org/10.1007/978-3-031-59835-7_7
4. Grover, S., Gupta, N., Khuller, S.: LP-based approximation for uniform capacitated facility location problem. Discret. Optim. **45**, 100723 (2022)
5. Kao, M.J. Improved approximation algorithm for capacitated facility location with uniform facility cost. In: Proceedings of the 34th International Symposium on Algorithms and Computation of the Conference, Kyoto, Japan (2023). https://doi.org/10.4230/LIPIcs.ISAAC.2023.45
6. Kao, M.J.: On the integrality gap of MFN relaxation for the capacitated facility location problem. In: Proceedings of the 2023 Annual ACM-SIAM Symposium on Discrete Algorithms of the Conference, Florence, Italy (2023). https://doi.org/10.1137/1.9781611977554.ch40
7. Levi, R., Shmoys, D.B., Swamy, C.: LP-based approximation algorithms for capacitated facility location. Math. Program. **131**(1), 365–379 (2012)
8. Li, Y., Du, D., Xiu, N., Xu, D.: Improved approximation algorithms for the facility location problems with linear/submodular penalties. Algorithmica **73**, 460–482 (2015)
9. Liu, X., Li, W.: An approximation algorithm for the-prize-collecting multicut problem in trees with submodular penalties. Math. Struct. Comput. Sci. **34**(3), 193–210 (2024)
10. Liu, X., Li, W.: Combinatorial approximation algorithms for the submodular multicut problem in trees with submodular penalties. J. Comb. Optim. **44**(3), 1964–1976 (2022)
11. Liu, X., Li, W.: Approximation algorithms for the multiprocessor scheduling with submodular penalties. Optim. Lett. **15**(6), 2165–2180 (2021). https://doi.org/10.1007/s11590-021-01724-1
12. Liu, X., Dai, H., Li, S., Li, W.: The k-prize-collecting minimum power cover problem with submodular penalties on a plane. Sci. Sin. Inform **52**(6), 947–959 (2022)
13. Liu, X., Li, W., Dai, H.: Approximation algorithms for the minimum power cover problem with submodular/linear penalties. Theoret. Comput. Sci. **923**, 256–270 (2022)
14. Miao, R., Wu, C., Yuan, J.: LP-rounding based algorithm for capacitated uniform facility location problem with soft penalties. Tsinghua Sci. Technol. **30**(1), 279–289 (2024)
15. Xiao, H., Zhang, J., Zhang, Z., Li, W.: A survey of approximation algorithms for the universal facility location problem. Mathematics **13**(7), 1023 (2025)
16. Xu, G., Xu, J.: An improved approximation algorithm for uncapacitated facility location problem with penalties. J. Comb. Optim. **17**(4), 424–36 (2009)

Prediction-Augmented Mechanism Design for Weighted Facility Location

Yangguang Shi[1,2(✉)] and Zhenyu Xue[1]

[1] School of Computer Science and Technology, Shandong University, Qingdao, China
shiyangguang@sdu.edu.cn, xuezhenyu@mail.sdu.edu.cn

[2] Department of Economics, University of Bath, Claverton Down, Bath, UK

Abstract. Facility location is fundamental in operations research, mechanism design, and algorithmic game theory, with applications ranging from urban infrastructure planning to distributed systems. Recent research in this area has focused on augmenting classic strategyproof mechanisms with predictions to achieve an improved performance guarantee against the uncertainty under the strategic environment. Previous work has been devoted to address the trade-off obstacle of balancing the consistency (near-optimality under accurate predictions) and robustness (bounded inefficiency under poor predictions) primarily in the unweighted setting, assuming that all agents have the same importance. However, this assumption may not be true in some practical scenarios, leading to research of weighted facility location problems.

The major contribution of the current work is to provide a prediction augmented algorithmic framework for balancing the consistency and robustness over strategic agents with non-uniform weights. In particular, through a reduction technique that identifies a subset of *representative* instances and maps the other given locations to the representative ones, we prove that there exists a *strategyproof* mechanism achieving a bounded consistency guarantee of $\frac{\sqrt{(1+c)^2 W_{\min}^2 + (1-c)^2 W_{\max}^2}}{(1+c)W_{\min}}$ and a bounded robustness guarantee of $\frac{\sqrt{(1-c)^2 W_{\min}^2 + (1+c)^2 W_{\max}^2}}{(1-c)W_{\min}}$ in weighted settings, where c can be viewed as a parameter to make a trade-off between the consistency and robustness and $W_{\min}$ and $W_{\max}$ denote the minimum and maximum agents' weight. We also prove that there is no strategyproof deterministic mechanism that reach 1-consistency and $O\left(n \cdot \frac{W_{\max}}{W_{\min}}\right)$-robustness in weighted FLP, even with fully predictions of all agents.

Keywords: Weighted Facility Location · Prediction-Augmented Mechanism · Consistency and Robustness Trade-off · Upper Bound

1 Introduction

Facility Location Problem (FLP) is a classic problem in operations research, computer science, and economics [8]. Its goal is to choose one or more facility

The full version of this paper is available at https://arxiv.org/abs/2507.06509.

M. Li et al. (Eds.): TAMC 2025, LNCS 16084, pp. 262–274, 2026.
https://doi.org/10.1007/978-981-95-4839-2_20

locations within a given space to optimize a global cost function between users (referred to as agents) and facilities. If only one facility needs to place, we say it single FLP[1]. In the problem, there are n agents in a metric space $\mathcal{M}$, and each agent i has her preferred location p_i and must report a location $\hat{p}_i \in \mathcal{M}$ of the facility. We are required to collect this information to choose where to locate the facility $f \in \mathcal{M}$. Once we choose the facility location f, the individual cost of each agent i is given by the distance $d(p_i, f)$ between p_i and f, where $d(\cdot)$ is the distance function defined on $\mathcal{M}$. Our objective is to optimize a global function $g(P; f)$, where P is the location profile of all the agents. Naturally, here comes to a question: *What kind of global cost function we should consider?* Motivated by various applications in practical scenarios (see [8] for a more detailed overview), researchers in this area are often interested in the global objectives of (1) *egalitarian social cost* and (2) *utilitarian social cost*, where *egalitarian social cost* is to the maximum cost over all the agents, and *utilitarian social cost* is the average cost [1,28]. The study of the former is more difficult as applying predictions on any existing strategyproof mechanisms is hard, for the mechanisms focus on the bounding box of P that prediction can hardly get involved. So the current work is devoted to the study of the latter.

Under our setting, agents are selfish, so an agent i may misreport her preferred location to minimize her individual cost, which means the reported location $\hat{p}_i$ is not necessarily equivalent to her real preferred location p_i. This leads to the *strategic facility location problem*, where the focus shifts from purely optimizing a cost function to designing mechanisms that are strategyproof, i.e., mechanisms in which agents have no incentive to tell lie about their true preferences [6,25]. This problem is first studied by Procaccia and Tennenholtz [25], focusing on the egalitarian social cost, and Dekel [11] investigates how to optimize the utilitarian social cost. Procaccia and Tennenholtz [25] prove that no deterministic and strategyproof mechanism can achieve better than a 2-approximation, when the goal is the egalitarian social cost, and Meir [22] propose the coordinate median (CM) mechanism that reaches $\sqrt{n}$-approximation in n-dimensional case while optimizing the utilitarian social cost. These results establish the framework for subsequent research.

Research work on FLP has been applied in establishment of public services such as schools and hospitals [10,16]. Existing work is often based on the unweighted assumption on the agents, which means that all the agents in the environment are equivalent in importance. Although this assumption reflects equal treatment in public services, this setting is uncommon in reality, where agents often differ in importance. Therefore, the unweighted assumption rarely holds in practice. For example, in the field of healthcare, existing work suggests that the older adults are often associated with a relatively higher weights [2,23]. Agents with distinct importance (weights) also exist in other applications including logistics planning [5], disaster assistance [7], etc. Obviously, the unweighted setting of FLP cannot be adapted to these environments. That is why strategic

[1] If not mentioned on purpose, in the following text, we say the *FLP* as single FLP by default.

FLP with non-uniform weights has attracted the attention of researchers [29,32]. Under the weighted setting, we assume that different weight to quantifies her importance to the decision maker: agents with higher weights take greater social cost and vice versa.

Existing work on strategic FLP (unweighted or weighted) usually adopts the research paradigm of worst-case analysis. This paradigm focuses on extreme scenarios that rarely occur in practice, leading to algorithm design that is overly conservative for the average case. On the other hand, average-case analysis must deal with the uncertainty in real-world problems, such as the lack of knowledge about the true distribution of agent's preferred location in FLP. To address this issue, a novel research paradigm, *prediction-augmented algorithm design* has recently emerged. It allows decision-makers to access external prediction sources to obtain forecasts about the uncertainty in the problem setting, and allows them to decide how much they trust the forecasts [13,14,22,32]. When the prediction is accurate, we can trust it more and get a better resulting guarantee than using worst-case analysis, i.e., *consistency* of the mechanism. On the other hand, when the prediction is arbitrarily wrong, we would like to ensure that the wrong prediction does not worsen the result too much and obtains a solid performance guarantee, i.e., *robustness* of the mechanism [20]. More precisely speaking, in the framework of prediction-augmented mechanism design of strategic FLP, given the agents' reported locations P, the designer is provided a predicted location $\hat{o}$, regarding the real optimal location $o(P)$. The designer can use the information to make a mechanism, whose result is $f(P, \hat{o})$. Given a social cost function C, the mechanism is α-consistent when the prediction is accurate ($\hat{o} = o(P)$) if

$$\max_{P} \left\{ \frac{C(f(P, o(P)), P)}{C(o(P), P)} \right\} \leq \alpha$$

and β-robust even when the prediction is arbitrarily wrong if

$$\max_{P,\hat{o}} \left\{ \frac{C(f(P, \hat{o}), P)}{C(o(P), P)} \right\} \leq \beta.$$

1.1 Our Contributions

We investigate the prediction-augmented mechanism design of strategic FLP in a weighted setting, where the importance of the agents is non-uniform according to the weighted values. Our goal is to design deterministic, strategyproof mechanisms that incorporate a prediction of the optimal facility location, and achieve a tunable trade-off between consistency and robustness.

Our contributions are as follows:

Theorem 1. *There exists a deterministic and strategyproof mechanism in the weighted FLP with predictions that reaches* $\frac{\sqrt{(1+c)^2 W_{\min}^2 + (1-c)^2 W_{\max}^2}}{(1+c)W_{\min}}$*-consistency and* $\frac{\sqrt{(1-c)^2 W_{\min}^2 + (1+c)^2 W_{\max}^2}}{(1-c)W_{\min}}$*-robustness.*

$W_{\min}$ and $W_{\max}$ denote the minimum and maximum weights of agents, respectively. The mechanism uses a parameter $c \in [0, 1)$ to control the degree of reliance on the prediction. When c is large, the mechanism places more trust in the prediction, achieving better consistency; when c is small, it becomes more conservative, improving the robustness. In particular, we emphasize that our performance guarantee matches the existing results of Zhang and Li [32] by setting $c = 0$. Also, when $\frac{W_{\max}}{W_{\min}} = 1$, meaning that all agents' weights are equal, our results matches those from Agrawal [1] on the unweighted setting.

Theorem 2. *There is no deterministic and strategyproof mechanism in the weighted FLP with predictions that is* 1*-consistent and* $O\left(n \cdot \frac{W_{\max}}{W_{\min}}\right)$*-robust.*

Theorem 2 generalizes known impossibility results for the unweighted case and highlights how agents' weights fundamentally constrains achievable trade-offs. It indicates that the weighted case of facility location is more challenging than the unweighted one, as the performance guarantee of any mechanism designed for the weighted case can hardly get rid of the dependence on the input parameters $\frac{W_{\max}}{W_{\min}}$.

1.2 Related Work

Strategic Facility Location. Strategic FLP is a classic mechanism design problem that has been widely studied. For example, Procaccia and Tennenholtz [25] research the approximate mechanism design without money and take this problem as an example. For single facility location in one dimension, there exists a deterministic mechanism that is 2-approximate in utilitarian social cost and optimal in maximum social cost by placing the facility on the medium of all agents. This mechanism also reaches the best approximate ratio among all deterministic and strategyproof mechanisms. Zhang and Li [32] expand this theory into weighted cases, and show that the coordinate medium mechanism achieves $\frac{W_{\max}}{W_{\min}}$-approximation in utilitarian social cost. Also, it is the best approximation ratio.

A lot of research work focus on this framework along with strategic FLP [4,9]. Beyond the classic setting, extended researches include Obnoxious Facility Location (where agents seek maximum distance from facilities) [12,19] and Heterogeneous Facility Location (where facilities can be both desirable and undesirable, from the perspective of different agents). Results in this line of work allows agents to have different weights. However, the weight of each agent is constrained in a binary set $\{-1, 1\}$. In our work, the analysis is conducted precisely by removing this constraint.

Prediction-Augmented Algorithms. Prediction-augmented algorithms (also known as the learning-augmented algorithms), is an algorithm design paradigm aiming to surpass the worst-case performance guarantee by allowing the algorithm to access predictions. Lykouris and Vassilvtiskii [21] introduce consistency and robustness to analyze the performance of algorithms in this framework as the two primal metrics to measure a mechanism in this framework. After its release, many classic problems, like online paging problems [18,21], scheduling problems [26], ski-rental problems [30,31], and secretary problems [3] have been researched in this framework. Facility location problems are also in the list. Fotakis and Gergatsouli [15] study online facility location problem with prediction, and Jiang and Liu [17] introduce dynamic prediction update to amend the prediction models.

However, the problems mentioned above are not studied under a strategic setting. In strategic problems, the predictions focus on overcoming the information limitation of privately held information rather than future information [1]. Agrawal [1] first takes strategic FLP into research and designs Coordinate Medium (CM) mechanism with Prediction, also known as CMP mechanism. The CMP mechanism is simply adding cn copies of $\hat{o}$ into P to construct a new multiset P' to run the CM mechanism, which outputs the median of P' from each dimension separately. As is known that such median-point like mechanisms cannot be applied to general metric spaces [27], Agrawal's work is conducted over the two-dimensional Euclidean space. Our work can be viewed as the extension of their work by adopting the same assumption on the metric space, while allowing agents to have non-uniform weights, which is more common in practice.

2 Preliminaries

Facility Location Problem (FLP). The classic FLP considers a set of agents located in a metric space (the current work focuses on the two-dimensional Euclidean space $\mathbb{R}^2$). Each agent $i \in [n]$ has a preferred location $p_i \in \mathbb{R}^2$. Given the preferred location set $P = \{p_1, \ldots, p_n\}$, the goal of FLP is to choose a facility location $f \in \mathbb{R}^2$ that minimizes the average cost, measured as the average distance from all agents to the facility:

$$C^u(f, P) = \frac{1}{n} \sum_{p_i \in P} d(f, p_i),$$

where $d(\cdot, \cdot)$ denotes the Euclidean distance. This is referred to as the *utilitarian cost.* Given this social cost function, one can define and compute the optimal facility location, denoted as $o(P)$.

Strategic FLP. In strategic settings, the real preferred locations p_i are private information, and an agent may misreport it to reduce her individual cost $C(f, p_i) = d(f, p_i)$. A mechanism $\mathcal{M}$ takes reported locations $(\hat{p}_1, \ldots, \hat{p}_n)$ and outputs a facility location f. In order to prevent misreporting, we would like to ensure that the mechanism we design is *strategyproof.* A mechanism is said to be strategyproof if the optimal decision of any agent is reporting her real information, regardless of others' reports. In our work, the reported information is about the real preferred locations, so the strategyproof constraints can be formulated as follows:

$$C(f(P_{-i}, p_i), p_i) \leq C(f(P_{-i}, \hat{p}_i), p_i) \quad \forall P \in \mathbb{R}^{2n}, i \in [n], \hat{p}_i \in \mathbb{R}^2.$$

A well known mechanism designed for strategic FLP is the *Coordinate Median (CM)* mechanism [1], which selects the facility location by independently computing the median of all coordinates (if the number of agents n is even, we assume that the smaller of two medians is returned for all coordinates). It proves that such a deterministic mechanism is strategyproof in two-dimensional space, and it guarantees a bounded approximation ratio of $\sqrt{2}$ for the utilitarian objective [1]. The mechanism Generalized Coordinate Median (GCM) proposed by Agrawal [1] generalizes the classic version of CM by taking an independent and constant multiset P' of locations (often called the *phantom points*) as inputs parallel to the reported location set P. The location chosen by GCM for the facility is $CM(P \cup P')$. Similarly, the GCM mechanism is also proved to be deterministic and strategyproof [1].

Strategic FLP With Weights. In many practical applications, agents differ in demand, importance, or service sensitivity. This difference can be modeled by assigning each agent a public positive weight $w_i > 0$. In the weighted setting, the individual cost for agent i becomes:

$$C(f, p_i) = w_i \cdot d(f, p_i),$$

and the utilitarian cost becomes the average weighted distance:

$$C^u(f, P) = \frac{1}{n} \sum_{p_i \in P} w_i \cdot d(f, p_i).$$

The weighted case breaks many of the nice symmetry properties of the unweighted problem and presents additional difficulties for designing strategyproof mechanisms. For example, a single agent with an extremely high weight can significantly bias the optimal solution, or even dominate the outcome.

Prediction-Augmented Mechanism. For the weighted version of strategic FLP, we consider developing a prediction augmented mechanism that receives a prediction $\hat{o} \in \mathbb{R}^2$ of the optimal location. This prediction may comes from some external machine learning models trained by the history data of agents. Our mechanism can be viewed as a functions $f(P, \hat{o})$ that maps the reported preferred locations and the prediction to a location of the facility. The goal of the current paper is to ensure that the proposed mechanism is strategyproof, while achieving a competitive ratio as close to 1 as possible when the obtained predictions are highly accurate (i.e., providing a consistency guarantee), and maintaining an acceptable competitive ratio even when the prediction error rate is high (i.e., providing a robustness guarantee).

3 Prediction Augmented Mechanism and Upper Bounds Analysis

To prove Theorem 1, we first consider the CMP mechanism that is proved to be $\frac{\sqrt{2+2c^2}}{1+c}$-consistent and $\frac{\sqrt{2+2c^2}}{1-c}$-robust in the unweighted strategic FLP, and then try to extend it to the weighted case. The major contribution here is to prove that our extended version of CMP for the weighted setting has the same upper bound on the approximation ratio with the original version when choosing a proper value for c, a parameter for making trade-off between the consistency and robustness.

3.1 Extending CMP to Weighted Case

Vanilla CMP Mechanism. The vanilla CMP mechanism (i.e., the original version of CMP) is a simple way to combine agent reports with a predicted optimal location. It does this by taking P' as the multiset of some copies of predicted location and applying it to GCM mechanism. The number of copies is controlled by a confidence parameter $c \in [0, 1)$, more specifically, it adds cn copies of $\hat{o}$. The detailed vanilla CMP algorithm is shown in the full version.

If c is small, only a few copies of prediction are added, so the mechanism relies more on the reported information and is more robust to prediction errors. If c is large, more copies of prediction are added, and the mechanism more trusts the prediction.

Notably, when $c = 0$, no copies of prediction are added, and CMP becomes equivalent to the classic CM mechanism.

This design allows CMP to smoothly trade off between being robust (when the prediction may be wrong) and following the prediction (when it is accurate), while remaining deterministic and strategyproof.

In this work, we extend the Coordinate Median with Prediction (CMP) mechanism to the weighted strategic setting, and analyze its consistency and robustness.

Extension of CMP to Weighted Setting. We now analyze the Coordinate Median with Prediction (CMP) mechanism in the weighted setting introduced in our contribution section. Similar with the unweighted setting, here CMP still takes as input the reported agent locations and a predicted optimal location, adds phantom points at the prediction, and returns the coordinate-wise median over the augmented set. The procedure of our mechanism does not rely on the weights, thus remains deterministic and strategyproofness by construction.

Although sharing a similar description of procedure with the vanilla CMP, the extended CMP asks for novel analysis techniques to incorporate the non-uniform weights. In particular, each agent i is assigned a public weight $w_i > 0$, and the cost function becomes:

$$C^u(f, P) = \frac{1}{n} \sum_{p_i \in P} w_i \cdot d(f, p_i),$$

reflecting the average weighted distance across agents.

Our main result, as stated in Theorem 1, shows that CMP achieves a consistency-robustness trade-off parameterized by $c \in [0, 1)$, with explicit performance bounds depending on the weight ratio $W_{\max}/W_{\min}$. When the weights are identical, these bounds reduce to the known results for the unweighted case [1]. Particularly, when c is set to 0, the extended CMP reduces to the vanilla version and recovers the robustness bounds in [32].

We next analyze CMP's behavior on a structured input family, known as the Clusters-on-Axes (COA) construction proposed by Agrawal et al. [1], to derive its consistency and robustness bounds and demonstrate why these bounds are tight.

3.2 Consistency Analysis

We begin our performance analysis of CMP in the weighted setting by focusing on its consistency: how well the mechanism performs when the prediction is accurate. Recall that consistency is measured by the worst-case ratio between the mechanism's output cost and the optimal cost, assuming that the prediction is exactly the optimal location.

To analyze this, we consider a structured class of worst-case instances, called *Clusters-and-OPT-on-Axes* (COA), where some agents are at the top of the y-axis, and others split symmetrically on the left and right sides of the x-axis. These instances have been shown to yield tight bounds in the unweighted setting [1], and we extend them here to match the weighted setting.

Definition 1 (Weighted COA Instances). *Given confidence parameter $c \in [0, 1)$, a weighted COA instance consists of a reported location set P and a prediction $\hat{o}$ that outputs $f(P, \hat{o}, c) = (0, 0)$ and optimal location $o(P) = (0, 1)$. Its agents positioned at $(0, 1)$, $(x, 0)$, or $(-x, 0)$ for some $x \geq 0$. Agents located at $(0, 1)$ are assigned weight $W_{\max}$, and agents at $(\pm x, 0)$ are assigned weight $W_{\min}$.*

This setup is designed to achieve the largest possible approximation ratio by concentrating high-weight agents near the optimal point and low-weight agents near the prediction. The following result formalizes that this setting indeed leads to the worst-case consistency.

Theorem 3. *Among all weight distributions in COA instances, assigning weight $W_{\max}$ to agents at $(0, 1)$ and $W_{\min}$ to agents at $(\pm x, 0)$ maximizes the approximation ratio of CMP.*

Sketch of Proof. Let m be the number of agents at $(0, 1)$. We compare the total cost under any weight $w \in (W_{\min}, W_{\max})$ and the extreme values $W_{\min}$ or $W_{\max}$, and observe that the ratio is maximized when weights are polarized as described. The detailed algebraic comparisons are given in the full version.

We now compute the consistency bound achieved by CMP under this worst-case weighted COA structure.

Theorem 4. *For CMP with confidence $c \in [0, 1)$, the worst-case consistency in weighted COA instances is*

$$\frac{\sqrt{(1+c)^2 W_{\min}^2 + (1-c)^2 W_{\max}^2}}{(1+c) W_{\min}}.$$

Sketch of Proof. To ensure the CMP output remains at $(0, 0)$, at most $\frac{(1-c)n}{2}$ agents can be placed at $(0, 1)$. Under this constraint, we compute the mechanism cost using average weighted cost, optimize the ratio with respect to x, and show that the maximum occurs at

$$x = \frac{(1 + c)W_{\min}}{(1 - c)W_{\max}}.$$

Substituting back gives the stated consistency bound. Full derivation is provided in the full version.

This result demonstrates how the consistency of CMP degrades as the weight imbalance $\frac{W_{\max}}{W_{\min}}$ increases. In the special case where all agents have equal weights, the bound reduces to the known result in the unweighted case. Now we begin to prove why COA can achieve this bound.

In order for CMP to output the facility at $(0, 0)$, there must be a sufficient number of agents near the origin to counterbalance the influence of phantom points at the prediction. The weighted COA structure achieves this by placing low-weight agents symmetrically on the horizontal axis, which allows for maximum geometric leverage while contributing minimally to the cost.

Meanwhile, high-weight agents are concentrated at the optimal point $(0, 1)$, ensuring that the optimal facility location has minimal cost. This separation between the mechanism output and the optimal location—combined with the weight imbalance—leads to a maximal approximation ratio.

Theorem 5. *Among all instances where CMP outputs the facility at* $(0, 0)$ *and the prediction is accurate, the weighted COA construction achieves the highest possible consistency bound.*

Sketch of Proof. Let f be the CMP output and $o(P)$ the true optimum. To achieve large approximation ratio $C^u(f, P)/C^u(o(P), P)$, we aim to maximize the numerator (cost at f) and minimize the denominator (cost at $o(P)$). The numerator is maximized by placing as many agents as possible far from f, but with low weights so that they do not influence the median. The denominator is minimized by concentrating high-weight agents at the optimal point.

Any other configuration either (i) increases the weight of nearby agents (reducing the cost at f), or (ii) spreads high-weight agents away from the optimum (increasing the cost at $o(P)$). Therefore, no alternative weight or location arrangement can achieve a higher approximation ratio than the weighted COA construction.

A formal derivation and comparison argument is provided in the full version.

3.3 Robustness Analysis

We now analyze the robustness of CMP in the weighted setting. Robustness captures how poorly the mechanism can perform when the prediction is arbitrarily wrong. Following the same weighted COA construction as Definition 1, we now assume that the prediction $\hat{o}$ is at $(0, 0)$, simulating the wrong prediction. The CMP mechanism still outputs $(0, 0)$.

The agents are arranged as before: some are placed at $(0, 1)$, and the rest are split symmetrically on $(x, 0)$ and $(-x, 0)$. Agents with high weight are assigned to the optimal location $(0, 1)$, and those with low weight are set at x-axis. This setting ensures that the optimal solution incurs low cost, while the output of mechanism takes higher cost due to its reliance on the wrong prediction.

Theorem 6. *For CMP with confidence $c \in [0, 1)$, the worst-case robustness in weighted COA instances is*

$$\frac{\sqrt{(1-c)^2 W_{\min}^2 + (1+c)^2 W_{\max}^2}}{(1-c) W_{\min}}.$$

Sketch of Proof. To fix the CMP output at $(0, 0)$, the number of agents at $(0, 1)$ must be limited so that the phantom points at $(0, 0)$ dominate the median on y-axis. Given this configuration, we compute the cost incurred by CMP and compare it with the optimal cost at $(0, 1)$. By adjusting the parameter x, we maximize the approximation ratio and find that it reaches a maximum when

$$x = \frac{(1-c) W_{\min}}{(1+c) W_{\max}}.$$

Substituting this into the ratio yields the stated robustness bound. Full derivation is provided in the full version.

This result looks symmetric to the consistency bound, but with the roles of c and $-c$ reversed, illustrating the trade-off: as the mechanism places more trust in the prediction (larger c), consistency improves while robustness deteriorates.

3.4 Trade-Off Between Consistency and Robustness

Our previous results show that consistency and robustness cannot be optimized at the same time. As the confidence parameter increases, the mechanism relies more on the predicted location, resulting in improved performance when the prediction is accurate. However, this same reliance can lead to degraded performance when the prediction is far from the true optimum.

To make this trade-off concrete, consider a simple instance in $\mathbb{R}^2$ with three agents. One agent with weight 4 is located at $(0, 1)$, while two agents with weight 1 are placed at $(-1, 0)$ and $(1, 0)$. The optimal location that minimizes the utilitarian social cost is clearly $(0, 1)$, where the highest-weight agent locates.

If the prediction is accurate, $\hat{o} = (0, 1)$, then the phantom points introduced by CMP is the optimal point. The larger c is, the more phantom points are introduced, and the mechanism outputs a facility location closer to $(0, 1)$. Or more specifically, when $0 < c < 1/3$, no phantom points are introduced, the CMP outputs $(0, 0)$, and the consistency is $\frac{3}{\sqrt{2}}$; when $1/3 \leq c < 2/3$, only one phantom point is introduced, the CMP also outputs $(0, 0)$, and the consistency is $\frac{3}{\sqrt{2}}$; when $2/3 \leq c < 1$, two phantom points are introduced, the CMP outputs $(0, 1)$, and the consistency is 1. In this case, the consistency decreases when c get larger.

In contrast, if the prediction is wrong, we consider two cases: $\hat{o} = (0, -10)$ and $\hat{o} = (0, 1/2)$. In the former case, the CMP always outputs $(0, 0)$, which means CMP has a bottommost performance guarantee when the prediction is extremely wrong. In the latter case, the CMP outputs $(0, 0)$ when $0 < c < 2/3$, and $(0, 1/2)$ when $2/3 \leq c < 1$, in which case the robustness decreases to $\frac{2+\sqrt{5}}{2\sqrt{2}}$. It shows that when the prediction is slightly wrong, CMP can still provide a meaningful robustness guarantee as long as the confidence is high enough.

This example highlights the trade-off: placing more confidence on the predictions improves consistency under accurate predictions, but may worsen robustness when predictions are wrong. No fixed value of c can optimize both objectives simultaneously, showing a limitation of deterministic mechanisms in the weighted setting.

In the next section, we formalize this limitation and prove that no deterministic mechanism can simultaneously achieve perfect consistency and bounded robustness in the weighted setting.

3.5 Impossibility of Simultaneous Optimal Trade-Off

We now demonstrate a fundamental impossibility result: *No mechanism can simultaneously achieve perfect consistency and bounded robustness* in the weighted facility location problem with prediction.

Generalized Coordinate-wise Median (GCM) mechanism proposed by Procaccia et al. [25] assumes that the locations are in the (low-dimensional) Euclidean space and computes the median of the reported locations along each coordinate. With Corollary 3 in [24], we know that any deterministic, strategyproof, anonymous, and unanimous mechanism can be expressed as a GCM mechanism with $n - 1$ constant points in P.

We construct two specific scenarios in $\mathbb{R}^2$, where all agents lie on the vertical line $x = 0$:

Instance A: 4 agents at $(0, 10)$, each with weight 1; 1 agent at $(0, 20)$, with weight 5.

Instance B: 4 agents at $(0, 10)$, each with weight 5; 1 agent at $(0, 20)$, with weight 1.

In Instance A, to reach 1-consistency, all the 4 constant point must be at (x, y) where $y > 20$, otherwise, the mechanism would choose $(0, 10)$ and consistency would be $\frac{5}{4}$.

However, in Instance B, the mechanism chooses $(0, 20)$, the robustness become:

$$\frac{200}{10} = 20.$$

We can easily imagine that this ratio grows *linearly* with the $\frac{W_{\max}}{W_{\min}}$. More generally, if the agent at $(0, 20)$ has weight $W_{\min}$ and the others have weight

$W_{\max}$, the robustness becomes:

$$\frac{(n-1) \cdot W_{\max} \cdot d}{W_{\min} \cdot d} = (n-1) \cdot \frac{W_{\max}}{W_{\min}}.$$

This example is exactly what Theorem 2 says: *There is no deterministic and strategyproof mechanism in the weighted FLP with predictions that is* 1-*consistent and* $O\left(n \cdot \frac{W_{\max}}{W_{\min}}\right)$*-robust.* It demonstrates that requiring consistency $= 1$ inevitably leads to unbounded robustness as $\frac{W_{\max}}{W_{\min}} \to \infty$. Therefore, no deterministic mechanism can simultaneously ensure perfect consistency and bounded robustness in the presence of adversarial prediction error.

Acknowledgments. The work of Yangguang Shi was partially supported at Shandong University by the Science Fund Program of Shandong Province for Distinguished Oversea Young Scholars (Grant No. 2023HWYQ-006), by the National Natural Science Foundation of China (Grant No. 62302273), and by the Shandong Province Sponsored Overseas Study Program (Grant No. 10000082163106).

References

1. Agrawal, P., Balkanski, E., Gkatzelis, V., Ou, T., Tan, X.: Learning-augmented mechanism design: leveraging predictions for facility location. In: Proceedings of the 23rd ACM Conference on Economics and Computation, pp. 497–528 (2022)
2. Ahmadi-Javid, A., Seyedi, P., Syam, S.S.: A survey of healthcare facility location. Comput. Oper. Res. **79**, 223–263 (2017)
3. Antoniadis, A., Gouleakis, T., Kleer, P., Kolev, P.: Secretary and online matching problems with machine learned advice. Adv. Neural. Inf. Process. Syst. **33**, 7933–7944 (2020)
4. Balkanski, E., Gkatzelis, V., Shahkarami, G.: Randomized strategic facility location with predictions. Adv. Neural. Inf. Process. Syst. **37**, 35639–35664 (2024)
5. Boonmee, C., Arimura, M., Asada, T.: Facility location optimization model for emergency humanitarian logistics. Int. J. Disaster Risk Reduction **24**, 485–498 (2017)
6. Brandt, F., Conitzer, V., Endriss, U., Lang, J., Procaccia, A.D.: Handbook of Computational Social Choice. Cambridge University Press (2016)
7. Carnero Quispe, M.F., Chambilla Mamani, L.D., Yoshizaki, H.T.Y., Brito Junior, I.D.: Temporary facility location problem in humanitarian logistics: a systematic literature review. Logistics **9**(1), 42 (2025)
8. Chan, H., Filos-Ratsikas, A., Li, B., Li, M., Wang, C.: Mechanism design for facility location problems: a survey. arXiv preprint arXiv:2106.03457 (2021)
9. Chen, Q., Gravin, N., Im, S.: Strategic facility location via predictions. arXiv preprint arXiv:2410.07497 (2024)
10. Daskin, M.: Network and discrete location: models, algorithms and applications. J. Oper. Res. Soc. **48**(7), 763–764 (1997)
11. Dekel, O., Fischer, F., Procaccia, A.D.: Incentive compatible regression learning. J. Comput. Syst. Sci. **76**(8), 759–777 (2010)
12. Drezner, T., Drezner, Z., Kalczynski, P.: Multiple obnoxious facilities location: a cooperative model. IISE Trans. **52**(12), 1403–1412 (2020)

13. Escoffier, B., Gourvès, L., Kim Thang, N., Pascual, F., Spanjaard, O.: Strategy-proof mechanisms for facility location games with many facilities. In: Brafman, R.I., Roberts, F.S., Tsoukiàs, A. (eds.) ADT 2011. LNCS (LNAI), vol. 6992, pp. 67–81. Springer, Heidelberg (2011). https://doi.org/10.1007/978-3-642-24873-3_6
14. Feldman, M., Wilf, Y.: Strategyproof facility location and the least squares objective. In: Proceedings of the Fourteenth ACM Conference on Electronic Commerce, pp. 873–890 (2013)
15. Fotakis, D., Gergatsouli, E., Gouleakis, T., Patris, N.: Learning augmented online facility location. arXiv preprint arXiv:2107.08277 (2021)
16. Jamshidi, M.: Median location problem. Facility Location: Concepts, Models, Algorithms and Case Studies, pp. 177–191 (2009)
17. Jiang, S.H.C., Liu, E., Lyu, Y., Tang, Z.G., Zhang, Y.: Online facility location with predictions. In: International Conference on Learning Representations (2022)
18. Jiang, Z., Panigrahi, D., Sun, K.: Online algorithms for weighted paging with predictions. ACM Trans. Algorithms (TALG) **18**(4), 1–27 (2022)
19. Lam, A., Aziz, H., Li, B., Ramezani, F., Walsh, T.: Proportional fairness in obnoxious facility location. In: Proceedings of the 23rd International Conference on Autonomous Agents and Multiagent Systems, pp. 1075–1083 (2024)
20. Lykouris, T., Vassilvitskii, S.: Competitive caching with machine learned advice. In: Proceedings of the 35th International Conference on Machine Learning (ICML), pp. 3296–3305. PMLR (2018)
21. Lykouris, T., Vassilvitskii, S.: Competitive caching with machine learned advice. J. ACM (JACM) **68**(4), 1–25 (2021)
22. Meir, R.: Strategyproof facility location for three agents on a circle. In: International Symposium on Algorithmic Game Theory, pp. 18–33. Springer (2019)
23. Miç, P., Antmen, Z.F.: A healthcare facility location selection problem with fuzzy topsis method for a regional hospital. Avrupa Bilim ve Teknoloji Dergisi **16**, 750–757 (2019)
24. Peters, H., van der Stel, H., Storcken, T.: Range convexity, continuity, and strategy-proofness of voting schemes. Z. Oper. Res. **38**, 213–229 (1993)
25. Procaccia, A.D., Tennenholtz, M.: Approximate mechanism design without money. ACM Trans. Econ. Comput. (TEAC) **1**(4), 1–26 (2013)
26. Purohit, M., Svitkina, Z., Kumar, R.: Improving online algorithms via ml predictions. In: Advances in Neural Information Processing Systems, vol. 31 (2018)
27. Schummer, J., Vohra, R.V.: Strategy-proof location on a network. J. Econ. Theory **104**(2), 405–428 (2002)
28. Sumit, G., Wade, H.C.: Coordinate-wise median: not bad, not bad, pretty good. arXiv preprint arXiv:2007.00903 (2020)
29. Todo, T., Iwasaki, A., Yokoo, M.: False-name-proof mechanism design without money. In: The 10th International Conference on Autonomous Agents and Multiagent Systems-Volume 2, pp. 651–658 (2011)
30. Wang, S., Li, J., Wang, S.: Online algorithms for multi-shop ski rental with machine learned advice. Adv. Neural. Inf. Process. Syst. **33**, 8150–8160 (2020)
31. Wei, A., Zhang, F.: Optimal robustness-consistency trade-offs for learning-augmented online algorithms. Adv. Neural. Inf. Process. Syst. **33**, 8042–8053 (2020)
32. Zhang, Q., Li, M.: Strategyproof mechanism design for facility location games with weighted agents on a line. J. Comb. Optim. **28**, 756–773 (2014)

Max-Min Four-Dispersion Problems

Shin-ichi Nakano(✉)

Gunma University, Maebashi 371-8510, Japan
nakano@gunma-u.ac.jp

Abstract. Given a set P of n points on which facilities can be placed and an integer k,we want to place k facilities on k points in P so that the minimum distance between facilities is maximized. The problem is called the k-dispersion problem. In this paper, we consider the 4-dispersion problem when P is a set of points on a plane (2-dimensional space). Note that the solution of the 2-dispersion problem corresponds to the diameter of P. We give an $O(n^3)$ time algorithm to solve the 4-dispersion problem in the L_∞ metric, and an $O(n^3)$ time algorithm to solve the 4-dispersion problem in the L_1 metric. Also, we give an $O(n^3 \log^2 n)$ time algorithm to solve the 4-dispersion problem in the L_2 metric.

Also, we give faster (but complicated) $O(n^2 \log n)$-time algorithms to solve the max-min 4-dispersion problem for L_∞ metric and the L_1 metric.

Keywords: algorithms · dispersion problem · facility location

1 Introduction

The facility location problem and many of its variants have been studied. See good textbooks [10,11]. Typically, given a set of points on which facilities can be placed and an integer k, we want to place k facilities on k points in P so that a designated function on distance is minimized. By contrast in the *dispersion problem*, we want to place facilities so that a designated function on distance is maximized.

The intuition of the problem is as follows. Assume that we are planning to open several coffee shops in a city. We wish to locate the shops mutually far away from each other to avoid self-competition. We want to find k points so that the minimum distance among the coffee shops is maximized. For more applications, including *result diversification*, see [8,19,20].

Now we formally define the *max-min k-dispersion problem*. Given a set P of n points, a distance function d for each pair of points (we assume that d is a symmetric nonnegative function satisfying $d(p,p) = 0$ for all $p \in P$), and an integer k with $k \ll n$, we want to find a subset $S \subset P$ with $|S| = k$ such that the cost $cost(S) = \min_{\{u,v\} \subset S}\{d(u,v)\}$ is maximized. Such a set S is called a k-dispersion of P. Note that a 2-dispersion of P on a plane (2-dimensional space) with L_2 metric corresponds to the diameter of P. Thus a k-dispersion is a generalization of the diameter, which is one of basic concept in geometry. (Here the diameter of P is the maximum distance between two points in P.)

M. Li et al. (Eds.): TAMC 2025, LNCS 16084, pp. 275–282, 2026.
https://doi.org/10.1007/978-981-95-4839-2_21

This is the max-min version of the k-dispersion problem [19,21]. For the max-sum version, see [5–9,13,18,19], and for a variety of related problems, see [5,9]. The max-min k-dispersion problem is NP-hard even when the triangle inequality is satisfied [12,21]. An exponential-time exact algorithm for the max-min k-dispersion problem is known [2]. The running time is $O(n^{\omega k/3} \log n)$, where $\omega < 2.373$ is the matrix multiplication exponent.

If P is a set of n points on a line (1-dimensional space) and the order of points in P on the line is given, the k-dispersion problem can be solved in $O(kn)$ time [21]. The running time was improved to $O(n \log\log n)$ [3] by the sorted matrix search method [14]. See a good survey for the sorted matrix search method in [1, Section 3.3].Later it was improved to $O(n)$ [2] by a reduction to the path partitioning problem [14]. Even if the order of points in P on the line is not given, the k-dispersion problem can be solved in $O(n)$ time [4] if k is a constant.

If P is a set of n points on a plane (2-dimensional space) the k-dispersion problem is NP-hard [21].

Ravi et al. [19] proved that the max-min k-dispersion problem cannot be approximated within any constant factorin polynomial time, and cannot be approximated within a factor of two in polynomial time when the distance satisfies the triangle inequality, unless P = NP. They also gave a polynomial-time algorithm with approximation ratio two when the triangle inequality is satisfied.

If $k = 2$ one can solve the k-dispersion problem (diameter problem) in $O(n \log n)$ time [17].

If $k = 3$ and P is a set of points on a plane (2-dimensional space)one can solve the 3-dispersion problem in $O(n)$ time when d is the L_∞ metric or$d$ is the L_1 metric, and in $O(n^2 \log n)$ time when d is the L_2 metric [15,16].

In this paper we consider for $k = 4$ and P is a set of n points on a plane. First we design a simple $O(n^3)$ time algorithm to compute the 4-dispersion of P when d is the L_∞ metric. Then we design a similar $O(n^3)$ time algorithm to compute the 4-dispersion of P when d is the L_1 metric. Then we study the case where d is the L_2 metric. We give an algorithm to compute a 4-dispersion of P in $O(n^3 \log^2 n)$ time. Then we design a faster (but complicated) $O(n^2 \log n)$ time algorithms to solve the max-min 4-dispersion problem for either L_∞ metric or the L_1 metric.

In this paper we use the following notations. P is a set of n points on a plane. For $p \in P$, we write their $x-, y-$ coordinates as $x(p)$ and $y(p)$. A point $p' \in P$ is a farthest point from $p'' \in P$ if $d(p', p'') \geq d(p, p'')$ for every $p \in P$.

The remainder of this paper is organized as follows. Section 2 gives an $O(n^3)$ time algorithm to solve the 4-dispersion problem if d is the L_∞ metric. Section 3 gives an $O(n^3)$ time algorithm to solve the 4-dispersion problem if d is the L_1 metric. Section 4 gives an $O(n^3 \log^2 n)$ time algorithm to solve the 4-dispersion problem if d is the L_2 metric. Section 5 gives faster (but complicated) $O(n^2 \log n)$ timealgorithms to solve the max-min 4-dispersion problem foreither L_∞ metric or L_1 metric. Finally Sect. 6 is a conclusion.

2 4-Dispersion in L_∞ Metric

Let P is a set of n points on a plane and d is the L_∞ metric. We assume that no three points are on a line. One can solve the 4-dispersion problem n $O(n^4)$ time by brute force manner. We design a faster $O(n^3)$ time algorithm to solve the 4-dispersion problem.

Let $\{s, t, u, v\}$ be a 4-dispersion of P. We have the following two cases.

Case 1: Each point of $\{s, t, u, v\}$ is an extreme point of the convex hull of $\{s, t, u, v\}$.

One can assume that s, t, u, v appear on the contour of the convex hull of $\{s, t, u, v\}$in this order, and $d(s, u) \leq d(t, v)$. One can find 4-dispersion as follows.

For every possible choice of s and u let ℓ be the line containing s and u.Let P_L be the set of points locating on or to the left of $\ell$and P_R be the set of points locating on or to the right of ℓ.(If ℓ is a horizontal line then we rotate ℓ a little.) Note that both P_L and P_R contains s and u.Choose t from P_L so that $cost(s, u, t)$ is maximized. Also choose v from P_R so that $cost(s, u, v)$ is maximized. We need $O(n)$ time by brute force method, respectively. Let $\{s, t, u, v\}$ be the resultant four points for s and u.

Among all possible choices of s and $u$we choose the $\{s, t, u, v\}$ attaining the maximum $cost(s, t, u, v)$. The number of choices of s and u is at most n^2 and we need $O(n)$ time to choose t and v for each choice of s and t, so we can compute the four points,which is a possible 4-dispersion,in $O(n^3)$ time.

Case 2: Otherwise. One can assume that s, t, u appear on the contour of the convex hull of $\{s, t, u, v\}$counterclockwise in this order, and v in not an extreme point of the convex hull.(So v is located in the triangles, t, u.)

We divide the plane into eight regions by the following eight line containing v.Those are the horizontal line, the straight line with 45°C slope, the vertical line and the straight line with -45°C slope. See Fig. 1.

We have the following three subcases depending on the location of s, t, u in the eight regions. For every possible choice of v among $|P| = n$ points, we compute a possible 4-dispersion$\{s, t, u, v\}$ containing v in $O(n^2)$ time. Since the number of cases is a constant we need $O(n^3)$ time in total.

Case 2(a): Four consecutive regions contain none of s, t, u. See Fig. 1(a).

Then v is an extreme point of the convex hull of $\{s, t, u, v\}$. A contradiction.

Case 2(b): There are three consecutive regions containing none of s, t, u.

We have the following five subsubcases depending on the location of s, t, u in the remaining five regions. See Fig. 1(b1)–(b5) (Symmetrical cases are omitted.)

Case 2(b1): There are one more three consecutive regions containing none of s, t, u, as shown in Fig. 1(b1).

Now $d(s, v) = |y(s) - y(v)|$, $d(s, v) \leq d(s, t)$ and $d(s, v) \leq d(s, u)$ hold. So we can assume s is the point with the maximum $|y(s)|$. Thus we can choose s in $O(n)$ time. Also we can choose t and u by brute force method in $O(n^2)$ time.

Case 2(b2): There are one more three consecutive regions containing none of s, t, u, as shown in Fig. 1(b2).

Now $d(u,v) = |y(u) - y(v)|$, $d(u,v) \leq d(u,s)$ and $d(u,v) \leq d(u,t)$ hold. So we can assume u is the point with the maximum $|y(u)|$. Thus we can choose u in $O(n)$ time. Also we can choose s and t by brute force method in $O(n^2)$ time.

Case 2(b3): There are two consecutive regions containing none of s, t, u, as shown in Fig. 1(b3).

Similar to Case2(b2).

Case 2(b4): There are two consecutive regions containing none of s, t, u, as shown in Fig. 1(b4).

Similar to Case2(b1).

Case 2(b5): There are no two consecutive regions containing none of s, t, u, as shown in Fig. 1(b5).

Similar to Case2(b1).

Case 2(c): There are no three consecutive regions contain none of s, t, u.

We have the following two subsubcases depending on the location of s, t, u. (Symmetrical cases are omitted.) See Fig. 1(c1)–(c2).

Case 3(c1): s, t, u are located as shown in Fig. 1(c1).

Similar to Case2(b1).

Case 3(c2): s, t, u are located as shown in Fig. 1(c2).

Similar to Case2(b2).

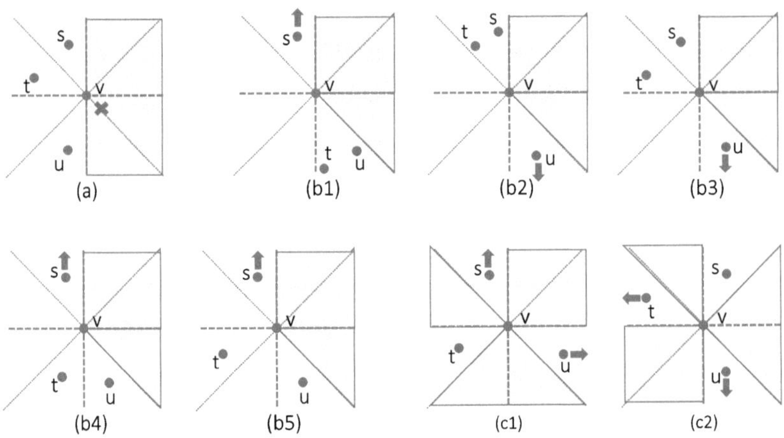

Fig. 1. Illustrations for Case 2.

Theorem 1. *If P is a set of n points on a plane and d is the L_∞ metric, then one can solve the max-min 4-dispersion problem in $O(n^3)$ time.*

3 4-Dispersion in L_1 Metric

Let P is a set of n points on a plane and d is the L_1 metric. We assume that no three points are on a line. We design an $O(n^3)$ time algorithm to solve the 4-dispersion problem.

Let $\{s,t,u,v\}$ be a 4 dispersion of P. We have the following two cases.

Case 1: Each point of $\{s,t,u,v\}$ is an extreme point of the convex hull of $\{s,t,u,v\}$.

One can assume that s,t,u,v appear on the contour of the convex hull of $\{s,t,u,v\}$ in this order, and $d(s,u) \leq d(t,v)$. One can find 4-dispersion as follows.

For every possible choice of s and u let ℓ be the line containing s and u. Let P_L be the set of points locating on or to the left of ℓ and P_R be the set of points locating on or to the right of ℓ. (If ℓ is a horizontal line then we rotate ℓ a little.) Note that both P_L and P_R contains s and u. Choose t from P_L so that $cost(s,u,t)$ is maximized. Also choose v from P_R so that $cost(s,u,v)$ is maximized. We need $O(n)$ time by brute force method, respectively. Let $\{s,t,u,v\}$ be the resultant four points for s and u.

Among all possible choices of s and u we choose the $\{s,t,u,v\}$ attaining the maximum $cost(s,t,u,v)$. The number of choices of s and u is at most n^2 and we need $O(n)$ time to choose t and v for each choice of s and t, so we can compute the four points, which is a possible 4-dispersion, in $O(n^3)$ time.

Case 2: Otherwise. One can assume that s,t,u appear on the contour of the convex hull of $\{s,t,u,v\}$ counterclockwise in this order, and v in not an extreme point of the convex hull. (So v is located in the triangle s,t,u.)

We divide the plane into four regions by the horizontal line and the vertical line containing v. See Fig. 2.

We have the following three subcases depending on the location of s,t,u in the four regions. For every possible choice of v among $|P| = n$ points, we compute a possible 4-dispersion $\{s,t,u,v\}$ containing v in $O(n^2)$ time. Since the number of cases is a constant we need $O(n^3)$ time in total.

Case 2(a): Two consecutive regions contain none of s,t,u. See Fig. 2(a).

Then v is an extreme point of the convex hull of $\{s,t,u,v\}$. A contradiction.

Case 2(b): There are two regions containing none of s,t,u, as shown in Fig. 2(b). (Symmetrical cases are omitted.)

Now $d(u,v) \leq d(u,s)$ and $d(u,v) \leq d(u,t)$ hold. So we can assume u is the point with the maximum $|x(u)+y(u)|$. Thus we can choose u in $O(n)$ time. Also we can choose s and t by brute force method in $O(n^2)$ time.

Case 2(c): There are exactly one regions containing none of s,t,u.

We have the following four subsubcases depending on the relative location of t to s and u. See Fig. 2(c1)-(c4). (Symmetrical cases are omitted.)

Case 2(c1): $y(s) \leq y(t)$ and $x(u) \leq x(t)$. See Fig. 1(c1).

Now $d(t,s) = x(t) - x(s) + y(t) - y(s)$, $d(u,s) = x(u) - x(s) + y(u) - y(s)$ and $d(v,s) = x(v) - x(s) + y(v) - y(s)$ hold. So we can assume s is the point with the maximum $|x(s) + y(s)|$. Thus we can choose s in $O(n)$ time. Also we can choose t and u by brute force method in $O(n^2)$ time.

Case 2(c2): $y(s) \leq y(t)$ and $x(u) > x(t)$. See Fig. 1(c2).

Now $d(t, s) = x(t) - x(s) + y(t) - y(s)$, $d(u, s) = x(u) - x(s) + y(u) - y(s)$ and $d(v, s) = x(v) - x(s) + y(v) - y(s)$ hold. So we can assume s is the point with the maximum $|x(s) + y(s)|$. Thus we can choose s in $O(n)$ time. Also we can choose t and u by brute force method in $O(n^2)$ time.

Case 2(c3): $y(s) > y(t)$ and $x(u) \leq x(t)$. See Fig. 1(c3).

Now $d(s, t) = x(t) - x(s) + y(s) - y(t)$, $d(u, t) = x(t) - x(u) + y(u) - y(t)$ and $d(v, t) = x(t) - x(v) + y(v) - y(t)$ hold. So we can assume t is the point with the maximum $|x(t) - y(t)|$.Thus we can choose t in $O(n)$ time. Also we can choose s and u by brute force method in $O(n^2)$ time.

Case 2(c4): $y(s) > y(t)$ and $x(u) > x(t)$. See Fig. 1(c4).

Now $d(s, u) = x(u) - x(s) + y(u) - y(s)$, $d(t, u) = x(u) - x(t) + y(u) - y(t)$ and $d(v, u) = x(u) - x(v) + y(u) - y(v)$ hold. So we can assume u is the point with the maximum $|x(u) + y(u)|$. Thus we can choose u in $O(n)$ time. Also we can choose s and t by brute force method in $O(n^2)$ time.

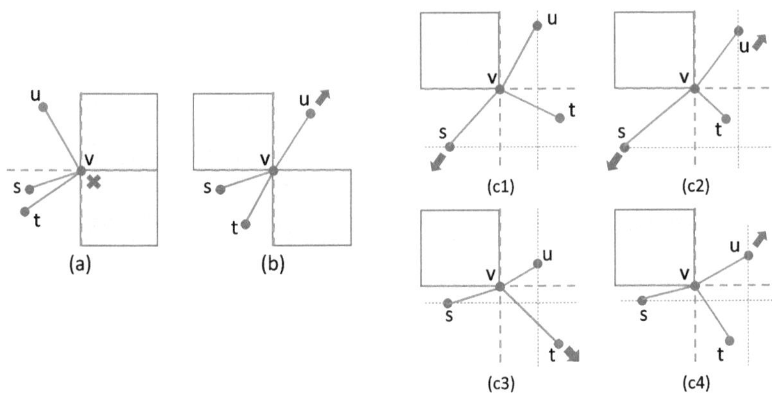

Fig. 2. Illustrations for Case 2.

Theorem 2. *If P is a set of n points on a plane and d is the L_1 metric, then one can solve the max-min 4-dispersion problem in $O(n^3)$ time.*

4 4-Dispersion in L_2 Metric

In this section, we design an $O(n^3 \log^2 n)$ time algorithm to solve the 4-dispersion problem when P is a set of n points on a plane (2-dimensional space) and d is the L_2 metric.

Fix a point $p_a \in P$. Let $P_i = \{p_1, p_2, \cdots, p_i\}$ be the subset of P consisting of the i farthest points from p_a, and p_i the i-th farthest point from p_a. By using an algorithm in [15,16] one can compute the 3-dispersion of P_i in $O(n^2 \log n)$ time. By binary search for i one can find a possible 4-dispersion containing p_a with cost $\min\{d(p_a, p_i), cost_3(P_i)\}$, where $cost_3(P_i)$ is the cost of 3-dispersion of P_i, in $O(n \log n + n^2 \log n * \log n) = O(n^2 log^2 n)$ time. (Because $d(p_a, p_i)$ is a monotonic increasing function in i, while $cost(P_i)$ is a monotonic decreasing function in i, if there exists a point i where the function $d(p_a, p_i) - cost(P_i)$ changes sign, this point can be found efficiently in $logn$ rounds using binary search.)

By computing above for every point p_a in P and choose the possible 4-dispersion with the maximum cost, one can compute the 4-dispersion of P in $O(n^3 log^2 n)$ time.

We have the following theorem.

Theorem 3. *If P is a set of n points on a plane and d is the L_2 metric, then one can solve the max-min 4-dispersion problem in $O(n^3 \log^2 n)$ time.*

5 Faster Algorithms

By a similar manner to the algorithm in the last Section one can design faster (but complicated) algorithms for 4-dispersion for L_∞ metric and L_1 metric, as follows.

Fix a point $p_a \in P$. Let $P_i = \{p_1, p_2, \cdots, p_i\}$ be the subset of P consisting of the i farthest points from p_a, and p_i the i-th farthest point from p_a. One can sort P with respect to the distance to p_a in $O(n \log n)$ time. One can compute the 3-dispersion of P_i in $O(n)$ time [15,16]. By binary search for i in $\log n$ stages, one can find a possible 4-dispersion containing p_a with cost $\min\{d(p_a, p_i), cost_3(P_i)\}$, where $cost_3(P_i)$ is the cost of 3-dispersion of P_i, in $O(n) \times \log n = O(n \log n)$ time.

Thus,by checking for each $p_a \in P$, one can compute a 4-dispersion in $O(n^2 \log n)$ time in total.

Theorem 4. *If P is a set of n points on a plane and d is either L_∞ metric or the L_1 metric, then one can solve the max-min 4-dispersion problem in $O(n^2 \log n)$ time.*

6 Conclusion

In this paper we have designed several algorithms to compute a 4-dispersion of a set of points on a plane.

Can we design a faster algorithm to compute the 4-dispersion?

References

1. Agarwal, P., Sharir, M.: Efficient algorithms for geometric optimization. ACM Comput. Surv. **30**, 412–458 (1998)
2. Akagi, T., et al.: Exact Algorithms for the Max-Min Dispersion Problem. In: Chen, J., Lu, P. (eds.) FAW 2018. LNCS, vol. 10823, pp. 263–272. Springer, Cham (2018). https://doi.org/10.1007/978-3-319-78455-7_20
3. Akagi. T., Nakano. S.: Dispersion on the line. IPSJ SIG Technical Reports, 2016-AL-158-3 (2016)
4. Araki, T., Nakano, S.: Max-Min Dispersion on a Line. In: Kim, D., Uma, R.N., Zelikovsky, A. (eds.) COCOA 2018. LNCS, vol. 11346, pp. 672–678. Springer, Cham (2018). https://doi.org/10.1007/978-3-030-04651-4_45
5. Baur, C., Fekete, S.P.: Approximation of geometric dispersion problems. In: Jansen, K., Rolim, J. (eds.) APPROX 1998. LNCS, vol. 1444, pp. 63–75. Springer, Heidelberg (1998). https://doi.org/10.1007/BFb0053964
6. Birnbaum, B., Goldman, K.J.: An improved analysis for a greedy remote-clique algorithm using factor-revealing LPS. Algorithmica **50**, 42–59 (2009)
7. Cevallos. A., Eisenbrand. F., Zenklusen. R.: Max-sum diversity via convex programming. Proc. of SoCG 2016, pp. 26:1–26:14 (2016)
8. Cevallos, A., Eisenbrand, F., Zenklusen, R.: Local search for max-sum diversification. Proc. of SODA **2017**, 130–142 (2017)
9. Chandra, B., Halldorsson, M.M.: Approximation algorithms for dispersion problems. J. of Algorithms **38**, 438–465 (2001)
10. Drezner, Z.: Facility location: a survey of applications and methods. Springer (1995)
11. Drezner, Z., Hamacher, H.W.: Facility location: applications and theory. Springer (2004)
12. Erkut, E.: The discrete p-dispersion problem. Eur. J. Oper. Res. **46**, 48–60 (1990)
13. Fekete, S.P., Meijer, H.: Maximum dispersion and geometric maximum weight cliques. Algorithmica **38**, 501–511 (2004)
14. Frederickson, G.: Optimal algorithms for tree partitioning. Proc. of SODA **1991**, 168–177 (1991)
15. Horiyama, T., et al.: Max-Min 3-Dispersion Problems. In: Du, D.-Z., Duan, Z., Tian, C. (eds.) COCOON 2019. LNCS, vol. 11653, pp. 291–300. Springer, Cham (2019). https://doi.org/10.1007/978-3-030-26176-4_24
16. Horiyama. T., et al.: Max-min 3-dispersion problems. IEICE Trans. Fundam. Electron. Commun. Comput. Sci., 104-A(9):1101–1107 (2021)
17. Preparata, F.P., Shamos, M. I.: Computational geometry: an introduction. Springer-Verlag (1985)
18. Rubinstein, S., Hassin, R., Tamir, A.: Approximation algorithms for maximum dispersion. Oper. Res. Lett. **21**, 133–137 (1997)
19. Ravi, S.S., Rosenkrantz, D.J., Tayi, G.K.: Heuristic and special case algorithms for dispersion problems. Oper. Res. **42**, 299–310 (1994)
20. Sydow, M.: Approximation guarantees for max sum and max min facility dispersion with parameterised triangle inequality and applications in result diversification. Math. Appl. **42**, 241–257 (2014)
21. Wang, D.W., Kuo, Y.-S.: A study on two geometric location problems. Inf. Process. Lett. **28**, 281–286 (1988)

Generalized Last Open-End Bin Packing Problem

Zhikang Zhang(✉) and Weidong Li

School of Mathematics and Statistics, Yunnan University, Kunming, PR China
zhikang2024@163.com

Abstract. In this paper, we study the online version of generalized last open-end bin packing problem. In this problem items arrive sequentially and need to be packed into a bin with a load strictly less than 1. If no such bin exists a new one needs to be opened. We introduce a unit overload cost c. Each bin has a unit capacity and unit cost. If the total size of the items packed into the bin is greater than 1, an overload cost is needed for the overloaded portion, and the objective changes from packing all arriving items as required using the minimized number of bins to minimizing the total cost of all bins used. In this paper, we use an algorithm based on the First Fit strategy for this problem, and proved that the asymptotic competitive ratio of this algorithm is at most $\frac{2}{1+c}$ when $0 \le c \le 1$ and at most $\frac{2c}{1+c}$ when $c > 1$.

Keywords: Open-end bin packing · Online algorithm · Asymptotic competitive ratio

1 Introduction

In the field of combinatorial optimization, there are numerous problems, such as the bin packing problem [1,2], the machine scheduling problem [17–19], the allocation problem [12], the mobile crowdsensing services problem [26], the cloud-edge collaborative computation offloading problem [22], and the facility location problem [23], among others. The bin packing problem is one of the most fundamental problems in combinatorial optimization and is the cornerstone of approximation algorithms. In this paper, we consider an important variant of the bin packing problem, defined as follows:

Given a set $O = \{o_1, o_2, \ldots, o_n\}$ of n items, each item $o_j \in O$ has a size $s(o_j) \in (0, 1]$. A subset $B_i = \{o_{i_1}, o_{i_2}, \ldots, o_{i_k}\} \subseteq O$ is called valid or feasible if there is a permutation $\sigma : \{i_1, i_2, \ldots, i_k\} \mapsto \{i_1, i_2, \ldots, i_k\}$ with $\{i_1, i_2, \ldots, i_k\} = \{\sigma(i_1), \sigma(i_2), \ldots, \sigma(i_k)\}$ such that

$$s(B_i \backslash \{o_{i_l}\}) = \sum_{j:j\in[k], j\neq l} s(o_{i_j}) < 1, \text{ where } i_l = \max_{j:j\in[k]} \sigma(i_j) \text{ and } [k] = \{1, 2, \ldots, k\}.$$

M. Li et al. (Eds.): TAMC 2025, LNCS 16084, pp. 283–291, 2026.
https://doi.org/10.1007/978-981-95-4839-2_22

Motivating by the Hong Kong subway riding strategy, the last open-end bin packing (LOEBP) problem [14,27] is to pack the n items into bins of unit capacity such that the number of used bins is minimized, where the set B_i of items packed in bin i is valid. Note that the asymptotic competitive ratio is usually seen as a more interesting than absolute competitive ratio for various variants of bin packing problems. For the offline case, Leung et al. [14] proved that the LOEBP problem is strong NP-hard, and presented an asymptotic fully polynomial time approximation scheme (AFPTAS). For the online case, items are presented one by one to be packed and we can only use the permutation $\sigma(i) = i$ for any online algorithm. For the offline form, LOEBP problem only needs to satisfy that the load of the bin after removing one item (or the last item packed into the bin) is strictly less than 1. Leung et al. [14] presented a best possible online algorithm with an asymptotic worst-case competitive ratio of 2. Zhang [27] studied a parameterized version where the size of each item is in $(0, \frac{1}{m}]$, and gave a best possible online algorithms with a competitive ratio of $1 + \frac{1}{m}$.

The first important variant of LOEBP is the maximum open-end bin packing (Max-OEBP) problem, where the permutation σ is determined by the item size, *i.e.*

$$\sigma(i_1) < \sigma(i_2) \text{ implies that } s(o_{i_1}) \leq s(o_{i_2}).$$

For the offline case, the result in [14] implies that Max-OEBP admits an AFPTAS. For the online case, Epstein [10] presented an online algorithm with a competitive ratio of 2, and analysed several classical online algorithms for the parametric setting.

The second important variant of LOEBP is the minimum open-end bin packing (Min-OEBP) problem [8] which is also called strong open-end bin packing, where the permutation σ is determine by the item size, *i.e.*

$$\sigma(i_1) < \sigma(i_2) \text{ implies that } s(o_{i_1}) \geq s(o_{i_2}).$$

For the offline case, Epstein and Levin [8] proved that Min-OEBP is equivalent to the lazy bin covering problem [11,15,16] and presented an AFPTAS. For the online case, Epstein and Levin [9] gave a lower bound of $\frac{7}{6}$ on the asymptotic competitive ratio of any online algorithm and designed a modified harmonic type algorithm whose asymptotic competitive ratio is $\frac{124}{105}$.

The third important variant of LOEBP is the ordered open-end bin packing (OOEBP) problem [3,6–8,24], where the permutation $\sigma(i) = i$ is given in the input and any feasible solution must obey the permutation σ. For the offline form, the bins used for packing in the OOEBP must satisfy the definition of feasible, which is the difference from the LOEBP. Epstein and Levin [8] proved that OOEBP is strongly NP-hard. Balogh et al. [3] proposed an asymptotic polynomial time approximation scheme (APTAS). For the online case, Yang and Leung [24] gave a lower bound of 1.6303 on the asymptotic competitive ratio of any online algorithm. Balogh et al. [3] proposed an online algorithm with an asymptotic competitive ratio of 1.6916 and an improved lower bound of 1.6305.

In this paper, we introduce a new problem, called generalized last open-end bin packing, which generalized both the LOEBP problem [14,27] and the classical bin packing problems [1,21], and present an online algorithm.

2 Preliminaries

Given a set $O = \{o_1, o_2, \ldots, o_n\}$ of n items and unit overload cost $c \geq 0$, each item $o_j \in O$ has a size $s(o_j) \in (0, 1]$. A subset $B_i = \{o_{i_1}, o_{i_2}, \ldots, o_{i_k}\} \subseteq O$ with $i_1 < i_2 < \cdots < i_k$ is called feasible if $\sum_{j=1}^{k-1} s(o_{i_j}) < 1$. Moreover, as in [4,5,20], the cost of nonempty bin B_i is defined as

$$c(B_i) = 1 + c \cdot \max\{0, s(B_i) - 1\},$$

where $s(B_i)$ is the load of bin B_i. The objective of online generalized last open-end bin packing (GLOEBP) problem is to pack all arrived items so that the total cost of all bins used is minimized, where each item must be packed into a bin before the next item appears.

If $c = 0$, the GLOEBP problem is exactly the LOEBP problem [14,27]. If $c = +\infty$, the size of each bin in the optimal solution is no more than 1, and the GLOEBP problem is exactly the classical bin packing problem [1,2,21].

GLOEBP is very close to the extensive bin packing (EBP) or surgery scheduling problems [4,5,20,25], as the loads of used bin can be larger than 1 in both problems. The main difference is that each used bin in the EBP problem cannot have a size larger than 2. However, in the GLOEBP problem, the load of a bin is strictly less than 1 after removing the last item packed into the bin, implying that the load of the used bin is strictly less than 2.

Li et al. [19] studied another related problem. They introduced a novel scheduling criterion called a discounted profit, denoted as $DP_j = X_j + \delta Y_j$ for job $J_j (1 \leq j \leq n)$, where $X_j(Y_j)$ is the early (late) work of this job, and $0 \leq \delta < 1$ is a discount factor. More precisely, the early work X_j (late work Y_j) of J_j is the part of this job executed before (after) its due date, and the goal of the scheduling models with such a criterion is to maximize the total discounted profit of all jobs, i.e., $DP = \sum_{j=1}^{n} DP_j$.

The current best results about the asymptotic competitive ratios are depicted as the following table.

Unit cost	Upper bound	Lower bound
$c - 0$	2 [14]	2 [14]
$c \in (0, 1]$	$\frac{2}{1+c}$ (This paper)	$\frac{2}{1+c}$ (This paper)
$c \in (1, 4]$	$\frac{2c}{1+c}$ (This paper)	$\frac{36}{c^2-8c+43}$(This paper)
$c \in (4, +\infty)$	$\frac{2c}{1+c}$ (This paper)	$\frac{4}{3}$(This paper)
$c = +\infty$	1.578 [1]	1.543 [2]

3 Online Algorithms for the GLOEBP Problem

Luo and Spieksma [20] studied the online Bin Packing problem with linear Overload Cost (BPOC), gave lower bounds for the problem with different overload costs, and proposed an algorithm based on the First-Fit strategy, called First-Fit Algorithm with Fixed Overload (FFO), which is optimal under certain overload costs.

The algorithm we give next is based on [20] with the addition of a simple condition. First, the algorithm is based on the First-Fit strategy, which is to pack arrived items into the first bin that does not violate the capacity constraints. The difference is that the FFO algorithm has an overload capacity for a given overload cost c. Let $f(c)$ be the new capacity of the bin at the overload cost c. However, this alone cannot be applied to the problem we study, and an additional condition needs to be added: item o_j can put into bin B_i only if $s(B_i) < 1$ and $s(B_i) + s(o_j) \leq f(c)$. We define the overload capacity for different unit overload cost as follows

$$f(c) = \begin{cases} 1+1, & \text{if } 0 \leq c \leq 1, \\ 1+\frac{1}{c}, & \text{if } 1 < c \leq +\infty. \end{cases}$$

Algorithm 1: First-Fit Algorithm with Fixed Overload (FFO)

Input: Overload cost c.
Output: A packing π.
Initialization: $f(c)$, $k = 1$, $B_1 = \emptyset$, $s(B_1) = 0$.
for *each item* o_j **do**
    for $i = 1$ *to* k **do**
        $d_i = f(c) - (s(B_i) + s(o_j))$;
    if $\exists i \in \{1, 2, \ldots k\}$, $d_i \geq 0$ *and* $s(B_i) < 1$ **then**
        pack o_j into bin B_i, where $i = \min\{i | d_i \geq 0, s(B_i) < 1\}$;
        set $B_i \leftarrow B_i \cup \{o_j\}$, and $s(B_i) \leftarrow s(B_i) + s(o_j)$;
    else
        $h = h + 1$, and set $B_k = \emptyset$;
        pack o_j into bin B_k;
        set $B_k \leftarrow B_k \cup \{o_j\}$, and $s(B_k) \leftarrow s(B_k) + s(o_j)$;
return the packing $\pi = \{B_1, B_2, \ldots, B_k\}$;

For convenience of analysis, we let C be the set of bins used in the packing scheme obtained by the algorithm and $k = |C|$. When $c > 1$, we have $\frac{f(c)}{2} < 1$. Let C_1 be the set of bins with load no less than 1 and $k_1 = |C_1|$; let C_2 be the set of bins with load in $\left[\frac{f(c)}{2}, 1\right)$ and $k_2 = |C_2|$; let C_3 be the set of bins with load strictly less than $\frac{f(c)}{2}$ and $k_3 = |C_3|$. It follows from the above definition that $\{C_1, C_2, C_3\}$ is a partition of C. Let W be the total size of all items, f_0 be the total overloaded size, and f_1 be the average overloaded size, then we have $f_1 < 1$ and $f_0 = k_1 \cdot f_1$.

Let $cost_{OPT}$ and $cost_{\mathcal{A}}$ be the cost of the optimal packing OPT and the algorithm $\mathcal{A}$ for the output packing, respectively. Next we discuss the asymptotic competition ratio of FFO algorithm depending on the value of c.

3.1 The Case $c \in [0, 1]$

In this section, we prove that the asymptotic competitive ratio of the FFO algorithm is consistent with the general lower bound when $0 \le c \le 1$.

Theorem 1. *If $0 \le c \le 1$, the asymptotic competitive ratio of any online algorithm for GLOEBP problem is not less than $\frac{2}{1+c}$.*

Proof. We construct an instance where the adversary releases N items of size 1, followed by N^2 items of size $\frac{1}{N}$, where N is a large integer. Any online algorithm $\mathcal{A}$ uses at least $2N$ bins, with the least cost being the case where the first N bins each contain an item of size 1, and the next N bins each contain N items of size $\frac{1}{N}$. These $2N$ bins are all loaded with 1, then we have $cost_{\mathcal{A}} \ge 2N$. Furthermore, we can construct a packing scheme using $(N+1)$ bins, where the first N bins contain $(N-1)$ items of size $\frac{1}{N}$ and one of size 1, and the last bin contains N items of size $\frac{1}{N}$. Thus there is $cost_{OPT} \le \left[1 + (N-1) \cdot c \cdot \frac{1}{N}\right] \cdot N + 1 = N + 1 + (N-1) \cdot c$ and we have

$$\frac{cost_{\mathcal{A}}}{cost_{OPT}} \ge \frac{2N}{N + 1 + (N-1) \cdot c} = \frac{2}{1 + c + \frac{1-c}{N}} \to \frac{2}{1+c} \quad (N \to \infty).$$

Lemma 1. *When $0 \le c \le 1$, the cost of the optimal packing is at least $\frac{W}{2}(1+c)$.*

By Lemma 1 we obtain a lower bound for the $cost_{OPT}$. Next we will analyze the asymptotic competitive ratio of the algorithm through this lower bound.

Theorem 2. *When $0 \le c \le 1$, the competitive ratio of the FFO algorithm for GLOEBP is at most $\frac{2}{1+c}$.*

Proof. Notice that $f(c) = 2$ for $0 \le c \le 1$, which indicates that for the same batch of items the FFO algorithm outputs the same packing scheme as the FF algorithm outputs without the overload cost, and then there are at least $(k-1)$ bins in C that have load not less than 1. We have $cost_{FFO} = k + k_1 \cdot c \cdot f_1$, and

$$W \ge k + k_1 \cdot f_1 > k - 1 + k_1 \cdot c \cdot f_1 = cost_{FFO} - 1.$$

By Lemma 1 we have

$$cost_{OPT} \ge \frac{cost_{FFO} - 1}{2}(1+c) \Rightarrow cost_{FFO} \le \frac{2}{1+c} \cdot cost_{OPT} + 1.$$

3.2 The Case $c \in (1, +\infty)$

In this section we give a lower bound on the asymptotic competitive ratio of any online algorithm and prove that the asymptotic competitive ratio of the FFO algorithm is at most $\frac{2c}{1+c}$ when $c > 1$.

Theorem 3. *For GLOEBP problem, the asymptotic competitive ratio of any online algorithm is at least* $\frac{36}{c^2-8c+43}$ *if* $1 < c \leq 4$ *and at least* $\frac{4}{3}$ *if* $c > 4$.

Proof. Consider two lists L_1 and L_2 of items. L_1 contains N items of size $\frac{1}{2}+\varepsilon$ and L_2 contains N items of size $\frac{1}{2}-\varepsilon$, where $2\varepsilon \cdot c \leq 1$ and $2\left(\frac{3}{2}-3\varepsilon-1\right)\cdot c > 1$. Assume that online algorithm $\mathcal{A}$ uses $\frac{N}{2} \leq N_1 \leq N$ bins to pack the items in list L_1. This means that there are $N - N_1$ bins containing two items and $2N_1 - N$ bins containing one item of size $\frac{1}{2}+\varepsilon$. The optimal packing of L_1 uses $\frac{N}{2}$ bins each containing two items of size $\frac{1}{2}+\varepsilon$. The asymptotic competitive ratio of algorithm $\mathcal{A}$ on list L_1 is $\frac{N_1+(N-N_1)\cdot 2\varepsilon\cdot c}{\frac{N}{2}(1+2\varepsilon\cdot c)}$.

After list L_1, algorithm $\mathcal{A}$ continues to process list L_2. The best result is to pack $2N_1 - N$ items of size $\frac{1}{2}-\varepsilon$ into the previous $2N_1 - N$ bin that contains one item of size $\frac{1}{2}+\varepsilon$. The remaining $2(N-N_1)$ items are packed together in pairs to get $N - N_1$ bins. The optimal packing under the combination of lists L_1 and L_2 uses N bins each containing one item of size $\frac{1}{2}+\varepsilon$ and one item of size $\frac{1}{2}-\varepsilon$. The asymptotic competitive ratio of algorithm $\mathcal{A}$ on the combination of lists L_1 and L_2 is $\frac{N+(N-N_1)\cdot 2\varepsilon\cdot c}{N}$. Therefore, the asymptotic competition ratio of algorithm $\mathcal{A}$ is at least

$$\max\left\{\frac{N_1+(N-N_1)\cdot 2\varepsilon\cdot c}{\frac{N}{2}(1+2\varepsilon\cdot c)}, \frac{N+(N-N_1)\cdot 2\varepsilon\cdot c}{N}\right\}.$$

The above equation is minimized for $N_1 = \left(1+\frac{t-1}{2(t^2-t+1)}\right)$ where $t = \varepsilon c$, which implies that $1+\frac{t(1-t)}{(t^2-t+1)}$ is a lower bound on the asymptotic competitive ratio of any online algorithm. Based on the item size setting we know $t \leq \min\{\frac{1}{2}, \frac{c-1}{6}\}$. $1+\frac{t(1-t)}{(t^2-t+1)}$ increases as t increases when $t \leq \frac{1}{2}$ and decreases as t increases when $t > \frac{1}{2}$. Therefore, if $1 < c \leq 4$ then $1+\frac{t(1-t)}{(t^2-t+1)}$ reaches the maximum value $\frac{36}{c^2-8c+43}$ by taking $t = \frac{c-1}{6}$ and if $c > 4$ then $1+\frac{t(1-t)}{(t^2-t+1)}$ reaches the maximum $\frac{4}{3}$ by taking $t = \frac{1}{2}$.

In order to analyse the asymptotic competitive ratio of the FFO algorithm, we first give two lemmas about the load of bins.

Lemma 2. *When* $c > 1$, *at least* $(k-1)$ *of the* k *bins in* C *have a load greater than* $\frac{f(c)}{2}$, *i.e.* $k_3 \leq 1$.

Lemma 3. *When* $k_2 \geq 1$, *the total size of the items in* $C_2 \cup C_3$ *is at least* $\frac{f(c)}{2}\cdot(k-k_1)$.

Theorem 4. *When $1 < c < +\infty$, the asymptotic competitive ratio of the FFO algorithm for this problem is at most $\frac{2c}{1+c}$.*

Proof. Notice that $cost_{FFO} = k + k_1 \cdot c \cdot f_1$ and $cost_{OPT} \geq W$. We next discuss the classification of k_2.

Case 1 $k_2 \geq 1$.

By Lemma 3, it follows that $cost_{OPT} \geq W \geq k_1 + k_1 \cdot f_1 + \frac{f(c)}{2} \cdot (k - k_1)$. Then we have

$$\begin{aligned} cost_{FFO} &= k + k_1 \cdot c \cdot f_1 \\ &\leq W \cdot \frac{k + k_1 \cdot c \cdot f_1}{k_1 + k_1 \cdot f_1 + \frac{f(c)}{2} \cdot (k - k_1)} \\ &\leq W \cdot \frac{k + k_1}{k_1(1 + \frac{1}{c}) + \frac{1}{2}(1 + \frac{1}{c})(k - k_1)} \\ &\leq \frac{2c}{1+c} \cdot cost_{OPT}. \end{aligned}$$

Case 2 $k_2 = 0$.

Case 2.1 $k_1 = 0$. In this case, since $W < 1$, we have $cost_{FFO} = cost_{OPT} = 1$.

Case 2.2 $k_1 \geq 1$. In this case there is $cost_{FFO} = k_1 + k_3 + k_1 \cdot c \cdot f_1$. Then we have

$$\begin{aligned} cost_{FFO} &\leq k_1 + 1 + k_1 \cdot c \cdot f_1 \\ &\leq W \cdot \frac{k_1 + k_1 \cdot c \cdot f_1}{k_1 + k_1 \cdot f_1} + 1 \\ &\leq W \cdot \frac{2c}{1+c} + 1 \\ &\leq \frac{2c}{1+c} \cdot cost_{OPT} + 1. \end{aligned}$$

4 Conclusion

In this paper, we study the generalized open-end bin packing problem with linear overload cost and analysis the competitive ratios of FFO algorithm under different overload cost c. Haouari and Mhiri [13] studied the bin packing problem with concave cost functions, gave three lower bounds as well as two heuristic algorithms. This also gives us a direction for future research, which is to introduce nonlinear cost functions.

Acknowledgments. The work was supported in part by Yunnan Fundamental Research Projects (Grants No. 202501AS070076, No. 202501AS070170), and Research and Innovation Program of Yunnan University (Grants No. TM-23236819).

References

1. Balogh, J., Bekesi, J., Dosa, G., Epstein, L., Levin, A.: A new and improved algorithm for online bin packing. In: Azar, Y., Bast, H., Herman, G. (eds.) 26th Annual European Symposium on Algorithms, ESA 2018. LIPIcs, vol. 112, pp. 5:1–5:14. Helsinki, Finland (2018). https://doi.org/10.4230/LIPIcs.ESA.2018.5
2. Balogh, J., Békési, J., Dósa, G., Epstein, L., Levin, A.: A new lower bound for classic online bin packing. Algorithmica **83**, 2047–2062 (2021)
3. Balogh, J., Epstein, L., Levin, A.: More on ordered open end bin packing. J. Sched. **24**(6), 589–614 (2021). https://doi.org/10.1007/s10951-021-00709-3
4. Bansal, A., Richard, J.P., Berg, B.P., Huang, Y.L.: A sequential follower refinement algorithm for robust surgery scheduling. INFORMS J. Comput. **36**(3), 918–937 (2024)
5. Berg, B.P., Denton, B.T.: Fast approximation methods for online scheduling of outpatient procedure centers. INFORMS J. Comput. **29**(4), 631–644 (2017)
6. Ceselli, A., Righini, G.: An optimization algorithm for the ordered open-end bin-packing problem. Oper. Res. **56**(2), 425–436 (2008)
7. Dell'Amico, M., Díaz, J.C.D., Iori, M.: The bin packing problem with precedence constraints. Oper. Res. **60**(6), 1491–1504 (2012)
8. Epstein, L., Levin, A.: Asymptotic fully polynomial approximation schemes for variants of open-end bin packing. Inf. Process. Lett. **109**(1), 32–37 (2008)
9. Epstein, L., Levin, A.: A note on a variant of the online open end bin packing problem. Oper. Res. Lett. **48**(6), 844–849 (2020)
10. Epstein, L.: Open-end bin packing: new and old analysis approaches. Discret. Appl. Math. **321**, 220–239 (2022)
11. Gai, L., Zhang, G.: Hardness of lazy packing and covering. Oper. Res. Lett. **37**(2), 89–92 (2009)
12. Guo, H., Li, W., Deng, B.: A survey on fair allocation of chores. Mathematics **11**(16), 3616 (2023)
13. Haouari, M., Mhiri, M.: Lower and upper bounding procedures for the bin packing problem with concave loading cost. Eur. J. Oper. Res. **312**(1), 56–69 (2024)
14. Leung, J.Y.T., Dror, M., Young, G.H.: A note on an open-end bin packing problem. J. Sched. **4**(4), 201–207 (2001)
15. Lin, M., Yang, Y., Xu, J.: Improved approximation algorithms for maximum resource bin packing and lazy bin covering problems. Algorithmica **57**(2), 232–251 (2010)
16. Lin, M., Yang, Y., Xu, J.: On lazy bin covering and packing problems. Theoret. Comput. Sci. **411**(1), 277–284 (2010)
17. Li, W., Ou, J.: Approximation algorithms for scheduling parallel machines with an energy constraint in green manufacturing. Eur. J. Oper. Res. **314**(3), 882–893 (2024)
18. Li, W., Ou, J.: Machine scheduling with restricted rejection: an application to task offloading in cloud–edge collaborative computing. Eur. J. Oper. Res. **314**(3), 912–919 (2024)
19. Li, W., Yang, Y., Xiao, M., Chen, X., Sterna, M., Błażewicz, J.: Scheduling with a discounted profit criterion on identical machines. Discret. Appl. Math. **367**, 195–209 (2025)
20. Luo, K., Spieksma, F.C.R.: Online bin packing with overload cost. In: Mudgal, A., Subramanian, C.R. (eds.) Algorithms and Discrete Applied Mathematics. CALDAM 2021. LNCS, vol. 12601, pp. 3–15. Springer, Cham (2021). https://doi.org/10.1007/978-3-030-67899-9_1

21. Seiden, S.S.: On the online bin packing problem. J. ACM **49**(5), 640–671 (2002)
22. Su, Q., Zhang, Q., Li, W., Zhang, X.: Primal-dual-based computation offloading method for energy-aware cloud-edge collaboration. IEEE Trans. Mob. Comput. **23**(2), 1534–1549 (2023)
23. Xiao, H., Zhang, J., Zhang, Z., Li, W.: A survey of approximation algorithms for the universal facility location problem. Mathematics **13**(7), 1023 (2025)
24. Yang, J., Leung, J.Y.T.: The ordered open-end bin-packing problem. Oper. Res. **51**(5), 759–770 (2003)
25. Yang, Y., Xiao, M., Li, W.: Semi-Online algorithms for the hierarchical extensible bin-packing problem and early work problem. Computation **12**(4), 68 (2024)
26. Zhang, J., Zhang, Y., Wu, H., Li, W.: An ordered submodularity-based budget-feasible mechanism for opportunistic mobile crowdsensing task allocation and pricing. IEEE Trans. Mob. Comput. **23**(2), 1278–1294 (2022)
27. Zhang, G.: Parameterized on-line open-end bin packing. Computing **60**, 267–273 (1998)

On the Windy k-Traveling Salesman Problem

Jianping Li[1(✉)], Xiaoxiao Yang[1], Junran Lichen[2], Wei Li[1], Runtao Xie[1], and Ping Yang[1]

[1] School of Mathematics and Statistics, Yunnan University, University Town, Chenggong District, Kunming 650504, P.R. China
jianping@ynu.edu.cn, yx698896@163.com, 1379715961@qw.com, xieruntao7@163.com, 1573395725@qq.com

[2] School of Mathematics and Physics, Beijing University of Chemical Technology, North Third Ring East Road, Chaoyang District, Beijing 100029, P.R. China
J.R.Lichen@buct.edu.cn

Abstract. In this paper, we address three variations of the windy k-traveling salesman problem (the k-WTSP), which are related to the asymmetric traveling salesman problem. Concretely, given a weighted connected graph $G = (V, E; w; r)$ of order n and an integer $k \geq 1$, where $r \in V$ and a windy cost function $w : A \rightarrow \mathbb{R}^+$, where $A = \{(v_i, v_j), (v_j, v_i) \mid v_iv_j \in E\}$, *i.e.*, for each edge $v_iv_j \in E$, we denote $w(v_i, v_j)$ to be the cost to traverse this edge v_iv_j from v_i to v_j and $w(v_j, v_i)$ to be the cost to traverse the same edge v_iv_j from v_j to v_i, respectively, the k-WTSP is asked to find k circuits, starting and ending at r, such that each vertex of G is contained in at least one of these k circuits, the objective is to minimize the total costs of these k circuits. In addition, the minmax windy k-traveling salesman problem (the minmax k-WTSP) is asked to find k circuits as mentioned-above, the objective is to minimize the maximum cost of these k circuits. Finally, the minmax windy k-circuit cover problem (the minmax k-WCCP) is asked to find k circuits such that each vertex of G is contained in at least one of these k circuits, the objective is to minimize the maximum cost of these k circuits. We have three key contributions as follows. (1) We design a $\frac{3}{2}(\alpha + 1)$-approximation algorithm to solve the k-WTSP in time $O(n^3)$, where $\alpha = \max\{\frac{w(v_i,v_j)}{w(v_j,v_i)} \mid v_iv_j \in E\}$; (2) We present a $(\frac{3}{2}(\alpha + 1) + 2(1 - \frac{1}{k}))$-approximation algorithm to solve the minmax k-WTSP in time $O(n^3)$; (3) We provide a $(\frac{3}{2}(\alpha + 1) + k)$-approximation algorithm to solve the minmax k-WCCP in time $O(n^3)$.

This study is supported by several funding sources, including the Natural Science Foundation of China [Nos.12361066, 12101593], the Yunnan Education Department [No.2025Y0087], the Science Research and Innovation Fund of Yunnan University [No.KC-24248783], and the Fundamental Research Funds for the Central Universities [No.buctrc202219]. Additional support is provided by the Project of Yunling Scholars of Yunnan Province [No.K264202011820].

M. Li et al. (Eds.): TAMC 2025, LNCS 16084, pp. 292–303, 2026.
https://doi.org/10.1007/978-981-95-4839-2_23

Keywords: Combinatorial optimization · Windy k-traveling salesman routings · Minmax k-circuit covers · Approximation algorithms · Complexity

1 Introduction and Problem Descriptions

The routing problem can be divided into the visiting vertex routing problem and the traversing edge or arc routing problem. The traveling salesman problem (TSP) [15] is one of the classical routing problems for visiting vertices, and the Chinese postman problem (CPP) [10] is one of the classic routing problems for traversing edges. They are widely used in routing optimization, logistics distribution, road maintenance and other fields. The metric TSP is defined as follows. Given a complete graph $G = (V, E; w)$, where $w : E \to \mathbb{R}^+$ is a cost function satisfying the triangle inequality, it is asked to find a cycle such that each vertex of G is visited exactly once by that cycle, the objective is to minimize the cost of that cycle. The most classic and effective algorithm for solving the metric TSP is the $\frac{3}{2}$-approximation algorithm designed in 1976 by Christofides [3]. In contrast, the CPP is asked to find a circuit such that each edge in E is traversed at least once by that circuit, the objective is to minimize the cost of that circuit. Guan [10] in 1962 presented a sufficient and necessary condition to produce an optimal edge routing for the CPP. Edmonds and Johnson [5] in 1973 designed a first optimal algorithm for solving the CPP in polynomial time, based on the perfect matching algorithm for odd-degree vertices [4].

For each edge $e = v_iv_j \in E$, when the cost function satisfies $w(v_i, v_j) = w(v_j, v_i)$ in the TSP, this problem is called as the symmetric TSP (commonly referred to as the TSP). When $w(v_i, v_j) \neq w(v_j, v_i)$ holds in general for each edge $e = v_iv_j \in E$, this problem is called as the asymmetric traveling salesman problem (ATSP) [9]. Concretely, given a strongly weighted connected digraph $D = (V, A; w)$, where $w : A \to \mathbb{R}^+$ is a cost function, it is asked to find a circuit such that each vertex in V is visited by at least once of that circuit, the objective is to minimize the cost of that circuit. Another definition of the ATSP is to find a Hamilton cycle that visits each vertex in V exactly once, the objective is to minimize the cost of that cycle. When the cost function $w(\cdot)$ satisfies the triangle inequality, these two definitions are equivalent [17].

Frieze et al. [8] in 1982 designed the first approximation algorithm with an approximation ratio $O(\log_2 n)$ for solving the ATSP. Asadpour et al. [1] in 2010 reduced the approximation ratio to $O(\log n/ \log\log n)$ by relaxation of the Euler-property condition. Svensson et al. [17] in 2020 proposed the first constant factor approximation algorithm for solving the ATSP and achieved an approximation ratio 506.

In real-world scenarios, due to terrain slopes, uphill traversal often incurs higher costs than downhill. So this problem, where the cost is related to the traversal direction, is called as the windy problem, for example, the windy postman problem (WPP) [13]. The WPP is an interesting generalization of the classic CPP and it is defined as follows. Given a weighted connected graph

$G = (V, E; w)$, $w : A \to \mathbb{R}^+$, where $A = \{(v_i, v_j), (v_j, v_i) \mid v_iv_j \in E\}$, *i.e.*, for each edge $e = v_iv_j \in E$, we denote $w(v_i, v_j)$ to be the cost to traverse this edge v_iv_j from v_i to v_j, and $w(v_j, v_i)$ to be the cost to traverse this edge v_iv_j (also v_jv_i) from v_j to v_i, respectively, it is asked to find a circuit to traverse each edge in E at least once, the objective is to minimize the cost of that circuit. The WPP has been proven to be *NP*-complete [11], Win [18] in 1989 designed a 2-approximation algorithm for solving the WPP.

In real life applications, multi-vehicle routing problems are frequently encountered. Frederickson et al. [6] in 1978 studied the min-max k-traveling salesman problem (the minmax k-TSP). Concretely, given a weighted complete graph $G = (V, E; w; r)$ and an integer $k \geq 1$, where $r \in V$ and a cost function $w : E \to \mathbb{R}^+$ satisfies the triangle inequality, the minmax k-TSP asks to find k cycles, starting and ending at r, collectively visiting all vertices in V, the objective is to minimize the maximum cost of these k cycles. The min-max k-Chinese postman problem (the minmax k-CPP) is similarly described. Concretely, given a weighted connected graph $G = (V, E; w; r)$ with an integer $k \geq 1$, $r \in V$ and a cost function $w : E \to \mathbb{R}^+$, it is asked to find k circuits, starting and ending at r, such that each edge of E is traversed by one of these k circuits at least once, the objective is to minimize the maximum cost of these k circuits. Frederickson et al. [6] in 1978 proposed a $(\frac{5}{2} - \frac{1}{k})$-approximation algorithm for solving the minmax k-TSP and designed a $(2 - \frac{1}{k})$-approximation algorithm for solving the minmax k-CPP, respectively. In addition, Frieze [7] in 1983 considered other two optimization problems, *i.e.*, the k-traveling salesman problem (the k-TSP) and the k-Chinese postman problem (the k-CPP), which is asked to find k circuits as mentioned-above, the objective is to minimize the total cost of these k circuits. Frieze [7] modified the algorithm for solving the TSP based on Christofides [3] and designed a $\frac{3}{2}$-approximate algorithm for solving the k-TSP. Pearn et al. [14] in 1994 showed that the k-CPP for some special versions is polynomial-time solvable.

Motivated by many arc routing problems as mentioned-above, we address the windy k-traveling salesman problem (the k-WTSP). Concretely, given a weighted connected graph $G = (V, E; w; r)$ of order n and an integer $k \geq 1$, where $r \in V$ and a windy cost function $w : A \to \mathbb{R}^+$, $A = \{(v_i, v_j), (v_j, v_i) \mid v_iv_j \in E\}$, *i.e.*, for each edge $v_iv_j \in E$, we denote $w(v_i, v_j)$ to be the cost to traverse this edge v_iv_j from v_i to v_j and $w(v_j, v_i)$ to be the cost to traverse the same edge v_iv_j from v_j to v_i, respectively, the k-WTSP is asked to find k circuits $\mathcal{C} = \{C_1, C_2, \ldots, C_k\}$, starting and ending at r, such that each vertex of G is visited at least once by these k circuits in $\mathcal{C}$, the objective is to minimize the total costs of these k circuits in $\mathcal{C}$, *i.e.*, $w(\mathcal{C}) = \sum_{j=1}^{k} w(C_j)$. In addition, the minmax windy k-traveling salesman problem (the minmax k-WTSP) asks to find k circuits as mentioned-above, the objective is to minimize the maximum cost of these k circuits in $\mathcal{C}$, *i.e.*, $w(\mathcal{C}) = \max\{w(C_j) \mid j = 1, 2, \ldots, k\}$. Finally, if there is no common vertex r such that k circuits cover all vertices in V, we propose the minmax windy k-circuit cover problem (the minmax k-WCCP), in details, it is asked to find k circuits $\mathcal{C} = \{C_1, C_2, \ldots, C_k\}$ such that each vertex of G is visited at least once by one of these k circuits, the objective is to minimize the maximum cost of these k circuits in $\mathcal{C}$, *i.e.*, $w(\mathcal{C}) = \max\{w(C_j) \mid j = 1, 2, \ldots, k\}$.

The remainder of this paper is organized into the following sections. In Sect. 2, we provide some terminologies, notations and fundamental lemmas to ensure the correctness of our algorithms. In Sect. 3, we design a $\frac{3}{2}(\alpha + 1)$-approximation algorithm to solve the k-WTSP in time $O(n^3)$, where $\alpha = \max\{\frac{w(v_i,v_j)}{w(v_j,v_i)} \mid v_iv_j \in E\}$. In Sect. 4, we present a $(\frac{3}{2}(\alpha + 1) + 2(1 - \frac{1}{k}))$-approximation algorithm to solve the minmax k-WTSP in time $O(n^3)$. In Sect. 5, we provide a $(\frac{3}{2}(\alpha+1)+k)$-approximation algorithm to solve the minmax k-WCCP in time $O(n^3)$. In Sect. 6, we summarize our conclusions and put forward some interesting related problems that will be worth of considering in the further research.

2 Terminologies, Notations and Fundamental Lemmas

In order to make this paper understood clearly, we present some related terminologies, notations and fundamental lemmas. Other ones not defined in the sequel will be found in the literature [2,12,16].

Given a windy weighted connected graph $G = (V, E; w)$ with order n and size m, we denote $w : A \to \mathbb{R}^+$ to be a windy cost function of G, where $A = \{(v_i, v_j), (v_j, v_i) \mid v_iv_j \in E\}$, *i.e.*, for any edge $e = v_iv_j \in E$, $w(v_i, v_j)$ is the cost to traverse this edge v_iv_j from v_i to v_j and $w(v_j, v_i)$ is the cost to traverse the same edge v_iv_j from v_j to v_i, respectively. A walk $W = (v_{i_0}e_{i_1}v_{i_1}e_{i_2} \cdots v_{i_{t-1}}e_{i_t}$ $v_{i_t})$ to connect v_{i_0} and v_{i_t} is a sequence of alternating vertices and edges, satisfying $e_{i_k} = v_{i_{k-1}}v_{i_k}$ for each $k = 1, 2, \ldots, t$, we denote this walk as $W = (v_{i_0}v_{i_1} \cdots v_{i_{t-1}}v_{i_t})$, where edges and vertices may be permitted to repeat, and each edge $e_{i_k} = v_{i_{k-1}}v_{i_k}$ is called as an internal edge for $k = 2, 3, \ldots, t-1$ and v_{i_k} is called as an internal vertex for $k = 1, 2, \ldots, t-1$ of this walk W. When the internal vertices of this walk W are all distinct, this walk W is called as a path to connect v_{i_0} and v_{i_t}, and we refer this path as $P(v_{i_0}, v_{i_t})$, simply, a v_{i_0}-v_{i_t} path. In addition, if a walk W satisfies $v_{i_0} = v_{i_t}$, this walk W is called as a circuit, denoted by C. Further, when the vertices of this circuit C are all distinct, this circuit C is called as a cycle.

Given a windy weighted connected graph $G = (V, E; w)$, since the cost of an edge is related to the direction of traversal of that edge, we denoted $\alpha = \max\{\frac{w(v_i,v_j)}{w(v_j,v_i)} \mid v_iv_j \in E\}$.

In a windy weighted connected graph $G = (V, E; w)$, given two vertices u, v in V, if $P(u, v)$ is a path from u to v, we denote its cost as $w(P(u, v)) = \sum_{xy \in P(u,v)} w(x, y)$, where the value $w(x, y)$ means that $P(u, v)$ traverses its internal edge xy from x to y. In addition, we denote $s(u, v)$ to be the cost of a shortest path from u to v in G.

If C is a circuit (of G) with a certain orientation, where all edges of C are either in a clockwise direction or in an anticlockwise direction, for any two distinct vertices $v_{i_s}, v_{i_t} \in V(C)$, we denote $v_{i_s} \prec v_{i_t}$ to indicate that v_{i_s} is traversed or "visited" before v_{i_t} along such certain orientation of C. In addition, we denote $P_C(v_{i_s}, v_{i_t})$ to present the path (on this circuit C with its certain orientation) from v_{i_s} to v_{i_t} along such certain orientation of C, and its cost is denoted as $w(P_C(v_{i_s}, v_{i_t}))$.

Before we present some algorithms to solve these problems as mentioned-above, we need the following result, due to Yu et al. [19].

Lemma 1. [19] *Given a connected graph $G = (V, E; w)$, an edge subset $E' \subseteq E$, and constants $B > 0$, $\beta \geq 0$, let C be a feasible solution, i.e., a circuit, to the rural postman problem, and $\max_{e \in E'} w(e) \leq \beta$. Then this circuit C can be divided into $l(\leq \lceil \frac{w(C)}{B} \rceil)$ walks, whose cost does not exceed $B + \beta$, and all walks cover the vertices in $V(C)$ and the edges in E', with a time complexity of $O(m)$.*

Using Lemma 1, we can obtain the following lemma.

Lemma 2. *Given a connected graph $G = (V, E; w)$, two constants $B > 0$, $\beta \geq 0$, and C as a feasible solution to the WTSP on G, we may assume $v_{i'} \prec v_{j'}$ follows in C along its certain orientation of C, $s(v_{j'}, v_{i'})$ is a shortest path from $v_{j'}$ to $v_{i'}$ in G, and suppose $\max_{v_{i'} \prec v_{j'} \in C} w(s(v_{j'}, v_{i'})) \leq \beta$. Then this circuit C can be divided into l $(\leq \lceil \frac{w(C)}{B} \rceil)$ sub-circuits in G, whose cost is at most $B + \beta$, and these sub-circuits cover all vertices in $V(C)$ and all edges in $E(C)$, with a time complexity of $O(n^3)$.*

Proof. Suppose that C is a feasible solution to the WTSP. For any walk $W_i = (v_{i_0} e_{i_1} v_{i_1} e_{i_2} \cdots e_{i_t}\ v_{i_t})$, if $w(W_i) \leq B$, we connect the vertex v_{i_t} to vertex v_{i_0} via a directed shortest path to form a sub-circuit C_i, and the cost is at most $B+\beta$. If $w(W_i) > B$, but $w(W_{i'}) = w(v_{i_0} e_{i_1} v_{i_1} \cdots v_{i_{t-1}}) < B$, we similarly connect $v_{i_{t-1}}$ to v_{i_0} via a directed shortest path $s(v_{i_{t-1}}, v_{i_0})$ to form a sub-circuit C_i with cost:

$$w(C_i) = \sum_{e_i \in W_{i'}} w(e_i) + w(s(v_{i_{t-1}}, v_{i_0})) \leq B + \max_{v_{i'} \prec v_{j'}} w(s(v_{j'}, v_{i'})) \leq B + \beta.$$

The construction of sub-circuit C_i, for each $i = 1, 2, \ldots, l$, requires computing directed shortest paths by using the Dijkstra algorithm [12], and its time complexity is $O(n^3)$. Thus, the time complexity is $O(n^3)$. □

3 A $\frac{3}{2}(\alpha + 1)$-Approximation Algorithm for the k-WTSP

In this section, we first consider the WTSP, where $k = 1$ for the k-WTSP, and we design an algorithm $\mathcal{A}_{WTSP}$ in the following setting.

Algorithm: $\mathcal{A}_{WTSP}$
Input: A weighted connected graph $G = (V, E; w)$, where $w : A \rightarrow \mathbb{R}^+$ is a windy cost function.
Output: A windy traveling salesman circuit.
Begin
Step 1. Construct a graph $G_1 = (V, E; \bar{w})$, where we may define $\bar{w}(v_i, v_j) = \min\{w(v_i, v_j), w(v_j, v_i)\}$ for each edge $e = v_i v_j \in E$.

Step 2. Use the Prim algorithm [12] to find a minimum spanning tree $T_1 = (V, E_{T_1})$ in $G_1 = (V, E; \bar{w})$, according to the function $\bar{w}(\cdot)$.
Step 3. Construct a graph $G_o = (V_o, E_o; w_o)$, where $V_o = \{x_i \mid d_{T_1}(v_i) \text{ is odd}\}$, $E_o = \{x_i x_j \mid \text{there exists a shortest path } P(v_i, v_j) \text{ to connect } v_i \text{ and } v_j \text{ in } G_1, \text{ where } x_i, x_j \in V_o\}$, $w_o : E_o \to \mathbb{R}^+$ is a cost function, having its cost of $w_o(x_i, x_j)$ as the cost of a shortest path to connect v_i and v_j in G_1.
Step 4. Use the Edmonds algorithm [4] to find a minimum cost perfect matching M in $G_o = (V_o, E_o; w_o)$.
Step 5. Construct a graph $G_M = (V, E_{T_1} \cup E_M)$, where $E_M = \bigcup_{x_i x_j \in M} \{(v_i v_j)_M \in E \mid (v_i v_j)_M \text{ is an edge of the shortest path } P(v_i, v_j) \text{ in } G_1\}$, *i.e.*, corresponding to each edge $x_i x_j \in M$, copy each edge $(v_i v_j)_M$ of the shortest path $P(v_i, v_j)$ connecting v_i and v_j in G_1 once and then superimpose it into the edge set E_{T_1}.
Step 6. Use the Win algorithm [18] to produce a directed Eulerian circuit C on $G_M = (V, E_{T_1} \cup E_M; w)$.
Step 7. Output a feasible windy traveling salesman circuit C.
End

Using the algorithm $\mathcal{A}_{WTSP}$, we can obtain the following lemma.

Lemma 3. *The algorithm $\mathcal{A}_{WTSP}$ is a $\frac{3}{2}(\alpha + 1)$-approximation algorithm to solve the WTSP, and this algorithm runs in time $O(n^3)$.*

Proof. Let C^* represent the optimal solution, OPT denote the optimal value, C represent the algorithm output solution, and OUT indicate the output value.

Since the circuit C is optimal solution for G_M, let C_1 be a feasible solution of WPP, which is obtained by traversing the edge set $E_{T_1} \cup E_M$ once in each direction, then the cost of C_1 can be expressed as: $w(C_1) = \sum_{v_i v_j \in E_{T_1} \cup E_M} [w(v_i, v_j) + w(v_j, v_i)]$, and we obtain $w(C) \leq w(C_1)$. Thus, we have the following

$$w(C) \leq w(C_1) = \sum_{v_i v_j \in E_{T_1} \cup E_M} [w(v_i, v_j) + w(v_j, v_i)] \leq (\alpha+1) \sum_{v_i v_j \in E_{T_1} \cup E_M} \bar{w}(v_i, v_j)$$

where $\bar{w}(\cdot)$ is the function defined at Step 1 in the algorithm $\mathcal{A}_{WTSP}$. The first inequality holds because C is an optimal solution to the WPP problem on the graph G_M, the second inequality holds because, for each edge, the cost of traversing this edge will not exceed α-times the smaller cost.

Since $G_M = (V, E_{T_1} \cup E_M)$ is an Eulerian graph produced at Step 5, recalling $\bar{w}(v_i, v_j) = \min\{w(v_i, v_j),\ w(v_j, v_i)\}$ for each edge $e = v_i v_j \in E$, we obtain that an Eulerian routing C_{cp} on the graph G_M, having its value $\sum_{v_i v_j \in E_{T_1} \cup E_M} \bar{w}(v_i, v_j)$, is an optimal solution on G_M of the CPP, according to the function $\bar{w}(\cdot)$. Thus, we have the following

$$w(C_{cp}) = \sum_{v_i v_j \in E_{T_1} \cup E_M} \bar{w}(v_i, v_j) = \sum_{v_i v_j \in E_{T_1}} \bar{w}(v_i, v_j) + \sum_{v_i v_j \in E_M} \bar{w}(v_i, v_j)$$

However, by deleting at least one edge from the optimal solution C^*, we can obtain a spanning tree T_1. Because T_1 is a minimum spanning tree of G under the cost $\bar{w}(\cdot)$, we obtain $\sum_{v_i v_j \in E_{T_1}} \bar{w}(v_i, v_j) \leq OPT$.

Now, we find all vertices in V_o of the first occurrence in C^*. We may suppose the order of occurrence of odd-degree vertices is $x_1, x_2, \ldots, x_{|V_o|}$, we connect the odd-degree vertices with the adjacent odd-degree vertices by the shortest path to construct a cycle C_o, *i.e.*, $C_o = \{p(x_1, x_2), p(x_2, x_3), \ldots, p(x_{|V_o|}, x_{|V_o|+1})\}$, where we denote $x_{|V_o|+1} = x_1$. For each path $p(x_i, x_{i+1})$, where $i = 1, 2, \ldots, |V_o|$, its cost is less than that of a path from x_i to x_{i+1} in the optimal solution C^*, implying that $w(C_o) \leq OPT$. Since C_o can be viewed as the cycle formed by two different perfect matchings, and however, M is the minimum cost perfect matching, then we obtain $\sum_{v_i v_j \in E_M} \bar{w}(v_i, v_j) \leq \frac{1}{2} w(C_o) \leq \frac{1}{2} OPT$. Thus, we have the following

$$w(C_{cp}) = \sum_{v_i v_j \in E_{T_1}} \bar{w}(v_i, v_j) + \sum_{v_i v_j \in E_M} \bar{w}(v_i, v_j) \leq OPT + \frac{1}{2} OPT \leq \frac{3}{2} OPT$$

To sum up, we have the following

$$OUT = w(C) \leq w(C_1) \leq (\alpha + 1) \sum_{v_i v_j \in E_{T_1} \cup E_M} \bar{w}(v_i, v_j) \leq \frac{3}{2}(\alpha + 1) OPT$$

The time complexity of the algorithm is determined in the following way. (1) At Step 1, constructing graph G_1 takes $O(m)$; (2) Applying the Prim algorithm [12] takes $O(n^2)$ in Step 2; (3) Constructing graph G_o requires $O(m)$ in Step 3; (4) Applying the Edmonds algorithm [4] takes $O(n^3)$ in Step 4; (5) At Step 5, constructing the Eulerian graph G_M requires $O(m)$; (6) Applying the Win algorithm [18] takes $O(n^3)$ in Step 6. To sum up, the time complexity of the algorithm $\mathcal{A}_{WTSP}$ is $O(n^3)$. □

From now on, we consider the k-WTSP. Building on the Frieze algorithm [8] for solving the k-TSP, our strategies to solve the k-WTSP are as follows.

(1) Construct a complete graph G_c under a smaller cost.
(2) The Prim algorithm [12] is used to find a minimum spanning tree with r as the root having degree $2k$ in the complete graph G_c.
(3) Apply the minimum cost perfect matching to construct an Eulerian graph, then utilize the Win algorithm [18] to compute a directed Eulerian circuit.

Algorithm: $\mathcal{A}_{k\text{-}WTSP}$
Input: A weighted connected graph $G = (V, E; w; r)$ and an integer $k \geq 1$, where $r \in V$ and $w : A \to \mathbb{R}^+$ is a windy cost function.
Output: A set $\mathcal{C}_1 = \{C_1, C_2, \ldots, C_k\}$ of circuits, starting and ending at r.
Begin
Step 1. Construct a graph $G_1 = (V, E; \bar{w})$, where we define the cost $\bar{w}(v_i, v_j) = \min\{w(v_i, v_j), w(v_j, v_i)\}$ for each edge $e = v_i v_j \in E$.
Step 2. Construct a complete graph $G_c = (V, E_c; w_c)$ as follows:
2.1 Set $E_c = \{v_i v_j | v_i, v_j \in V, v_i \neq v_j\}$;

2.2 For each edge $e = v_iv_j \in E_c$, define the cost

$$w_c(v_i, v_j) = \begin{cases} \bar{w}(v_i, v_j) & \text{if } v_iv_j \in E \\ \bar{w}(P(v_i, v_j)) & \text{otherwise.} \end{cases}$$

Step 3. Execute the Prim algorithm [12] to find a minimum spanning tree T_1 in G_c, where the degree of r in T_1 is $2k$.

Step 4. Construct an auxiliary graph $G_o = (V_o, E_o; w_o)$, where $V_o = \{x_i | d_{T_1}(v_i)$ is odd$\}$, $E_o = \{x_ix_j|$ there exists a shortest path $P(v_i, v_j)$ to connect v_i and v_j in G_c, where $x_i, x_j \in V_o\}$, w_o is a cost function, satisfying for any edge $(x_i, x_j) \in E_o$, $w_o(x_i, x_j)$ is the cost of a shortest path connecting v_i and v_j in G_c.

Step 5. Use the Edmonds algorithm [4] to find a minimum cost perfect matching M in G_o.

Step 6. Construct an Eulerian graph $G_M = (V, E_{T_1} \cup E_M)$, where $E_M = \bigcup_{x_ix_j \in M}\{(v_iv_j)_M \in E \mid (v_iv_j)_M$ is an edge of the shortest path $P(v_i, v_j)$ in $G_1\}$, *i.e.*, corresponding to each edge $x_ix_j \in M$, copy each edge $(v_iv_j)_M$ of the shortest path $P(v_i, v_j)$ connecting v_i and v_j in G_1 once and then superimpose it into the edge set E_{T_1}.

Step 7. Use the Win algorithm [18] to produce a directed Eulerian circuit C on $G_M = (V, E_{T_1} \cup E_M; w)$.

Step 8. Divide the circuit C into k circuits starting and ending at r, output k circuits $\mathcal{C}_1 = \{C_1, C_2, \ldots, C_k\}$.

End

Using the algorithm $\mathcal{A}_{k\text{-}WTSP}$, we can obtain the following result.

Theorem 1. *The algorithm $\mathcal{A}_{k\text{-}WTSP}$ is a $\frac{3}{2}(\alpha + 1)$-approximation algorithm to solve the k-WTSP, and this algorithm runs in time $O(n^3)$.*

Proof. Given a weighted connected graph $G = (V, E; w, r)$ and an integer $k \geq 1$, we may assume that $\mathcal{C}^* = \{C_1^*, C_2^*, \cdots, C_k^*\}$ is an optimal solution of G for the k-WTSP and OPT is the optimal value. For the same instance, the algorithm $\mathcal{A}_{k\text{-}WTSP}$ produces a feasible solution $\mathcal{C}_1 = \{C_1, C_2, \ldots, C_k\}$ with output value OUT.

Let $\mathcal{C}_1$ be the optimal windy postman circuit on the Eulerian graph G_M. A feasible tour C_f, traversing each edge once in each direction, satisfies $w(C_f) = \sum_{v_iv_j \in E_{T_1} \cup E_M}(w(v_i, v_j) + w(v_j, v_i))$, and $w(\mathcal{C}_1) \leq w(C_f)$. Thus, we have the following

$$w(\mathcal{C}_1) \leq w(C_f) = \sum_{v_iv_j \in E_{T_1} \cup E_M} (w(v_i, v_j) + w(v_j, v_i)) \leq (\alpha + 1) \sum_{v_iv_j \in E_{T_1} \cup E_M} \bar{w}(v_i, v_j)$$

It is easy to see that $\sum_{v_iv_j \in E_{T_1} \cup E_M} \bar{w}(v_i, v_j)$ corresponds to the CPP optimal solution C_{cp} on G_M, which traverses each edge in $E_{T_1} \cup E_M$ exactly once, implying that we have the following

$$w(C_{cp}) = \sum_{v_iv_j \in E_{T_1}} \bar{w}(v_i, v_j) + \sum_{v_iv_j \in E_M} \bar{w}(v_i, v_j)$$

$$w(C_{cp}) = \sum_{v_iv_j \in E_{T_1} \cup E_M} \bar{w}(v_i, v_j) = \sum_{v_iv_j \in E_{T_1}} \bar{w}(v_i, v_j) + \sum_{v_iv_j \in E_M} \bar{w}(v_i, v_j)$$

Since T_1 is a minimum spanning tree under the cost function $\bar{w}(\cdot)$, we have $\sum_{v_iv_j \in E_{T_1}} \bar{w}(v_i, v_j) \leq OPT$. For the optimal solution $\mathcal{C}^*$, we can find the vertices that appear in order for the first time in the odd-degree vertex set V_o. We may suppose that the odd-degree vertices in $\mathcal{C}^*$ appear in the order x_1, x_2, ..., $x_{|V_o|}$, connecting the odd-degree vertices with the adjacent odd-degree vertices in the form of the shortest path to construct a cycle C_o, *i.e.*, $C_o = \{P(x_1, x_2), P(x_2, x_3), \ldots, P(x_{|V_o|}, x_{|V_o|+1})\}$, where we denote $x_{|V_o|+1} = x_1$.

For each $i = 1, 2, \ldots, |V_o|$, since the cost of each shortest path $P(x_i, x_{i+1})$ is less than or equal to the cost a path from x_i to x_{i+1} in the aforementioned k-WTSP optimal solution of G, then we have $w(C_o) \leq OPT$. Since C_o can be regarded as a cycle composed of two different perfect matchings, and however, M is the minimum cost perfect matching in G_o such that we have $\sum_{v_iv_j \in E_M} \bar{w}(v_i, v_j) \leq \frac{1}{2}w(C_o) \leq \frac{1}{2}OPT$. Finally, we obtain the following

$$w(C_{cp}) = \sum_{v_iv_j \in E_{T_1}} \bar{w}(v_i, v_j) + \sum_{v_iv_j \in E_M} \bar{w}(v_i, v_j) \leq OPT + \frac{1}{2}OPT \leq \frac{3}{2}OPT$$

implying that $OUT = w(\mathcal{C}_1) \leq \frac{3}{2}(\alpha + 1)OPT$.

The time complexity of the algorithm $\mathcal{A}_{k\text{-}WTSP}$ can be determined as follows. (1) At Step 1, constructing the graph G_1 requires $O(m)$; (2) At Step 2, constructing the complete graph G_c takes $O(n^3)$; (3) Using the Prim algorithm [12] takes $O(n^2)$ in Step 3; (4)Constructing a graph G_o requires $O(m)$ in Step 4; (5) At Step 5, it executes the Edmonds algorithm [4] in time $O(n^3)$; (6) At Step 6, constructing the Eulerian graph takes $O(m)$; (7) Applying the Win algorithm [18] requires $O(n^3)$ in Step 7. Thus, the algorithm $\mathcal{A}_{k\text{-}WTSP}$ runs in total time $O(n^3)$. □

4 A $(\frac{3}{2}(\alpha + 1) + 2(1 - \frac{1}{k}))$-Approximation Algorithm for the Minmax k-WTSP

In this section, we consider the minmax k-WTSP. Motivated by the algorithm for solving the k-CPP, we design an algorithm to solve the minmax k-WTSP as follows.

Algorithm: $\mathcal{A}_{minmax\ k\text{-}WTSP}$

Input: A weighted connected graph $G = (V, E; w; r)$ and an integer $k \geq 1$, where $r \in V$ and $w : A \to \mathbb{R}^+$ is a windy cost function.

Output: A set $\mathcal{C}_2 = \{C_1, C_2, \ldots, C_k\}$ of circuits, starting and ending at r.

Begin

Step 1. Execute the algorithm $\mathcal{A}_{WTSP}$ (in Sect. 3) on G to obtain a feasible WTSP solution $C_0 = (rv_{i_1}v_{i_2} \cdots v_{i_t})$, having total cost L, where $r = v_{i_t}$.

Step 2. For each $j = 1, 2, \ldots, k-1$ do:
2.1 Let $L_j = \frac{j}{k}(L - 2s_{\max}) + s_{\max}$, where $s_{\max} = \max\{p(r, v_j), p(v_j, r) \mid v_j \in V\backslash\{r\}\}$.
2.2 Along a certain orientation of the circuit C_0, find the last vertex $v_{p(j)}$, starting from r, which satisfies $w(P_{C_0}(r, v_{p(j)})) \leq L_j$.
Step 3. Define these walks: $C'_1 = (r\, v_{i_1} \cdots v_{p(1)})$, $C'_2 = (v_{p(1)+1} \cdots v_{p(2)})$, ... , $C'_k = (v_{p(k-1)+1} \cdots r)$.
Step 4. Construct directed graph $\overleftrightarrow{G} = (V, A; c)$, where $A = \{(v_i, v_j), (v_j, v_i) \mid v_i v_j \in E\}$, $c(v_i, v_j) = w(v_i, v_j)$, $c(v_j, v_i) = w(v_j, v_i)$.
Step 5. For $j = 1, 2, 3, \ldots, k$ do:
Use the Dijkstra algorithm [12] on $\overleftrightarrow{G}$ to find two shortest paths from r to the starting vertex of C'_j and from the ending vertex of C'_j to r, respectively. Connect these two paths and the walk C'_j to form sub-circuit C_j.
Step 6. Output k circuits $C_1, C_2, \ldots, C_k$, starting and ending at r.
End

Using the algorithm $\mathcal{A}_{minmax\, k\text{-}WTSP}$, we can obtain the following result.

Theorem 2. *The algorithm $\mathcal{A}_{minmax\, k\text{-}WTSP}$ is a $(\frac{3}{2}(\alpha + 1) + 2(1 - \frac{1}{k}))$-approximation algorithm to solve the minmax k-WTSP, and this algorithm runs in time $O(n^3)$.*

Proof. Let $\mathcal{C}^* = \{C^*_1, C^*_2, \ldots, C^*_k\}$ be the optimal solution, OPT the optimal value, $\mathcal{C}_2 = \{C_1, C_2, \ldots, C_k\}$ output solution produced by the algorithm, OUT denote its output value of $\mathcal{C}_2$, and C^*_1 the optimal solution for the WTSP.

We consider the jth sub-circuit C_j. Since every vertex must be visited, the optimal cost to visit any vertex v_j is at least the sum of a shortest path cost from r to v_j and a shortest path cost from v_j to r. Thus, we obtain $s_{\max} = \max\{p(r, v_j), p(v_j, r) | v_j \in V\backslash r\} \leq OPT$.

By the algorithm, each cost of C'_1 and C'_k is at most $\frac{1}{k}(L - 2s_{\max}) + s_{\max}$. For $j = 2, 3, \ldots, k-1$, the cost of C'_j is at most $\frac{1}{k}(L - 2s_{\max})$. Thus, for each $j = 1, 2, 3, \ldots, k$, the cost $w(C_j) \leq s_{\max} + \frac{1}{k}(L - 2s_{\max}) + s_{\max}$.

Let C^*_1 be the optimal solution for the WTSP. We have $\frac{1}{k}w(C^*_1) \leq OPT$, and $w(C_0) = L \leq \frac{3}{2}(1 + \alpha)w(C^*_1)$, and combining $s_{\max} \leq OPT$, we have the following

$$OUT \leq 2s_{\max} + \frac{1}{k}(L - 2s_{\max}) \leq 2s_{\max}(1 - \frac{1}{k}) + \frac{L}{k} = [\frac{3}{2}(1 + \alpha) + 2(1 - \frac{1}{k})] \cdot OPT$$

The time complexity of the algorithm $\mathcal{A}_{minmax\, k\text{-}WTSP}$ can be determined in the following ways. (1) At Step 1, it requires time $O(n^3)$ to execute the algorithm $\mathcal{A}_{WTSP}$; (2) At Step 2, it needs time at most $O(m)$; (3) At Step 4, it needs time $O(n^2)$ to execute the Dijkstra algorithm [12] to find the shortest paths. Thus, the algorithm $\mathcal{A}_{minmax\, k\text{-}WTSP}$ runs in total time $O(n^3)$. □

5 A ($\frac{3}{2}(\alpha + 1) + k$)-Approximation Algorithm for the Minmax k-WCCP

In this section, we consider the minmax k-WCCP, and we design an algorithm in the following setting.

Using the algorithm $\mathcal{A}_{minmax\ k\text{-}WCCP}$, we can obtain the following result.

Theorem 3. *The algorithm* $\mathcal{A}_{minmax\ k\text{-}WCCP}$ *is a* $(\frac{3}{2}(1+\alpha)+k)$*-approximation algorithm to solve the minmax k-WCCP, and it runs in time* $O(n^3)$.

Proof. Let OPT be the optimal value, OUT the output value produced by the algorithm, and C_1^* the optimal circuit of G for the WTSP, having the $w(C_1^*) \leq k \cdot OPT$.

At Step 1, use the algorithm $\mathcal{A}_{WTSP}$ to obtain the WTSP feasible solution of G, then we have $w(C_0) \leq \frac{3}{2}(1+\alpha)w(C_1^*) \leq \frac{3}{2}(1+\alpha) \cdot k \cdot OPT$.

By using Lemma 2, we have the following

$$B = \frac{w(C_0)}{k} \leq \frac{\frac{3}{2}(1+\alpha)w(C_1^*)}{k} \leq \frac{\frac{3}{2}(1+\alpha) \cdot k \cdot OPT}{k} = \frac{3}{2}(1+\alpha) \cdot OPT$$

By the fact that $\beta = \max_{v_{i'} \prec v_{j'}\, in\, C_0} w(P(v_{j'}, v_{i'}))$ is less than or equal to the cost from $v_{j'}$ to $v_{i'}$ in C_1^*, we obtain $\beta = max_{v_{i'} \prec v_{j'}\, in\, C_0} w(P(v_{j'}, v_{i'})) \leq w(C_1^*) \leq k \cdot OPT$, implying that the cost of each sub-circuit is at most $B + \beta$. Thus we have the following

$$OUT \leq B + \beta \leq [\frac{3}{2}(1+\alpha) + k] \cdot OPT$$

The time complexity of the algorithm $\mathcal{A}_{minmax\ k\text{-}WCCP}$ can be determined as follows. (1) At Step 1, it requires time $O(n^3)$ to execute the algorithm $\mathcal{A}_{WTSP}$; (2) At Step 2, it requires at most time $O(n^3)$ to execute an algorithm in Lemma 3. Thus, the algorithm $\mathcal{A}_{minmax\ k\text{-}WCCP}$ runs in total time $O(n^3)$. □

6 Conclusions

In this paper, we consider three variations of the windy k-traveling salesman problem, *i.e.*, the k-WTSP, the minmax k-WTSP, and the minmax k-WCCP, and we design three approximation algorithms to solve these three problems.

The further research work, we will intend to design new algorithms to reduce dependence on α, such as based on linear programming methods or other combinatorial methods to design algorithms independent of α. On the other hand, we may consider some other windy problems related to the windy k-traveling salesman problem.

Acknowledgement. These authors are grateful to the three anonymous reviewers for their insightful comments and for their suggested changes that improve the current presentation greatly.

Declarations. The authors declare that they have no conflict of interest.

References

1. Asadpour, A., Goemans, M.X., Madry, A., Gharan, S.O., Saberi, A.: An $O(\log n/\log\log n)$-approximation algorithm for the asymmetric traveling salesman problem. Oper. Res. **65**(4), 1043–1061 (2017)
2. Bang-Jensen, J., Gutin, G.Z.: Digraphs: Theory, Algorithms and Applications (2 edn). Springer Monographs in Mathematics (2010)
3. Christofides, N.: Worst-case analysis of a new heuristic for the travelling salesman problem. Report 388, Carnegie Mellon University (1976)
4. Edmonds, J.: Maximum matching and a polyhedron with (0,1)-vertices. J. Res. Natl. Bur. Stand. B **69**, 125–130 (1965)
5. Edmonds, J., Johnson, E.L.: Matching, Euler tours and the Chinese postman. Math. Program. **5**, 88–124 (1973)
6. Frederickson, G.N., Hecht, M.S., Kim, C.E.: Approximation algorithms for some routing problems. SIAM J. Comput. **7**(2), 178–193 (1978)
7. Frieze, A.M.: An extension of Christofides heuristic to the k-person travelling salesman problem. Discret. Appl. Math. **6**, 79–83 (1983)
8. Frieze, A.M., Galbiati, G., Maffioli, F.: On the worst-case performance of some algorithms for the asymmetric traveling salesman problem. Networks **12**(1), 23–39 (1982)
9. Glover, F., Gutin, G., Yeo, A., Zverovich, A.: Construction heuristics for the asymmetric TSP. Eur. J. Oper. Res. **129**(3), 555–568 (2001)
10. Guan, M.G.: Graphic programming using odd or even points. Chin. Math. **1**, 237–277 (1962)
11. Guan, M.G.: On the windy postman problem. Discret. Appl. Math. **9**(1), 41–46 (1984)
12. Korte, B.H., Vygen, J.: Combinatorial Optimization: Theory and Algorithms. Springer, Berlin (2011)
13. Minieka, E.: The Chinese postman problem for mixed networks. Manag. Sci. **25**(7), 643–648 (1979)
14. Pearn, W.L.: Solvable cases of the k-person Chinese postman problem. Oper. Res. Lett. **16**(4), 241–244 (1994)
15. Reinelt, G.: The Traveling Salesman: Computational Solutions for TSP Applications. Springer, Berlin (2003)
16. Schrijver, A.: Combinatorial Optimization: polyhedra and efficiency (Volume A C), Algorithms and Combinatorics, vol 24. Springer (2004)
17. Svensson, O., Tarnawski, J., Végh, L.A.: A constant-factor approximation algorithm for the asymmetric traveling salesman problem. J. ACM **67**(6), 1–53 (2020)
18. Win, Z.: On the windy postman problem on Eulerian graphs. Math. Program. **44**, 97–112 (1989)
19. Yu, W., Liu, Z., Bao, X.: Approximation algorithms for some min-max postmen cover problems. Ann. Oper. Res. **300**, 267–287 (2021)

Approximating Total Effective Resistance Minimization with Small Budget

Yun Li and Hong Zhou(✉)

School of Mathematics and Statistics, Fuzhou University, Fuzhou, China
{230320066,hong.zhou}@fzu.edu.cn

Abstract. Total effective resistance (Kirchhoff index) is a fundamental graph metric with applications in network robustness, consensus algorithms, and graph neural networks. We study Total Effective Resistance Minimization (TERM): Given a graph G and a budget k, find a subgraph H with at most k edges to minimize the sum of pairwise effective resistances. While efficient approximation algorithms exist for spanning trees ($k = n - 1$) and large budgets ($k \gg n$), the challenging small-budget regime ($k - n = o(n)$) lacks non-trivial approximations. We bridge this gap with an $O(\sqrt{n})$-approximation algorithm for all $k \geq n - 1$, achieved through a greedy edge deletion framework that combines solutions for extreme budget regimes. Further, we study a natural Single-Source TERM variant (STERM), which minimizes the sum of effective resistances from a source vertex. Surprisingly, while TERM is NP-hard for spanning trees, STERM admits polynomial-time algorithms in this regime. We establish new connections: an α-approximation for STERM yields a 2α-approximation for TERM. In addition, we make progress toward resolving STERM's complexity status by proving APX-hardness via a reduction from 3-Dimensional Matching.

Keywords: Approximation Algorithm · Effective Resistance · Network Design · Kirchhoff Index

1 Introduction

Network design problems typically involve finding a minimum-cost subgraph satisfying connectivity constraints [1,6,10]. Effective resistance, a measure of "closeness" between vertices in a graph, provides a natural alternative metric for network design. Given a weighted undirected graph $G = (V, E)$ with a weight $w_e > 0$ on each edge $e \in E$, the *effective resistance* Reff_{uv} between vertices $u, v \in V$ is the potential difference induced by sending a unit electrical flow from u to v through the graph G, where each edge e is treated as a single resistor with resistance $1/w_e$ (or with conductance w_e). Informally, Reff_{uv} is the resistance

This work was supported in part by National Key R&D Program of China (Grant No. 2023YFA1010202) and Natural Science Foundation of Fujian Province of China (Grant No. 2023J01414).

M. Li et al. (Eds.): TAMC 2025, LNCS 16084, pp. 304–316, 2026.
https://doi.org/10.1007/978-981-95-4839-2_24

between u and v when we treat the whole resistance network G as a big resistor. Effective resistance forms a metric over the vertex set (see, *e.g.*, [13]), thus it serves as a measure of "connectivity" between vertices in the graph. Therefore, it is natural to incorporate effective resistance into network design problems.

The *total effective resistance* (or Kirchhoff index), denoted as $\mathrm{Reff}_{\mathrm{tot}}$, is the sum of all pairwise effective resistances (*i.e.*, $\mathrm{Reff}_{\mathrm{tot}} = \sum_{u<v} \mathrm{Reff}_{uv}$), which can serve as a measure on the robustness of a graph. Total effective resistance finds wide applications range from network connectivity analysis [19] and social recommendation systems [20] to graph neural networks [4] and consensus algorithms [17,18,23].

In this paper, we consider a combinatorial optimization variant of the Total Effective Resistance Minimization (TERM) problem introduced by Ghosh, Boyd, and Saberi [7]. Given an undirected graph $G = (V, E)$ and a budget k, TERM seeks a subgraph H with at most k edges to minimize $\mathrm{Reff}_{\mathrm{tot}}(H)$. Formally, let $x \in \{0,1\}^m$ indicate edges in H. Then, x_e can be treated as the conductance on edge e, and the total effective resistance of H is a function of x, denoted by $\mathrm{Reff}_{\mathrm{tot}}(x)$. TERM can be formulated as the following integer programming.

$$\begin{aligned} \underset{x \in \mathbb{R}^m}{\text{minimize}} \quad & \mathrm{Reff}_{\mathrm{tot}}(x) \\ \text{subject to} \quad & \sum_{e \in E} x_e \le k \\ & x_e \in \{0,1\}, \forall e \in E. \end{aligned} \tag{1}$$

TERM finds applications in minimizing average commute time of a random walk on the graph, or average Elmore delay of a RC circuit, or the total time constant of an averaging network (see [7] for more details and more applications). TERM is also a special case of the A-Optimal Experiment Design problem [7], which is a classical problem in statistics and recently found wide applications from machine learning to numerical linear algebra, to signal processing etc. [9].

As a combinatorial optimization problem, TERM is NP-hard [11] even for the special case $k = n - 1$ (also known as the *minimum routing cost spanning tree* problem). Researchers have been focusing on designing approximation algorithms for TERM. For the minimum routing cost spanning tree problem, a polynomial time approximation scheme (PTAS) is known [22], but the techniques are highly limited to deal with spanning trees and cannot be applied to the general TERM problem. On the other hand, when we have a large budget $k \gg n$, the problem is also well-understood. In particular, there are $(1+\epsilon)$-approximation algorithms for TERM when $k > \Omega(n/\epsilon)$ [15], matching the integrality gap of a natural convex programming relaxation of (1). However, these algorithms cannot provide good approximation ratios when the budget is small. In particular, no non-trivial approximation algorithms are known in the challenging small-budget regime where $k - n = o(n)$.

1.1 Our Results

In this paper, we advance toward bridging this gap with a greedy edge deletion framework, combining techniques for $k = n - 1$ and large $k \gg n$.

Theorem 1. *There exists a polynomial-time* $O(\min\{k - n + 2, n/(k - n + 1)\})$*-approximation algorithm for TERM. This yields* $O(\sqrt{n})$*-approximation algorithm for all* $k \geq n - 1$.

The algorithm in Theorem 1 has non-trivial approximation guarantees in the small-budget regime, which goes beyond the integrality gap of a widely used convex programming relaxation for TERM (see Sect. 2 for more details).

Then, in Sect. 4, we consider a natural variant *Single-Source Total Effective Resistance Minimization* (STERM), which has been mentioned in [7] as an extension to TERM. Given a graph G, a source vertex s, and a budget k, find a subgraph H with at most k edges such that the sum of the effective resistance from s to all other vertices, *i.e.*, $\sum_{u:u \neq s} \mathrm{Reff}_{us}(H)$, is minimized.

We explore the connections between TERM and STERM by showing that approximation for STERM implies approximation for TERM.

Theorem 2. *If there exists an* α*-approximation algorithm to STERM, then we have a* 2α*-approximation algorithm for TERM.*

Furthermore, we can show that the approximation algorithm in Theorem 1 can be extended to STERM as well and obtain the same approximation guarantee as in Theorem 1. As the extension is relatively straight forward, we omit this part due to the space limit.

Although closely related, TERM and STERM exhibit distinct complexities. While TERM is NP-hard when $k = n-1$, STERM reduces to shortest path trees, which is solvable in polynomial time via Dijkstra's algorithm. In this paper, we establish APX-hardness of STERM via a reduction from 3-Dimensional Matching.

Theorem 3. *The problem STERM is APX-hard.*

The proof of Theorem 3 is the most technical part in this paper. Due to the space constraints, we only highlight some key ides in Sect. 4.2, and postpone the complete proof to the full version of this work.

1.2 Related Work

The continuous optimization version of TERM was studied by Ghosh, Boyd, and Saberi [7] where the goal is to allocate a fixed amount of conductance to the edges of the input graph to minimize the total effective resistance. They formulated the problem as an SDP and provided an interior-point algorithm. However, they did not address the combinatorial optimization variant of the problem.

As for the combinatorial version of TERM, it has also been studied in the setting of matrix column subset selection [3,8]. However, this line of work do not provide approximation algorithms, instead they measure the algorithm performance by comparing the returned solution (a submatrix) to the original input matrix. We remark that the greedy edge deletion idea used in this paper was inspired by these work on column subset selection [3,8].

In terms of approximation algorithms for TERM, besides the earlier mentioned PTAS proposed by Wu et al. [22] for the special case with $k = n - 1$, most of the recent algorithmic progresses were made in the large-budget setting $k \gg n$ under the more general context of A-optimal experiment design problem. Allen-Zhu et al. [2] provided a regret minimization based framework to a family of experimental design problems with different objectives, which achieves $(1+\epsilon)$-approximation when $k \geq \Omega(n/\epsilon^2)$ for the A-optimal design. Nikolov, Singh, and Tantipongpipat improved the result with a proportional volume sampling technique to achieve $(1+\epsilon)$-approximation when $k \geq \Omega\left(\frac{n}{\epsilon} + \frac{1}{\epsilon^2}\log\frac{1}{\epsilon}\right)$. Lau and Zhou further improved the result using a local search framework to achieve $(1+\epsilon)$-approximation when $k \geq \Omega(\frac{n}{\epsilon})$.

Besides TERM and STERM, Ghosh, Boyd, and Saberi [7] also mentioned another variant where the goal is to minimize the effective resistance between a specific pair of vertices. The combinatorial version of this problem was studied by Chan et al. [5] who provided a constant approximation algorithm.

As for some more recent related work, Kooij and Achterberg [14] proved that it is NP-hard to add a specific number of edges to a given graph in order to minimize the total effective resistance. Zhou, Zehmakan, and Zhang [24] proposed two algorithms for TERM: a traditional greedy algorithm and a gradient-based greedy algorithm. However, both algorithms do not have worst-case approximation guarantees as their approximation ratios may depend on the structure of input graphs.

2 Preliminaries

We denote $M \succeq 0$ when M is positive semidefinite (PSD). For PSD matrices $A, B \in \mathbb{R}^{n\times n}$, their Frobenius inner product is $\langle A, B\rangle := \mathrm{tr}(A^\top B)$.

Let $G = (V, E)$ be a weighted undirected graph with $n = |V|$ vertices and $m = |E|$ edges, where each edge $e \in E$ has conductance $w_e > 0$ (equivalently, resistance $1/w_e$). The *Laplacian matrix* $L_G \in \mathbb{R}^{n\times n}$ is defined as $L_G = D - A$, where D is the weighted degree matrix and A is the weighted adjacency matrix.

Let $\lambda_1 \leq \lambda_2 \leq \cdots \leq \lambda_n$ be the eigenvalues of L_G. It is well known that L_G is a PSD matrix with the minimum eigenvalue $\lambda_1 = 0$ ($\mathbf{1}$ is the corresponding eigenvector), and $\lambda_2 > 0$ for connected graphs. Let $L_G = \sum_{k=2}^{n} \lambda_k u_k u_k^\top$ be the eigen-decomposition of L_G, where $u_k \in \mathbb{R}^n$ is a unit eigenvector corresponding to λ_k for $k = 2, \ldots, n$, then the *Moore-Penrose pseudoinverse* of L_G is defined as $L_G^\dagger := \sum_{k=2}^{n} \frac{1}{\lambda_k} u_k u_k^T$.

For $i, j \in V$, define $b_{ij} \in \mathbb{R}^n$ with $b_{ij}(i) = 1$, $b_{ij}(j) = -1$, and 0 elsewhere. The effective resistance between i and j in G can be written as (see, e.g., [7])

$$\mathrm{Reff}_{ij}(G) = b_{ij}^T L_G^\dagger b_{ij} = \langle b_{ij} b_{ij}^T, L_G^\dagger\rangle.$$

Therefore, the total effective resistance can be reformulated as

$$\mathrm{Reff}_{\mathrm{tot}}(G) = \sum_{i<j} \mathrm{Reff}_{ij}(G) = \sum_{i<j} \left\langle b_{ij} b_{ij}^T, L_G^\dagger \right\rangle = \left\langle nI - \mathbf{1}\mathbf{1}^T, L_G^\dagger \right\rangle = n \cdot \mathrm{tr}\left(L_G^\dagger\right),$$

where the last equality holds since $L_G^\dagger \mathbf{1} = 0$.

We adapts the well-known Sherman-Morrison formula to Laplacian systems.

Lemma 1 (Sherman-Morrison formula for graph Laplacians). *Let L_G be the Laplacian matrix of a connected graph G, and let $w_{ij} b_{ij} b_{ij}^T$ be a weighted rank-1 perturbation to the edge (i,j) with $w_{ij} \geq 0$. The perturbed pseudoinverse is given by:*

$$(L_G - w_{ij} b_{ij} b_{ij}^T)^\dagger = L_G^\dagger + \frac{w_{ij} L_G^\dagger b_{ij} b_{ij}^T L_G^\dagger}{1 - w_{ij} b_{ij}^T L_G^\dagger b_{ij}}.$$

Then, we recall some key results for TERM that we need to use in this paper.

When $k = n-1$, TERM is exactly the minimum routing cost spanning tree problem, for which there exists a PTAS.

Lemma 2 ([22]). *For any fixed small constant $\epsilon > 0$, there is a polynomial time algorithm that returns a $(1+\epsilon)$-approximate solution to TERM when $k = n-1$.*

To deal with TERM with large budget $k \gg n$, most of the recent algorithms are based on the following natural relaxation of (1).

$$\begin{aligned} \underset{x \in \mathbb{R}^m}{\text{minimize}} \quad & \mathrm{Reff}_{\mathrm{tot}}(x) = n \, \mathrm{tr}\left(L_x^\dagger\right) \\ \text{subject to} \quad & \sum_{e \in E} x_e \leq k \\ & x_e \in [0,1], \forall e \in E, \end{aligned} \tag{2}$$

where L_x is the Laplacian matrix of the input graph with edge weight $x_e w_e$ on each edge. Ghosh, Boyd, and Saberi [7] showed that (2) is a convex programming relaxation[1] of (1) (as the $\mathrm{Reff}_{\mathrm{tot}}(x)$ is a convex function in x), and (2) can be reformulated by an SDP and can be solved efficiently.

After obtaining an optimal fractional solution x to (2), we use the regret minimization based framework proposed by Allen-Zhu et al. [2][2] to round x to an integer solution with the following guarantees.

[1] We can show that the integrality gap of (1) is at least $\Omega(n)$ in the worst case. This follows by slightly modifying the integrality gap example for a related convex programming relaxation [16] for A-optimal experiment design. We omit the details due to space constraints.

[2] Instead of the algorithm in [2], we may also use the randomized local search algorithm in [15], which has better guarantees ($k \geq \Omega(n/\epsilon)$ vs $k \geq \Omega(n/\epsilon^2)$). However, this does not affect our result in Theorem 1, as we did not try to optimize the constant in the approximation ratio.

Lemma 3 ([2]). *Let $x \in [0,1]^m$ with $\sum_{e \in E} x_e \leq k$ and let $\epsilon \in (0, \frac{1}{3}]$, if $k \geq 5n/\epsilon^2$ then there exists a polynomial time deterministic algorithm to find an integer solution $z \in \{0,1\}^m$ such that*

$$L_z \succeq (1-3\epsilon)L_x \quad \text{and} \quad \sum_{e \in E} z_e \leq k.$$

Since $L_z \succeq (1-3\epsilon)L_x$ implies $\operatorname{tr}(L_z^\dagger) \leq (1+O(\epsilon))\operatorname{tr}(L_x^\dagger)$, Lemma 3 immediately implies the following corollary, which provides good approximation for TERM with large budget.

Corollary 1. *For some small constant $\epsilon > 0$, if $k \geq \Omega(n/\epsilon^2)$, then there exists a convex relaxation based polynomial time rounding algorithm that finds a $(1+\epsilon)$-approximate solution to TERM. In particular, if $k \geq cn$ for some large enough constant c, then we can find an $O(1)$-approximate solution in polynomial time.*

3 Greedy Edge Deletion for TERM

We present a nontrivial approximation algorithm for TERM with a small budget, which goes beyond the limit of the integrality gap of (2). Our approach combines two key ideas that bridge the gap between the extreme budget regimes ($k = n-1$ and $k \gg n$): (1) Starting from an initial approximate solution for large k (Corollary 1), we iteratively remove edges using a greedy criterion; (2) When the greedy solution is suboptimal, we fall back to an approximate minimum routing cost spanning tree (Lemma 2).

When $k \geq 10^3 n$ (we did not try to optimize the constant 10^3), Corollary 1 already provides an $O(1)$-approximation. Thus, we focus on the small-budget case where $k < 10^3 n$. Algorithm 1 formalizes this approach, which clearly runs in polynomial time. We now proceed to analyze its approximation guarantee.

Algorithm 1. Greedy Edge Deletion for TERM

Input: a weighted graph $G = (V, E)$ with n vertices and a budget $n-1 \leq k < 10^3 n$
Output: a subgraph H of G with at most k edges

```
Apply Corollary 1 with budget k' = 10^3 n to get a subgraph S_0 with k' edges;
Let S ← S_0 be the initial solution;
while |E(S)| > k do
    Greedily find an edge e ∈ E(S) such that e = argmin_{e∈E(S)} tr(L†_{S\{e}});
    Set S ← S\{e};
end while
Let T be a spanning tree returned by applying the algorithm in Lemma 2 with
some constant ϵ, e.g., ϵ = 1;
if Reff_tot(T) < Reff_tot(S) then
    return H ← T;
end if
return H ← S.
```

Let H^* be an optimal solution to TERM with budget $k < 10^3 n$. We first show that the objective value of the initial solution S_0 of the while loop is not much worse than H^*.

Lemma 4. *Let S_0 be the initial solution of the while loop with $10^3 n > k$ edges, then it holds that* $\mathrm{Reff}_{\mathrm{tot}}(S_0) \leq O(1) \cdot \mathrm{Reff}_{\mathrm{tot}}(H^*)$

Proof. Let $\mathrm{OPT}(k')$ be the optimal objective value of TERM with budget $k' = 10^3 n$. By Corollary 1,

$$\mathrm{Reff}_{\mathrm{tot}}(S_0) \leq O(1) \cdot \mathrm{OPT}(k') \leq O(1) \cdot \mathrm{Reff}_{\mathrm{tot}}(H^*),$$

where the last inequality holds as $k' > k$ and the objective value of the optimal solution is non-increasing as the budget grows. □

The core of our algorithm relies on controlling resistance growth during the greedy edge removal process.

Lemma 5. *Let G be a connected graph with n vertices and $m \geq n$ edges, then there exists an edge $e \in E(G)$ such that*

$$\mathrm{Reff}_{\mathrm{tot}}(G \backslash \{e\}) \leq \frac{m-n+2}{m-n+1} \cdot \mathrm{Reff}_{\mathrm{tot}}(G).$$

Proof. By Sherman-Morison formula (Lemma 1), removing edge e (with weight w_e) updates the pseudoinverse as

$$L_{G\backslash\{e\}}^{\dagger} = (L_G - w_e b_e b_e^T)^{\dagger} = L_G^{\dagger} + \frac{w_e L_G^{\dagger} b_e b_e^T L_G^{\dagger}}{1 - w_e b_e^T L_G^{\dagger} b_e}. \tag{3}$$

The new total effective resistance is

$$\begin{aligned}\mathrm{Reff}_{\mathrm{tot}}(G\backslash\{e\}) &= \left\langle \sum_{i<j} b_{ij} b_{ij}^T, L_{G\backslash\{e\}}^{\dagger} \right\rangle = \left\langle \sum_{i<j} b_{ij} b_{ij}^T, L_G^{\dagger} + \frac{w_e L_G^{\dagger} b_e b_e^T L_G^{\dagger}}{1 - w_e b_e^T L_G^{\dagger} b_e} \right\rangle \\ &= \mathrm{Reff}_{\mathrm{tot}}(G) + \left\langle \sum_{i<j} b_{ij} b_{ij}^T, \frac{w_e L_G^{\dagger} b_e b_e^T L_G^{\dagger}}{1 - w_e b_e^T L_G^{\dagger} b_e} \right\rangle.\end{aligned}$$

Notice that

$$\sum_{e' \in E(G)} (1 - w_{e'} b_{e'}^T L_G^{\dagger} b_{e'}) = |E(G)| - \left\langle \sum_{e' \in E(G)} w_{e'} b_{e'} b_{e'}^T, L_G^{\dagger} \right\rangle = m - n + 1 \geq 0,$$

thus, we can define a deletion probability distribution over edges $e \in E(G)$:

$$\Pr[\text{removing } e] = \frac{1 - w_e b_e^T L_G^{\dagger} b_e}{\sum_{e' \in E(G)} (1 - w_{e'} b_{e'}^T L_G^{\dagger} b_{e'})} = \frac{1 - w_e b_e^T L_G^{\dagger} b_e}{m - n + 1}.$$

The expected total resistance after removal is

$$\begin{aligned}
\mathbb{E}[\mathrm{Reff}_{\mathrm{tot}}(G\backslash\{e\})] &= \mathrm{Reff}_{\mathrm{tot}}(G) + \sum_{e\in E(G)} \frac{1-w_e b_e^T L_G^\dagger b_e}{m-n+1} \cdot \left\langle \sum_{i<j} b_{ij}b_{ij}^T, \frac{w_e L_G^\dagger b_e b_e^T L_G^\dagger}{1-w_e b_e^T L_G^\dagger b_e} \right\rangle \\
&= \mathrm{Reff}_{\mathrm{tot}}(G) + \frac{1}{m-n+1} \cdot \left\langle \sum_{i<j} b_{ij}b_{ij}^T, \sum_{e\in E(G)} L_G^\dagger (w_e b_e b_e^T) L_G^\dagger \right\rangle \\
&= \mathrm{Reff}_{\mathrm{tot}}(G) + \frac{1}{m-n+1} \cdot \left\langle \sum_{i<j} b_{ij}b_{ij}^T, L_G^\dagger \right\rangle \\
&= \frac{m-n+2}{m-n+1} \cdot \mathrm{Reff}_{\mathrm{tot}}(G).
\end{aligned}$$

Therefore, there exists some edge e achieves the claimed bound. □

Iterating Lemma 5 bounds the solution after greedy deletion.

Lemma 6. *After removing edges from initial solution S_0 with $k' = 10^3 n$ edges down to k edges, we obtain a solution S with objective value*

$$\mathrm{Reff}_{\mathrm{tot}}(S) \le \frac{k'-n+2}{k-n+1} \cdot \mathrm{Reff}_{\mathrm{tot}}(S_0).$$

Proof. According to the greedy edge deletion algorithm, in each iteration we remove an edge from the current solution to minimize the total effective resistance of the remaining graph. Thus, it holds by Lemma 5 that

$$\mathrm{Reff}_{\mathrm{tot}}(S) \le \left(\prod_{i=k}^{k'} \frac{i-n+2}{i-n+1} \right) \cdot \mathrm{Reff}_{\mathrm{tot}}(S_0) = \frac{k'-n+2}{k-n+1} \cdot \mathrm{Reff}_{\mathrm{tot}}(S_0),$$

where we note that the product telescopes. □

Finally, we relate the optimal minimum routing cost spanning tree T^* to H^*.

Lemma 7. *Let G be the input graph, and T^* be an optimal minimum routing cost spanning tree, and H^* be an optimal solution to TERM with budget k. Then, it holds that*

$$\mathrm{Reff}_{\mathrm{tot}}(T^*) \le (k-n+2) \cdot \mathrm{Reff}_{\mathrm{tot}}(H^*).$$

Proof. Greedily remove edges from H^* (with k edges) down to a spanning tree $\hat{H}$ with $n-1$ edges as in Algorithm 1. Note that, H^* is unknown, and the greedy deletion process here is for analysis only.

Applying Lemma 5 repeatedly, we obtain

$$\mathrm{Reff}_{\mathrm{tot}}(\hat{H}) \le \left(\prod_{i=n}^{k} \frac{i-n+2}{i-n+1} \right) \cdot \mathrm{Reff}_{\mathrm{tot}}(H^*) = (k-n+2) \cdot \mathrm{Reff}_{\mathrm{tot}}(H^*).$$

Since T^* minimizes $\mathrm{Reff}_{\mathrm{tot}}$ among spanning trees, the result follows. □

Now, we are ready to prove the main theorem in this section.

Proof (Proof of Theorem 1). As mentioned earlier, it suffices to consider the small budget case where $k < 10^3 n$. By Lemma 4 and Lemma 6, the solution S after greedy deletion has total effective resistance at most

$$\mathrm{Reff}_{\mathrm{tot}}(S) \le \frac{10^3 n - n + 2}{k - n + 1} \cdot \mathrm{Reff}_{\mathrm{tot}}(S_0) \le \frac{O(n)}{k - n + 1} \cdot \mathrm{Reff}_{\mathrm{tot}}(H^*).$$

As for the spanning tree T from Lemma 2 with ϵ being constant, it holds that $\mathrm{Reff}_{\mathrm{tot}}(T) \le O(1) \cdot \mathrm{Reff}_{\mathrm{tot}}(T^*)$, where T^* is the optimal minimum routing cost spanning tree. Combining with Lemma 7, it holds that

$$\mathrm{Reff}_{\mathrm{tot}}(T) \le O(k - n + 2) \cdot \mathrm{Reff}_{\mathrm{tot}}(H^*).$$

Taking the better solution among S and T, the approximation ratio of the algorithm is at most $O(\min\{k-n+2, n/(k-n+1)\})$. The worst case happens when $k = n + \Theta(\sqrt{n})$, which yields $O(\sqrt{n})$-approximation. □

4 Single-Source Total Effective Resistance Minimization

We study the Single-Source Total Effective Resistance Minimization (STERM) problem. We first show that STERM and TERM are closely related in Sect. 4.1. Then, we establish the APX-hardness of STERM in Sect. 4.2.

4.1 Approximation-Preserving Reduction from TERM to STERM

We show that STERM approximability implies TERM approximability with a factor-2 loss. This reduction extends Wong's approach for minimum routing cost spanning tree [21] to STERM.

Proof (Proof of Theorem 2). Given a graph F, let $\mathrm{Reff}_v(F) := \sum_{u:u\neq v} \mathrm{Reff}_{uv}(F)$ denote the sum of effective resistance from vertex v to all other vertices. Fix a source vertex s, we denote $\mathrm{Reff}_s^* := \min_{F\subseteq G, |E(F)|\le k} \mathrm{Reff}_s(F)$ as the optimal value for STERM with budget k. Let $s^* = \arg\min_{s\in V} \mathrm{Reff}_s^*$ be the best source vertex in terms of STERM value, and F_{s^*} be the optimal solution achieving STERM value $\mathrm{Reff}_{s^*}^*$.

We first claim that F_{s^*} is a 2-approximate solution for TERM with budget k. Notice that

$$\begin{aligned}\mathrm{Reff}_{\mathrm{tot}}(F_{s^*}) = \frac{1}{2} \cdot \sum_{i\neq j} \mathrm{Reff}_{ij}(F_{s^*}) &\le \frac{1}{2} \cdot \sum_{i\neq j} (\mathrm{Reff}_{is^*}(F_{s^*}) + \mathrm{Reff}_{js^*}(F_{s^*})) \\ &= (n-1) \cdot \mathrm{Reff}_{s^*}(F_{s^*}),\end{aligned} \quad (4)$$

where the inequality follows as the effective resistance forms a metric over the vertex set of F_{s^*}.

Meanwhile, suppose H^* is an optimal solution to TERM with budget k. Then, for any fixed vertex $s \in V$, let F_s be the optimal solution to STERM with source vertex s and budget k, it holds that

$$\sum_{u:u \neq s} \mathrm{Reff}_{us}(H^*) \geq \sum_{u:u \neq s} \mathrm{Reff}_{us}(F_s) = \mathrm{Reff}_s(F_s) \geq \mathrm{Reff}_{s^*}(F_{s^*}), \tag{5}$$

where the first inequality follows as F_s is an optimal solution to STERM with source vertex s, and the second inequality follows by the definition of s^*. Summing over all vertex $s \in V$ on both sides of (5), we obtain

$$2\mathrm{Reff}_{\mathrm{tot}}(H^*) = \sum_{u \neq s} \mathrm{Reff}_{us}(H^*) \geq n \cdot \mathrm{Reff}_{s^*}(F_{s^*}). \tag{6}$$

Combine (4) and (6) together, we establish that F_{s^*} is a 2-approximate solution for TERM, as claimed.

Now, suppose we have an α-approximation algorithm for STERM, then we run the algorithm on n different STERM instances, each with the same input graph G but with a different source vertex $v \in V$. Let H_v be the solution returned by the approximation algorithm when v is the source vertex, then $\mathrm{Reff}_v(H_v) \leq \alpha \cdot \mathrm{Reff}^*_v$, as H_v is an α-approximate solution.

Let $s' = \arg\min_{s \in V} \mathrm{Reff}_s(H_s)$ be the source vertex such that $H_{s'}$ has the smallest outgoing effective resistance among all returned subgraphs. Note that

$$\sum_{u:u \neq s'} \mathrm{Reff}_{us'}(H_{s'}) \leq \mathrm{Reff}_{s^*}(H_{s^*}) \leq \alpha \cdot \mathrm{Reff}_{s^*}(F_{s^*}),$$

where s^* is the best source vertex in terms of STERM value, and the first inequality holds by the definition of s', and the second inequality holds as H_{s^*} is an α-approximate solution to STERM with source vertex s^*. Then, a similar argument as in (4), (5), and (6) can show that $H_{s'}$ is a 2α-approximate solution to TERM. □

4.2 Hardness of STERM

We establish APX-hardness of STERM, thus proving Theorem 3, via a reduction from 3-Dimensional Matching (3DM), which is a well-known APX-hard problem [12]. Since this is the most technical part of this paper, we will outline key ideas in this subsection, and defer the rigorous proof to the full version of this work due to the space limit.

Given a 3DM instance (*i.e.* a 3-uniform hypergraph), Let m be the number of hyperedges (triples) and n be the number of vertices. The 3DM instance can be represented as a bipartite graph $G_M = (T, L, E_M)$, where T corresponds to the set of hyperedges (referred to as *triple vertex*), and L corresponds to the vertices of the hypergraph (referred to as *element vertex*). Each triple vertex $t = (v_1, v_2, v_3) \in T$ has exactly three neighbors $v_1, v_2, v_3 \in L$ in G_M.

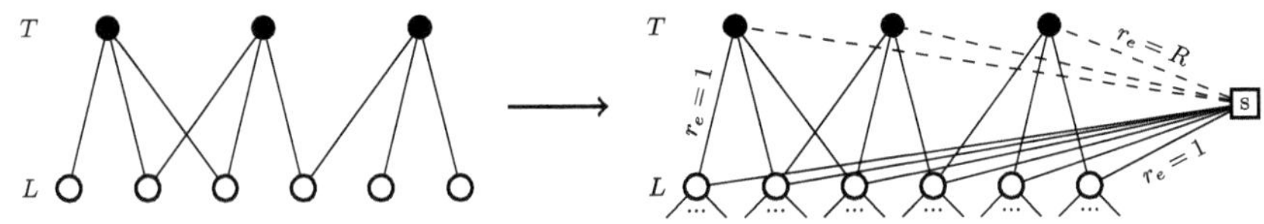

Fig. 1. A simple demonstration of the reduction.

In the reduction (see Fig. 1 for a demonstration), we first construct an intermediate instance $G = (V, E)$, where $V = T \cup L \cup \{s\}$, $E = E_M \cup E_T \cup E_L$ and $E_T = \{(s,t)\,|t \in T\}$, $E_L = \{(s,v)\,|v \in L\}$. In other words, we add a source vertex s to the vertex set and connect s with every other vertex in G_M. The edges in $E_M \cup E_L$ have resistance 1, and the edges in E_T have resistance $R \in (0,1)$, which will be some small enough constant.

In the intermediate instance G, we would like to find a subgraph H with at most $m + 2n$ edges to minimize the sum of effective resistances from s to all vertices in L, *i.e.* $\mathrm{Reff}_{s,L}(H) := \sum_{v \in L} \mathrm{Reff}_{sv}(H)$. This is not exactly a STERM instance, as we ignored those vertices in T. However, there is a simple trick of adding dummy vertices and edges to further reduce to a STERM instance (see the full version of this work).

Finally, to prove APX-hardness of the intermediate problem, the main key observation is that, if there is a perfect matching in the 3DM instance, then we can pick a solution to be an edge-disjoint union of many copies of book graphs as shown in Fig. 2. This is the most efficient structure to minimize effective resistances from s to vertices in L. Otherwise, if the maximum matching of the 3DM instance is far from perfect, then a significant portion of vertices $v \in L$ need to involve more edges than the book graph structure to achieve the same effective resistance. However, under the budget constraint, this forces many vertices in L to become degree-1, significantly increasing the effective resistance $\mathrm{Reff}_{s,L}(H)$. The gap between the efficient and inefficient structure implies APX-hardness.

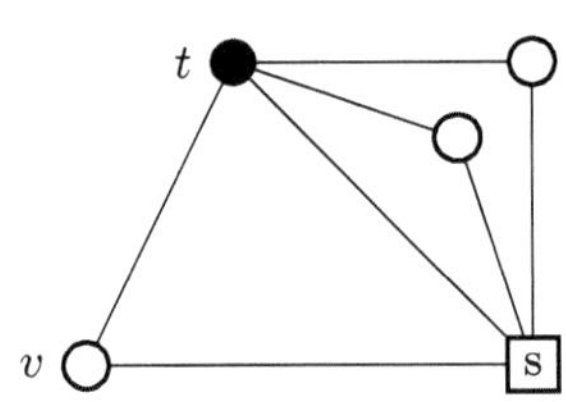

Fig. 2. Desired book graph structure, where those white vertices are element vertices and the black vertex is a triple vertex.

Concluding Remarks

This paper presents an $O(\sqrt{n})$-approximation algorithm for TERM, advancing the state-of-the-art for the challenging small-budget regime ($k - n = o(n)$).

While this work addresses a key gap in the literature, the approximation ratio remains far from optimal, leaving substantial room for improvement. A natural direction would be to augment the greedy edge deletion framework with greedy edge addition starting from an approximate minimum routing cost spanning tree. Unfortunately, we do not know how to rigorously analyze the greedy addition operation yet.

While we have established APX-hardness for STERM, our reduction only applies when $k = cn$ for a constant c in a certain range, leaving open the critical question: At what budget level does STERM shift from being computationally tractable to intractable? Specifically, we do not know whether it becomes NP-hard for $k = n$.

Acknowledgments. We thank Lap Chi Lau for helpful discussions at the initial stage of this work.

References

1. Agrawal, A., Klein, P., Ravi, R.: When trees collide: an approximation algorithm for the generalized steiner problem on networks. In: Proceedings of the Twenty-Third Annual ACM Symposium on Theory of Computing, pp. 134–144 (1991). https://doi.org/10.1145/103418.103437
2. Allen-Zhu, Z., Li, Y., Singh, A., Wang, Y.: Near-optimal discrete optimization for experimental design: a regret minimization approach. Math. Program. **186**, 439–478 (2021). https://doi.org/10.1007/s10107-019-01464-2
3. Avron, H., Boutsidis, C.: Faster subset selection for matrices and applications. SIAM J. Matrix Anal. Appl. **34**(4), 1464–1499 (2013). https://doi.org/10.1137/120867287
4. Black, M., Wan, Z., Nayyeri, A., Wang, Y.: Understanding oversquashing in GNNs through the lens of effective resistance. In: International Conference on Machine Learning, pp. 2528–2547. PMLR (2023)
5. Chan, P.H., Lau, L.C., Schild, A., Wong, S.C.W., Zhou, H.: Network design for S-T effective resistance. ACM Trans. Algorithms (TALG) **18**(3), 1–45 (2022). https://doi.org/10.1145/3522588
6. Gabow, H.N., Goemans, M.X., Tardos, É., Williamson, D.P.: Approximating the smallest k-edge connected spanning subgraph by LP-rounding. Netw. Int. J. **53**(4), 345–357 (2009). https://doi.org/10.1002/net.20289
7. Ghosh, A., Boyd, S., Saberi, A.: Minimizing effective resistance of a graph. SIAM Rev. **50**(1), 37–66 (2008). https://doi.org/10.1137/050645452
8. de Hoog, F.R., Mattheij, R.M.M.: Subset selection for matrices. Linear Algebra Appl. **422**(2–3), 349–359 (2007). https://doi.org/10.1016/j.laa.2006.08.034
9. Huan, X., Jagalur, J., Marzouk, Y.: Optimal experimental design: formulations and computations. Acta Numer **33**, 715–840 (2024). https://doi.org/10.1017/S0962492924000023
10. Jain, K.: A factor 2 approximation algorithm for the generalized steiner network problem. Combinatorica **21**, 39–60 (2001). https://doi.org/10.1007/s004930170004
11. Johnson, D.S., Lenstra, J.K., Kan, A.R.: The complexity of the network design problem. Networks **8**(4), 279–285 (1978). https://doi.org/10.1002/net.3230080402

12. Kann, V.: Maximum bounded 3-dimensional matching is MAX SNP-complete. Inform. Process. Lett. **37**(1), 27–35 (1991). https://doi.org/10.1016/0020-0190(91)90246-E
13. Klein, D.J., Randić, M.: Resistance distance, vol. 12, pp. 81–95 (1993), applied graph theory and discrete mathematics in chemistry (Saskatoon, SK, 1991)
14. Kooij, R.E., Achterberg, M.A.: Minimizing the effective graph resistance by adding links is np-hard. Oper. Res. Lett. **51**(6), 601–604 (2023). https://doi.org/10.1016/j.orl.2023.10.002
15. Lau, L.C., Zhou, H.: A local search framework for experimental design. SIAM J. Comput. **51**(4), 900–951 (2022). https://doi.org/10.1137/20M1386542
16. Nikolov, A., Singh, M., Tantipongpipat, U.: Proportional volume sampling and approximation algorithms for A-optimal design. Math. Oper. Res. **47**(2), 847–877 (2022). https://doi.org/10.1287/moor.2021.1129
17. Patterson, S., Bamieh, B.: Consensus and coherence in fractal networks. IEEE Trans. Control Netw. Syst. **1**(4), 338–348 (2014). https://doi.org/10.1109/TCNS.2014.2357552
18. Qi, Y., Zhang, Z., Yi, Y., Li, H.: Consensus in self-similar hierarchical graphs and sierpiński graphs: convergence speed, delay robustness, and coherence. IEEE Trans. Cybern. **49**(2), 592–603 (2018). https://doi.org/10.1109/TCYB.2017.2781714
19. Tizghadam, A., Leon-Garcia, A.: Autonomic traffic engineering for network robustness. IEEE J. Sel. Areas Commun. **28**(1), 39–50 (2009). https://doi.org/10.1109/JSAC.2010.100105
20. Wong, F.M.F., Liu, Z., Chiang, M.: On the efficiency of social recommender networks. IEEE/ACM Trans. Networking **24**(4), 2512–2524 (2015). https://doi.org/10.1109/TNET.2015.2475616
21. Wong, R.T.: Worst-case analysis of network design problem heuristics. SIAM J. Algebraic Discrete Methods **1**(1), 51–63 (1980). https://doi.org/10.1137/0601008
22. Wu, B.Y., Lancia, G., Bafna, V., Chao, K.M., Ravi, R., Tang, C.Y.: A polynomial-time approximation scheme for minimum routing cost spanning trees. SIAM J. Comput. **29**(3), 761–778 (2000). https://doi.org/10.1137/S009753979732253X
23. Yi, Y., Zhang, Z., Patterson, S.: Scale-free loopy structure is resistant to noise in consensus dynamics in complex networks. IEEE Trans. Cybern. **50**(1), 190–200 (2018). https://doi.org/10.1109/TCYB.2018.2868124
24. Zhou, X., Zehmakan, A.N., Zhang, Z.: Efficient algorithms for minimizing the kirchhoff index via adding edges. IEEE Trans. Knowl. Data Eng. (2025). https://doi.org/10.1109/TKDE.2025.3552644

Modified Greedy Algorithm for Monotone Submodular Maximization with Knapsack and Partition Matroid Constraints

Dongkai Xu[1,2], Jianhua Yuan[1,2], and Zhongzheng Tang[1,2(✉)]

[1] School of Mathematical Sciences, Beijing University of Posts and Telecommunications, Beijing 100876, China
{xudongkai,jianhuayuan,tangzhongzheng}@bupt.edu.cn

[2] Key Laboratory of Mathematics and Information Networks (Beijing University of Posts and Telecommunications), Ministry of Education, Beijing 100876, China

Abstract. In this paper, we study the monotone submodular maximization problem with a knapsack constraint and a partition matroid constraint, where the ground set is partitioned into k disjoint classes. We propose the maximum imbalance parameter λ, defined as the maximum ratio of cost values between any two elements within the same class. We design a modified greedy algorithm which returns the better solution between the constructed greedy solution and a single element. We show that this algorithm can achieve an approximation ratio of $\min\{\frac{1}{1+\lambda}(1-e^{-\frac{1+\lambda}{2}}), \frac{1}{2+\lambda}\}$. In particular, when all elements within each class have identical cost values, i.e., $\lambda = 1$, the algorithm obtains an approximation guarantee of $\frac{1}{2}(1-e^{-1}) \approx 0.316$.

Keywords: Greedy Algorithm · Submodular Maximization · Approximation Guarantee

1 Introduction

Submodularity is a concept that has a significant prevalence in a wide range of scientific and technological disciplines. Submodular functions emerge naturally in diverse domains, such as combinatorics, graph theory, probability theory, and economics. Many combinatorial optimization problems can be fundamentally attributed to monotone submodular maximization. Examples of such problems include the maximum coverage problem [3,7], the maximum facility location problem [1], and the maximum entropy sampling problem [8,13].

Furthermore, an increasing number of problems in real-world applications are also shown to be monotone submodular maximization. These problems include

This work was partially supported by the National Natural Science Foundation of China (No. 12101069 and No. 12171052), the Beijing Natural Science Foundation (No. Z220004).

M. Li et al. (Eds.): TAMC 2025, LNCS 16084, pp. 317–329, 2026.
https://doi.org/10.1007/978-981-95-4839-2_25

sensor placement [9], data summarization [6,10], image segmentation [2], data subset selection [17], network optimization [14], etc.

After reviewing the applications of submodular maximization, let us introduce related definitions for monotone submodular maximization.

A set function $f : 2^V \to \mathbb{R}$ is submodular if for all $S,\ T \subseteq V$, it holds that

$$f(S) + f(T) \geq f(S \cup T) + f(S \cap T).$$

In addition, we assume that f is normalized, i.e., $f(\varnothing) = 0$. Alternatively, defining

$$f(v \mid S) := f(S \cup \{v\}) - f(S)$$

as the marginal gain of adding an element $v \in V$ to a set $S \subseteq V$, and equivalent definition [11] of a submodular set function f is that for all $S \subseteq T \subseteq V$ and $v \in V \setminus T$, we have

$$f(v \mid S) \geq f(v \mid T).$$

The latter form of definition describes the concept of diminishing return in economics. The function f is monotone nondecreasing if and only if $f(S) \leq f(T)$ for all $S \subseteq T \subseteq V$ (or equivalently $f(v \mid S) \geq 0$, for every $S \subseteq V$ and $v \in V \setminus S$). Additionally, we also consider the marginal gain of a set $T \subseteq V$ with respect to another set $S \subseteq V$, which is defined as $f(T \mid S) := f(T \cup S) - f(S)$.

1.1 Related Work

Nemhauser et al. [11] conducted the seminal work on submodular optimization, establishing its foundational significance in the field. They studied monotone submodular maximization with a cardinality constraint and proposed a greedy heuristic that achieves an approximation factor of $(1 - e^{-1})$, which is tight.

When dealing with the problem of maximizing a monotone submodular function subject to a knapsack constraint, the Greedy+Singleton Algorithm returns the better solution between the greedy solution constructed and a single element, which is a well-known and intuitive algorithmic framework for finding solutions. The approximation ratio of 0.357 for this algorithm was first proven by Wolsey [18] in 1982. Khuller et al. [7] obtained an approximation ratio of 0.316 in 1999. In addition, Sviridenko [15] first enumerated 3 elements and then applied the greedy algorithm based on this, achieving an approximation ratio of $(1-e^{-1})$, with the algorithm requiring $O(n^5)$ function value computations. Jing Tang et al. [16] showed that the Greedy+Singleton Algorithm achieves an improved constant approximation ratio of 0.405 in 2021. Recently, Feldman et al. [4] showed that the approximation ratio of the algorithm is within $[0.427, 0.462]$.

In addition to the knapsack constraints, the matroid constraints are also considered in the problem of submodular maximization. Fisher [5] showed that greedy heuristic has approximation ratios at least $\frac{1}{P+1}$ with the intersection of P matroids, where P is any positive integer. In the case where the number of matroids is one, i.e., $P = 1$, the greedy heuristic achieves an approximation factor of $\frac{1}{2}$. To maximize a monotone submodular function subject to a knapsack and k matroid constraints (where $k > 1$), Sarpatwar et al. [12] proposed an algorithm that combines a greedy approach with local search. This algorithm attains approximation ratios of $\frac{1-e^{-(k+1)}}{k+1}$ with the algorithm requiring $n^{O(k)}$ function value computations. In the case of having only a single knapsack constraint and a single matroid constraint, the approximation ratio of this algorithm is $\frac{1-e^{-2}}{2}$ with $\tilde{O}(n^6)$ evaluations of the function, which is relatively high.

In this paper, we study monotone submodular maximization with a knapsack constraint and a partition matroid constraint. In this problem, the input consists of a nonnegative monotone submodular function $f : 2^V \to \mathbb{R}_{\geq 0}$, a positive cost function $c : V \to \mathbb{R}_{>0}$, and a positive budget b. The objective of the problem is to find a set maximizing f among the sets $S \subseteq V$ whose cost is at most b (i.e., $c(S) = \sum_{v \in S} c(v) \leq b$), which is defined as follows:

$$\begin{aligned} &\max_{S \subseteq V} f(S) \\ &\text{s.t.}\ \ c(S) \leq b,\ |S \cap V_i| \leq 1, \end{aligned} \tag{1}$$

where $V = V_1 \sqcup V_2 \sqcup \cdots \sqcup V_k$, $i \in [k] = 1, 2, \cdots, k$. Without loss of generality, we may assume that $c(v) \leq b$ for all $v \in V$, since elements with a cost greater than b do not belong to any feasible solution.

For the ground set V and the indices of classes $V_1, V_2, \cdots, V_k$, where $V = V_1 \sqcup V_2 \sqcup \cdots \sqcup V_k$, we define the function $\varphi : V \to [k]$ such that $\varphi(u) = i$ if and only if $u \in V_i$, where $u \in V$ and $i \in [k]$.

2 The Modified Greedy Algorithm

Next, we will solve problem (1) using the greedy method, we propose a modified greedy algorithm, as shown in Algorithm 1.

Algorithm 1 The Modified Greedy Algorithm

initialize $S_g \leftarrow \varnothing$, $V' \leftarrow V$.
while $V' \neq \varnothing$ **do**
  find $u \leftarrow \arg\max_{v \in V'} \{\frac{f(v|S_g)}{c(v)}\}$;
  if $c(S_g) + c(u) \leq b$ **then**
    $S_g \leftarrow S_g \cup \{u\}$,
    $V' \leftarrow V' \setminus V_{\varphi(u)}$;
  else
    $V' \leftarrow V' \setminus \{u\}$;
  end if
end while
$u^* \leftarrow \arg\max_{v \in V} f(v)$;
$S_q \leftarrow \arg\max_{S \in \{\{u^*\}, S_g\}} f(S)$;
return S_q.

2.1 Notations and Definitions

To characterize Algorithm 1, we denote $u_i \in V_i$ as the i-th element added to the greedy solution S_g by the greedy heuristic. Denote $S_i := \{u_1, u_2, \cdots, u_i\}$ as the first i added elements for any $0 \le i \le |S_g| = m$. Clearly, $S_0 = \varnothing,\ S_m = S_g$.

Let o be the first element in the optimal solution OPT considered by the greedy heuristic but not added to the set of elements S_g due to budget violation, and let Q be the set of elements constructed until o considered by the greedy heuristic. Note that $Q \subseteq S_g$, then $f(Q) \le f(S_g)$ due to the monotonicity of f. We define λ as

$$\lambda = \max_{i\in[k]} \max_{a,b\in V_i} \frac{c(a)}{c(b)},$$

which is the maximum cost ratio between two elements of the same classes V_i for all $i \in [k]$, and we refer to λ as the maximum imbalance coefficient. Clearly, $\lambda \ge 1$.

2.2 The Approximation Guarantee

The main result of this paper is to establish that Algorithm 1 achieves the approximation ratio of $\min\{\frac{1}{1+\lambda}(1-e^{-\frac{1+\lambda}{2}}), \frac{1}{2+\lambda}\}$.

Theorem 1. *The Modified Greedy Algorithm achieves an approximation factor of*

$$\min\{\frac{1}{1+\lambda}(1-e^{-\frac{1+\lambda}{2}}), \frac{1}{2+\lambda}\},$$

where $\lambda \ge 1$ is the maximum imbalance coefficient in V. When all elements within each class have identical cost values, i.e., $\lambda = 1$, the approximation factor can reach $\frac{1}{2}(1-e^{-1}) \approx 0.316$.

3 Proof of Theorem 1

In this section we prove the approximation guarantee stated in Theorem 1. First, we establish the following preliminary results.

Lemma 1. *For any nondecreasing submodular set function f, given any set $X \subseteq V$, the marginal function $f(S \mid X) = f(S \cup X) - f(X)$ in X is also a monotone nondecreasing submodular set function with respect to S.*

Proof. It follows that $f(S \mid X)$ is monotone since for any $S \subseteq T \subseteq V$,

$$f(S \mid X) = f(S \cup X) - f(X) \le f(T \cup X) - f(X) = f(T \mid X).$$

Furthermore, by the definition of a submodular set function f, we have

$$\begin{aligned} f(S \mid X) + f(T \mid X) &= f(S \cup X) - f(X) + f(T \cup X) - f(X) \\ &\ge f((S \cup X) \cup (T \cup X)) + f((S \cup X) \cap (T \cup X)) - 2f(X) \\ &= f((S \cup T) \cup X) - f(X) + f((S \cap T) \cup X) - f(X) \\ &= f(S \cup T \mid X) + f(S \cap T \mid X), \end{aligned}$$

which shows the submodularity. □

Lemma 2. *For any nondecreasing submodular (and nonnegative) set function* f*, given any two sets* S*,* T *with* $|S \cap V_i| \leq 1$*,* $|T \cap V_i| \leq 1$ *and* $S \subseteq T$*, we have*

$$f(T) - f(S) \leq \sum_{u \in T \setminus S} f(u \mid S).$$

Proof. Let $t = |T| - |S|$. Without loss of generality, we may assume that $T \setminus S = \{u_1, u_2, \cdots, u_t\}$, where $u_i \in V_i$, $i \in [t]$. Define arbitrary sets $\{S_j\}_{j=0}^{t}$ such that (1) $S_0 = S$; (2) $|S_j \setminus S_{j-1}| = 1$; (3) $S_t = T$. Then we have

$$\begin{aligned}
f(T) - f(S) &= \sum_{j=0}^{t-1} f(u_{j+1} \mid S_j) \\
&= f(u_1 \mid S_0) + f(u_2 \mid S_1) + \cdots + f(u_t \mid S_{t-1}) \\
&\leq f(u_1 \mid S) + f(u_2 \mid S) + \cdots + f(u_t \mid S) \\
&= \sum_{j=0}^{t-1} f(u_{j+1} \mid S) = \sum_{u \in T \setminus S} f(u \mid S).
\end{aligned}$$

This completes the proof. □

Subsequently, we study the problem (1) without the knapsack constraint, over a subset $V' = \bigcup_{i \in I} V_i$, for $I \subseteq [k]$. Without loss of generality, let $I = \{1, 2, \cdots, s\}$. We fix an arbitrary ordering for the classes in V'. Furthermore, we can assume that they are ordered as $V_1, V_2, \cdots, V_s$.

Next, we consider a greedy algorithm, which sequentially adds the element with the highest marginal density in each class according to the order of elements class, as shown in Algorithm 2.

Algorithm 2 The Greedy Algorithm without Knapsack Constraint

```
1: initialize G ← ∅.
2: for i = 1 to s do
3:     find v_i ← arg max_{v∈V_i} { f(v|G) / c(v) },
4:     G ← G ∪ {v_i}.
5: end for
6: return G.
```

We assume that the Algorithm 2 processes the items in the order $v_1 \in V_1, v_2 \in V_2, \cdots, v_s \in V_s$, and let $G := \{v_1, v_2, \cdots, v_s\}$ be the greedy solution returned.

Let $T := \{v_1^*, v_2^*, \cdots, v_s^*\}$, $v_i^* \in V_i$, $i \in [s]$ be an optimal solution that reaches the maximum value of the function on V', where the indexing follows G.

We define $G_j := \{v_1, v_2, \cdots, v_j\}$ as the first j added elements, where $G_j \subseteq G$, $j \in [s]$. Furthermore, let T_j be the set obtained by replacing the first j elements of T with G_j, i.e., $T_j := (T \setminus \{v_1^*, \cdots, v_j^*\}) \cup G_j$ for $j = 0, 1, \cdots, s$. Observe that $T_0 = T$, $T_s = G$.

The following lemma derives several upper bounds for $f(T)$.

Lemma 3. *For* $t = 0, 1, \cdots, s$, *we have*

$$f(T) \leq (1+\lambda) f(G_t) + \sum_{v \in T_t \setminus G_t} f(v \mid G_t),$$

where $\lambda \geq 1$ *is the maximum imbalance coefficient in* V.

Proof. The lemma holds directly when $t = 0$. In the following, we consider $t \in [s]$. For $j = 0, 1, \cdots, t-1$, we introduce an intermediate

$$P_j := T_j \setminus \{v_{j+1}^*\} = T_{j+1} \setminus \{v_{j+1}\}.$$

Namely, P_j contains $s-1$ items (excluding items with index $j+1$), where the indices of items $v_1, v_2, \cdots v_j$ are the same as those in G, and the indices of other items are the same as those in T. Evidently, $G_j \subseteq P_j$. Then, we have

$$\begin{aligned} f(T_j) &= f(P_j) + f(v_{j+1}^* \mid P_j), \\ f(T_{j+1}) &= f(P_j) + f(v_{j+1} \mid P_j). \end{aligned}$$

Due to the submodularity of f and the greedy rule, we have

$$\begin{aligned} f(T_j) - f(T_{j+1}) &= f(v_{j+1}^* \mid P_j) - f(v_{j+1} \mid P_j) \\ &\leq f(v_{j+1}^* \mid P_j) \\ &\leq f(v_{j+1}^* \mid G_j) \\ &= \frac{f(v_{j+1}^* \mid G_j)}{c(v_{j+1}^*)} \cdot c(v_{j+1}^*) \\ &\leq \frac{f(v_{j+1} \mid G_j)}{c(v_{j+1})} \cdot c(v_{j+1}^*) \\ &\leq \lambda f(v_{j+1} \mid G_j). \end{aligned}$$

Summing up the above inequalities from $j = 0$ to $t-1$, we get

$$\begin{aligned} f(T_0) - f(T_t) &\leq \lambda \cdot \sum_{j=0}^{t-1} f(v_{j+1} \mid G_j) \\ &= \lambda (f(G_t) - f(G_0)). \end{aligned}$$

Equivalently, we have

$$f(T) - f(T_t) \leq \lambda f(G_t), \; t \in [s].$$

Since we have $G_t \subseteq T_t$, then by Lemma 2, we get

$$\begin{aligned} f(T) &\leq \lambda f(G_t) + f(T_t) \\ &\leq \lambda f(G_t) + f(G_t) + \sum_{v \in T_t \setminus G_t} f(v \mid G_t) \\ &= (1+\lambda) f(G_t) + \sum_{v \in T_t \setminus G_t} f(v \mid G_t). \end{aligned}$$

This completes the proof. □

Now, let us return to the problem (1) and let OPT be the optimal solution of the problem (1). The next lemma provides a series of upper bounds for the optimal value $f(\text{OPT})$.

Lemma 4. *Let $|Q| = l$. Then for $t \in [l]$, we have*

$$f(OPT) \leq (1+\lambda) f(S_{t-1}) + b \cdot \frac{f(u_t \mid S_{t-1})}{c(u_t)}, \ \lambda \geq 1.$$

Proof. Fixing any $t = 1, \cdots, l$, we consider the Algorithm 2 in

$$V^* = (\bigcup_{u \in S_{t-1}} V_{\varphi(u)}) \cup (\bigcup_{v \in \text{OPT}} V_{\varphi(v)}).$$

Since Algorithm 2 works for any order of the classes, we can assume that Algorithm 2 considers the classes in an order of $V_1, V_2, \cdots, V_{t-1}, \cdots$, i.e., the first $t-1$ classes coincide those considered by Algorithm 1, and for the remaining classes with indices in $\{\varphi(v) \mid v \in \text{OPT}\} \setminus \{\varphi(u) \mid u \in S_{t-1}\}$, Algorithm 2 selects them in an arbitrary class order.

Recall the notations of G_j and T_j, and note that $G_j = S_j$ for $j = 1, \cdots, t-1$, i.e., the partial greedy solutions in Algorithm 1 also coincide those in Algorithm 2.

Denote by T the optimal solution of the function f on V^*, then we apply Lemma 3 to bound $f(T)$. By Lemma 3 and greedy rule, we have

$$\begin{aligned} f(\text{OPT}) \leq f(T) &\leq (1+\lambda) f(S_{t-1}) + \sum_{u \in T_{t-1} \setminus S_{t-1}} f(u \mid S_{t-1}) \\ &= (1+\lambda) f(S_{t-1}) + \sum_{u \in T_{t-1} \setminus S_{t-1}} (c(u) \cdot \frac{f(u \mid S_{t-1})}{c(u)}) \\ &\leq (1+\lambda) f(S_{t-1}) + \sum_{u \in T_{t-1} \setminus S_{t-1}} (c(u) \cdot \frac{f(u_t \mid S_{t-1})}{c(u_t)}) \\ &\leq (1+\lambda) f(S_{t-1}) + b \cdot \frac{f(u_t \mid S_{t-1})}{c(u_t)}, \end{aligned}$$

where the last inequality holds since $c(T_{t-1} \setminus S_{t-1}) \leq c(T) \leq b$. This completes the proof. □

Lemma 5. *[18] If P and D are arbitrary positive integers, $(\rho_i)_{i=1}^{P}$ are arbitrary nonnegative reals with $\rho_1 > 0$, then*

$$\frac{\sum_{i=1}^{P} \rho_i}{\min_{t=1,\cdots,P}(\sum_{i=1}^{t-1} \rho_i + D\rho_t)} \geq 1 - (1 - \frac{1}{D})^P \geq 1 - e^{-\frac{P}{D}}.$$

This is a valuable conclusion, and combining it with Lemma 4 yields lower bounds on the value of the intermediate greedy solution in the next lemma.

Lemma 6. *For any $t \in [l]$, we have*

$$f(S_t) \geq \frac{1}{1+\lambda}(1 - e^{-\frac{(1+\lambda)c(S_t)}{b}}) \cdot f(OPT),\ \lambda \geq 1.$$

Proof. Let $b_t = c(S_t) = \sum_{\tau=1}^{t} c(u_\tau)$ for $t \in [l]$. We define $\theta_j := \frac{f(u_j|S_{j-1})}{c(u_j)}$ as the marginal density in the j-th step for $j \in [l]$. For $i = 1, \cdots, b_l$, we define

$$\rho_i = \theta_j,\ i = b_{j-1}+1, \cdots, b_j.$$

Specifically, it holds that

$$\begin{aligned}
&\rho_1 = \cdots = \rho_{b_1} = \theta_1,\\
&\rho_{b_1+1} = \cdots = \rho_{b_2} = \theta_2,\\
&\cdots\\
&\rho_{b_{l-1}+1} = \cdots = \rho_{b_l} = \theta_l.
\end{aligned}$$

Using this definition, we obtain $f(S_j) = \sum_{\tau=1}^{j} c(u_\tau)\theta_\tau = \sum_{i=1}^{b_j} \rho_i$. Then, we have

$$\begin{aligned}
&\min_{s=1,\cdots,b_t}\{\sum_{i=1}^{s-1}\rho_i + \frac{b}{1+\lambda}\cdot\rho_s\}\\
&= \min_{j=0,\cdots,t-1}\{\sum_{i=1}^{b_j}\rho_i + \frac{b}{1+\lambda}\cdot\rho_{b_j+1}\}\\
&= \min_{j=0,\cdots,t-1}\{f(S_j) + \frac{b}{1+\lambda}\cdot\theta_{j+1}\}.
\end{aligned}$$

By Lemma 4, we have

$$f(\mathrm{OPT}) \leq \min_{j=0,\cdots,t-1}\{(1+\lambda)f(S_j) + b\cdot\frac{f(u_{j+1} \mid S_j)}{c(u_{j+1})}\}.$$

Therefore,

$$\begin{aligned}
\frac{f(S_t)}{f(\mathrm{OPT})} &= \frac{\sum_{i=1}^{b_t}\rho_i}{f(\mathrm{OPT})}\\
&\geq \frac{\sum_{i=1}^{b_t}\rho_i}{\min_{j=0,\cdots,t-1}\{(1+\lambda)f(S_j) + b\cdot\frac{f(u_{j+1}|S_j)}{c(u_{j+1})}\}}\\
&= \frac{\sum_{i=1}^{b_t}\rho_i}{(1+\lambda)\cdot\min_{j=0,\cdots,t-1}\{f(S_j) + \frac{b}{1+\lambda}\cdot\theta_{j+1}\}}\\
&= \frac{\sum_{i=1}^{b_t}\rho_i}{(1+\lambda)\cdot\min_{s=1,\cdots,b_t}\{\sum_{i=1}^{s-1}\rho_i + \frac{b}{1+\lambda}\cdot\rho_s\}}\\
&\geq \frac{1}{1+\lambda}(1 - e^{-\frac{(1+\lambda)b_t}{b}})\\
&= \frac{1}{1+\lambda}(1 - e^{-\frac{(1+\lambda)c(S_t)}{b}}),
\end{aligned}$$

where the last inequality holds by Lemma 5. Rearranging it immediately completes the proof. □

While the above lemma presents lower bounds for the value of the intermediate greedy solution, the next lemma focuses on the lower bounds of $f(S_g)$.

Lemma 7. *Given any element set T with $|T \cap V_i| \leq 1$, $i \in [k]$, the greedy heuristic returns S_g subject to a budget constraint b satisfying*

$$f(S_g) \geq \frac{b - c(T)}{(1+\lambda)b - \lambda \cdot c(T)} \cdot f(T),\ \lambda \geq 1.$$

Proof. The lemma is trivial when $T \subseteq S_g$. Suppose $T \setminus S_g \neq \varnothing$. Let $S_l = \{u_1, u_2, \cdots, u_l\}$ be the set of elements constructed by greedy heuristic when the first element from T is considered but not added to S_l due to budget violation. We construct a set T' satisfying $T \subseteq T'$ and T' includes at least one representative from each equivalence class occurring in S_l.

Let T_l be the set obtained by replacing the first l elements of T' with S_l, i.e., $T_l := (T' \setminus \{u_1^*, \cdots, u_l^*\}) \cup S_l$. Due to submodularity of f and the greedy rule, we have

$$\begin{aligned}&\frac{f(u_1 \mid S_0)}{c(u_1)} \geq \cdots \geq \frac{f(u_l \mid S_{l-1})}{c(u_l)}\\ &\geq \max_{v \in T_l \setminus S_l} \frac{f(v \mid S_l)}{c(v)} \geq \frac{\sum_{v \in T_l \setminus S_l} f(v \mid S_l)}{c(T_l \setminus S_l)} \geq \frac{f((T_l \setminus S_l) \mid S_l)}{c(T_l \setminus S_l)}.\end{aligned}$$

The last inequality holds since Lemma 2. Then, by the proof process of Lemma 3 and $T \subseteq T'$, we have

$$\begin{aligned}f(T) \leq f(T') &\leq \lambda f(S_l) + f(T_l)\\ &= (1+\lambda) f(S_l) + f((T_l \setminus S_l) \mid S_l).\end{aligned}$$

Meanwhile, we get

$$\begin{aligned}f(S_l) &= \sum_{i=1}^{l} f(u_i \mid S_{i-1})\\ &= \sum_{i=1}^{l} (c(u_i) \cdot \frac{f(u_i \mid S_{i-1})}{c(u_i)})\\ &\geq \sum_{i=1}^{l} (c(u_i) \cdot \frac{f((T_l \setminus S_l) \mid S_l)}{c(T_l \setminus S_l)})\\ &= c(S_l) \cdot \frac{f((T_l \setminus S_l) \mid S_l)}{c(T_l \setminus S_l)}.\end{aligned}$$

By the algorithm definition, note that

$$c(S_l) + c(T_l \setminus S_l) > b,$$
$$\text{and } c(T_l \setminus S_l) \leq c(T).$$

Therefore,

$$\begin{aligned} f(T) &\leq (1+\lambda)f(S_l) + f((T_l \setminus S_l) \mid S_l) \\ &\leq (1+\lambda)f(S_l) + f(S_l)\cdot \frac{c(T_l \setminus S_l)}{c(S_l)} \\ &< (1+\lambda+\frac{c(T_l \setminus S_l)}{b-c(T_l \setminus S_l)})\cdot f(S_l) \\ &\leq (1+\lambda+\frac{c(T)}{b-c(T)})\cdot f(S_l) \\ &\leq \frac{(1+\lambda)b-\lambda\cdot c(T)}{b-c(T)}\cdot f(S_g). \end{aligned}$$

Equivalently, we have

$$f(S_g) \geq \frac{b-c(T)}{(1+\lambda)b-\lambda\cdot c(T)}\cdot f(T).$$

It completes the proof. □

Based on Lemma 7, a relation between the greedy solution S_g and the optimal solution OPT can be derived as follows.

Lemma 8. *Let* $OPT' := OPT \setminus (Q \cup \{o\})$*, and* $c(Q) \leq \frac{b}{2}$*. Then*

$$f(S_g) \geq f(Q) + \frac{b-2c(Q)}{(1+\lambda)b-(1+2\lambda)c(Q)}\cdot f(OPT' \mid Q), \quad \lambda \geq 1.$$

Proof. According to Lemma 1, we know that $f(S \mid Q)$ is a monotone nondecreasing submodular set function with respect to S. Then, by Lemma 7, we have

$$\begin{aligned} f(S_g) &= f(Q) + f((S_g \setminus Q) \mid Q) \\ &\geq f(Q) + \frac{b-c(Q)-c(\mathrm{OPT}')}{(1+\lambda)(b-c(Q))-\lambda\cdot c(\mathrm{OPT}')}\cdot f(\mathrm{OPT}' \mid Q). \end{aligned}$$

Furthermore, by the algorithm definition, we know that

$$\begin{aligned} &c(Q)+c(o) > b \\ \text{and}\ \ &c(\mathrm{OPT}')+c(o) \leq c(\mathrm{OPT}) \leq b. \end{aligned}$$

Thus, $c(\mathrm{OPT}') < c(Q)$. Putting it together, we have

$$\begin{aligned} f(S_g) &\geq f(Q) + \frac{b-c(Q)-c(\mathrm{OPT}')}{(1+\lambda)(b-c(Q))-\lambda\cdot c(\mathrm{OPT}')}\cdot f(\mathrm{OPT}' \mid Q) \\ &\geq f(Q) + \frac{(b-c(Q))-c(Q)}{(1+\lambda)(b-c(Q))-\lambda\cdot c(Q)}\cdot f(\mathrm{OPT}' \mid Q) \\ &= f(Q) + \frac{b-2c(Q)}{(1+\lambda)b-(1+2\lambda)c(Q)}\cdot f(\mathrm{OPT}' \mid Q), \end{aligned}$$

where the last inequality holds since the function $t(x) = \frac{(b-c(Q))-x}{(1+\lambda)(b-c(Q))-\lambda x}$ is monotonically decreasing. Thus, the proof is complete. □

In the following, we can derive a lower bound on the worst-case approximation of the Algorithm 1 by solving the following optimization problem.

Lemma 9. *We have* $f(S_q) \geq \alpha^* \cdot f(OPT)$*, where* α^* *is the minimum of the following optimization problem with respect to* α, x_1, x_2, x_3*.*

$$\begin{aligned} \min \quad & \alpha \\ \text{s.t.} \quad & \alpha \geq x_1, & \text{(2a)} \\ & \alpha \geq 1 - x_1 - x_3, & \text{(2b)} \\ & \alpha \geq x_1 + x_3 \cdot \frac{1 - 2x_2}{(1+\lambda) - (1+2\lambda)x_2}, & \text{(2c)} \\ & x_1 \geq \frac{1}{\lambda+1}(1 - e^{-(1+\lambda)x_2}), & \text{(2d)} \\ & \alpha, x_1, x_2, x_3 \in [0,1]. & \text{(2e)} \end{aligned}$$

Proof. To simplify the notations, we define $f_Q = \frac{f(Q)}{f(\text{OPT})}$, $f_o = \frac{f(o|Q)}{f(\text{OPT})}$, $c_Q = \frac{c(Q)}{b}$, $f' = \frac{f(\text{OPT}'|Q)}{f(\text{OPT})}$ and $\alpha_q = \frac{f(S_q)}{f(\text{OPT})}$. We show that $\alpha = \alpha_q$, $x_1 = f_Q$, $x_2 = c_Q$, and $x_3 = f'$ are always feasible to the optimization problem defined in the lemma, which indicates that $f(S_q) \geq \alpha^* \cdot f(\text{OPT})$. By the algorithm definition,

$$\alpha_q \geq \frac{f(S_g)}{f(\text{OPT})} \geq f_Q. \qquad \text{(Constraint (2a))}$$

Due to monotonicity, submodularity of f and the greedy rule,

$$\begin{aligned} f(\text{OPT}) &\leq f(Q \cup \{o\} \cup \text{OPT}') \\ &= f(Q) + f(o \mid Q) + f(\text{OPT}' \mid Q \cup \{o\}) \\ &\leq f(Q) + f(o \mid Q) + f(\text{OPT}' \mid Q). \end{aligned}$$

Hence, we have

$$f_Q + f_o + f' \geq 1.$$

Then by the algorithm definition, we get

$$\alpha_q \geq \frac{f(u^*)}{f(\text{OPT})} \geq f_o.$$

Putting it together yields

$$\alpha_q \geq f_o \geq 1 - f_Q - f'. \qquad \text{(Constraint (2b))}$$

By Lemma 8, we have

$$\alpha_q \geq f_Q + \frac{1 - 2c_Q}{(1+\lambda) - (1+2\lambda)c_Q} \cdot f'. \qquad \text{(Constraint (2c))}$$

By Lemma 6, let $S_t = Q$, then $t = |Q| = l$, thus we have

$$f(Q) \geq \frac{1}{1+\lambda}(1 - e^{-\frac{(1+\lambda)c(Q)}{b}}) \cdot f(\text{OPT}).$$

Namely,

$$f_Q \geq \frac{1}{1+\lambda}(1 - e^{-(1+\lambda)c_Q}). \qquad \text{(Constraint (2d))}$$

Finally, by definitions of α, x_1, x_2, x_3,

$$\alpha, x_1, x_2, x_3 \in [0,1]. \qquad \text{(Constraint (2e))}$$

As can be seen, all constraints are satisfied, and hence the proof is done. □

Lemma 10. *We have*

$$\alpha^* \geq \min\{\frac{1}{1+\lambda}(1 - e^{-\frac{1+\lambda}{2}}), \frac{1}{2+\lambda}\},$$

where $\lambda \geq 1$ is the maximum imbalance coefficient in V.

Proof of Theorem 1. The approximation factor of the Algorithm 1 is $\frac{f(S_q)}{f(\text{OPT})}$. Then combining Lemmas 9 and 10, we have

$$\frac{f(S_q)}{f(\text{OPT})} \geq \alpha^* \geq \min\{\frac{1}{1+\lambda}(1 - e^{-\frac{1+\lambda}{2}}), \frac{1}{2+\lambda}\}.$$

In particular, when all elements within each class have identical cost values, i.e., $\lambda = 1$, we can get

$$\begin{aligned}\frac{f(S_q)}{f(\text{OPT})} &\geq \alpha^* \\ &\geq \min\{\frac{1}{2}(1 - e^{-1}), \frac{1}{3}\} \\ &= \frac{1}{2}(1 - e^{-1}) \approx 0.316.\end{aligned}$$

This establishes the theorem. □

References

1. Ageev, A.A., Sviridenko, M.I.: An 0.828-approximation algorithm for the uncapacitated facility location problem. Discret. Appl. Math. **93**(2-3), 149–156 (1999)
2. Delong, A., Veksler, O., Osokin, A., Boykov, Y.: Minimizing sparse high-order energies by submodular vertex-cover. In: Advances in Neural Information Processing System, vol. 25 (2012)
3. Feige, U.: A threshold of $\ln n$ for approximating set cover. J. ACM (JACM) **45**(4), 634–652 (1998)
4. Feldman, M., Nutov, Z., Shoham, E.: Practical budgeted submodular maximization. Algorithmica **85**(5), 1332–1371 (2023)

5. Fisher, M.L., Nemhauser, G.L., Wolsey, L.A.: An Analysis of Approximations for Maximizing Submodular Set Functions—II. Springer, Cham (1978)
6. Kazemi, E., et al.: Data summarization with privacy and fairness constraints. In: Proceedings of the 35th International Conference on Machine Learning, pp. 1–10 (2018)
7. Khuller, S., Moss, A., Naor, J.S.: The budgeted maximum coverage problem. Inf. Process. Lett. **70**(1), 39–45 (1999)
8. Ko, C.W., Lee, J., Queyranne, M.: An exact algorithm for maximum entropy sampling. Oper. Res. **43**(4), 684–691 (1995)
9. Krause, A., Leskovec, J., Guestrin, C., VanBriesen, J., Faloutsos, C.: Efficient sensor placement optimization for securing large water distribution networks. J. Water Resour. Plan. Manag. **134**(6), 516–526 (2008)
10. Mitrovic, M., Kazemi, E., Zadimoghaddam, M., Karbasi, A.: Data summarization at scale: a two-stage submodular approach. In: International Conference on Machine Learning, pp. 3596–3605. PMLR (2018)
11. Nemhauser, G.L., Wolsey, L.A., Fisher, M.L.: An analysis of approximations for maximizing submodular set functions–I. Math. Program. **14**, 265–294 (1978)
12. Sarpatwar, K.K., Schieber, B., Shachnai, H.: Constrained submodular maximization via greedy local search. Oper. Res. Lett. **47**(1), 1–6 (2019)
13. Shewry, M.C., Wynn, H.P.: Maximum entropy sampling. J. Appl. Stat. **14**(2), 165–170 (1987)
14. Son, K., Kim, H., Yi, Y., Krishnamachari, B.: Base station operation and user association mechanisms for energy-delay tradeoffs in green cellular networks. IEEE J. Sel. Areas Commun. **29**(8), 1525–1536 (2011)
15. Sviridenko, M.: A note on maximizing a submodular set function subject to a knapsack constraint. Oper. Res. Lett. **32**(1), 41–43 (2004)
16. Tang, J., Tang, X., Lim, A., Han, K., Li, C., Yuan, J.: Revisiting modified greedy algorithm for monotone submodular maximization with a knapsack constraint. Proc. ACM Measur. Anal. Comput. Syst. **5**(1), 1–22 (2021)
17. Wei, K., Iyer, R., Bilmes, J.: Submodularity in data subset selection and active learning. In: International Conference on Machine Learning, pp. 1954–1963. PMLR (2015)
18. Wolsey, L.A.: Maximising real-valued submodular functions: primal and dual heuristics for location problems. Math. Oper. Res. **7**(3), 410–425 (1982)

Random Greedy Deployment of Heterogeneous UAVs

Yang Lv[1], Fengmin Wang[2], Xiankun Yu[1], Xin Li[3], Dachuan Xu[1], and Ruiqi Yang[1](✉)

[1] Beijing Institute for Scientific and Engineering Computing, Beijing University of Technology, Beijing 100124, China
{lvyang,22062213}@emails.bjut.edu.cn, {xudc,yangruiqi}@bjut.edu.cn
[2] Jinghang Research Institute of Computing and Communication, Beijing 100074, China
[3] College of Economics and Management, Beijing University of Technology, Beijing 100124, China
lixinyz@bjut.edu.cn

Abstract. Unmanned Aerial Vehicles (UAVs) have been extensively utilized in environmental modeling tasks, including forest monitoring, agricultural mapping, and disaster assessment. These applications benefit from UAVs' ability to collect high-resolution data for purposes such as deforestation detection, crop management, and damage evaluation. However, the common assumption that UAVs are homogeneous leads to significant inaccuracies due to differences in UAV platforms and sensor characteristics. This paper investigates the deployment of heterogeneous UAVs equipped with various sensors in fire-affected regions, modeled via a point of interest framework. The goal is to minimize measurement errors by explicitly leveraging UAV heterogeneity. We introduce a weighted frame potential metric based on matrix orthogonality, and develop a randomized greedy algorithm to optimize UAV positioning. The proposed algorithm guarantees at least a 50% approximation of the optimal solution and offers a theoretical error bound on the metric. Experimental results demonstrate that the proposed method outperforms joint greedy and determinant-based algorithms in reducing measurement errors.

Keywords: UAV measurement · Approximation algorithm · Weighted frame potential · Submodular optimization

1 Introduction

UAVs are valuable tools for environmental management due to their mobility, adaptability, and sensor-carrying capabilities [20]. In environmental monitoring, UAV-mounted sensors are deployed to collect diverse data in various scenarios. However, key challenges remain, such as determining optimal UAV positions to improve detection accuracy.

M. Li et al. (Eds.): TAMC 2025, LNCS 16084, pp. 330–341, 2026.
https://doi.org/10.1007/978-981-95-4839-2_26

To better understand the deployment problem in UAV-based sensing, we first review related work in traditional sensor networks. In traditional sensor networks [1–17], sensors are typically considered homogeneous and independent, especially in linear observation models that solve linear inverse problems for reconstructing physical fields. A substantial amount of research focuses on selecting k points from N candidate locations to maximize information gain and minimize errors. This forms a combinatorial optimization problem, which has led to the development of numerous heuristic and approximation algorithms. Most studies aim to design various metrics to assess and minimize computational error, and then develop optimization algorithms that exploit the properties of these metrics. While significant progress is made in coupling metrics with algorithms [3], research on heterogeneous UAV measurements remains limited, with most studies addressing homogeneous sensor issues. UAV heterogeneity stems from differences in sensor models, often caused by variations in procurement timing or manufacturing inconsistencies. These variations lead to distinct sensor errors, and treating all UAVs as homogeneous results in a model that poorly reflects real-world conditions. Effectively coordinating these heterogeneous UAVs for accurate detection remains a challenging task [21].

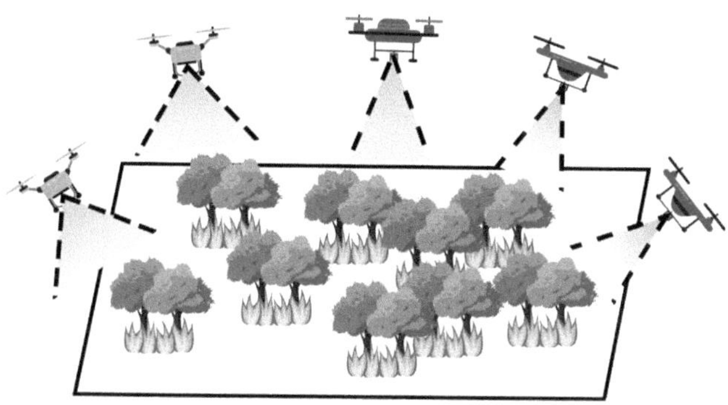

Fig. 1. Heterogeneous UAV Measurement Challenge: This study addresses a scenario involving multiple distinct UAV types. In the figure, the fleet consists of two yellow UAVs, two blue UAVs and one red UAV, each type exhibiting different accuracy levels. The objective is to ascertain the optimal deployment of these UAVs to reduce detection errors as much as possible in monitoring forest fires. (Color figure online)

We investigate how UAVs equipped with sensors of varying accuracies can collaboratively measure fire-affected areas and reconstruct the fire scene to identify hazardous zones. This approach is also applicable to real-time forest fire warnings. We assume a set of predefined points of interest, where UAVs hover to collect measurements. The collected data is transmitted to a central system, which reconstructs the overall fire scenario, as illustrated in Fig. 1. Each UAV group is subject to Gaussian noise with a distinct variance. The goal is to determine an optimal UAV deployment strategy that solves a linear inverse problem while minimizing reconstruction error. Due to the challenges introduced by sensor heterogeneity, new metrics are needed. Classical criteria such as mutual information or A-optimality fail to explicitly capture sensor heterogeneity. To address this issue, we propose the *weighted frame potential (WFP)*, which fully accounts for the accuracy differences among heterogeneous sensors. The proposed metric is submodular, enabling the use of a random greedy algorithm. The submodularity of WFP ensures that the random greedy algorithm can provide provable approximation guarantees for this otherwise intractable selection problem, while maintaining high computational efficiency.

The main contributions of this paper are summarized as follows:

- We address the problem of environmental modeling using heterogeneous UAVs, particularly in scenarios involving multiple UAV groups.
- Despite differences in UAV models and factory configurations, we show that effective collaborative detection can still be achieved.
- We introduce the weighted frame potential metric and apply a random greedy algorithm, which guarantees a solution that is at least 1/2 of the optimal. Our method provides a provable performance bound.
- We further provide an upper bound on the solution quality based on the weighted frame potential.

2 Problem Statement

2.1 UAVs Model

A forest fire has occurred, requiring heterogeneous UAVs to investigate the affected area. The fire is assumed to be located on a two-dimensional plane. On this plane, there are N candidate locations that require observation, denoted by the set $\mathcal{N} = \{1, 2, \ldots, N\}$. Each candidate point is more suitable for specific types of UAVs. Thus, the set $\mathcal{N}$ can be partitioned into subsets $S_1, S_2, \ldots, S_m \subseteq \mathcal{N}$, where S_i represents the points assigned to UAVs of type i, for $i = 1, 2, \ldots, m$. The objective is to select appropriate locations from the candidate set to deploy the UAVs, such that the estimation of the fire scene is as accurate as possible.

Since the number of UAVs available for each type is fixed, any feasible deployment $\mathcal{L} \subseteq \mathcal{N}$ must satisfy the UAV-type constraints. Specifically, if M_i UAVs of type i are available, then the solution must satisfy $|\mathcal{L} \cap S_i| = M_i$. The total number of deployed UAVs is given by $|\mathcal{L}| = \sum_{i=1}^{m} M_i$. In other words, all UAVs must be assigned to their designated subsets, and their total usage matches the available inventory.

2.2 UAVs Placement Problem

Suppose the goal is to reconstruct the temperature at K selected points. The corresponding true values are denoted by $\boldsymbol{\alpha} \in \mathbb{R}^K$. At each PoI, the temperatures at the K detection points will influence the measurement at this PoI. In addition, each PoI measurement is corrupted by Gaussian noise associated with the UAV's sensor. Following the sensing model in [9], the aggregated measurements across all candidate points can be expressed as

$$\boldsymbol{f} = \boldsymbol{\Psi}\boldsymbol{\alpha} + \boldsymbol{\eta}, \tag{1}$$

The differences in each row of $\boldsymbol{\Psi} \in \mathcal{R}^{N\times K}$ reflect variations in the spatial distribution of UAVs, where more similar vectors indicate closer distances. The differences in each row of $\boldsymbol{\Psi}$ reflect variations in the spatial distribution of UAVs. $\boldsymbol{\eta}$ represents additive measurement noise. The errors across all UAVs stem from an N-dimensional vector $\boldsymbol{\eta} \in \mathbb{R}^N$, where $\boldsymbol{\eta}$ is additive zero-mean white measurement noise with a variance matrix $\boldsymbol{C}$.

The covariance matrix $\boldsymbol{C}$ is a diagonal matrix $\boldsymbol{C} = \mathbf{diag}(\sigma_1^2, \sigma_2^2, ..., \sigma_N^2)$, where UAVs of the same type have the same variance. Furthermore, outstanding UAVs are characterized by smaller variances. For all UAVs belonging to the j-th type, σ_i^2 remains consistent for any $i \in S_j$.

If the set of observation positions $\mathcal{L}$ for the UAV is determined, the vector obtained from the measurements is denoted by $\boldsymbol{f}_{\mathcal{L}} \in \mathbb{R}^{|\mathcal{L}|}$, where $\boldsymbol{f}_{\mathcal{L}}$ represents the sub-vector of $\boldsymbol{f}$ indexed by $\mathcal{L}$. Simultaneously, $\boldsymbol{\eta}_{\mathcal{L}}$ signifies the sub-vector of $\boldsymbol{\eta}$ indexed by $\mathcal{L}$. $\boldsymbol{\Psi}_{\mathcal{L}} \in \mathbb{R}^{|\mathcal{L}|\times K}$ corresponds to the sub-matrix of $\boldsymbol{\Psi}$, formed by the rows indexed by $\mathcal{L}$. The resultant sub-formula from observations is given by:

$$\boldsymbol{f}_{\mathcal{L}} = \boldsymbol{\Psi}_{\mathcal{L}}\boldsymbol{\alpha} + \boldsymbol{\eta}_{\mathcal{L}} \tag{2}$$

Because the number of UAVs is limited and cannot cover all selected points, the size of the measurement results from $\boldsymbol{f}_{\mathcal{L}}$ is smaller than that of $\boldsymbol{f}$. This reduction in scale decreases the precision in describing the physical scene $\boldsymbol{\alpha}$. Therefore, the central challenge of this paper is to effectively utilize the limited heterogeneous UAVs to enhance the description of $\boldsymbol{\alpha}$.

2.3 Weighted Frame Potential (WFP)

Reconstructing the temperature typically requires different criteria to evaluate the effectiveness of the references. Previous works have proposed various criteria such as MSE [22], VCE [13], and WCEV [14] to assess performance. Benedetto et al. [15] introduce an approach centered around the concept of frame potential. Building on this foundation, subsequent research extended the frame potential metric to incorporate heterogeneous metrics, leading to the development of the weighted frame potential. Let the chosen set be denoted as $\mathcal{L}$. Define $\boldsymbol{\Psi}_{\mathcal{L}} = \{\boldsymbol{\psi}_1, \boldsymbol{\psi}_2, \ldots, \boldsymbol{\psi}_L\}$, where $\boldsymbol{\psi}_i$ represents the i-th row vector of $\boldsymbol{\psi}_{\mathcal{L}}$. The weighting coefficients are given by $\omega_i = \sigma_i$.

$$WFP(\boldsymbol{\Psi}_{\mathcal{L}}) = \sum_{i,j\in\mathcal{L}} \omega_i\omega_j \frac{\langle \boldsymbol{\psi}_i, \boldsymbol{\psi}_j \rangle^2}{\|\boldsymbol{\psi}_i\|_2^2 \cdot \|\boldsymbol{\psi}_j\|_2^2}. \tag{3}$$

Since the primary objective is to minimize the weighted frame potential, it becomes evident that larger variances in UAV measurement errors lead to poorer evaluations of the WFP's performance metric, so the logical strategy is to increase the corresponding ω values. Research [6] findings indicate that in scenarios involving two groups of UAVs, $\omega_i = \frac{1}{1+\exp(-(\sigma_i-\overline{\sigma}))}$ performs better, wherein $\overline{\sigma}$ symbolizes the average error across UAVs. In scenarios involving multiple groups of UAVs, this paper discovers that setting $\omega_i = \sigma_i$ leads to favorable performance. In other words, selecting ω_i as the standard deviation corresponding to index i yields superior results.

Minimizing the WFP presents significant challenges in algorithm design. To address this, a monotone, normalized, and submodular function f is constructed. The core principle behind constructing f is to identify the complement of the solution set and exclude row vectors that minimally decrease the value of WFP. These positions are deemed unsuitable for UAV placement. Through this construction, the function f becomes

$$f(\mathcal{L}) = WFP(\boldsymbol{\Psi}_{\mathcal{N}}) - WFP(\boldsymbol{\Psi}_{\mathcal{N}\setminus\mathcal{L}}). \tag{4}$$

We start by proving the normalization of function f. When $\mathcal{L} = \emptyset$, the following holds:

$$\begin{aligned} f(\mathcal{L}) &= WFP(\boldsymbol{\Psi}_{\mathcal{N}}) - WFP(\boldsymbol{\Psi}_{\mathcal{N}\setminus\emptyset}) \\ &= WFP(\boldsymbol{\Psi}_{\mathcal{N}}) - WFP(\boldsymbol{\Psi}_{\mathcal{N}}) \\ &= 0. \end{aligned} \tag{5}$$

This confirms that when the set $\mathcal{L}$ is empty, $f(\mathcal{L})$ evaluates to 0, thereby satisfying the normalization property.

Based on the definition of WFP in Equation (3), observing that the summation terms of WFP are positive and the number of terms increases as $\mathcal{L}$ grows, it can be deduced that WFP is a monotonically increasing function. Therefore, the complement of the set denoted as $WFP(\boldsymbol{\Psi}_{\mathcal{N}\setminus\mathcal{L}})$, is a monotonically decreasing function. Consequently, the expression $WFP(\boldsymbol{\Psi}_{\mathcal{N}}) - WFP(\boldsymbol{\Psi}_{\mathcal{N}\setminus\mathcal{L}})$ represents a monotonically increasing function.

Next, we establish the submodularity property for any two sets, $\mathcal{L}_1$ and $\mathcal{L}_2$, within $\mathcal{N}$, where $\mathcal{L}_1 \subseteq \mathcal{L}_2 \subseteq \mathcal{N}$. For any element $e \notin \mathcal{L}_2$, we aim to prove the submodularity condition.

$$\begin{aligned} & f(\mathcal{L}_1 \cup \{e\}) - f(\mathcal{L}_1) \\ =& \left(WFP(\boldsymbol{\Psi}_{\mathcal{N}}) - WFP(\boldsymbol{\Psi}_{\mathcal{N}\setminus(\mathcal{L}_1\cup\{e\})})\right) - \left(WFP(\boldsymbol{\Psi}_{\mathcal{N}}) - WFP(\boldsymbol{\Psi}_{\mathcal{N}\setminus\mathcal{L}_1})\right) \\ =& WFP(\boldsymbol{\Psi}_{\mathcal{N}\setminus\mathcal{L}_1}) - WFP(\boldsymbol{\Psi}_{\mathcal{N}\setminus(\mathcal{L}_1\cup\{e\})}) \\ =& \sum_{i\in\mathcal{N}\setminus(\mathcal{L}_1\cup\{e\})} \omega_i\omega_e \frac{\langle \boldsymbol{\psi}_i, \boldsymbol{\psi}_e\rangle^2}{\|\boldsymbol{\psi}_i\|_2^2 \cdot \|\boldsymbol{\psi}_e\|_2^2} \\ &+ \sum_{j\in\mathcal{N}\setminus(\mathcal{L}_1\cup\{e\})} \omega_e\omega_j \frac{\langle \boldsymbol{\psi}_e, \boldsymbol{\psi}_j\rangle^2}{\|\boldsymbol{\psi}_e\|_2^2 \cdot \|\boldsymbol{\psi}_j\|_2^2} + \omega_e\omega_e \frac{\langle \boldsymbol{\psi}_e, \boldsymbol{\psi}_e\rangle^2}{\|\boldsymbol{\psi}_e\|_2^2 \cdot \|\boldsymbol{\psi}_e\|_2^2} \end{aligned} \tag{6}$$

Since $\mathcal{L}_1$ is contained within $\mathcal{L}_2$, the number of terms in the given expression reduces. We can establish the submodularity property by demonstrating that: $f(\mathcal{L}_1 \cup \{e\}) - f(\mathcal{L}_1) \geq f(\mathcal{L}_2 \cup \{e\}) - f(\mathcal{L}_2)$.

By taking these steps, we have demonstrated that f is normalized, monotone, and submodular. Based on these properties, we can present an approximation algorithm for f.

3 Algorithm and Theoretical Foundation

In this section, the heterogeneous UAV placement problem is tackled by solving a submodular maximization problem. The algorithm employed here is the random greedy algorithm, which offers a straightforward way to demonstrate the theoretical guarantees of this approach.

3.1 Algorithm

f is a normalized, monotone submodular function. Given that the number of UAV of the i-th model is denoted as M_i, the constraints of the problem can be expressed as $|(\mathcal{N} \setminus \mathcal{L}) \cap S_i| = M_i$. Assuming $|S_i|$ represents the number of elements in S_i, the problem can be simplified as $|\mathcal{L} \cap S_i| = |S_i| - M_i$, where let us denote $M_i' = |S_i| - M_i$, $L' = \sum_{i=1}^{m} M_i'$ for simplicity. This can be framed as an optimization problem:

$$\begin{aligned} &\arg\max_{\mathcal{L} \subseteq \mathcal{N}} f(\mathcal{L}), \\ &s.t. |\mathcal{L} \cap S_i| = M_i' \quad \text{for } i \text{ from 1 to } m \end{aligned} \tag{7}$$

For such a problem, when $m = 1$, it becomes a classic submodular function maximization problem [19]. Utilizing a greedy algorithm, the problem can be solved to yield a value of the f function that is $1 - 1/e$ times the optimal solution. Next, the aim is to extend this algorithm to the scenario where m is an arbitrary constant.

Algorithm 1 The random greedy algorithm for submodular maximization problems

Require: submodular function $f : 2^{\mathcal{N}} \to \mathbb{R}$, ground set $S_1, S_2, ..., S_m$, restrictions $M_1', M_2', ..., M_m'$

Create a multi-element index list T containing M_1' instances of 1, M_2' instances of 2, $\ldots$, M_m' instances of m. {This list is used to determine the order of sensor placement in a greedy manner.}
Randomly permute this list
for each element j in the list, taken from the head of the list **do**
  $v_i = \arg\max_{v_i \in S_j} f(T \cup \{v_i\}) - f(T)$
  $T \leftarrow T \cup \{v_i\}$
end for
return T

A greedy algorithm is employed to address this issue. Although the subsequent approximation ratio analysis applies to all greedy algorithms, numerical experiments have shown that the random greedy algorithm yields the best performance. Therefore, we only introduce the random greedy algorithm here. The algorithm first randomly shuffles the selection order of M_1' elements from S_1, M_2' elements from S_2, and so on before execution. Each time, the UAV is placed at the position with the highest marginal gain among the corresponding elements.

3.2 Algorithm Approximation Guarantee

Given that f is a submodular function, the marginal gain of set A over set B according to the function f is defined as $f(A|B) = f(A \cup B) - f(B)$. If A is an element rather than a set, then $f(A|B) = f(A \cup B) - f(B)$. Let us establish the proof that for any chosen subset $V \subseteq \mathcal{N}$, the function $f(\cdot|V)$ exhibits submodularity.

$$\begin{aligned}
&f(A|V) + f(B|V) \\
=&f(A \cup V) - f(V) + f(B \cup V) - f(V) \\
=&(f(A \cup V) + f(B \cup V)) - 2f(V) \\
\geq&f(A \cup B \cup V) + f((A \cap B) \cup V) - 2f(V) \\
=&f(A \cup B|V) + f(A \cap B|V)
\end{aligned} \tag{8}$$

The inequality is based on the properties of the submodular function of f.

We define the optimal solution of the submodular function as OPT and introduces the definitions $OPT_1 = OPT \cap S_1$, $OPT_2 = OPT \cap S_2$, ..., $OPT_m = OPT \cap S_m$. Here, OPT_i represents the elements within S_i that are part of the optimal solution OPT. Let T_i denote the outcome of set T after completing the i-th round of iterations. Given that OPT is a valid set of solutions, we have $|OPT_i| = M_i'$ for all i from 1 to m.

We establish a collection of bijections from $T \cap S_j$ to OPT_j. If an element $e_j \in (T \cap S_j) \cap OPT_j$, we map e_j to itself. The remaining elements from $T \cap S_j$ lie in $(T \cap S_j) \setminus OPT_j$, and we randomly construct a bijection from $(T \cap S_j) \setminus OPT_j$ to $OPT_j \setminus (T \cap S_j)$. This way, a mapping is established from $v_i \in T \cap S_j$ to $o_i^j \in OPT_j$. Next, we show that $f(v_i|S_{i-1}) \geq f(o_i^j|S_{i-1})$. If v_i is in the set $(T \cap S_j) \cap OPT_j$, it is evident that v_i and o_i^j are the same element. T_0 is defined as the empty set. If v_i lies in the set $(T \cap S_j) \setminus OPT_j$, and in the i-th round of greedy selection, v_i is chosen while o_i^j is not, it implies $f(v_i|T_{i-1}) \geq f(o_i^j|T_{i-1})$.

Based on the above discussion, this inequality holds.

$$
\begin{aligned}
f(T) &= \sum_{j=1}^{m} \sum_{i \in A_j} f(v_i|T_{i-1}) \\
&\geq \sum_{j=1}^{m} \sum_{i \in A_j} f(o_i^j|T_{i-1}) \geq \sum_{j=1}^{m} \sum_{i \in A_j} f(o_i^j|T) \\
&\geq \sum_{j=1}^{m} f(OPT_j|T) \geq f\left(\bigcup_{j=1}^{m} OPT_j|T\right) \\
&= f(OPT|T) \geq f(OPT) - f(T)
\end{aligned}
\tag{9}
$$

The first inequality follows from the constructed bijections mentioned earlier. The second inequality is a consequence of the submodularity of f and the fact that $T_{i-1} \subseteq T$. The third and fourth inequalities stem from the submodularity of $f(\cdot|T)$. The fifth inequality holds due to the monotonicity property of f, where $OPT \subseteq (OPT \cup T)$. By simplifying (9), we arrive at the result $f(T) \geq \frac{1}{2} f(OPT)$.

Theorem 1. *For any heterogeneous UAV placement problem, assuming* $\overline{OPT} = \arg\min_{\mathcal{A} \subseteq \mathcal{N}, |\mathcal{A} \cap S_i| = M_i} WFP(\mathcal{A})$, *and considering the sub-optimal solution found using the greedy algorithm as* $\mathcal{N} \setminus T$, *the relationship between* $\boldsymbol{\Psi}_{\mathcal{N} \setminus T}$ *and* $\boldsymbol{\Psi}_{\overline{OPT}}$ *is as follows:*

$$
WFP(\boldsymbol{\Psi}_{\mathcal{N} \setminus T}) \leq \gamma WFP(\boldsymbol{\Psi}_{\overline{OPT}}) \tag{10}
$$

where $\gamma = \frac{1}{2}\left(1 + WFP(\boldsymbol{\Psi})\frac{K}{L_{min}^2}\right)$ *is the approximation factor and* $L_{min} = \min\limits_{|\mathcal{L}|=L} \sum_{i \in L} \sigma_i^2 \left\|\boldsymbol{\psi}_i\right\|^2$.

Proof. The detailed proof of this result is omitted here for brevity and will be included in an extended version.

This establishes the theoretical guarantee for the upper bound of the approximation algorithm's performance measured by the WFP metric. As the values of the WFP in $\boldsymbol{\Psi}_{\mathcal{N} \setminus \mathcal{L}}$ decrease, $\boldsymbol{\Psi}_{\mathcal{N} \setminus \mathcal{L}}$ approaches orthogonality, resulting in a smaller error.

4 Numerical Results

In this section, we analyze the performance of the algorithm proposed in the paper and compare it with algorithms from other studies through numerical experiments. This section focuses on a large-scale scenario to test the algorithms.

We compare our random greedy algorithm with three other algorithms as follows:

Random Greedy Algorithm (RGA): The WFP and random greedy algorithm are utilized to select points where UAVs cannot be placed, and then their complements are determined. The random greedy algorithm randomly selects a category at each step and removes an element from it. The final result must precisely satisfy the specified constraints.

Joint Greedy Algorithm (JGA) [6]**:** The algorithm identifies the largest potential point from all available potential points that do not exceed the constraints and then determines its complement for UAV placement. Similarly, it shares the same theoretical upper bound as RGA.

medskip**Determinant Algorithm (DA)** [11]**:** The calculation of UAVs with different accuracies is conducted in stages. In each step of the process, the algorithm identifies the value of j corresponding to the maximum marginal gain, which is given by $\log\det\left(\overline{\sigma}^{-2}\sum_{i\in S}\boldsymbol{\psi}_i\tilde{\boldsymbol{\psi}}_i+\boldsymbol{\psi}_j\tilde{\boldsymbol{\psi}}_j+\boldsymbol{\Sigma}^{-1}\right)$. This algorithm is solely dependent on $\overline{\sigma}$ and does not take into account the heterogeneity among UAVs.

Random Algorithm (RA): Potential points that meet the problem requirements are selected from each category of UAV locations for UAV placement.

In large-scale fires, dispatching UAVs for disaster site observation is of utmost importance. The selected area for fire observation is 1km × 2km, with the average temperature ranging from 30 to 400 °C. The fire data are obtained via interpolation of regression models based on a set of points. From Fig. 2, the distribution of the fire is quite distinct, with two prominent high-temperature regions. The algorithm should strive to depict these two hotspots as accurately as possible.

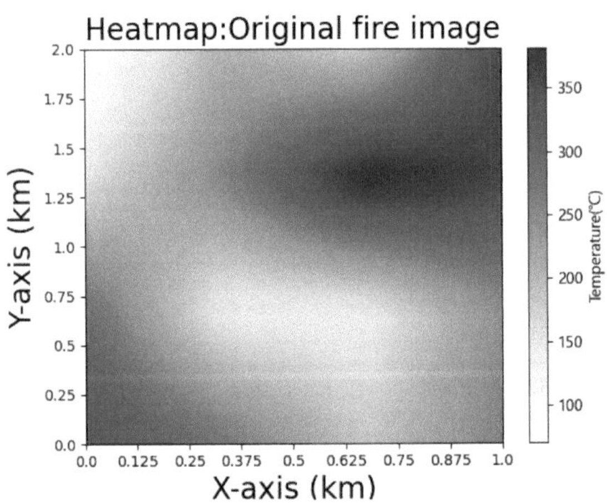

Fig. 2. A simulated fire scene heatmap measuring 2km × 1km.

In this scenario, two types of UAVs are deployed: one type comprises 20 UAVs with 80 potential locations, and the other type consists of 55 UAVs with 200 potential locations. The UAVs have variance values of 2 and 6 respectively. To reconstruct the fire disaster scene, the UAVs primarily observe the temperatures at 75 specific points on the fire site. The overall fire disaster scene is approximated using interpolation techniques. The matrix $\boldsymbol{\Psi}$ is designed as a 280×75 matrix, which serves as a sub-matrix of a 280-dimensional orthogonal matrix. For the algorithmic approach, only the RGA, JGA, DA and RA from this study are considered.

The algorithmic results reveal that the execution times for RGA, JGA, DA, and RA are 0.007 s, 0.007 s, 34.6 s, and 0.0 s, respectively. It is noticeable that RGA and JGA continue to maintain a rapid execution speed, while DA experiences a sharp increase in runtime due to the high time complexity associated with determinant calculations. Therefore, DA is not a favorable choice for large-scale scenarios, as its performance significantly degrades in such cases.

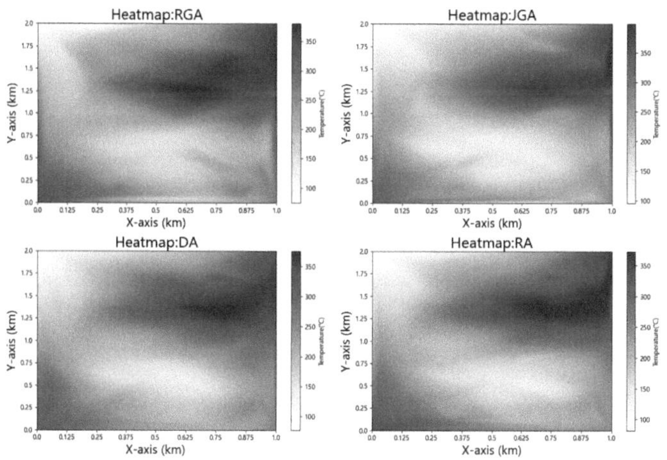

Fig. 3. Interpolated fire scene plots provided by the four algorithms: top-left for RGA, top-right for JGA, bottom-left for DA, and bottom-right for RA.

The measurement results of the algorithms are fitted using interpolation to obtain Fig. 3. Intuitively, it is evident that the fitting outcome of the RGA image in the top-left corner is the most favorable, clearly displaying the two hotspots on the image while achieving excellent results for other intermediate areas. The poor fitting performance of the other images can be attributed to inaccurate measurement results, which negatively impact the quality of the fit. Specific numerical values, as depicted in Fig. 4, reveal that RGA performs the best in terms of both MSE and average temperature difference in larger-scale data scenarios. The average temperature difference is improved by 4.0% compared to the DA algorithm. These numerical results align with the patterns observed in the previous images. Thus, RGA proves to be a fast and accurate algorithm in larger-scale scenarios.

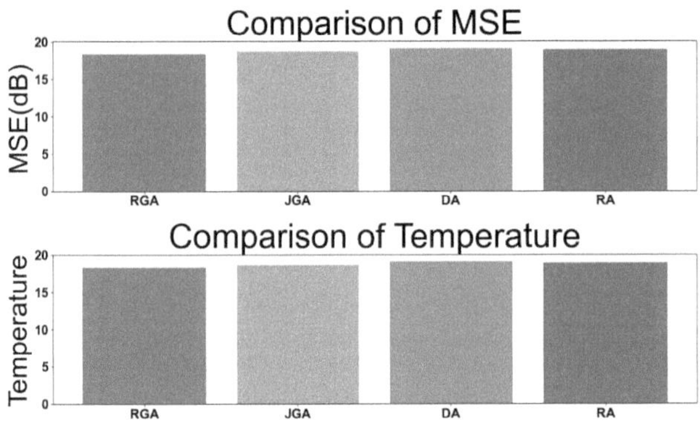

Fig. 4. Comparison chart of MSE and average temperature difference among the four algorithms in the large-scale fire detection scenarios.

5 Conclusion

This paper studies a measurement problem with multiple groups of heterogeneous UAVs. Each group has a limited number of UAVs. The goal is to reduce errors in solving linear inverse problems. A submodular maximization algorithm is proposed using the weighted frame potential. Its upper bound is rigorously proven. The algorithm is tested in large-scale fire scenarios. The results indicate that it outperforms other methods in terms of speed. Additionally, it reduces the average errors by 4.0% in large-scale fire observations. Future research could further explore the combination of UAV placement and dynamic target tracking [18] for improved disaster monitoring.

Acknowledgement. This work was partially supported by Beijing Natural Science Foundation (Nos. Z220004, IS24001, QY24081).

References

1. Ranieri, J., Chebira, A., Vetterli, M.: Near-optimal sensor placement for linear inverse problems. IEEE Trans. Signal Process. **62**(5), 1135–1146 (2014)
2. Bian, F., Kempe, D., Govindan, R.: Utility based sensor selection. In: Proceedings of the 5th International Conference on Information Processing in Sensor Networks, pp. 11–18. ACM, Nashville (2006)
3. Li, B., Liu, H., Wang, R.: Efficient sensor placement for signal reconstruction based on recursive methods. IEEE Trans. Signal Process. **69**, 1885–1898 (2021)
4. Zhou, H., Li, X., Cher, C.Y., Kursun, E., Qian, H., Yao, S.C.: An information-theoretic framework for optimal temperature sensor allocation and full-chip thermal monitoring. In: Proceedings of the 49th Annual Design Automation Conference, pp. 642–647. ACM, San Francisco (2012)
5. Rusu, C., Thompson, J., Robertson, N.M.: Sensor scheduling with time, energy, and communication constraints. IEEE Trans. Signal Process. **66**(2), 528–539 (2017)

6. Majumder, K., Pillai, S. B., Mulleti, S.: Greedy Selection for Heterogeneous Sensors. arXiv preprint arXiv:2307.00840 (2023)
7. Welch, W.J.: Branch-and-bound search for experimental designs based on D optimality and other criteria. Technometrics **24**(1), 41–48 (1982)
8. Krause, A., Singh, A., Guestrin, C.: Near-optimal sensor placements in Gaussian processes: theory, efficient algorithms and empirical studies. J. Mach. Learn. Res. **9**(2), 235–284 (2008)
9. Jamali-Rad, H., Simonetto, A., Leus, G.: Sparsity-aware sensor selection: centralized and distributed algorithms. IEEE Signal Process. Lett. **21**(2), 217–220 (2014)
10. Liu, S., Vempaty, A., Fardad, M., Masazade, E., Varshney, P.K.: Energy-aware sensor selection in field reconstruction. IEEE Signal Process. Lett. **21**(12), 1476–1480 (2014)
11. Joshi, S., Boyd, S.: Sensor selection via convex optimization. IEEE Trans. Signal Process. **57**(2), 451–462 (2008)
12. Chepuri, S.P., Leus, G.: Sparsity-promoting sensor selection for non-linear measurement models. IEEE Trans. Signal Process. **63**(3), 684–698 (2014)
13. Shamaiah, M., Banerjee, S., Vikalo, H.: Greedy sensor selection: leveraging submodularity. In: 49th IEEE Conference on Decision and Control, pp. 2572–2577. IEEE, Atlanta (2010)
14. Jiang, C., Soh, Y.C., Li, H.: Sensor placement by maximal projection on minimum eigenspace for linear inverse problems. IEEE Trans. Signal Process. **64**(21), 5595–5610 (2016)
15. Benedetto, J.J., Fickus, M.: Finite normalized tight frames. Adv. Comput. Math. **18**, 357–385 (2003)
16. Alonso, A.A., Kevrekidis, I.G., Banga, J.R., Frouzakis, C.E.: Optimal sensor location and reduced order observer design for distributed process systems. Comput. Chem. Eng. **28**(1–2), 27–35 (2004)
17. Mokhasi, P., Rempfer, D.: Optimized sensor placement for urban flow measurement. Phys. Fluids **16**(5), 1758–1764 (2004)
18. Lv, Y., Fan, G., Li, M., Liu, X., Liu, P., Zhang, Y.: UAV target tracking with bandit-based data fusion. In: International Conference on Combinatorial Optimization and Applications, Beijing, China, pp. 278–286 (2024)
19. Nemhauser, G.L., Wolsey, L.A., Fisher, M.L.: An analysis of approximations for maximizing submodular set functions—I. Math. Program. **14**, 265–294 (1978)
20. Fang, Z., Savkin, A.V.: Strategies for optimized UAV surveillance in various tasks and scenarios: a review. Drones **8**(5), 1–41 (2024)
21. Sultan, L., Anjum, M., Rehman, M., Murawwat, S., Kosar, H.: Communication among heterogeneous unmanned aerial vehicles (UAVs): classification, trends, and analysis. IEEE Access **9**, 118815–118836 (2021)
22. Jiang, C., Chen, Z., Su, R., Soh, Y.C.: Group greedy method for sensor placement. IEEE Trans. Signal Process. **67**(9), 2249–2262 (2019)

Parameterized Algorithms

A Parameterized Approximation Algorithm for the Diversity-Aware l-Centrum Problem

Junteng Song[1,2], Shuilian Liu[1,2], Yicheng Xu[1,2(✉)], and Yong Zhang[1,2]

[1] Shenzhen Institutes of Advanced Technology, Chinese Academy of Sciences, Shenzhen, China

[2] University of Chinese Academy of Sciences, Beijing, China

yc.xu@siat.ac.cn

Abstract. We propose and study a combination model of *l-centrum* and the *diversity-aware k-median* problem, called the *diversity-aware l-centrum* problem. In this problem, we are given a set of clients, a set of facilities in a metric space, integers k and l, and $r[i]$ ($i \in [t]$) for each facility group. The diversity constraint refers to the requirement that each group must contain at least $r[i]$ open facilities. The goal is to find a subset of facilities of size at most k so as to minimize the sum of the l largest distances from a client to its nearest open facility while satisfying the diversity constraint. As our main contribution, this work provides the first constant approximation algorithm for the *diversity-aware l-centrum* problem in FPT time, based on a combination of leader selection and coreset construction techniques.

Keywords: Fair clustering · Diversity-aware clustering · l-centrum · Fixed parameter tractability

1 Introduction

Clustering problems are fundamental in machine learning and algorithmic theory, with traditional models like the *k-median* and *k-center* aiming to partition data points by optimizing objective functions. The *k-median* problem minimizes the sum of distances from all clients to their nearest facilities, while the *k-center* problem minimizes the maximum distance from any client to a facility. In recent years, variants of classical clustering problems have also been widely studied. For example, the *ordered k-median* problem proposed by Chakrabarty and Swamy [1] alters the clustering objective of the *k-median* problem. After assigning clients to facilities, all client assignment costs are sorted in non-increasing order. We are also given non-increasing weightsÂăw_i ($i \in [n]$), where n is the number of clients. The objective of the problem is to minimize the sum of the products of each client's assignment cost and its corresponding weight, i.e., $w_1\cdot$(largest assignment cost) $+ w_2\cdot$(second-largest assignment cost) $+ \ldots + w_n\cdot$(n-th largest assignment cost). Note that setting $w_1 = 1 = w_2 = \ldots = w_n$ yields the *k-median* problem; setting $w_1 = 1, w_2 = 0 = w_3 = \ldots = w_n$ yields the *k-center* problem.

M. Li et al. (Eds.): TAMC 2025, LNCS 16084, pp. 345–353, 2026.
https://doi.org/10.1007/978-981-95-4839-2_27

A natural version of this model sets the first l weights to 1 and the remaining weights to 0, and the objective becomes calculating the sum of the l largest assignment costs. This model is referred to as the *l-centrum* problem by Tamir et al. [2]. The model has a wide range of applications in real life, as many clustering problems in practical scenarios are dominated by larger assignment costs. Aouad and Segev [3] obtained a logarithmic-approximation algorithm. Chakrabarty and Swamy [4] present approximation algorithms with approximation ratios $(12+\varepsilon)$ and $(18+\varepsilon)$ for *l-centrum* and *ordered k-median*, respectively. Subsequently, the results were improved to $(5+\varepsilon)$ [5]. Its variants such as *outliers* [6], *fault-tolerance* [7] have also been studied.

In the field of clustering, algorithmic fairness has emerged as a critical consideration, particularly in applications where resource allocation impacts diverse societal groups. Fairness often requires that no subgroup is systematically *under-represented* or *over-represented* in the clustering outcome. For instance, in public service provision (e.g., healthcare facilities or transportation hubs), unequal representation of facilities across demographic or geographic groups can perpetuate disparities. Motivated by the need to address *under-representation* in clustering, Thejaswi et al. [8] first introduced diversity constraints into the *k-median* problem, named this model the *diversity-aware k-median* problem and shown its NP-hardness, polynomial-time inapproximability, and fixed-parameter intractability. In this model, facilities are partitioned into several groups, and each group is associated with a threshold that serves as a lower bound on the number of facilities selected. This ensures fair representation of each facility group in the solution, balancing cost optimization with fairness considerations. Subsequently, Thejaswi et al. [9] presented fixed-parameter approximation algorithms with approximation ratios of $(1+2/e+\varepsilon)$ and $(1+8/e+\varepsilon)$ for the *diversity-aware k-median* and *diversity-aware k-means* problems, respectively.

In algorithmic complexity,Âăfixed-parameter tractable (FPT) algorithms offer a powerful framework for solving combinatorial problems efficiently when parameters (such as the number of facilities and groups, or the dimension) are small relative to the input size. FPT algorithms ensure that the running time depends polynomially on the input size while allowing exponential growth in parameters, i.e., $f(k)poly(n)$, where k is a small parameter relative to n. Cohen-Addad et al. [10] introduced FPT algorithms with approximation ratios ofÂă$(1+2/e+\varepsilon)$ÂăandÂă$(1+8/e+\varepsilon)$Âăfor theÂă*k-median* andÂă*k-means* problems, respectively. Their work demonstrates that FPT algorithms can achieve near-optimal solutions for *k-median* andÂă*k-means* in metric spaces while maintaining polynomial running time in the input size. This approach is particularly valuable for applications whereÂăkÂăis a manageable parameter. Additional research devoted to designing algorithms for clustering problems in FPT time includes, e.g., [11–13].

In the domain of clustering, coresets have emerged as a fundamental technique to address the computational challenges of large-scale data by constructing a compact summary (*coreset*) that approximates the original dataset while preserving key geometric properties. Coresets are particularly valuable in large-scale data scenarios, as they enable efficient approximation algorithms with rigorous theoretical guarantees.

Our Result: In this paper, we construct a new model by incorporating diversity constraints into the *l-centrum* problem, which we term the *div-l-centrum* (short for *diversity-aware l-centrum*) problem. Specifically, we replace the clustering objective of the aforementioned *diversity-aware k-median* problem with that of the *l-centrum* problem. The objective of this model is to minimize the sum of the l largest assignment costs while satisfying the diversity constraint. Our contribution is a $(3+\varepsilon)$-approximation algorithmÂăfor the *div-l-centrum* problem running in time $f(k, t, dim)poly(n)$. We employ a matching-based inequality scaling method for the analysis, and our technique is a combination of the leader selection idea from Thejaswi et al. [9] and the coreset construction approach from Osadnik [14].

2 Preliminaries

We give the definition of the *l-centrum* problem. In this problem, we are given a metric space, a set of clients, a set of facilities, a budget k, and an integer l. We find a subset S of facilities of size k. After assigning all clients to the facilities in solution S, we denote by $\boldsymbol{d}^{\downarrow}$ the vector of all clients' assignment costs sorted in non-increasing order. That is, we have $\boldsymbol{d}_1^{\downarrow} \geq \boldsymbol{d}_2^{\downarrow} \geq \cdots \geq \boldsymbol{d}_n^{\downarrow}$. The objective is to minimize

$$cost(l, \boldsymbol{d}) = \sum_{i=1}^{l} \boldsymbol{d}_i^{\downarrow}$$

Upon introducing diversity constraints, facilities are partitioned into several groups, and each group has a lower bound on the number of facilities selected. The objective is to minimize the cost of the *l-centrum*. We provide the precise definition of the *diversity-aware l-centrum* problem below.

Definition 1. ***(Diversity-aware l-centrum****(div-l-centrum)). Given a metric space $(U \subseteq \mathcal{R}^{dim}, d)$, a set $C \subseteq U$ of clients, a set $F \subseteq U$ of facilities, a collection (groups) $G' = \{G_1, \ldots, G_t\}$ of facility sets $G_i \subseteq F$ $(i \in [t])$, a budget $k \leq |F|$, an integer $l \leq |C|$, and a vector of requirements $\boldsymbol{r} = (r[1], \ldots, r[t])$ that serves as the lower bound on the number of facilities selected from each group. The goal is to find a subset of facilities $S \subseteq F$ of size k. After assigning all clients to the facilities in solution S, we denote by $\mathbf{d}^{\downarrow}$ the vector of all clients' assignment costs sorted in non-increasing order($\mathbf{d}_1^{\downarrow} \geq \mathbf{d}_2^{\downarrow} \geq \cdots \geq \mathbf{d}_n^{\downarrow}$). We aim to minimize the sum of the l largest assignment costs $cost(l, \mathbf{d}) = \sum_{i=1}^{l} \mathbf{d}_i^{\downarrow}$. An instance of div-l-centrum is denoted as $I = ((U \subset \mathcal{R}^{dim}, d), F, C, G', \boldsymbol{r}, k, l)$*

Our algorithm employs coreset construction methods to reduce the enumeration time. Roughly speaking, coreset is used to reduce the number of clients such that for a solution of the clustering problem, the cost distortion on set C of clients and coreset C' of clients is bounded by a multiplicative factor $(1 \pm \varepsilon)$, for some $\varepsilon \geq 0$. Below is the formal definition of *coreset* for k-clustering problems.

Definition 2. *(Coreset). Given an instance $I = ((U \subseteq \mathcal{R}^{dim}, d), C, F, k)$ of the k-clustering problem and a constant $\varepsilon \geq 0$. for any subset of facilities $S \subseteq F$ of size k, we denote by $cost(S, C)$ the cost of the k-clustering problem on the client set C. a (strong) coreset is a subset of clients $C' \subseteq C$ with associated weights $\{w_c : c \in C'\}$ such that*

$$(1 - \varepsilon) \cdot cost(S, C) \leq cost(S, C') \leq (1 + \varepsilon) \cdot cost(S, C)$$

3 Algorithm Design and Analysis

In our algorithm, we combine the leader selection idea from Thejaswi et al. [9] with the coreset construction approach from Osadnik [14]. In the analysis, we employ a matching-based inequality scaling approach to analyze the approximation ratio.

3.1 Leader Selection and Coreset Construction

Since there may be overlaps between facility groups, we regroup the facilities using the characteristic vector of each facility. For an instance $I = ((U \subseteq \mathcal{R}^{dim}, d), F, C, G', \boldsymbol{r}, k, l)$ of *div-l-centrum*, where $G' = \{G_1, ..., G_t\}$. For each facility $f \in F$ we denote by $\boldsymbol{v} \in \{0,1\}^t$ the characteristic vector of f with respect to G'. Specifically, $\boldsymbol{v}_f[i] = 1$ if $f \in G_i$, 0 otherwise, for all $i \in [t]$. We put the facilities with the same characteristic vector into the same group $E(\boldsymbol{v})$, let $E' = \{E(\boldsymbol{v}) : \boldsymbol{v} \in \{0,1\}^t\}$. In this way, the facility set F is partitioned into at most $O(2^t)$ subsets and there is no overlap among the facility groups in E'.

To satisfy diversity constraints, we pick a k-multiset $E_k = \{E(\boldsymbol{v}_1), ..., E(\boldsymbol{v}_k)\}$ from E' and calculate the element-wise sum of the characteristic vectors $\{\boldsymbol{v}_1, ..., \boldsymbol{v}_k\}$, that is, $\sum_{i \in [k]} \boldsymbol{v}_i$. We call a k-multiset feasible if $\sum_{i \in [k]} \boldsymbol{v}_i \geq \boldsymbol{r}$, where the inequality is taken element-wise. In other words, if we pick exactly one facility from each set $E(\boldsymbol{v}_i)$ from feasible k-multiset E_k, we can obtain a feasible solution of *div-l-centrum*. We can prove that for an instance of *div-l-centrum*, we can enumerate all feasible k-multisets in FPT time.

Lemma 1. *Given an instance $I = ((U \subseteq \mathcal{R}^{dim}, d), F, C, G', \boldsymbol{r}, k, l)$ of the div-l-centrum problem, where $|U| = n$. We can enumerate all the feasible k-multisets in time $O(2^{tk}tn)$.*

Proof. We construct the set E' in time $O(2^t|F|)$, since $|E'| \leq 2^t$. Then, we enumerate all possible feasible k-multisets of E'. Since there are $\binom{|E'|+k-1}{k}$ possible k-multisets of E', enumerating all feasible k-multisets can be done in $O(|E'|^k t)$ time, resulting in total running time $O(2^{tk}tn)$. □

Subsequently, we employ a coreset construction method designed for the *l-centrum* problem as stated in the following theorem. Observe that the coreset framework for the *l-centrum* problem can be adapted to address the *diversity-aware l-centrum* (*div-l-centrum*) problem.

Theorem 1. *([14]) For any $k, l \in \mathbb{Z}^+$, data set $X \subseteq \mathcal{R}^{dim}$, and any $\varepsilon \in (0,1)$, there is a weighted set D of size $O(\frac{k^2 \cdot dim}{\varepsilon^{dim+1}})$such that:*

$$\text{cost}_\ell(D, S) \in \big(1 \pm O(\varepsilon)\big) \cdot \text{cost}_\ell(X, S)$$

for all sets $S \subseteq \mathcal{R}^{dim}$ of size k and D is obtained in polynomial time.

For each feasible k-multiset $E_k = \{E(\boldsymbol{v}_1), ..., E(\boldsymbol{v}_k)\}$, we generate an instance $J_{E_k} = ((U, d), \{E(\boldsymbol{v}_1), ..., E(\boldsymbol{v}_k)\}, C', k, l)$, where C' is a coreset constructed by Theorem 1. Observe that there are at most $O(2^{tk})$ instances in total. Let $F^* = \{f_i^* \in E(\boldsymbol{v}_i)\}_{i \in [k]}$ be an optimal solution of instance J_{E_k}, and for each $f_i^* \in F^*$, let c_i^* be a closest client to f_i^* with $d(c_i^*, f_i^*) = \lambda_i^*$. We show how to find c_i^* and λ_i^* in FPT time, as shown below.

Find c_i^* and λ_i^* : We already have a coreset C' with size $O(\frac{k^2 \cdot dim}{\varepsilon^{dim+1}})$, so we can enumerate all k-multisets $\{c_i^*\}_{i \in [k]}$ in $O((\frac{k^2 \cdot dim}{\varepsilon^{dim+1}})^k)$ time. Then we employ the method of Cohen-Addad et al. [10], which builds on the work of Feldman and Langberg [15] to find λ_i^*. We would discretize the search space of λ_i^* (which is at most $\Delta = (\max_{x,y \in U} d(x,y))/(\min_{x \neq y \in U} d(x,y)) = poly(n)$) so that we can enumerate all possible choices in FPT time. Precisely, we discretize the interval $[\Delta]$ with the power of $(1+\varepsilon)$ so we can enumerate all k-multisets in at most $O(log^k n)$ time. The total time to find c_i^* and λ_i^* is $O((\frac{k^2 \cdot dim}{\varepsilon^{dim+1}} \cdot logn)^k)$ which is FPT.

Let $\prod_i^* \subseteq E(\boldsymbol{v}_i)_{i \in [k]}$ represent the set of facilities within the radius range centered on c_i^* and with λ_i^* as the radius, that is, for every $f_i \in \prod_i^*$, $d(f_i, c_i^*) \leq \lambda_i^*$. Obviously, $\prod_i^*$ contains f_i^*. We finally obtain the feasible solution for the *div-l-centrum* problem by arbitrarily picking exactly one facility from each $\prod_i^*$.

3.2 Analysis

Unlike the objective in the *diversity-aware k-median* problem, the *div-l-centrum* problem only considers the l largest assignment costs. Consequently, there may exist clients that contribute to the objective in a solution output by *ALG* (*div-l-centrum*) but not in the optimal solution. We call such clients *bad clients*. Before the analysis, we provide the definitions of *good client* and *bad client* below.

Definition 3. *(**good/bad client**). Given an instance $I = ((U \subseteq \mathcal{R}^{dim}, d), F, C, G', \boldsymbol{r}, k, l)$, let S be a solution output by ALG (div-l-centrum) and O be the optimal solution. In solution S, let $c_1, c_2, ..., c_l$ denote the clients corresponding to the l largest assignment costs, and define the set $c' = \{c_1, ..., c_l\}$. Similarly, in the optimal solution O, let $o_1, ..., o_l$ denote the clients corresponding to the l largest assignment costs, and define the set $o = \{o_1, ..., o_l\}$. We call a client $c \in c'$ good if $c \in o$ and bad otherwise. We use the function $\phi(c_i) = o_j$ $(i \in [l])$ to represent this correspondence, that is, good clients satisfy $i, j \in [1, l]$, bad clients satisfy $i \in [1, l]$ and $j \in [l+1, n]$.*

Algorithm 1: *ALG* (*div-l-centrum*)

Input: $I = ((U \subseteq \mathcal{R}^{dim}, d), F, C, G', \boldsymbol{r}, k, l)$, an instance of *div-l-centrum*
Output: S^*, subset of facilities

begin
  foreach $\boldsymbol{v} \in \{0,1\}^t$ **do**
    $E(\boldsymbol{v}) \leftarrow \{f \in F : \boldsymbol{v} = \boldsymbol{v}_f\}$
  end
  $E' \leftarrow \{E(\boldsymbol{v}) : \boldsymbol{v} \in \{0,1\}^t\}$
  $C' \leftarrow coreset((U \subseteq \mathcal{R}^{dim}, d), F, C, k, l, \varepsilon)$
  $S^* \leftarrow \emptyset$
  foreach *k-multiset* $\{E(\boldsymbol{v}_1), ..., E(\boldsymbol{v}_k)\} \subseteq E'$ **do**
    if $\sum_{i \in [k]} \boldsymbol{v}_i \geq \boldsymbol{r}$ **then**
      foreach $i \in [k]$ **do**
        Find c_i^* and λ_i^*
        $\prod_i^* \leftarrow \{f \in E(\boldsymbol{v}_i) : d(f, c_i^*) \leq \lambda_i^*\}$
      end
      $S \leftarrow$Pick exactly one facility f_i from each $\prod_i^*$
    else
  end
  $S^* \leftarrow S$ with smallest $cost(S, C)$
end
return S^*

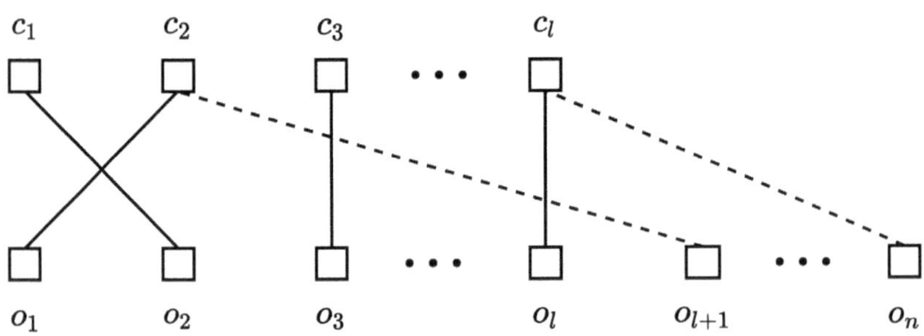

Fig. 1. An illustrative example of *good client* and *bad client*.

We analyze each of the l clients $c_1, c_2, ..., c_l$ that contribute to the objective in solution S. For a *good client* with a mapping $\phi(c_i) = o_j$ ($j \in [l]$), we directly apply the triangle inequality. Since *bad clients* have no contribution to the objective in the optimal solution, we cannot directly use the triangle inequality for them. To address this, we enlarge the distance of a *bad client* to the facility in the optimal solution to the smallest distance among the l largest distances in the optimal solution. Specifically, for a *bad client* with a mapping $\phi(c_{i'}) = o_j$ ($j \in [l+1, n]$), we enlarge its distance $d(c_{i'}(o_j), f^*)$ to $d(o_{j'}, f^*)$, which is the smallest distance among the l largest distances in the optimal solution and $o_{j'}$ has no corresponding c_i for $i \in [j]$. We then set $\phi(c_{i'}) = o_{j'}$. By doing so, we ensure $d(c_{i'}(o_j), f^*) \leq d(o_{j'}, f^*)$.

We provide an illustrative example (see Fig. 1). $c_1, c_2, ..., c_l$ denote the l clients contributing to the objective in solution S. Here, c_1 and c_3 are *good clients* with mappings $\phi(c_1) = o_2$ and $\phi(c_3) = o_3$, respectively, while c_2 and c_l are *bad clients* with mappings $\phi(c_2) = o_{l+1}$ and $\phi(c_l) = o_n$, respectively. For c_l, assuming $d(o_l, f^*)$ (o_l has no corresponding c_i, for $i \in [l]$) is the smallest distance among the l largest distances in the optimal solution, we enlarge $d(c_l(o_n), f^*)$ to $d(o_l, f^*)$, and set $\phi(c_l) = o_l$. Similarly, we enlarge $d(c_2(o_{l+1}), f^*)$ to $d(o_1, f^*)$, and set $\phi(c_2) = o_1$.

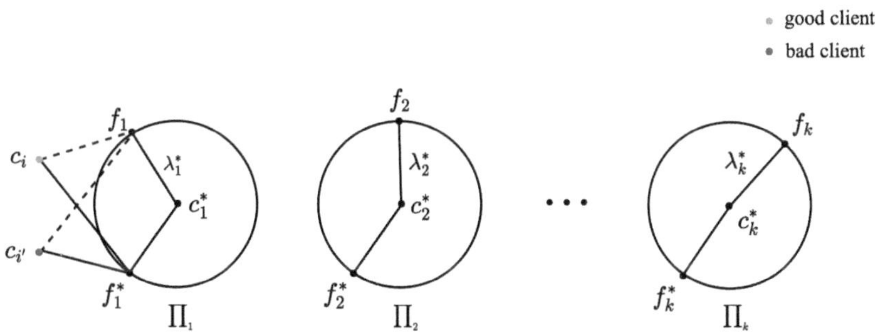

Fig. 2. An illustration of facility selection for *ALG* (*div-l-centrum*).

We employ the triangle inequality in our analysis (see Fig. 2). $f_1^*, ..., f_k^*$ denote the facilities in the optimal solution, and c_i^* and λ_i^* denote the closest client and its distance for each f_i^*, respectively, and $f_1, ..., f_k$ denote the facilities in a solution output by *ALG* (*div-l-centrum*). For *good client* $c_i \in C_{good}$, we have $d(c_i, f_i) \leq d(c_i, f_i^*) + d(f_i^*, c_i^*) + d(c_i^*, f_i)$, and for *bad client* $c_{i'} \in C_{bad}$, we have $d(c_{i'}, f_i) \leq d(o_{j'}, f_i^*) + d(f_i^*, c_i^*) + d(c_i^*, f_i)$, where $d(o_{j'}, f_i^*)$ is the scaled distance.

By the choice of c_i^* we have $d(f_i^*, c_i^*) + d(c_i^*, f_i) \leq 2\lambda_i^* \leq 2d(c_i, f_i^*)$. For *good clients*, we have $d(c_i, f_i) \leq 3d(c_i, f_i^*)$; for *bad clients*, we have $d(c_{i'}, f_i) \leq 3d(o_{j'}, f_i^*)$, which implies

$$
\begin{aligned}
cost(l, \boldsymbol{d}) &\leq \sum_{c_i \in C_{good}} d(c_i, f_i) + \sum_{c_{i'} \in C_{bad}} d(c_{i'}, f_i) \\
&\leq 3d(c_i, f_i^*) + 3d(c_{i'}, f_i^*) \\
&\leq 3cost(l, \boldsymbol{o})
\end{aligned}
$$

Combining the error ε introduced by the coreset construction and the discretization of the search space, along with the time complexity incurred by the enumeration process, we obtain the following theorem.

Theorem 2. *ALG (div-l-centrum) is a* $(3+\varepsilon)$*-approximation algorithm for the div-l-centrum problem running in time* $O((\frac{2^t \cdot k^3 \cdot dim \cdot logk}{\varepsilon^{dim+1}})^k)$*, for every* $\varepsilon > 0$.

4 Conclusion

In this paper, we introduce the *diversity-aware l-centrum* (*div-l-centrum*) problem, which integrates the *l-centrum* problem with diversity constraints. This new model addresses scenarios requiring both the minimization of the l largest assignment costs (as in *l-centrum*) and fair representation among facility groups (via diversity constraints). Our contribution is a $(3+\varepsilon)$-approximation algorithm for *div-l-centrum* running in FPT time with respect to parameters k, t, dim.

The diversity constraints only provide lower bounds on the number of facilities to be selected from each group. Furthermore, exploring variants that impose *fair range* constraints on the *l-centrum* objective (defining both lower bounds and upper bounds on the number of facilities selected from each group) remains an interesting direction.

Acknowledgements.
This work is supported by Guangdong Basic and Applied Basic Research Foundation 2024A1515030197, Natural Science Foundation of China 12371321 and Shenzhen key Laboratory of Intelligent Bioinformatics ZDSYS20220422103800001.

References

1. Puerto, J., Tamir, A.: Locating tree-shaped facilities using the ordered median objective. In: Mathematical Programming, pp. 313–338 (2005)
2. Tamir, A.: The k-centrum multi-facility location problem. In: Discrete Applied Mathematics, pp. 293–307 (2001)
3. Aouad, A., Segev, D.: The ordered k-median problem: surrogate models and approximation algorithms. In: Mathematical Programming, pp. 55–83 (2019)
4. Chakrabarty, D., Swamy, C.: Interpolating between k- median and k-center: approximation algorithms for ordered k-median. In: International Colloquium on Automata, Languages, and Programming, pp. 29:1–29:14 (2018)
5. Chakrabarty, D., Swamy, C.: Approximation algorithms for minimum norm and ordered optimization problems. In: Proceedings of Annual ACM SIGACT Symposium on Theory of Computing, pp. 126–137 (2019)
6. Deng, S., Zhang, Q.: Ordered k-median with outliers. In: Approximation, Randomization, and Combinatorial Optimization. Algorithms and Techniques, pp. 34:1–34:22 (2022)
7. Deng, S., Zhang, Q.: Ordered k-median with outliers and fault-tolerance. In: arXiv preprint arXiv:2011.04289 (2020)
8. Thejaswi, S., Ordozgoiti, B., Gionis, A.: Diversity-aware k-median: clustering with fair center representation. In: Machine Learning and Knowledge Discovery in Databases, pp. 765–780 (2021)
9. Thejaswi, S., Gadekar, A., Ordozgoiti, B., Osadnik, M.: Clustering with fair-center representation: parameterized approximation algorithms and heuristics. In: Proceedings of ACM SIGKDD Conference on Knowledge Discovery and Data Mining, pp. 1749–1759 (2022)

10. Cohen-Addad, V., Gupta, A., Kumar, A., Lee, E., Li, J.: Tight FPT approximations for k-median and k-means. In: International Colloquium on Automata, Languages, and Programming, pp. 42:1–42:14 (2019)
11. Cohen-Addad, V., Li, J.: On the fixed-parameter tractability of capacitated clustering. In: International Colloquium on Automata, Languages, and Programming, pp. 41:1–41:14 (2019)
12. Inamdar, T., Varadarajan, K.: Capacitated sum-of-radii clustering: an FPT approximation. In: Annual European Symposium on Algorithms, pp. 62:1–62:17 (2020)
13. Bandyapadhyay, S., Lochet, W., Saurabh, S.: FPT constant- approximations for capacitated clustering to minimize the sum of cluster radii. In: International Symposium on Computational Geometry, pp. 12:1–12:14 (2023)
14. Osadnik, M.: Fixed parameter tractable algorithm and coreset for the ordered k-median problem. In: Master's Thesis of Aalto University School of Science (2023)
15. Feldman, D., Langberg, M.: A unified framework for approximating and clustering data. In: Proceedings of ACM Symposium on Theory of Computing, pp. 569–578 (2011)

An Improved Approximation Algorithm for the k-Supplier Problem with Parameterized Triangle Inequality

Wei Ding[1(✉)], Guangting Chen[2], Ke Qiu[3], and Yu Zhou[4]

[1] Nanxun Innovation Institute, and School of Computer Science and Technology, Zhejiang University of Water Resources and Electric Power, Hangzhou 310018, China
dingweicumt@163.com

[2] Zhejiang University of Water Resources and Electric Power, Hangzhou 310018, China
gtchen@hdu.edu.cn

[3] Department of Computer Science, Brock University, St. Catharines, Canada
kqiu@brocku.ca

[4] College of Data Science, Taiyuan University of Technology, Taiyuan 030024, China
zhouyu@tyut.edu.cn

Abstract. Given an undirected connected graph $G = (V, E, \rho, \mathcal{X}, \mathcal{Y})$, where $\rho : E \to \mathrm{R}^+ \cup \{0\}$ is an edge-weight function, $\mathcal{X} \subset V$ is a subset of clients, and $\mathcal{Y} \subset V$ is a subset of candidates, and a positive integer $k < |\mathcal{Y}|$, the k**-Supplier Problem (**k**SP)** asks for an optimal subset of $\mathcal{Y}$ of cardinality at most k in G to minimize the radius from $\mathcal{X}$ to the subset. In this paper, we focus on the case of $\mathcal{X} \cap \mathcal{Y} \neq \emptyset, \mathcal{X}, \mathcal{Y}$, and consider the scenario where the shortest path distances $d(\cdot, \cdot)$ satisfy a *parameterized triangle inequality* between $\mathcal{X}$ and $\mathcal{Y}$, i.e., $d(v, w) + d(w, u) \geq \alpha \cdot d(v, u)$, $1 < \alpha \leq 2$, for any $v, u \in \mathcal{X}, v \neq u$ and $w \in \mathcal{Y}$.

We present a *two-stage* dual approximation algorithm KSP-PG for the kSP with parameterized triangle inequality between $\mathcal{X}$ and $\mathcal{Y}$. If KSP-PG stops at the end of Stage 1 then it obtains a $\frac{2}{\alpha}$-approximation, and if it stops at the end of Stage 2 then it obtains a $(\frac{2}{\alpha^2} + \frac{1}{\alpha})$-approximation. It is implied by $\frac{2}{\alpha} \in [1, 2)$ and $\frac{2}{\alpha^2} + \frac{1}{\alpha} \in [1, 3)$ that KSP-PG is better than the previously best 3-approximation algorithm of Hochbaum and Shmoys (J. ACM. **33**: 533–550, 1986). Regardless of the parameterized aspect, KSP-PG obtains a 2-approximation if it stops at the end of Stage 1 and a 3-approximation if it stops at the end of Stage 2, for the kSP.

Keywords: Supplier problem · Two-Stage · Dual approximation · Parameterized

1 Introduction

Let $G = (V, E, \rho)$ be an undirected connected graph where V is the vertex set, E is the edge set, and $\rho : E \to \mathrm{R}^+ \cup \{0\}$ is an edge-weight function. Given a subset

M. Li et al. (Eds.): TAMC 2025, LNCS 16084, pp. 354–366, 2026.
https://doi.org/10.1007/978-981-95-4839-2_28

$\mathcal{X} \subseteq V$ of *clients*, a subset $\mathcal{Y} \subseteq V$ of *candidates*, and a positive integer k, the **k-Supplier Problem (kSP)** aims to find at most k facilities from $\mathcal{Y}$ such that the maximum distance connecting client to the closest facility is minimized [15]. The **k-Center Problem (kCP)** is the special case of kSP where $\mathcal{X} = \mathcal{Y} = V$ [13,16]. Both of them have important applications in emergency facility location, machine learning, data science, etc.

1.1 Related Results

For the *classic* kSP, Hochbaum and Shmoys [15] gave a 3-approximation algorithm, and proved it is NP-hard to approximate within a factor of less than 3 unless NP = P for general metrics. For the classic kCP, Hochbaum and Shmoys [14], and Gonzalez [11] obtained 2-approximation algorithm, respectively. It was proved that it is NP-hard to approximate within a factor of less than 2 unless NP = P [14].

The *Euclidean* version of kSP and kCP were proved by Feder and Greene [10] to be NP-hard to approximate within a factor of less than $\sqrt{7}$ and $\sqrt{3}$, respectively. Recently, Nagarajan et al. [21] obtained a $(1 + \sqrt{3})$-approximation algorithm for the kSP in Euclidean metrics of arbitrary dimension, and an approximation algorithm with a factor of less than 3 for the kSP in Euclidean metrics of constant dimensions. In [19], Lee et al. discussed two variants, priority Euclidean kSP and Euclidean kSP with Outliers, and presented $(1 + \sqrt{3})$-approximation algorithms for both.

The *capacitated* version of kCP has been extensively studied. The kCP with identical capacity upper bounds was first proposed by Barilan et al. [4], where the number of clients each facility can serve is limited by an upper bound. Some approximation algorithms were obtained in [4,17]. The kCP with non-identical capacity upper bounds was studied in [3,7,18]. On the other hand, the kCP with capacity lower bounds was first proposed by Sweeney [22]. Aggarwal et al. [1] considered the kCP with identical capacity lower bounds and its variant with outliers. Ahmadian and Swamy [2] considered the kCP with non-identical capacity lower bounds and its variant with outliers. Later, Ding et al. [8] considered the two-sided capacitated kCP and kSP (with both the capacity upper bounds and lower bounds), and obtained approximation algorithms in a variety of settings. In [12], Goyal and Jaiswal studied various *constrained* versions of the kSP and kCP and obtained approximation algorithms.

Given $\mathcal{X} \subset V$ and $\mathcal{Y} \subset V$, there exist four possible intersection relationships of $\mathcal{X}$ and $\mathcal{Y}$ as follows, see Fig. 1.

$$\mathcal{X} \cap \mathcal{Y} \neq \emptyset, \mathcal{X}, \mathcal{Y}, \quad \mathcal{X} \cap \mathcal{Y} = \emptyset, \quad \mathcal{X} \subseteq \mathcal{Y}, \quad \mathcal{Y} \subseteq \mathcal{X}.$$

For the case of $\mathcal{X} = \mathcal{Y}$, we can adapt Hochbaum and Shmoys's algorithm [14] or Gonzalez's algorithm [11] to obtain a 2-approximation. For $\mathcal{X} \cap \mathcal{Y} = \emptyset$, we can adapt Hochbaum and Shmoys's algorithms [15]. For $\mathcal{X} \subseteq \mathcal{Y}$, we can adapt Ding and Qiu's algorithm [9] to select facilities from $\mathcal{X}$ to obtain a 2-approximation. For $\mathcal{Y} \subseteq \mathcal{X}$, we can adapt Liang's algorithm [20] to obtain a 2-approximation.

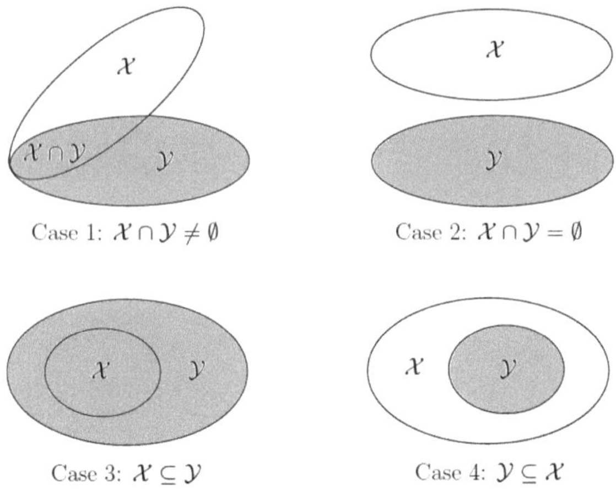

Fig. 1. Illustration for the intersection relationships of $\mathcal{X}$ and $\mathcal{Y}$.

1.2 Our Contributions

Let $G = (V, E, \rho, \mathcal{X}, \mathcal{Y})$ be an edge-weighted undirected connected graph, in which $\mathcal{X} \subset V$ is a subset of clients and $\mathcal{Y} \subset V$ is a subset of candidates admitting facilities, and let $d(v, u)$ denote the shortest path distance in G between v and u, for $v, u \in V$. We introduce a novel *parameterized triangle inequality*, quite different from those in [5,6,23], between $\mathcal{X}$ and $\mathcal{Y}$ instead of over the whole vertex set V, as follows: $d(v, w) + d(w, u) \geq \alpha \cdot d(v, u)$, for any $v, u \in \mathcal{X}$ and $w \in \mathcal{Y}$, where $1 < \alpha \leq 2$ is a parameter. This inequality arises from the real-world scenarios, e.g., V has many vertices, and both $\mathcal{X}$ and $\mathcal{Y}$ have (much) fewer vertices than V. In this paper, we study the kSP with parameterized triangle inequality between $\mathcal{X}$ and $\mathcal{Y}$, and focus on the case of $\mathcal{X} \cap \mathcal{Y} \neq \emptyset, \mathcal{X}, \mathcal{Y}$.

We design a *two-stage* dual approximation algorithm $\mathsf{KSP\text{-}PG}$. If it stops at the end of Stage 1 then it outputs client facilities $\mathcal{S}_1 \subset \mathcal{X} \cap \mathcal{Y}$ as a $\frac{2}{\alpha}$-approximation, and if it stops at the end of Stage 2 then it obtains $\mathcal{S}_2 \subset \mathcal{Y} \setminus \mathcal{S}_1$ in Stage 2 and outputs $\mathcal{S}_1 \cup \mathcal{S}_2$ as a $(\frac{2}{\alpha^2} + \frac{1}{\alpha})$-approximation. Because $1 \leq \frac{2}{\alpha} < 2$ and $1 \leq \frac{2}{\alpha^2} + \frac{1}{\alpha} < 3$, we can say $\mathsf{KSP\text{-}PG}$ is better than the previously best 3-approximation algorithm of Hochbaum and Shmoys [15] for the kSP with parameterized triangle inequality between $\mathcal{X}$ and $\mathcal{Y}$. Regardless the parameterized aspect of input, $\mathsf{KSP\text{-}PG}$ obtains a 2-approximation if it stops at the end of Stage 1 and a 3-approximation if it stops at the end of Stage 2, for the kSP.

Organization. In Sect. 2, we define the notations, k-Supplier Problem (kSP) and parameterized triangle inequality formally, and show several basic properties. In Sect. 3, we first design two separate test procedures and then merge them into an approximation algorithm. In Sect. 4, we conclude this paper. Due to the page limit, most proof details throughout this paper are omitted.

2 Preliminaries

2.1 Notations and Problem Statement

Let $G = (V, E, \rho, \mathcal{X}, \mathcal{Y})$ be an edge-weighted connected graph, where V is the set of n vertices, E is the set of m edges, $\rho : E \to \mathrm{R}^+ \cup \{0\}$ is an edge-weight function, $\mathcal{X} = \{x_1, x_2, \ldots, x_p\} \subseteq V$ is a subset of $p \leq n$ *clients*, and $\mathcal{Y} = \{y_1, y_2, \ldots, y_q\} \subseteq V$ is a subset of $q \leq n$ *candidates* admitting facilities. Let $d(v, u)$ denote the *shortest path distance (SPD)* in G from v to u, abbreviated as the v-to-u SPD, for any $v, u \in V$.

A set of cardinality g is called a *g-set*. Let $\mathcal{S} = \{s_1, s_2, \ldots, s_g\} \subseteq \mathcal{Y}$ be a g-set of $\mathcal{Y}$ with no duplicate. For any $1 \leq i \leq p$, the distance from x_i to $\mathcal{S}$, $d(x_i, \mathcal{S})$, is defined as the SPD in G from x_i to the closest facility in $\mathcal{S}$, i.e., $d(x_i, \mathcal{S}) = \min_{1 \leq j \leq g} d(x_i, s_j)$, abbreviated as the x_i-to-$\mathcal{S}$ distance. The maximum distance from a client to $\mathcal{S}$ is called the $\mathcal{X}$-to-$\mathcal{S}$ *radius* and denoted as $r(\mathcal{X}, \mathcal{S})$, i.e.,

$$r(\mathcal{X}, \mathcal{S}) = \max_{1 \leq i \leq p} d(x_i, \mathcal{S}). \tag{1}$$

Given $G = (V, E, \rho, \mathcal{X}, \mathcal{Y})$ and a positive integer $k < |\mathcal{Y}|$, the objective of the **k-Supplier Problem (kSP)** is to find an optimal subset of $\mathcal{Y}$ of cardinality at most k in G such that the radius is minimized. Let $\mathcal{S}^* \subset \mathcal{Y}, |\mathcal{S}^*| \leq k$ be an optimum solution to the kSP in G. We have

$$r(\mathcal{X}, \mathcal{S}^*) = \min_{\mathcal{S} \subset \mathcal{Y},\ |\mathcal{S}| \leq k} r(\mathcal{X}, \mathcal{S}). \tag{2}$$

The classic kCP is just a special case of the kSP where $\mathcal{X} = \mathcal{Y} = V$. This paper studies the other case of the kSP where $\mathcal{X} \subset V, \mathcal{Y} \subset V$ and $\mathcal{X} \cap \mathcal{Y} \neq \emptyset, \mathcal{X}, \mathcal{Y}$.

2.2 Parameterized Triangle Inequality

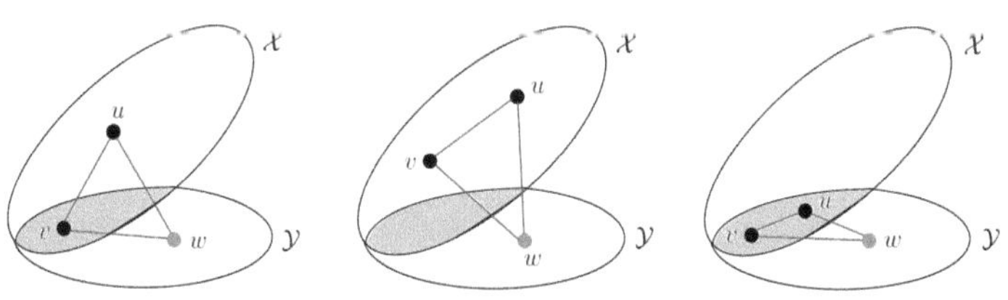

Fig. 2. Illustrating $\mathrm{PTI}^{\alpha}_{\mathcal{X},\mathcal{Y}}$ and $G^{\alpha}_{\mathcal{X},\mathcal{Y}}$.

A novel *parameterized triangle inequality* is defined as follows, which is quite different from those in [5,6,23].

Definition 1 (see Fig. 2.) *Given an edge-weighted connected graph, $G = (V, E, \rho, \mathcal{X}, \mathcal{Y})$, and a real number $1 < \alpha \leq 2$, if $d(\cdot, \cdot)$ satisfies that $d(v, w) + d(w, u) \geq \alpha \cdot d(v, u)$, for any $v, u \in \mathcal{X}, v \neq u$ and $w \in \mathcal{Y}$, then we say that $d(\cdot, \cdot)$ satisfies an α-**parameterized triangle inequality** (α-**PTI**) between $\mathcal{X}$ and $\mathcal{Y}$, denoted as* $\mathrm{PTI}^{\alpha}_{\mathcal{X},\mathcal{Y}}$.

Moreover, $G = (V, E, \rho, \mathcal{X}, \mathcal{Y})$ with $\mathrm{PTI}^{\alpha}_{\mathcal{X},\mathcal{Y}}$ is denoted as $G^{\alpha}_{\mathcal{X},\mathcal{Y}}$.

2.3 Properties of Radius

Lemma 1. *Given any subset $Y \subseteq \mathcal{Y}$, it always holds that $r(X_1, Y) \leq r(X_2, Y)$, for any pair of subsets, $X_1 \subseteq X_2 \subseteq \mathcal{X}$.*

Proof. By Eq. (1) and $X_1 \subseteq X_2$, we obtain

$$r(X_1, Y) = \max_{x \in X_1} d(x, Y) \leq \max_{x \in X_2} d(x, Y) = r(X_2, Y).$$

□

Lemma 2. *Given any subset $Y \subseteq \mathcal{Y}$ and any pair of subsets, $X_1 \subseteq X_2 \subseteq \mathcal{X}$, it always holds that*

$$r(X_1 \cup X_2, Y) = \max\{r(X_1, Y),\ r(X_2, Y)\}. \tag{3}$$

Proof. By Eq. (1), we obtain

$$\begin{aligned} r(X_1 \cup X_2, Y) &= \max_{x \in X_1 \cup X_2} d(x, Y) \\ &= \max\left\{\max_{x \in X_1} d(x, Y),\ \max_{x \in X_2} d(x, Y)\right\} \\ &= \max\{r(X_1, Y),\ r(X_2, Y)\}. \end{aligned}$$

□

Lemma 3. *Given two pairs of disjoint subsets, $X_1, X_2 \subseteq \mathcal{X}$ and $Y_1, Y_2 \subseteq \mathcal{Y}$, it always holds that*

$$r(X_1 \cup X_2, Y_1 \cup Y_2) \leq \max\{r(X_1, Y_1),\ r(X_2, Y_2)\}. \tag{4}$$

3 Parameterized-Factor Approximation to *k*SP in $G^{\alpha}_{\mathcal{X},\mathcal{Y}}$

In this section, we study the *k*SP in $G^{\alpha}_{\mathcal{X},\mathcal{Y}} = (V, E, \rho, \mathcal{X}, \mathcal{Y}, \alpha)$, and present a parameterized-factor approximation algorithm called KSP-PG. Let *opt* be the optimal value and $\mathcal{S}^{\mathrm{A}}$ be an algorithm solution of KSP-PG.

The main idea of KSP-PG is as follows (see Fig. 3): we first select facilities from $\mathcal{X} \cap \mathcal{Y}$ and then select facilities from the remaining candidates in $\mathcal{Y}$, and finally combine two parts into a solution to *k*SP in $G^{\alpha}_{\mathcal{X},\mathcal{Y}}$.

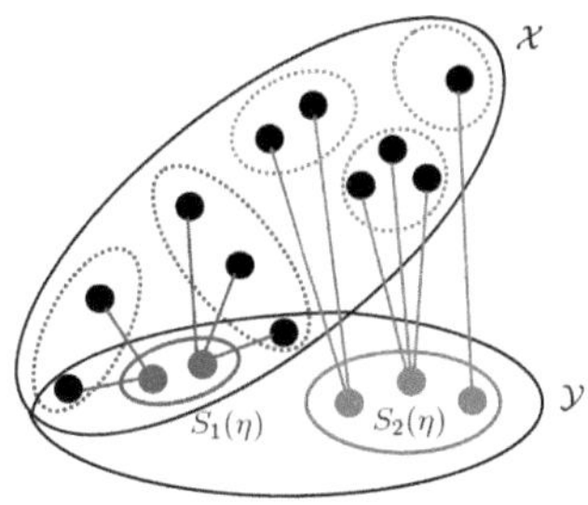

Fig. 3. Illustration for the idea of KSP-PG.

3.1 Stage 1: Selecting Facilities from $\mathcal{X} \cap \mathcal{Y}$

In this subsection, we design a procedure TEST_1 using the dual approximation method to determine whether or not we can select facilities from $\mathcal{X} \cap \mathcal{Y}$ and output them if it answers affirmatively. TEST_1 will serve as an important subprocedure of KSP-PG.

Let $S_1(\eta)$ be the subset of the facilities obtained by $\text{TEST}_1(G^{\alpha}_{\mathcal{X},\mathcal{Y}}, \eta)$ using a real number $\eta > 0$ as a parameter, and $\mathcal{X}_1 \subseteq \mathcal{X}$ be the subset of the clients labelled by TEST_1 when it ends. Accordingly, $\mathcal{X} \setminus \mathcal{X}_1$ denotes the subset of the unlabelled clients. Initially, $S_1(\eta) = \emptyset$ and $\mathcal{X}_1 = \emptyset$. Let $s^\bullet$ be the latest client which is arbitrarily selected as a facility from $(\mathcal{X} \setminus \mathcal{X}_1) \cap \mathcal{Y}$. Every time $s^\bullet$ is selected, all the unlabelled clients $x \in \mathcal{X} \setminus \mathcal{X}_1$ whose distances to $s^\bullet$ are no more than $\frac{2\eta}{\alpha}$ are labelled, i.e., satisfying $d(x, s^\bullet) \leq \frac{2\eta}{\alpha}$. This operation is repeated until all the clients in $\mathcal{X} \cap \mathcal{Y}$ are labelled, i.e., $\mathcal{X}_1 \cap \mathcal{Y} = \mathcal{X} \cap \mathcal{Y}$, or k facilities are obtained from $\mathcal{X} \cap \mathcal{Y}$, i.e., $|S_1(\eta)| = k$. Let $N(s^\bullet; \frac{2\eta}{\alpha})$ be the subset of the unlabelled clients in $\mathcal{X} \setminus \mathcal{X}_1$ which are labelled when $s^\bullet$ is selected. We have

$$N(s^\bullet; \frac{2\eta}{\alpha}) = \left\{ x \in \mathcal{X} \setminus \mathcal{X}_1 : d(x, s^\bullet) \leq \frac{2\eta}{\alpha} \right\}. \tag{5}$$

If TEST_1 ends with $|S_1(\eta)| = k$ and $\mathcal{X}_1 \cap \mathcal{Y} \neq \mathcal{X} \cap \mathcal{Y}$, then TEST_1 fails and returns NO, meaning that there exists no subset $H \subset \mathcal{Y}$ with $|H| \leq k$ satisfying $r(\mathcal{X}, H) \leq \eta$. If TEST_1 ends with $\mathcal{X}_1 \cap \mathcal{Y} = \mathcal{X} \cap \mathcal{Y}$, then TEST_1 returns YES, meaning that it can find a subset $S_1(\eta) \subseteq \mathcal{X} \cap \mathcal{Y}$ with $|S_1(\eta)| \leq k$ satisfying $r(\mathcal{X}_1, S_1(\eta)) \leq \frac{2\eta}{\alpha}$ (Fig. 4).

$\mathsf{TEST}_1(G^{\alpha}_{\mathcal{X},\mathcal{Y}}, \eta)$: Test procedure 1.
Input: $G^{\alpha}_{\mathcal{X},\mathcal{Y}} = (V, E, \rho, \mathcal{X}, \mathcal{Y}, \alpha)$ and $\eta > 0$; **Output:** YES with $S_1(\eta)$ and $\mathcal{X}_1$, or NO.

$S_1(\eta) \leftarrow \emptyset$; $\mathcal{X}_1 \leftarrow \emptyset$;
while $\mathcal{X}_1 \cap \mathcal{Y} \neq \mathcal{X} \cap \mathcal{Y}$ and $\|S_1(\eta)\| < k$ **do**
    Select a client $s^\bullet \in (\mathcal{X} \setminus \mathcal{X}_1) \cap \mathcal{Y}$ arbitrarily;
    $S_1(\eta) \leftarrow S_1(\eta) \cup \{s^\bullet\}$;
    $N(s^\bullet; \frac{2\eta}{\alpha}) \leftarrow \{x \in \mathcal{X} \setminus \mathcal{X}_1 : d(x, s^\bullet) \leq \frac{2\eta}{\alpha}\}$;
    $\mathcal{X}_1 \leftarrow \mathcal{X}_1 \cup N(s^\bullet; \frac{2\eta}{\alpha})$;
end
if $\mathcal{X}_1 \cap \mathcal{Y} = \mathcal{X} \cap \mathcal{Y}$ **then**
    Return YES with $S_1(\eta)$ and $\mathcal{X}_1$;
else Return NO; **endif**

Theorem 1. *Given $G^{\alpha}_{\mathcal{X},\mathcal{Y}} = (V, E, \rho, \mathcal{X}, \mathcal{Y}, \alpha)$, where $|\mathcal{X}| = p$, $|\mathcal{Y}| = q$, and $1 < \alpha \leq 2$, if $\mathcal{X} \cap \mathcal{Y} \neq \emptyset$ and there is a subset $H \subset \mathcal{Y}$ with $|H| \leq k$ satisfying $r(\mathcal{X}, H) \leq \eta$, then $\mathsf{TEST}_1(G^{\alpha}_{\mathcal{X},\mathcal{Y}}, \eta)$ output YES and two subsets, $\mathcal{X}_1 \subseteq \mathcal{X}$ and $S_1(\eta) \subseteq \mathcal{X} \cap \mathcal{Y}$, satisfying $|S_1(\eta)| \leq k$ and $r(\mathcal{X}_1, S_1(\eta)) \leq \frac{2\eta}{\alpha}$ in $O(kp)$ time for the kSP in $G^{\alpha}_{\mathcal{X},\mathcal{Y}}$, for any $\eta > 0$.*

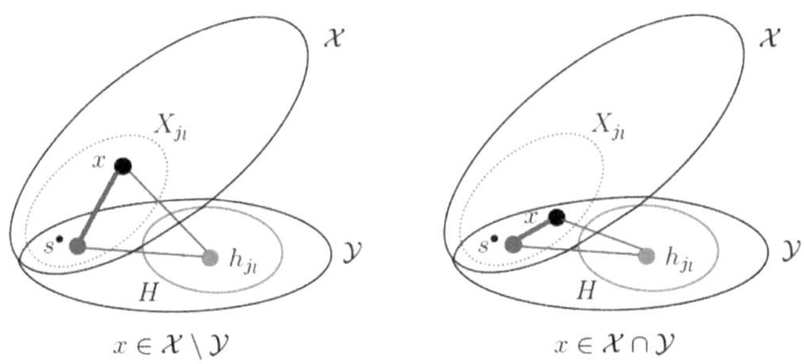

Fig. 4. Illustration for TEST_1 and the proof of Theorem 1 ($x \in \mathcal{X} \setminus \mathcal{Y}$ or $x \in \mathcal{X} \cap \mathcal{Y}$).

Proposition 1. *If $\mathcal{X}_1 = \mathcal{X}$ then $S_1(\eta)$ is a solution to the kSP in $G^{\alpha}_{\mathcal{X},\mathcal{Y}}$.*

Proof. It is implied by $\mathcal{X}_1 = \mathcal{X}$ that all the clients in $\mathcal{X}$ are labelled and $|S_1(\eta)| \leq k$ when $\mathsf{TEST}_1(G^{\alpha}_{\mathcal{X},\mathcal{Y}}, \eta)$ ends. $\square$

Let $H = \{h_1, h_2, \ldots, h_g\}$ be a g-set of $\mathcal{Y}$ satisfying that $r(\mathcal{X}, H) \leq \eta$. For $1 \leq j \leq g$, we let X_j denote the subset of $\mathcal{X}$, where h_j is the closest facility in H to the clients in X_j, i.e., all the clients in X_j connect to h_j. We say that h_j is the *center* of X_j.

Proposition 2. *If $\mathcal{X}_1 \subset \mathcal{X}$ then it always holds that $h_j \notin S_1(\eta)$ such that $h_j \in H$ is the center of the subset X_j having x, for any $x \in \mathcal{X} \setminus \mathcal{X}_1$.*

Proof. We discuss two cases as follows:

(1) If $h_j \in \mathcal{Y} \setminus \mathcal{X}$, then it follows from $S_1(\eta) \in \mathcal{X} \cap \mathcal{Y}$ that $h_j \notin S_1(\eta)$.

(2) If $h_j \in \mathcal{X} \cap \mathcal{Y}$, then we suppose that $h_j \in S_1(\eta)$. Obviously, all the clients in X_j are labelled by TEST_1. Due to $x \in X_j$, we claim that x is labelled and so $x \in \mathcal{X}_1$. This causes a contradiction. □

Proposition 3. $\mathsf{TEST}_1(G^{\alpha}_{\mathcal{X},\mathcal{Y}}, opt)$ *must return* YES, *for any* $1 < \alpha \leq 2$.

Proof. Recall that $\mathcal{S}^*$ is an optimum solution to the kSP in $G^{\alpha}_{\mathcal{X},\mathcal{Y}}$ and *opt* is the optimal value. So, $\mathcal{S}^*$ is a subset of $\mathcal{Y}$ with $|\mathcal{S}^*| \leq k$ and satisfies $r(\mathcal{X}, \mathcal{S}^*) \leq opt$. According to Theorem 1, we claim that $\mathsf{TEST}_1(G^{\alpha}_{\mathcal{X},\mathcal{Y}}, opt)$ must return YES. □

Corollary 1. *Given* $G^{\alpha}_{\mathcal{X},\mathcal{Y}} = (V, E, \rho, \mathcal{X}, \mathcal{Y}, \alpha)$, *where* $1 < \alpha \leq 2$, *if* $\mathsf{TEST}_1(G^{\alpha}_{\mathcal{X},\mathcal{Y}}, opt)$ *obtains* $\mathcal{X}_1 = \mathcal{X}$ *then it can output a* $\frac{2}{\alpha}$*-approximation to the* k*SP in* $G^{\alpha}_{\mathcal{X},\mathcal{Y}}$.

3.2 Stage 2: Selecting Facilities from $\mathcal{Y} \setminus S_1(\eta)$

Provided that TEST_1 outputs $\mathcal{X}_1 \subset \mathcal{X}$ and $|S_1(\eta)| < k$ in Stage 1, we need to select at most $k - |S_1(\eta)|$ facilities from the unlabelled candidates in $\mathcal{Y} \setminus S_1(\eta)$ in Stage 2. In this subsection, we design a procedure TEST_2 using a dual approximation method to determine whether or not we can select the required facilities from $\mathcal{Y} \setminus S_1(\eta)$ and output them if it answers affirmatively. TEST_2 will serve as an important sub-procedure of KSP-PG (Fig. 5).

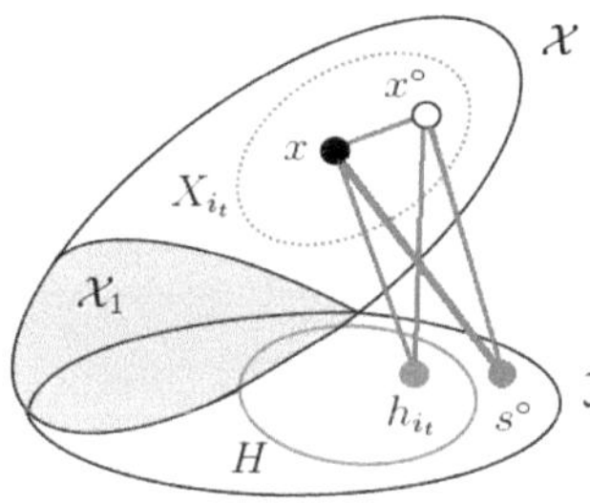

Fig. 5. Illustration for TEST_2 and the proof of Theorem 2.

Let $S_2(\eta)$ be the subset of the facilities obtained by TEST_2 using a real number $\eta > 0$ as a parameter, and $\mathcal{X}_2 \subseteq \mathcal{X}$ be the subset of the clients labelled by TEST_2 when it ends. Accordingly, $\mathcal{X} \setminus (\mathcal{X}_1 \cup \mathcal{X}_2)$ denotes the subset of the unlabelled clients. Let Y be the subset of the candidates in $\mathcal{Y} \setminus S_1(\eta)$ which are not selected as facilities in Stage 1. Initially, $S_2(\eta) = \emptyset$, $\mathcal{X}_2 = \emptyset$, and $Y = \mathcal{Y} \setminus S_1(\eta)$. Let x° be the latest client which is arbitrarily selected from the set $\mathcal{X} \setminus (\mathcal{X}_1 \cup \mathcal{X}_2)$ of unlabelled clients, and s° be the closest candidate in Y to x°. Every time s° is selected, all the unlabelled clients $x \in \mathcal{X} \setminus (\mathcal{X}_1 \cup \mathcal{X}_2)$ whose distance to s° do not exceed $(\frac{2}{\alpha^2} + \frac{1}{\alpha})\eta$ are labelled. This operation is repeated until all the clients in $\mathcal{X} \setminus \mathcal{X}_1$ are labelled, or $k - |S_1(\eta)|$ facilities are selected from $\mathcal{Y} \setminus S_1(\eta)$. Let $N(s^\circ; (\frac{2}{\alpha^2} + \frac{1}{\alpha})\eta))$ be the subset of the unlabelled clients in $\mathcal{X} \setminus (\mathcal{X}_1 \cup \mathcal{X}_2)$ that are labelled when s° is selected out. We have

$$N(s^\circ; (\frac{2}{\alpha^2} + \frac{1}{\alpha})\eta) = \left\{x \in \mathcal{X} \setminus (\mathcal{X}_1 \cup \mathcal{X}_2) : d(x, s^\circ) \leq (\frac{2}{\alpha^2} + \frac{1}{\alpha})\eta\right\}. \quad (6)$$

If TEST_2 ends with $|S_2(\eta)| = k - |S_1(\eta)|$ and $\mathcal{X} \setminus (\mathcal{X}_1 \cup \mathcal{X}_2) \neq \emptyset$, then TEST_2 fails and returns NO, meaning that there exists no subset $H \subset \mathcal{Y}$ with $|H| \leq k$ satisfying $r(\mathcal{X}, H) \leq \eta$. If TEST_2 ends with $\mathcal{X} \setminus (\mathcal{X}_1 \cup \mathcal{X}_2) = \emptyset$, then TEST_2 returns YES, meaning that it can find a subset $S_2(\eta)$ from $\mathcal{Y} \setminus S_1(\eta)$ satisfying $|S_2(\eta)| \leq k - |S_1(\eta)|$ and $r(\mathcal{X}_2, S_2(\eta)) \leq (\frac{2}{\alpha^2} + \frac{1}{\alpha})\eta$.

$\mathsf{TEST}_2(G^\alpha_{\mathcal{X},\mathcal{Y}}, \eta, \mathcal{X}_1, S_1(\eta))$: Test procedure 2.
Input: $G^\alpha_{\mathcal{X},\mathcal{Y}} = (V, E, \rho, \mathcal{X}, \mathcal{Y}, \alpha)$, $\eta > 0$, $\mathcal{X}_1$, and $S_1(\eta)$; **Output:** YES with $S_2(\eta)$ and $\mathcal{X}_2$, or NO.

$S_2(\eta) \leftarrow \emptyset$; $\mathcal{X}_2 \leftarrow \emptyset$; $Y \leftarrow \mathcal{Y} \setminus S_1(\eta)$;
while $\mathcal{X} \setminus (\mathcal{X}_1 \cup \mathcal{X}_2) \neq \emptyset$ and $\lvert S_2(\eta)\rvert < k - \lvert S_1(\eta)\rvert$ **do**
   Select a client $x^\circ \in \mathcal{X} \setminus (\mathcal{X}_1 \cup \mathcal{X}_2)$ arbitrarily;
   Select the closest candidate, s°, in Y to x°;
   $S_2(\eta) \leftarrow S_2(\eta) \cup \{s^\circ\}$; $Y \leftarrow Y \setminus \{s^\circ\}$;
   $N(s^\circ; (\frac{2}{\alpha^2} + \frac{1}{\alpha})\eta) \leftarrow \{x \in \mathcal{X} \setminus (\mathcal{X}_1 \cup \mathcal{X}_2) : d(x, s^\circ) \leq (\frac{2}{\alpha^2} + \frac{1}{\alpha})\eta\}$;
   $\mathcal{X}_2 \leftarrow \mathcal{X}_2 \cup N(s^\circ; (\frac{2}{\alpha^2} + \frac{1}{\alpha})\eta)$;
end
if $\mathcal{X} \setminus (\mathcal{X}_1 \cup \mathcal{X}_2) = \emptyset$ **then**
   Return YES with $S_2(\eta)$ and $\mathcal{X}_2$;
else Return NO; **endif**

Theorem 2. *Given $G^\alpha_{\mathcal{X},\mathcal{Y}} = (V, E, \rho, \mathcal{X}, \mathcal{Y}, \alpha)$, where $|\mathcal{X}| = p$, $|\mathcal{Y}| = q$, and $1 < \alpha \leq 2$, if $\mathcal{Y} \setminus \mathcal{X} \neq \emptyset, \mathcal{X}_1 \subset \mathcal{X}$ and there is a subset $H \subset \mathcal{Y}$ with $|H| \leq k$ satisfying $r(\mathcal{X}, H) \leq \eta$, then $\mathsf{TEST}_2(G^\alpha_{\mathcal{X},\mathcal{Y}}, \eta, \mathcal{X}_1, S_1(\eta))$ output YES with two subsets, $\mathcal{X}_2 = \mathcal{X} \setminus \mathcal{X}_1$ and $S_2(\eta) \subseteq \mathcal{Y} \setminus S_1(\eta)$, satisfying that $|S_2(\eta)| \leq k - S_1(\eta)$ and $r(\mathcal{X}_2, S_2(\eta)) \leq (\frac{2}{\alpha^2} + \frac{1}{\alpha})\eta$ in $O(k(p+q))$ time for the kSP in $G^\alpha_{\mathcal{X},\mathcal{Y}}$, for any $\eta > 0$.*

Proposition 4. *$\mathsf{TEST}_2(G^\alpha_{\mathcal{X},\mathcal{Y}}, opt, \mathcal{X}_1, S_1(opt))$ must return YES, for any $1 < \alpha \leq 2$.*

Corollary 2. *Given $G^\alpha_{\mathcal{X},\mathcal{Y}} = (V, E, \rho, \mathcal{X}, \mathcal{Y}, \alpha)$, where $1 < \alpha \leq 2$, if $\mathsf{TEST}_1(G^\alpha_{\mathcal{X},\mathcal{Y}}, opt)$ obtains $\mathcal{X}_1 \subset \mathcal{X}$ then the combination of the output of $\mathsf{TEST}_1(G^\alpha_{\mathcal{X},\mathcal{Y}}, opt)$ and the output of $\mathsf{TEST}_2(G^\alpha_{\mathcal{X},\mathcal{Y}}, opt, \mathcal{X}_1, S_1(opt))$ forms a $(\frac{2}{\alpha^2} + \frac{1}{\alpha})$-approximation to the kSP in $G^\alpha_{\mathcal{X},\mathcal{Y}}$.*

3.3 Merging Two Stages

In Stage 1, $\mathsf{TEST}_1(G^\alpha_{\mathcal{X},\mathcal{Y}}, \eta)$ selects facilities from $\mathcal{X} \cap \mathcal{Y}$. Given a parameter $\eta > 0$, if $\mathsf{TEST}_1(G^\alpha_{\mathcal{X},\mathcal{Y}}, \eta)$ outputs NO then we claim by Theorem 1 that there is no subset $H \subset \mathcal{Y}$ with $|H| \leq k$ satisfying that $r(\mathcal{X}, H) \leq \eta$. In other words, η is an invalid parameter. If $\mathsf{TEST}_1(G^{\alpha}_{\mathcal{X},\mathcal{Y}}, \eta)$ outputs YES and $\mathcal{X}_1 = \mathcal{X}$, then we claim by Proposition 1 that $S_1(\eta)$ is a solution to the kSP in $G^{\alpha}_{\mathcal{X},\mathcal{Y}}$. If $\mathsf{TEST}_1(G^{\alpha}_{\mathcal{X},\mathcal{Y}}, \eta)$ outputs YES and $\mathcal{X}_1 \subset \mathcal{X}$, then we claim that we need to find more facilities, i.e., $S_1(\eta)$ is only part of a solution to the kSP in $G^{\alpha}_{\mathcal{X},\mathcal{Y}}$. Next, we go to Stage 2.

In Stage 2, $\mathsf{TEST}_2(G^{\alpha}_{\mathcal{X},\mathcal{Y}}, \eta, \mathcal{X}_1, S_1(\eta))$ selects facilities from $\mathcal{Y} \setminus S_1(\eta)$. If TEST_2 outputs NO then we claim by Theorem 2 that there is no subset $H \subset \mathcal{Y}$ satisfying the requirements. If TEST_2 outputs YES then we claim by Theorems 1 and 2 that $S(\eta) = S_1(\eta) \cup S_2(\eta)$ is a solution.

These two stages can be merged as a complete test procedure TEST, which can be used to determine whether or not $\eta > 0$ is a valid parameter and if the answer is affirmative then a solution is output.

```
TEST(G^α_{X,Y}, η): Complete test procedure.
Input: G^α_{X,Y} = (V, E, ρ, X, Y, α) and η > 0;
Output: YES and S(η), or NO.
if TEST_1(G^α_{X,Y}, η) = [YES, S_1(η), X_1] then // Stage 1
  if X_1 = X then
    S(η) ← S_1(η); Return YES and S(η);
  else
    if TEST_2(G^α_{X,Y}, η, X_1, S_1(η)) = [YES, S_2(η), X_2] then // Stage 2
      S(η) ← S_1(η) ∪ S_2(η); Return YES and S(η);
    else Return NO; endif
  endif
else Return NO; endif
```

Proposition 5. $\mathsf{TEST}(G^{\alpha}_{\mathcal{X},\mathcal{Y}}, opt)$ *must return YES, for any* $1 < \alpha \leq 2$.

Proof. Because both $\mathsf{TEST}_1(G^{\alpha}_{\mathcal{X},\mathcal{Y}}, opt)$ and $\mathsf{TEST}_2(G^{\alpha}_{\mathcal{X},\mathcal{Y}}, opt, \mathcal{X}_1, S_1(opt))$ must return YES, we claim that $\mathsf{TEST}(G^{\alpha}_{\mathcal{X},\mathcal{Y}}, opt)$ also must return YES, according to procedure TEST. □

Theorem 3. *Given* $G^{\alpha}_{\mathcal{X},\mathcal{Y}} = (V, E, \rho, \mathcal{X}, \mathcal{Y}, \alpha)$, *where* $|\mathcal{X}| = p$, $|\mathcal{Y}| = q$, *and* $1 < \alpha \leq 2$, *if* $\mathsf{TEST}(G^{\alpha}_{\mathcal{X},\mathcal{Y}}, opt)$ *obtains YES and* $\mathcal{X}_1 = \mathcal{X}$ *then it can output a* $\frac{2}{\alpha}$*-approximation in* $O(kp)$ *time, and if* $\mathsf{TEST}(G^{\alpha}_{\mathcal{X},\mathcal{Y}}, opt)$ *obtains YES and* $\mathcal{X}_1 \subset \mathcal{X}$ *then it can output a* $(\frac{2}{\alpha^2} + \frac{1}{\alpha})$*-approximation in* $O(k(p+q))$ *time, for the kSP in* $G^{\alpha}_{\mathcal{X},\mathcal{Y}}$.

3.4 Parameterized-Factor Approximation Algorithm

In this section, using TEST as a complete test procedure, we develop a two-stage dual approximation algorithm KSP-PG for the kSP in $G^{\alpha}_{\mathcal{X},\mathcal{Y}}$.

Because each of $d(x_i, y_j), \forall x_i \in \mathcal{X}, y_j \in \mathcal{Y}$ is a potential value of radius, the number of all the potential values of radius is at most $|\mathcal{X}| \cdot |\mathcal{Y}| = pq$. We first delete the duplicates in them and then arrange the remaining M values into an increasing sequence, $z_1 < z_2 < \cdots < z_M$. Clearly, $M \leq pq$.

Proposition 5 shows that $\mathsf{TEST}(G^{\alpha}_{\mathcal{X},\mathcal{Y}}, opt)$ must return YES, and Theorem 3 shows that $\mathsf{TEST}(G^{\alpha}_{\mathcal{X},\mathcal{Y}}, opt)$ can output an approximation. We find the smallest value, z_{m^*}, in the sequence $z_1, z_2, \ldots, z_M$ such that TEST returns YES, and take z_{m^*} as the substitute of *opt*. In order to reduce the times of search, KSP-PG applies a *binary* search to determine z_{m^*} instead of a sequential search. It is sufficient to use TEST $O(\log M)$ times.

If $\mathsf{TEST}(G^{\alpha}_{\mathcal{X},\mathcal{Y}}, z_1)$ returns YES and $S(z_1)$, then $z_{m^*} = z_1$ and $\mathcal{S}^{\mathrm{A}} = S(z_1)$ is output as a solution, and otherwise a binary search is used to determine z_{m^*}. Let LB and UB be the lower bound and upper bound of m^*, respectively, and $\mathrm{B} = \lfloor \frac{\mathrm{LB}+\mathrm{UB}}{2} \rfloor$. Initialize $\mathrm{LB} = 1, \mathrm{UB} = M$. If $\mathsf{TEST}(G^{\alpha}_{\mathcal{X},\mathcal{Y}}, z_{\mathrm{B}})$ returns YES, then it takes B as a new upper bound and LB remains as a lower bound. If $\mathsf{TEST}(G^{\alpha}_{\mathcal{X},\mathcal{Y}}, z_{\mathrm{B}})$ returns NO, then it takes B as a new lower bound and UB remains as an upper bound. This operation is repeated until UB and LB become two consecutive integers. The final UB is just the index m^*, and $\mathcal{S}^{\mathrm{A}} = S(z_{\mathrm{UB}})$ is output as a solution by $\mathsf{TEST}(G^{\alpha}_{\mathcal{X},\mathcal{Y}}, z_{\mathrm{UB}})$.

KSP-PG($G^{\alpha}_{\mathcal{X},\mathcal{Y}}$): Algorithm for the kSP in $G^{\alpha}_{\mathcal{X},\mathcal{Y}}$.
Input: $G^{\alpha}_{\mathcal{X},\mathcal{Y}} = (V, E, \rho, \mathcal{X}, \mathcal{Y}, \alpha)$; **Output:** YES and $\mathcal{S}^{\mathrm{A}}$, or NO.

Delete the duplicates in $\{d(x_i, y_j) : x_i \in \mathcal{X}, y_j \in \mathcal{Y}\}$ and arrange the remaining values into $z_1 < z_2 < \cdots < z_M$.
if $\mathsf{TEST}(G^{\alpha}_{\mathcal{X},\mathcal{Y}}, z_1) = [\mathsf{YES}, S(z_1)]$ **then**
    $\mathcal{S}^{\mathrm{A}} \leftarrow S(z_1)$; Return YES and $\mathcal{S}^{\mathrm{A}}$;
else
    $\mathrm{LB} \leftarrow 1; \mathrm{UB} \leftarrow M$;
    while $\mathrm{UB} - \mathrm{LB} \neq 1$ **do**
        $\mathrm{B} \leftarrow \lfloor \frac{\mathrm{LB}+\mathrm{UB}}{2} \rfloor$;
        if $\mathsf{TEST}(G^{\alpha}_{\mathcal{X},\mathcal{Y}}, z_{\mathrm{B}}) = [\mathsf{YES}, S(z_{\mathrm{B}})]$ **then** $\mathrm{UB} \leftarrow \mathrm{B}$;
        else $\mathsf{TEST}(G^{\alpha}_{\mathcal{X},\mathcal{Y}}, z_{\mathrm{B}}) = \mathsf{NO}$ **then** $\mathrm{LB} \leftarrow \mathrm{B}$; **endif**
    end
    $\mathcal{S}^{\mathrm{A}} \leftarrow S(z_{\mathrm{UB}})$; Return YES and $\mathcal{S}^{\mathrm{A}}$;
endif

Theorem 4. *Given $G^{\alpha}_{\mathcal{X},\mathcal{Y}} = (V, E, \rho, \mathcal{X}, \mathcal{Y}, \alpha)$, where $|\mathcal{X}| = p$, $|\mathcal{Y}| = q$, and $1 < \alpha \leq 2$, if KSP-PG($G^{\alpha}_{\mathcal{X},\mathcal{Y}}$) obtains $\mathcal{X}_1 = \mathcal{X}$ then it can output a $\frac{2}{\alpha}$-approximation, and if it obtains $\mathcal{X}_1 \subset \mathcal{X}$ then it can output a $(\frac{2}{\alpha^2} + \frac{1}{\alpha})$-approximation. Also, KSP-PG($G^{\alpha}_{\mathcal{X},\mathcal{Y}}$) has a time cost of $O(p(k \log M + q \log p) + q(k \log M + p \log q))$, for the kSP in $G^{\alpha}_{\mathcal{X},\mathcal{Y}}$.*

Corollary 3. $\frac{2}{\alpha} \in [1, 2)$ *and* $\frac{2}{\alpha^2} + \frac{1}{\alpha} \in [1, 3)$.

3.5 Further Discussions

It is not difficult to examine that KSP-PG is also appliable to the cases of $\mathcal{X} \subseteq \mathcal{Y}$ and $\mathcal{Y} \subseteq \mathcal{X}$ and obtains a 2-approximation, respectively. Next, we take no account of the parameterized aspect of the triangle inequality in $G = (V, E, \rho, \mathcal{X}, \mathcal{Y})$. Taking $\alpha = 1$ into Theorem 4, we obtain the following corollary.

Corollary 4. *Regardless of the parameterized aspect of graphs, if KSP-PG stops at the end of Stage 1 then it can output a 2-approximation, and if KSP-PG stops at the end of Stage 2 then it can output a 3-approximation.*

4 Conclusions

It is also of interest to study other clustering problems in graphs with parameterized triangle inequality.

Acknowledgments. The first author was supported by Nanxun scholars program of ZJWEU (RC2023010481). We thank reviewers so much for their insightful comments which help to improve the presentation of this paper significantly.

References

1. Aggarwal, G., et al.: Achieving anonymity via clustering. ACM Trans. Alg. **6**(3), 49:1-49:19 (2010)
2. Ahmadian, S. and Swamy, C.: Approximation algorithms for clustering problems with lower bounds and outliers. In: Proceedings of 43rd ICALP, pp. 69:1-69:15 (2016)
3. An, H.C., Bhaskara, A., Chekuri, C., Gupta, S., Madan, V., Svensson, O.: Centrality of trees for capacitated k-center. Math. Program. **154**(1–2), 29–53 (2015)
4. Barilan, J., Kortsarz, G., Peleg, D.: How to allocate network centers. J. Algor. **15**(3), 385–415 (1993)
5. Bläser, M., Manthey, B., Sgall, J.: An improved approximation algorithm for the asymmetric TSP with strengthened triangle inequality. J. Disc. Alg. **4**, 623–632 (2006)
6. Chandran, L.S., Ram, L.S: On the relationship between ATSP and the cycle cover problem. Theor. Comput. Sci. **370**, 218–228 (2007)
7. Cygan, M., Hajiaghayi, M., Khuller, S.: LP rounding for k-Centers with non-uniform hard capacities. In: Proceedings of 53rd FOCS, pp. 273–282 (2012)
8. Ding, H., Hu, L., Huang, L., Li, J.: Capacitated center problems with two sided bounds and outliers. In: Proceedings of 15th WADS, pp. 325–336 (2017)
9. Ding, W., Qiu, K.: A 2-approximation algorithm and beyond for the minimum diameter k-Steiner forest problem. Theor. Comput. Sci. **840**, 1–15 (2020)
10. Feder, T., Greene, D.H.: Optimal algorithms for approximate clustering. In: Proceedings of 20th STOC, pp. 434–444 (1988)
11. Gonzalez, T.F.: Clustering to minimize the maximum intercluster distance. Theor. Comput. Sci. **38**, 293–306 (1985)
12. Goyal, D., Jaiswal, R.: Tight FPT approximation for constrained k-center and k-supplier. Theor. Comput. Sci. **940**, 190–208 (2023)
13. Hakimi, S.L.: Optimum locations of switching centers and the absolute centers and medians of a graph. Oper. Res. **12**(3), 450–459 (1964)
14. Hochbaum, D.S., Shmoys, D.B.: A best possible heuristic for the k-Center problem. Math. Oper. Res. **10**(2), 180–184 (1985)
15. Hochbaum, D.S., Shmoys, D.B.: A unified approach to approximation algorithms for bottleneck problems. J. ACM **33**, 533–550 (1986)

16. Kariv, O., Hakimi, S.L.: An algorithmic approach to network location problems. I: the p-Centers. SIAM J. Appl. Math. **37**(3), 513–538 (1979)
17. Khuller, S., Sussmann, Y.J.: The capacitated k-center problem. SIAM J. Disc. Math. **13**(3), 403–418 (2000)
18. Kociumaka, T., Cygan, M.: Constant factor approximation for capacitated k-center with outliers. In: Proceedings of 31st STACS, pp. 251–262 (2014)
19. Lee, E., Nagarajan, V., Wang, L.: On some variants of Euclidean k-supplier. Oper. Res. Lett. **50**(2), 115–121 (2022)
20. Liang, H.Y.: The hardness and approximation of the star p-hub center problem. Oper. Res. Lett. **41**, 138–141 (2013)
21. Nagarajan, V., Schieber, B., Shachnai, H.: The Euclidean k-supplier problem. Math. Oper. Res. **45**(1), 1–14 (2020)
22. Sweeney, L.: k-anonymity: a model for protecting privacy. Int. J. Uncertain. Fuzz. Knowl.-Based Syst. **10**(05), 557–570 (2002)
23. Zhang, T.Q., Li, W.D., Li, J.P.: An improved approximation algorithm for the ATSP with parameterized triangle inequality. J. Algor. **64**, 74–78 (2009)

A Quadratic Kernel for {Claw, Diamond}-free Deletion

Hanchun Yuan(✉)

Zhejiang Normal University, Jinhua, China

Abstract. The {claw, diamond}-free edge deletion problem admits an $O(k^3)$-vertex kernel [Theoretical Computer Science 2022]. In this paper, we improve it to $O(k^2)$ vertices. We use the modulator technique and then bound the number of vertices outside the modulator. Let M be an arbitrary modulator. We find that, the propagation between the maximal cliques in $G - M$ is always due to the claw graph. Based on the observation, the number of vertices in the maximal cliques untouched to M can be bounded by $O(k^2)$. Since $|M| = O(k)$ and the maximal cliques in $G - M$ touched to M is bounded by some normal kernel frameworks, we thus have the claimed kernelization size.

Keywords: Kernelization · Claw-free · {Claw · Diamond}-free · Parameterized Algorithms

1 Introduction

The graph modification problem is that given a graph G and a parameter k, whether one can apply at most k operations to make G satisfy some properties. The operation can be vertex deletion, edge deletion, edge addition, and edge editing which allows edge deletion and addition. A graph is H-free if it has no induced subgraph isomorphic to H. The H-free modification problem is a subclass of graph modification problems where the property is H-free. There are much research about H-free modification problems via parameterized algorithms view. In parameterized algorithms, we consider whether one can apply at most k modifications to obtain an H-free graph in $f(k) \cdot n^{O(1)}$ time. If a problem can be detected in $f(k) \cdot n^{O(1)}$, then the problem is a fixed-parameter tractable problem. The complexity class containing all fixed-parameter tractable problems is called FPT. If a problem is in FPT, then we will consider whether the problem admits a polynomial kernel, where a kernel is an equivalent instance whose size is bounded by $g(k)$ and the instance can be obtained in polynomial time [6]. The polynomial time algorithm for obtaining a kernel is called a *kernelization algorithm*. When H is a claw, see Fig. 1, whether claw-free edge deletion admits a polynomial kernel is an important open problem [5].

In this paper, we consider the {claw, diamond}-free edge deletion problem where a diamond is a K_4 minus one edge, see Fig. 1. We now describe the formal definition of the problem in the following.

M. Li et al. (Eds.): TAMC 2025, LNCS 16084, pp. 367–378, 2026.
https://doi.org/10.1007/978-981-95-4839-2_29

(a) claw. (b) diamond.

Fig. 1. The claw and diamond.

{Claw, Diamond}-free Edge Deletion	
Input:	An undirected graph $G = (V, E)$ and $k \in \mathbb{N}$.
Parameter:	k.
Question:	Is there a set $S \subseteq E(G)$ such that $G - S$ is {claw, diamond}-free and $\|S\| \leq k$?

Let us simplify {claw, diamond}-free edge deletion by {claw, diamond}-free deletion. One may wonder the kernelization algorithm about {claw, diamond}-free vertex deletion. This problem can reduce to the 4-hitting set, which admits an $O(k^3)$-vertex kernel, see, e.g., [1]. Actually, the main difficulty about edge modification is the 'propagation', which means that deleting or adding an edge may cause some new forbiddens. Note that {claw, diamond}-free graph is a subclass of claw-free graph. Moreover, {claw, diamond}-free graph is a subclass of line graph [13]. Line graph and {claw, diamond}-free graph are all subclasses of claw-free graph. There are thoroughly research about the kernelization algorithms for subclasses of claw-free graph, such as {claw, diamond}-free graphs [7,12], proper Helly circular-arc graphs [2,14], unit interval graphs [3,9,10], and line graphs [8]. Note that the above research contains the vertex deletion version and some edge modification versions. Specifically, for the {claw, diamond}-free graph, Cygan et al. proved that {claw, diamond}-free deletion admits an $O(k^{12})$-vertex kernel [7], and then Li et al. improved it to an $O(k^3)$-vertex kernel [12]. In this paper, we give an $O(k^2)$-vertex kernel by five reduction rules.

Theorem 1. *The {claw, diamond}-free deletion problem admits an $O(k^2)$-vertex kernel.*

1.1 Overview of the Kernelization Algorithm

We first build a modulator M: Greedy find a maximal set T of edge-disjoint claws and diamonds and let $M = V(T)$. Note that if $|T| > k$, then we have a trivially no-instance. Since each claw or diamond has four vertices, we have that $|M| = |V(T)| \leq 4k$. Then, $G - M$ is {claw, diamond}-free. Since each{claw, diamond}-free graph has a maximal clique decomposition, see Sect. 2.3, and so $G - M$ does. Let $\mathcal{K}$ be the decomposition of $G - M$.

Then, we apply some rules to bound the number of vertices in $G - M$. First, we can delete all the components that are {claw, diamond}-free in G. The main trick of our algorithm is that *if there is a forbidden (due to edge deletion) between* *two maximal cliques in $\mathcal{K}$, the forbidden must be a claw.* By the observation, we have that if a clique in $\mathcal{K}$ is also maximal in G, then for any optimal solution S, the resulting graph $K - S$ is a cluster. Then, the propagation of causing claw is stop and "most of" cliques in $\mathcal{K}$ have no edge in S. We can design a rule to simplify the graph structure (Rule 3). Moreover, if a clique in $\mathcal{K}$ is big enough, any optimal solution cannot have an edge in the clique (Rule 4). After the above rules, we already have an $O(k^3)$-vertex kernel. To have an $O(k^2)$-vertex kernel, we need to delve into the lower bound of the number of deleted edges in some special cliques and then design a rule to delete some redundant vertices (Rule 5).

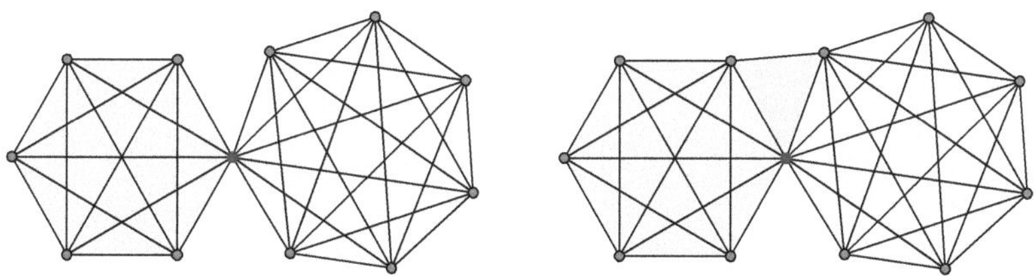

(a) A {claw, diamond}-free graph. (b) A line graph with a diamond.

Fig. 2. The right graph is a line graph, but it is not {claw, diamond}-free, since the violet vertex is contained in three maximal cliques with different colors.

2 Preliminaries

We now give some definitions and properties about {claw, diamond}-free and a maximal clique partition for a {claw, diamond}-free graph.

2.1 Basic Definitions

The vertex set and edge set of a graph G are denoted by $V(G)$ and $E(G)$. Given a vertex $v \in V(G)$, let $N(v)$ be the set of neighbors in G and $N[v] = \{v\} \cup N(v)$. Similarly, given a vertex set $V' \subset V(G)$, let $N(V') = \cup_{v \in V'} N(v)$ and $N[V'] = V' \cup N(V')$. An *induced subgraph* G' of G is a subgraph of G such that $uv \in E(G')$ if and only if $u, v \in V(G')$ and $uv \in E(G)$. For a vertex set X, we denote by $G - X$ the induced subgraph $G[V(G) \setminus X]$. When X is an edge set, $G - X$ is the subgraph of G by removing edges in X. A *clique* is a set of pairwise adjacent vertices and a *cluster* is a graph where each component is a clique. For basic definitions not included here, we refer the reader to [4].

2.2 The Structures of {Claw, Diamond}-free

We first give a characterization of {claw, diamond}-free graphs by maximal cliques, see Fig. 2 as an example.

Lemma 1 ([11,13]). *The graph G is {claw, diamond}-free if and only if every vertex is in at most two maximal cliques of G and every edge is in exactly one maximal clique.*

In the following, we give two pairwise inverse operations about the maximal cliques in G, see Fig. 3 as an example.

Definition 1 (Splitting). *Let K_1 and K_2 be two maximal cliques in G such that $K_1 \cap K_2 = \{v\}$ and there is no edge between K_1 and K_2. The splitting operation of K_1 and K_2, or say the splitting operation at v is that deleting v from G and then add two vertices v_1 and v_2 such that $N(v_1) = N(v) \cap K_1$ and $N(v_2) = N(v) \cap K_2$. Moreover, for a vertex v such that $N[v]$ is a maximal clique, the splitting operation at v is just adding one isolated vertex in G.*

Now we give an inverse operation of splitting, which is called gluing.

Definition 2 (Gluing). *Let K_1 and K_2 be two maximal cliques in G such that $K_1 \cap K_2 = \emptyset$ and there is no edge between K_1 and K_2. Let $v_1 \in K_1$ and $v_2 \in K_2$. The guling operation of K_1 and K_2, or say the splitting operation at $\{v_1, v_2\}$ is that deleting v_1 and v_2 from G and then adding one vertex v such that $N(v) = N(v_1) \cup N(v_2)$.*

Lemma 2. *Let G be a {claw, diamond}-free graph, and let $v \in V(G)$ satisfying one of the followinig conditions.*

1. *there are two maximal cliques such that $K_1 \cap K2 = \{v\}$ and there is no edge between them;*
2. *$N[v]$ is a maximal clique.*

After splitting at v, the new graph G' is also {claw, diamond}-free.

Lemma 3. *Let G' be a {claw, diamond}-free graph and K_1' and K_2' two maximal cliques such that $K_1' \cap K_2' = \emptyset$ and no edge between K_1' and K_2'. After gluing $v_1' \in K_1'$ and $v_2' \in K_2'$, the new graph G is also {claw, diamond}-free.*

2.3 The Structures of $\mathcal{K}$

We partition the graph by two parts, one is modulator M and another is $R = G - M$, where R is {claw, diamond}-free. By Lemma 1, in R, we can find a maximal clique decomposition $\mathcal{K}$ in polynomial time such that each element in $\mathcal{K}$ is a maximal clique.

We define some types of cliques in $\mathcal{K}$. First, A clique $K \in \mathcal{K}$ is *big* if it has at least $k + 3$ vertices and is *small* otherwise. For other types of cliques, see the following definition.

Definition 3 (Types of Cliques). *If there is a vertex $v \in M$ such that v is adjacent to all vertices of a clique $K \in \mathcal{K}$, then K is an* all-touched *clique. If a clique $K \in \mathcal{K}$ is not all-touched, and there is a vertex $v \in M$ such that*

– *v is adjacent to only one vertex, say k in K, and*
– *k is not in some all-touched clique.*

then K is a one-touched clique, *and v is called a* one-touched vertex. *All-touched cliques and one-touched cliques are called* uninvolved critical cliques. *An* involved critical clique *is a clique that is not uninvolved critical and is adjacent to some all-touched clique. An* uncritical clique *is a clique not critical.*

(a) Splitting v to v_1 and v_2. (b) Gluing v_1 and v_2 to v.

Fig. 3. Splitting and gluing are two pairwise inverse operations, which is only applied when there is no edge between the two cliques.

Proposition 1. *Let $K \in \mathcal{K}$ be an all-touched clique with size at least two and $B = \{b \in M \mid |N[b] \cap K| \geq 2$. Then for each $b \in B$, the vertex b is adjacent to all vertices in K and $G[B \cup K]$ is a clique.*

Proposition 2. *If $K \in \mathcal{K}$ is not all-touched, then K is maximal in G.*

In the following, we define some new types of cliques based on the clique intersection. The idea behind the definition is by the structure of {claw, diamond}-free, hence by Lemma 1.

Definition 4 (Perfect and Strongly Perfect). *For a clique $K \in \mathcal{K}$, let $\mathcal{X} = \{X_1, \ldots, X_t\}$ be the maximal cliques sharing a vertex with K in G. The clique K is perfect in G if K is maximal in G, and each $X \in \mathcal{X}$ satisfies that*

1. *$X \cap X' = \emptyset, \forall X' \in \mathcal{X} \setminus \{X\}$, and*
2. *$|X \cap K| = 1$, and*
3. *there is no edge between X and K.*

Moreover, if K is strongly perfect, then K is perfect and each $X \in \mathcal{X}$ is perfect.

There are closed relationships between Definition 3 and Definition 4.

Proposition 3. *If $K \in \mathcal{K}$ is not all-touched, then K is perfect; Moreover, if K is uncritical, then K is strongly perfect.*

Proof. If K has only one vertex, then K is an isolated vertex. The clique K is uncritical and is strongly perfect. Hence, in the following, we assume that K has at least two vertices. Since K is not all-touched, then by Proposition 2, clique K is maximal.

Let $\mathcal{X} = \{X_1, \ldots, X_t\}$ be the maximal cliques sharing a vertex with K in G and $\mathcal{K}'$ the not all-touched cliques in $\mathcal{K}$. Let $\mathcal{X}' = \mathcal{X} \cap \mathcal{K}'$. Then, each clique in $\mathcal{X}'$ is maximal by Proposition 2. Note that R is {claw, diamond}-free, by Lemma 1, we have that a) for each $X' \in \mathcal{X}'$, there is no edge between X' and K, b) $|X' \cap K| = 1$, and c) $X_1' \cap X_2' \neq \emptyset$ for any $X_1', X_2' \in \mathcal{X}'$.

Now let X' be a maximal clique in $\mathcal{X} \setminus \mathcal{X}'$. Then, the clique X' contains an all-touched clique, say K_1 in $\mathcal{K}$. Since R is {claw, diamond}-free, by Lemma 1, we have that a) $K_1 \cap K_1' = \emptyset, \forall K_1' \in \mathcal{X}'$, b) $|K_1 \cap K| = 1$, and c) there is no edge between K_1 and K. Let $T = X' \setminus K_1$. Since $T \subseteq M$ and $K_1 \cap K_1' = \emptyset, \forall K_1' \in \mathcal{X}'$, we have X' has no vertex in $V(\mathcal{X}')$. Since K is not all-touched and $T \subseteq M$, each $t \in T$ has at most one neighbor in K. If $t \in T$ has an edge incident to a vertex in K, then there is a diamond whose edge set is disjoint to M. Then, if $\mathcal{X}$ has only one clique not in $\mathcal{K}'$, the clique K is perfect.

Now that $\mathcal{X}$ has two cliques, say X_1 and X_2 not in $\mathcal{K}'$. Let K_1 and K_2 be the all-touched cliques contained in X_1 and X_2 respectively. Let $T_1 = X_1 \setminus K_1$ and $T_2 = X_2 \setminus K_2$. If $X_1 \cap X_2 \neq \emptyset$, then $T_1 \cap T_2 \neq \emptyset$, implying that K is all-touched. If X_1 or X_2 has two vertices in K, or there is an edge between X_1 or X_2 and K, then K is also all-touched.

If K is uncritical, since any clique $K_x \in \mathcal{X}$ is not all-touched, by the above, the clique K_x is perfect, hence K is strongly perfect. □

The perfect cliques have a nice property about any minimal solution, this is why we call them perfect. This property can help us delve into some other characterizations about optimal solutions, see Sect. 3.1.

Proposition 4. *Let K be a perfect clique in G and S a minimal edge set such that $G - S$ is {claw, diamond}-free. Let $\mathcal{X} = \{X_1, \ldots, X_t\}$ be the set of cliques sharing a vertex with K in G and $T = \{t_i \mid \{t_i\} = K \cap X_i, \forall X_i \in \mathcal{X}\}$. Let $G_{-S} = G - S$. For each $t_i \in T$, if $G_{-S}[N[t_i] \cap X_i]$ is a clique, then $S \cap K = \emptyset$.*

Proof. Let $K' = K \setminus T$. Since G_{-S} is {claw, diamond}-free, we have that $H_{-S} = G_{-S} - K'$ is also {claw,diamond}-free since H_{-S} is an induced subgraph of G_{-S}. Then, by Lemma 1, in H_{-S}, each vertex is in at most two maximal cliques and edge edge is in exaxtly one maximal cliques.

Now we add K' to H_{-S} and add all edges between K, and the graph denoted by G'. Observe that in G', we do not add any edge outside K. Then, we have $G'[N[t_i] \cap X_i]$ is also a clique for each $t_i \in T$. Since K is perfect in G, there is no edge between X_i and K, hence no edge between $G'[N[t_i] \cap X_i]$. Then, each vertex in G' is in at most two maximal cliques and each edge is in exactly one maximal clique. By Lemma 1, The graph G' is {claw, diamond}-free. Since G' is a super graph of G_{-S} and S is minimal, we have that $G' = G_{-S}$, implying that $S \cap K = \emptyset$.

If K is uncritical, then $\mathcal{X}$ has no all-touched cliques, implying that $\mathcal{X}$ are all perfect. Hence we have the proposition. □

3 Kernelization

In the section, we first give some propositions for feasiable solutions, and then give five rules to build our kernelization algorithm.

3.1 Propositions About Optimal Solutions

In this section, we always use S to denote an optimal solution. We will give some propositions which can help us design kernelization algorithm.

In the following, we give some new types based on an optimal solution.

Definition 5 (Propagation-free and Optimal-free). *Let $K \in \mathcal{K}$ be a clique that is not small-touched and S an optimal solution. We say that K is propagation-free if*

$$S \cap K = \bigcup_{v \in Q} \{vt \mid t \in K \setminus v\},$$

where $Q \subseteq K$ and each $q \in Q$ is in some small all-touched clique or q is a one-touched vertex; Moreoer, the clique K is optimal-free if $Q = \emptyset$.

Now we characterize which cliques are propagation-free or optimal-free. The following proposition shows us that the perfect cliques are propagation-free and the strongly perfect cliqeus are optimal-free.

Proposition 5. *If K is perfect, then K is propagation-free. If K is strongly perfect, then K is optimal-free.*

Proof. Let $\mathcal{X} = \{X_1, \ldots, X_t\}$ be the cliques share a vertex with K, and let $T = \{t_i \mid t_i = X_i \cap K\}$. Let S be an arbitrary optimal solution and $G_{-S} = G - S$.

Claim 1. *If K is perfect, then K is progagation-free.*

Proof. If K is perfect, then by Definition 4, we have that K is maximal, and each $X \in \mathcal{X}$ satisfies that

1. $|X \cap K| = 1$, and
2. there is no edge between X and K.

In the following, we let $S_1 = S \setminus E(K)$ and $S_2 = S \cap E(K)$. If $\forall t_i \in T$, the induced graph $G_{-S}[N[t_i] \cap X_i]$ is a clique, then by Proposition 4, we have $S \cap K = \emptyset$, which implies that $Q = \emptyset$, and we have done. Hence, in the following, we assume that at least one $t_x \in T$, the induced subgraph $G_{-S}[N[t_x] \cap X_x]$ is not a clique. Let

$$Q = \{t_i \mid G_{-S}[N[t_i] \cap X_i] \text{ is not a clique}\}, \text{ and}$$
$$E_Q = \bigcup_{v \in Q} \{vt \mid t \in K \setminus \{v\}\}.$$

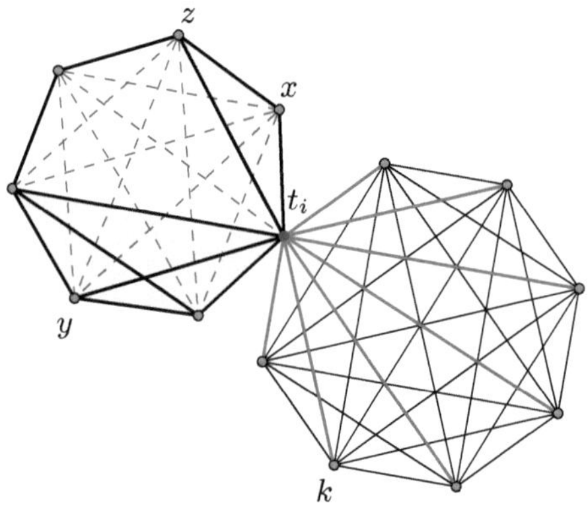

Fig. 4. The left clique is X_i and the right clique is K. In X_i, the blue dashed edges are contained in S_1. After deleting S_1, the resulting graph induced by $V(X_i)$ has four maximal cliques, in which two of them are colored and each of the others have two vertices. In K, the red edges are the edges in E_Q if $Q = \{t_i\}$. After deleting E_Q, the resulting graph induced by $V(K)$ is a cluster. In this example, if we do not delete some edge in E_Q, say t_ik, then there is a claw induced by $\{x, y, t_i, k\}$. However, in this figure, we only explain the symbols used in Proposition 5. In this example, the optimal solution contains $\{t_ix, t_iz\}$ and no need to delete any edge in E_Q, which is the intuition behind Rule 5. (Color figure online)

We will prove that $S_2 = E_Q$. Let $G_{-S_1} = G - S_1$. Note that for any t_x, we have that $G_{-S}[N[t_x] \cap X_x] = G_{-S_1}[N[t_x] \cap X_x]$. Hence, for each $t_i \in T$, the induced subgraph $G_{-S}[N[t_i] \cap X_i]$ is {claw, diamond}-free. Since for each $t_y \in Q$, the induced subgraph is not a clique, then we have that $G_{-S}[N[t_y] \cap X_y]$ is exactly two maximal cliques in G_{-S} by Lemma 1 and by that X_i is a maximal clique in G.

Suppose that there is an edge between $t_x \in Q$ and $k \in K$. Since $G_{-S}[N[t_x] \cap X_x]$ is exactly two maximal cliques, we let x and y be two vertices in the two maximal cliques su ch that $xy \notin E(G)$. Then $\{x, y, t_i, k\}$ induces a claw, see Fig. 4, which implies that

$$E_Q \subseteq S. \tag{1}$$

Since $K - E_Q$ is a cluster, we have that in $G_{-S_1} - E_Q$, each edge is in exactly one maximal clique and each vertex is in at most two maximal cliques. By Lemma 1, we have that $G_{-S_1} - E_Q$ is {claw, diamond}-free. By Eq. 1 and $S_1 \cup E_Q$ is solution, we have that $S = S_1 \cup E_Q$. ◁

Claim 2. *If K is strongly perfect, then K is optimal-free.*

Proof. If K is strongly perfect, then by Definition 4, we have that K is perfect and each $X \in \mathcal{X}$ is also perfect. By Claim 1, each $X \in \mathcal{X}$ in G_{-S} must be a cluster. Then for each vertex $k \in K$, the graph $G_{-S}[N[k] \cap X]$ is must be a clique. By Proposition 4, there is no need to delete any edge in K. ◁

By Claims 1 and 2, we have the proposition. □

3.2 Kernelization Algorithm

The kernelization algorithm contains five rules. The first two rules simlifies the structure of input graphs; Specially, the second rule is ready for the third rule. The third rule considers the uncritical cliques. The forth rule considers the big cliques in $\mathcal{K}$. The last rule considers the involved cliques.

Rule 1. *Delete all components that are {claw, diamond}-free.*

Lemma 4. *Rule 1 is safe and can be done in polynomial time.*

In the following, we set up a simple rule for the third rule.

Rule 2 *Let $K_1 \in \mathcal{K}$ and $K_2 \in \mathcal{K}$ be two big all-touched cliques such that $K_1 \cap K_2 \neq \emptyset$. Let $T_1 = N[K_1] \cap M$ and $T_2 = N[K_2] \cap$. Delete all edges between T_1 and T_2 and decrease k by the number of deleted edges.*

Lemma 5. *Rule 2 is safe and can be done in polynomial time.*

Proposition 6. *Let K_1' and K_2' be two cliques in $\mathcal{K}$ such that $K_1 \cap K_2 \neq \emptyset$. Let K_1 and K_2 be two maximal cliques containing K_1' and K_2' in G respectively. Then we have that $|K_1 \cap K_2| = 1$ and there is no edge between K_1 and K_2.*

By Proposition 6, we can do the splitting operation for two cliques of which none is small all-touched.

Rule 3 *For any two cliques $K_1, K_2 \in \mathcal{K}$, if none of them is small all-touched, then splitting up them.*

Lemma 6. *Rule 3 is safe and can be done in polynomial time.*

Proof. Since we can test whether a clique is all-touched or small, the reduction can be done in polynomial time. Let G' be the graph after separating up K_1 and K_2 and the splitted vertex in G is v. After the splitting operation, the new vertices replacing v are $v_1 \in K_1$ and $v_2 \in K_2$. We will prove that (G, k) is yes if and only if (G', k) is yes.

(G, k) is yes $\rightarrow$ (G', k) is yes. Let S be a feasiable solution of (G, k). In G', We use K_1' and K_2' to denote the cliques corresponding K_1 and K_2, which means that $K_i \setminus \{v\} = K_i' \setminus \{v_i\}$, $i \in \{1, 2\}$. We build an edge set

$$\begin{aligned} S' = &\{e \in S | V(e) \cap \{v\} = \emptyset\} \\ &\cup \{v_1 t | t \in M_1 \text{ and } vt \in S\} \\ &\cup \{v_2 t | t \in M_2 \text{ and } vt \in S\}. \end{aligned}$$

By the definition of splitting, we have that $|S| = |S'|$.

Now we prove that $G' - S'$ is {claw, diamond}-free. The edges in $K_1 - S$ and $K_2 - S$ are equal to the edges in $K_1' - S'$ and $K_2' - S'$, respectively. Then $G' - S'$ can be obtained by splitting v in $G - S$. Recall that the two maximal cliqeus K_1 and K_2 in G share only one vertex and there is no edge between them. Since

$G - S$ is {claw, diamond}-free, then $N[v]$ is in at most two maximal cliques. By Lemma 2, in $G - S$, splitting at v is also {claw, diamond}-free.

(G', k) is yes $\rightarrow$ (G, k) is yes. Let v_1 and v_2 be the vertices corresponding to the splitting vertex v. Note $N[v_1]$ and $N[v_2]$ are both one maximal cliques in G' such that the intersection of the two cliques is an emptyset and there is no edge between them. Let K_1' and K_2' are the maximal cliques containing v_1 and v_2 respectively. Moreover, in G, we use K_1 and K_2 to denote the cliques whose insection is exactly v.

Let S' be an optimal solution of (G', k), and Let

$$S = \{e \mid e \in S' \text{ and } e \cap \{v_1, v_2\} = \emptyset\} \cup \{vt \mid v_1t \in S' \text{ or } v_2t \in S'\}.$$

Note that the edges in $K_1 - S$ and $K_2 - S$ are equal to the edges in $K_1' - S'$ and $K_2' - S'$, respectively. Moreover, by the definition of splitting, we have that in $G' - S'$, the induced subgraphs $N[v_1]$ and $N[v_2]$ are all maximal cliques such that the intersection of them is empty and no edge between the two maximal cliques. Then, by Lemma 3, if we glue v_1 and v_2 in $G' - S'$, the graph, say G_1, is also {claw, diamond}-free. Note that $G - S$ is equal to G_1, and we have the lemma. □

After apply Rule 3, some cliques in $\mathcal{K}$ become a component that is {claw, diamond}-free. Hence, we need to apply Rule 1 again. Actually, In reduced instance, there is no uncritical cliques. Since the number of uninvolved critical cliques can be bounded easily, we only consider how to bound the size of an uninvolved critical clique and how to bound the number of vertices in involved critical cliques. The following rule consider the big uninvolved critical cliques.

Rule 4. *Let K be a big uninvolved critical clique, and $V_u \subset K$ the set of vertices each of which is in small all-touched cliques or is a one-touched vertex (see Definition 3). Let*

$$z = \max(0, k + 3 - |V_u|). \tag{2}$$

If $|K| \geq k + 3$, preserving z vertices in $K \setminus V_u$ and deleting others in $K \setminus V_u$.

Lemma 7. *Rule 4 is safe and can be done in polynomial time.*

Before we give the next rule, we now ready to prove that the number of vertices in uninvolved critical cliques is bounded by $O(k^2)$.

Lemma 8. *After applying Rules 1–4, the number of vertices in uninvolved critical cliques can be bounded by $O(k^2)$.*

Now we give the last rule which bound the number of vertices in involved critical cliques. Let K be an all-touched clique. For all splits (A, B) of K satisfying

$$(|A| - 1) \times (|B| - 1) \leq k, \tag{3}$$

we let

$$x = \max[\min(|A|, |B|) + 1]. \tag{4}$$

Rule 5. *Let K be a small all-touched clique. For each vertex $v \in K$ in some involved clique Y_v, mark x vertices in Y_v that are not in any other all-touched cliques.*

Lemma 9. *Rule 5 is safe and can be done in polynomial time.*

Proof. Since we can detect whether a clique is small all-touched, the running time is polynomial. For convenience, we prove that in Rule 5, deleting the vertices one by one is safe. Note that Y_v is perfect by Proposition 3. After deleting v, we let $Y'_v = Y_v \setminus \{v\}$.

Let (G', k) be the reduced instance. We will prove that (G, k) is yes if and only if (G', k) is yes. Let S be an optimal solution of (G, k). If (G, k) is yes, then $G' - S$ is also {claw, diamond}-free due to the induced property.

Now that (G', k) is yes and let S' be an optimal solution of (G', k). Let $G'_{-S'} = G' - S'$. Suppose for contradiction that $G - S'$ is not {claw, diamond}-free. Note that Y'_v is perfect in G' by Proposition 3. By Proposition 5, the clique Y'_v is propagation-free. Then, the subgraph $G'_{-S'}[Y'_v]$ is a cluster. Let C_v be the clique containing v in $G'_{-S'}$. Note that $|C_v| \geq x$. We will prove that $G'_{-S'}[Y'_v]$ is a clique. If the subgraph $G'_{-S'}[Y'_v]$ is not a cluster, then there is a vertex $x \in V(Y'_v)$ such that the set

$$E_x = \{xt \mid t \in V(Y'_v \setminus \{x\}\}$$

is a subset of S'. Let K' be the maximal clique such that $K' \cap Y'_v = \{x\}$. Note that $G'_{-S'}[N[x] \cap K']$ is exactly two cliques, say C_1 and C_2; Otherwise, we can add the edges between x and C_v, contradicting the optimal of S'. However, since $|C_v| \geq x$ and by Eq. 4, we can add the edges between x and C_v and delete the edges between x and C_1 to obtain a smaller solution than S', a contradiction.

Then, the subgraph $G'_{-S'}[Y'_v]$ is exactly a maximal clique in $G'_{-S'}$. After adding v, the graph $G - S'$ is also {claw, diamond}-free by Lemma 1. □

The Proof of Theorem 1. By Lemma 8, we have bounded the number of vertices in uninvolved critical cliques. Now we bound the number of vertices in involved critical cliques. We will prove that for each small all-touched clique, the number of marked vertices in involved cliques is $O(k)$.

For each small all-touched clique K, by Rule 5, we mark at most $|K| \times x$ vertices where $x = \max[\min(|A|, |B|) + 1$ for all splits (A, B) of K satisfying $(|A| - 1) * (|B| - 1) \leq k$, see Eq. 3.

Let (A', B') be the split such that $x = \min(|A'|, |B'|) + 1$. Without loss of generality, we can assume that $|A'| \geq |B'|$. Then we have

$$\begin{aligned}
|K| \times x &= (|A'| + |B'|) \times [\min(|A'|, |B'|) + 1] && (5)\\
&\leq (|A'| + |A'|) \times (|B'| + 1) && (6)\\
&= 2\,A'| \times |B'| + 2\,A'| &&\\
&= 2[(|A'| - 1) \times (|B'| - 1) + (|A'| + |B'|) - 1] + 2\,A'| && (7)\\
&\leq 6k + 6. && (8)
\end{aligned}$$

In the above equations, Eq. 5 is from the definition of x, see Eq. 4. Equation 6 is from the assumption that $|A'| \geq |B'|$. Equation 7 is a deformation to use Eq. 3. Equation 8 can obtained by Eq. 3 and by Rule 2. □

Acknowledgments. This research was supported by Zhejiang Provincial Natural Science Foundation of China under Grant No. ZCLQN25F0205.

References

1. Abu-Khzam, F.N.: A kernelization algorithm for d-hitting set. J. Comput. Syst. Sci. **76**(7), 524–531 (2010)
2. Agrawal, A., Jana, S., Sahu, A.: A polynomial kernel for proper Helly circular-arc vertex deletion. In: Soto, J.A., Wiese, A. (eds.) LATIN 2024: Theoretical Informatics - 16th Latin American Symposium, Puerto Varas, Chile, 18–22 March 2024, Proceedings, Part II. Lecture Notes in Computer Science, vol. 14579, pp. 208–222. Springer, Heidelberg (2024). https://doi.org/10.1007/978-3-031-55601-2_14
3. Bessy, S., Perez, A.: Polynomial kernels for proper interval completion and related problems. Inf. Comput. **231**, 89–108 (2013)
4. Bondy, J.A., Murty, U.S.R.: Graph Theory. Graduate Texts in Mathematics. Springer, Heidelberg (2008)
5. Crespelle, C., Drange, P.G., Fomin, F.V., Golovach, P.A.: A survey of parameterized algorithms and the complexity of edge modification. Comput. Sci. Rev. **48**, 100556 (2023)
6. Cygan, M., et al.: Parameterized Algorithms. Springer, Heidelberg (2015)
7. Cygan, M., Pilipczuk, M., Pilipczuk, M., van Leeuwen, E.J., Wrochna, M.: Polynomial kernelization for removing induced claws and diamonds. Theory Comput. Syst. **60**(4), 615–636 (2017)
8. Eiben, E., Lochet, W.: A polynomial kernel for line graph deletion. In: Grandoni, F., Herman, G., Sanders, P. (eds.) 28th Annual European Symposium on Algorithms, ESA 2020, Pisa, Italy (Virtual Conference), 7–9 September 2020. LIPIcs, vol. 173, pp. 42:1–42:15. Schloss Dagstuhl - Leibniz-Zentrum für Informatik (2020)
9. Fomin, F.V., Saurabh, S., Villanger, Y.: A polynomial kernel for proper interval vertex deletion. SIAM J. Disc. Math. **27**(4), 1964–1976 (2013)
10. Ke, Y., Cao, Y., Ouyang, X., Li, W., Wang, J.: Unit interval vertex deletion: fewer vertices are relevant. J. Comput. Syst. Sci. **95**, 109–121 (2018)
11. Kloks, T., Kratsch, D., Müller, H.: Dominoes. In: Mayr, E.W., Schmidt, G., Tinhofer, G. (eds.) Graph-Theoretic Concepts in Computer Science, 20th International Workshop, WG '94, Herrsching, Germany, 16–18 June 1994, Proceedings. Lecture Notes in Computer Science, vol. 903, pp. 106–120. Springer, Heidelberg (1994)
12. Li, W., Peng, H., Yang, Y.: Improved kernel and algorithm for claw and diamond free edge deletion based on refined observations. Theor. Comput. Sci. **906**, 83–93 (2022)
13. Metelsky, Y., Tyshkevich, R.: Line graphs of Helly hypergraphs. SIAM J. Disc. Math. **16**(3), 438–448 (2003)
14. Yuan, H., Zhang, Z.: Kernel for proper Helly circular-arc vertex deletion: smaller and simpler via graph isomorphism. In: Du, D., Han, L., Xu, D. (eds.) Combinatorial Optimization and Applications - 17th International Conference, COCOA 2024, Beijing, China, 6–8 December 2024, Proceedings, Part I. Lecture Notes in Computer Science, vol. 15434, pp. 388–400. Springer, Heidelberg (2024). https://doi.org/10.1007/978-981-96-4445-2_32

Continuous Optimization and Lattice Optimization

Inertial Bregman Proximal Alternating Linearized Minimization Method with Line Search for Non-convex Split Feasibility Problem

Jiayi Song, Biao Qu(✉), and Danping Yang

Institute of Operations Research, Qufu Normal University, Rizhao, China
qubiao001@163.com

Abstract. Split feasibility problem where the sets C and Q are nonconvex is considered in this paper. This class of problems are widely applied in compressed sensing, matrix factorization, outlier detection and many other fields. In this setting, split feasibility problem can be formulated as a non-convex and non-smooth problem with variable separation, so that alternating minimization algorithm can be used. We focus on the alternating minimization algorithm with Bregman distance studied in Chao et al. [7]. To further enhance efficiency, Armijo-type search method and linearization technique are employed. Under suitable assumptions on the sets involved, we show that, the sequence generated by our algorithm is bounded and globally convergent when applied to the nonconvex split feasibility problem. Some mild conditions that ensure the sequence converges to the solution of problem are proposed in this paper.

Keywords: Non-convex split feasibility problem · Proximal alternating linearized minimization · Bregman distance · Kurdyka-Lojasiewicz property · Convergence

1 Introduction

Split feasibility problem (SFP) introduced by Censor and Elfving [6] in 1994 is to find $x \in C$ with $Ax \in Q$, where $C \subseteq \Re^n$ and $Q \subseteq \Re^m$ are two non-empty, closed and convex sets, respectively, and A is an $m \times n$ real matrix. The SFP arises in many real-world applications, such as signal processing, image reconstruction, machine learning and various other fields, see, (Suantai et al. [23]; Censor and Elfving [6]). Many scholars have investigated the convex split feasibility problem (CSFP) and developed various algorithms, such as the classical CQ algorithm (Byrne [5]; Qu and Xiu [21]). However, these methods do not address modern applications with nonconvex constraints, including compressed sensing (Bolte et al. [3]), matrix factorization and outlier detection (Chen et al. [9]; López G et al. [17]), among others.

M. Li et al. (Eds.): TAMC 2025, LNCS 16084, pp. 381–395, 2026.
https://doi.org/10.1007/978-981-95-4839-2_30

In this paper, we consider non-convex split feasibility problem(NCSFP) of the following form:

$$\text{find } x \in \Re^n \text{ such that } x \in C, Ax \in Q, \tag{1}$$

where A is an $m \times n$ real matrix, $C \subseteq \Re^n$ and $Q \subseteq \Re^m$ are two non-empty, closed and non-convex sets. In 2020, utilizing indicator function and distance function, Gibali et al. [12] proposed four optimization models for non-convex split feasibility problem and provided their corresponding algorithms, such as alternating minimization (AM) method. These algorithms can be regarded as extensions of the CQ algorithm to non-convex sets. Chen et al. [9] used technique of Difference-of-Convex (DC) to rewrote objective function and applied the NPG_{major} to extend the CQ algorithm in the non-convex setting.

Follow from Gibali et al. [12], NCSFP can also be solved by transforming it into a non-convex optimization problem with variable separation as follow:

$$\min_{x \in \Re^n, y \in \Re^m} \{F(x, y) = \delta_C(x) + \delta_Q(y) + h(x, y)\}, \tag{2}$$

where $h(x, y) = \frac{1}{2}\|Ax - y\|^2$. $\delta_C(x), \delta_Q(y)$ denote the indicator functions of C and Q. A point $x \in C$ is a solution of problem (1) if and only if it serves as a minimizer for problem (2) with an optimal value 0. Consequently, we can solve problem (1) by solving problem (2).

For solving non-convex optimization problems with variable separation, the alternating minimization algorithm is a powerful tool. To relax the strict convexity of the function, Auslender et al. [2] proposed the proximal alternating minimization approach (PAM). PAM requires finding an exact solution at each step, however, the subproblems may be difficult to solve. The linearization technique is an useful method to tackle this drawbacks. Bolte et al. [3] combined the linearization technique with alternating minimization algorithm. With the help of Kurdyka–Lojasiewicz property, they proved that each bounded sequence generated by the algorithm globally converges to the critical point of problem.

In recent years, the inertial extrapolation technique starting from the heavy ball method of Polyak [20] have been widely used to accelerate algorithms for convex and nonconvex optimizations (Chao er al. [11]; Dong et al. [8]). The main feature of inertial method is that the next iteration is established by utilizing the previous two or more iterations. Many inertial algorithms have been proposed, such as Dang et al. [10] proposed inertial version of CQ algorithm for solving CSFP, Shehu et al. [22] introduced a relaxed CQ algorithm with alternating inertial extrapolation steps.

To further enhance the performance of PALM, unlike the typical inertial step, Gao et al. [13] introduced a Gauss-Seidel type inertial proximal alternating linearized minimization (GiPALM) scheme for solving non-convex optimization problems with variable separation of the form: $\min_{x,y}\{f(x)+g(y)+h(x,y)\}$, where f and g are proper lower semicontinuous functions, and $h(x, y)$ is a continuous differentiable function. Numerical results indicated the acceleration of GiPALM over PALM.

Another acceleration technique commonly used in recent years is replacing the Euclidean distance with the more general Bregman distance. Bregman distance introduced by Bregman in [4], describes the difference between a function and its linear approximation. Many well-known concepts of distance, such as Euclidean distance, Kullback–Leibler distance and Itakura–Saito distance are special cases of the Bregman distances. Bregman distance gives us more flexibility in the selection of regularization term of proximal point algorithm. Iterative algorithms involving Bregman distance have been widely used to solve optimization problems. For instance, Zhao et al. [16] proposed a general CQ algorithm to solve CSFP by combining the Bregman projection with classical CQ algorithm. Liu et al. [18] improved ADMM and its linearized version in the nonconvex setting by using Bregman distance; and Zhao et al. [14] proposed a two-step inertial proximal Bregman alternative algorithm involving the previous three iterates, and so on (Xu and Chao [24]; Mukkamala et al. [19]; Hanzely et al. [15]).

Combining with the Bregman distance and inertial, Chao et al. [7] proposed an inertial alternating minimization with Bregman distance (BIAM) algorithm. The regularization term of the AM algorithm is replaced by a Bregman distance according to a carefully chosen function.

Motivated by the results mentioned above, in this paper, our algorithm is an adaptation of the inertial alternating minimization with Bregman algorithm proposed in Chao [7]. To enhance the computational efficiency, Armijo-type search method and linearization technique are employed. In Chao [7], as in most of the literature, when solving non-convex optimization problems, it is only to ensure that the cluster points of the sequence are critical points. It cannot guarantee that it is a solution of the problem. In Sect. 3, we prove that under certain conditions, the proposed algorithm can converge to a solution of NCSFP. Due to the properties of the indicator function, the convergence proof of our algorithm is relatively simpler.

Furthermore, non-convex optimization problems frequently have numerous local optimum. Using a fixed step size may cause the algorithm to get stuck or oscillate around a local optimum. Thus, we introduce the algorithm combined with the Armijo-type search method proposed by Qu et al. [21] to avoid the high computational costs and can converge to the global optimal point. Some mild conditions that ensure the sequence generated by the algorithm converges to the solution of non-convex split feasibility problem are presented in this paper.

2 Basic Notations and Preliminaries

In this paper, let $\Re^n$ denote the n-dimensional Euclidean space with inner product and associated norm denoted by $\langle \cdot, \cdot \rangle$ and $\| \cdot \|$, respectively.

An extended real valued function $f : \Re^n \to (-\infty, \infty]$ is said to be proper if $\operatorname{dom} f := \{x : f(x) < \infty\} \neq \emptyset$, and closed if it is lower semicontinuous. When C is a nonempty closed set, the indicator function of C is both lower semicontinuous and proper.

Definition 1. *[4] Let G be a σ-strongly convex ($\sigma > 0$), differentiable function and such that ∇G is Lipschitz continuous with the modulus $L_{\nabla G} > 0$. The Bregman distance of G is defined as: $D_G(x, y) = G(x) - G(y) - \langle \nabla G(y), x - y \rangle$.*

Combing Definition 2.1 with descent lemma, we can get the following inequality:

$$\frac{\sigma}{2}\|x - y\|^2 \leq D_G(x, y) \leq \frac{L_{\nabla G}}{2}\|x - y\|^2. \tag{3}$$

The Kurdyka–Łojasiewicz property which plays a role in our analysis.

Definition 2. *[3] Let $f : \Re^n \to \Re \cup \{+\infty\}$ be proper and lower semicontinuous function. (i)The function f is said to have the Kurdyka-Lojasiewicz(KL) property at $\bar{u} \in dom\partial f$ if there exists $\eta \in (0, +\infty]$, a neighborhood U of $\bar{u}$ and a continuous concave function $\varphi \in \Phi_\eta$ that satisfy the following inequality*

$$\varphi^{'}(f(u) - f(\bar{u}))dist(0, \partial f(u)) \geq 1,$$

for all $u \in U \cap [f(\bar{u}) < f(u) < f(\bar{u}) + \eta]$. (ii) If f satisfies the KL property at each point of $dom\partial f$ then f is referred to as KL function.

Definition 3. *[3] Let Ω be a compact set and $f : \Re^n \to \Re \cup \{+\infty\}$ be a proper and low semicontinuous function. Assume that f is constant on Ω and satisfies the KL property at each point of Ω. Then, there exist $\varepsilon > 0, \eta > 0$, and $\varphi \in \Phi_\eta$ such that for all $\bar{u}$ in Ω and all u in the following intersection:*

$$\{u \in \Re^n : dist(u, \Omega) < \varepsilon\} \cap [f(\bar{u}) < f(u) < f(\bar{u}) + \eta], \tag{4}$$

one has,

$$\varphi^{'}(f(u) - f(\bar{u})) \cdot dist(0, \partial f(u)) \geq 1.$$

Remark 1. [1]
(i)The indicator function of semi-algebraic sets are semi-algebraic. (ii)Finite sums of semi-algebraic sets remain semi-algebraic.

3 Algorithm and Convergence Analysis

We formally give the inertial proximal alternating linearized minimization algorithm with Bregman distance, which is a linearized version of the algorithm in Chao [7], enhanced with the Armijo-type search technique. Then the algorithm is applied to nonconvex split feasibility problem.

Algorithm 3.1

---

step 0 Let $x^0 = \hat{x}^0 \in C$, $y^0 = \hat{y}^0 \in Q$, $\alpha \in [0, 1)$, $\beta \in [0, 1)$, $l \in (0, 1)$, $\mu \in (0, 1)$. Set $k := 0$.

step 1 Compute

$$x^{k+1} \in \arg\min_x \{\delta_C(x) + \alpha_k \langle A^T(A\hat{x}^k - \hat{y}^k), x - \hat{x}^k \rangle + \lambda_k D_{G_1}(x, \hat{x}^k)\}, \tag{5}$$

where $\alpha_k = l^{m_k}$ and m_k is the smallest nonnegative integer such that

$$\|\nabla_x h(x^{k+1}, \hat{y}^k) - \nabla_x h(\hat{x}^k, \hat{y}^k)\| \leq \frac{\mu\|x^{k+1} - \hat{x}^k\|}{\alpha_k}. \tag{6}$$

step 2 Compute

$$\hat{x}^{k+1} = x^{k+1} + \alpha(x^{k+1} - \hat{x}^k). \tag{7}$$

step 3 Compute

$$y^{k+1} \in \arg\min_y\{\delta_Q(y) + \beta_k\langle\hat{y}^k - A\hat{x}^{k+1}, y - \hat{y}^k\rangle + \mu_k D_{G_2}(y, \hat{y}^k)\}, \tag{8}$$

where $\beta_k = l^{n_k}$ and n_k is the smallest nonnegative integer such that

$$\|\nabla_y h(\hat{x}^{k+1}, y^{k+1}) - \nabla_y h(\hat{x}^{k+1}, \hat{y}^k)\| \leq \frac{\mu\|y^{k+1} - \hat{y}^k\|}{\beta_k}. \tag{9}$$

step 4 Compute

$$\hat{y}^{k+1} = y^{k+1} + \beta(y^{k+1} - \hat{y}^k). \tag{10}$$

The following assumptions are necessary for the convergence analysis of our algorithm.

Assumption 1. *(i)The Bregman distance function G_i (i=1,2) are $\sigma_i -$ strongly convex and ∇G_i is $L_i -$ Lipschitz continuous.*

(ii)There exists $\lambda^-, \lambda^+, \mu^-, \mu^+ > 0$ such that $\lambda^- < \lambda_k < \lambda^+$,$\mu^- < \mu_k < \mu^+$ and satisfy

$$\lambda^-\sigma_1 > \mu + (1+\alpha)^2(1 + \mu + \|A\|^2 + \|B\|^2) + \lambda^+ L_1\alpha^2,$$

$$\mu^-\sigma_2 > \mu + \beta(1 + \beta + 2\beta\|B\|^2) + \beta(1+\beta)^2(2 + \mu + \|C\|^2) + \mu^+ L_2\beta^2.$$

(iii) The sets C and Q are two compact semi-algebraic sets.

The following lemma shows that the function $\nabla_x h(x, y)$ and $\nabla_y h(x, y)$ are globally Lipschitz, satisfies the assumption 3.1 (i) and (iii) for the BIAM in [7].

Lemma 1. *Suppose $C = (A, -I) \in \Re^{m\times(n+m)}$, $B = A^T C \in \Re^{n\times(n+m)}$, $z^{k+1} = (x^{k+1}, y^{k+1})$, $z^k = (x^k, y^k)$, where I is an $m \times m$ identity matrix, all vectors are column vectors.*

Then

$$\|\nabla_x h(z^{k+1}) - \nabla_x h(z^k)\| \leq \|B\|\|z^{k+1} - z^k\|,$$
$$\|\nabla_y h(z^{k+1}) - \nabla_x h(z^k)\| \leq \|C\|\|z^{k+1} - z^k\|.$$

Proof. By the definition of $h(x, y)$, we have

$$\|\nabla_x h(z^{k+1}) - \nabla_x h(z^k)\| = \|B(z^{k+1}) - B(z^k)\| \leq \|B\| \|z^{k+1} - z^k\| = \|B\| \|z^{k+1} - z^k\|.$$

Similarly, $\|\nabla_y h(z^{k+1}) - \nabla_x h(z^k)\| \leq \|C\| \|z^{k+1} - z^k\|$. The proof is completed.

Remark 2. (1)For any fixed y, $\nabla_x h(x, y)$ is globally Lipschitz with modulus $\|A\|^2$, i.e., $\|\nabla_x h(x^k, y) - \nabla_x h(x^{k+1}, y)\| \leq \|A\|^2 \|x^k - x^{k+1}\|$.

(2)For any fixed x, $\nabla_y h(x, y)$ is globally Lipschitz with modulus 1, i.e., $\|\nabla_y h(x, y^k) - \nabla_x h(x, y^{k+1})\| = \|y^k - y^{k+1}\|$.

The following lemma shows that the line search procedure is well-defined.

Lemma 2. *For any $k = 0, 1, ...$, the Armijo-like search rules defined in (6) and (9) are well-defined.*

Proof. When $\alpha_k > 0$ is sufficiently small and satisfy $\alpha_k \leq \frac{\mu}{\|A\|^2}$, then $\|A\|^2 \leq \frac{\mu}{\alpha_k}$. Therefore,

$$\|\nabla_x h(x^{k+1}, \hat{y}^k) - \nabla_x h(\hat{x}^k, \hat{y}^k)\| = \|A^T A(x^{k+1} - \hat{x}^k)\| \leq \|A\|^2 \|x^{k+1} - \hat{x}^k\| \leq \frac{\mu \|x^{k+1} - \hat{x}^k\|}{\alpha_k}.$$

Similarly, when $\beta_k > 0$ is sufficiently small and satisfy $\beta_k \leq \mu$, then

$$\|\nabla_y h(\hat{x}^{k+1}, y^{k+1}) - \nabla_y h(\hat{x}^{k+1}, \hat{y}^k)\| = \|y^{k+1} - \hat{y}^k\| \leq \frac{\mu \|y^{k+1} - \hat{y}^k\|}{\beta_k}.$$

The proof is completed.

Remark 3. Let α_k and β_k be the sequences generated by Algorithm (3.1). Then: $\frac{\mu l}{\|A\|^2} < \alpha_k \leq 1$, $\mu l < \beta_k \leq 1$.

Proof. If $\alpha_k = 1$, by the definition of α_k, it is trivial.

If $\alpha_k < 1$, follow from (6), we have

$$\|\nabla_x h(x^{k+1}, \hat{y}^k) - \nabla_x h(\hat{x}^k, \hat{y}^k)\| > \frac{\mu l \|x^{k+1} - \hat{x}^k\|}{\alpha_k}.$$

Combine with Remark 2 (1), we obtain $\alpha_k > \frac{\mu l}{\|A\|^2}$. Thus, $\frac{\mu l}{\|A\|^2} < \alpha_k \leq 1$. Similarly, we have $\mu l < \beta_k \leq 1$. The proof is completed.

To make the notation more convenient, in the rest of this paper, let $\{z^k = (x^k, y^k)\}_{k \in N}$ be the sequence generated by Algorithm 3.1 and $\omega^k = (x^k, y^k, \hat{x}^k, \hat{y}^k)$.

Combine with the proof strategy of the Lemma 3.1 in Chao [7] and the inequality (29) and (30) derived from (7) and (10), which is fully shown in Gao et al. [13], the following property of the sequence $\{(x^k, y^k)\}_{k \in N}$ generated by Algorithm 3.1 can be derived.

$$\|x^{k+1} - x^k\|^2 \leq \frac{1+\alpha}{\alpha} \|x^k - \hat{x}^k\|^2 + \frac{1+\alpha}{\alpha^2} \|\hat{x}^{k+1} - x^{k+1}\|^2, \tag{11}$$

$$\|y^{k+1} - y^k\|^2 \leq \frac{1+\beta}{\beta}\|y^k - \hat{y}^k\|^2 + \frac{1+\beta}{\beta^2}\|\hat{y}^{k+1} - y^{k+1}\|^2. \tag{12}$$

Lemma 3. *If Assumption 3.1 holds, then*

$$F(x^{k+1}, y^{k+1}) + c_1\|x^{k+1} - \hat{x}^{k+1}\|^2 + c_2\|y^{k+1} - \hat{y}^{k+1}\|^2 \leq F(x^k, y^k) + c_3\|x^k - \hat{x}^k\|^2 + c_4\|y^k - \hat{y}^k\|^2,$$

where

$$c_1 = \frac{\lambda^- \sigma_1 - \mu - (1+\alpha)(\|A\|^2 + 1 + \|B\|^2 + \mu)}{2\alpha^2},$$

$$c_2 = \frac{\mu^- \sigma_2 - \mu - (1+\beta)(2 + \mu + \|C\|^2)}{2\beta^2},$$

$$c_3 = \frac{\lambda^+ L_1}{2} + \frac{(1+\alpha)(\|A\|^2 + 1 + \|B\|^2 + \mu)}{2\alpha},$$

$$c_4 = \frac{\mu^+ L_2}{2} + \frac{(1+\beta)(3 + \mu + \|C\|^2) + \|B\|^2\beta}{2\beta}.$$

Proof. The proof is in the appendix.

We construct an auxiliary function below, which helps us to deduce the convergence of Algorithm 3.1, $H(x, y, \hat{x}, \hat{y}) = F(x, y) + c_3\|x - \hat{x}\| + c_4\|y - \hat{y}\|$.

Using the above result, the following corollary establishes the monotonicity and convergence of the sequence $\{H(\omega^k)\}_{k\in N}$.

Corollary 1. *The sequence $\{H(\omega^k)\}_{k\in N}$ is monotonically nonincreasing and convergent.*

Moreover, it holds that

$$H(\omega^{k+1}) + \rho_1(\|x^{k+1} - \hat{x}^{k+1}\|^2 + \|y^{k+1} - \hat{y}^{k+1}\|^2) \leq H(\omega^k), \tag{13}$$

where, $\rho_1 = min\{c_1 - c_3, c_2 - c_4\}$.

Proof. It follows from Lemma 3, we have

$$H(\omega^{k+1}) + (c_1 - c_3)\|x^{k+1} - \hat{x}^{k+1}\|^2 + (c_2 - c_4)\|y^{k+1} - \hat{y}^{k+1}\|^2 \leq H(\omega^k).$$

From the assumption, we can obtain $c_1 - c_3 > 0, c_2 - c_4 > 0$, hence $H(\omega^{k+1}) \leq H(\omega^k)$, which, together with the property that $\{F(z^k)\}_{k\in N}$ is bounded below, shows that $\{H(\omega^k)\}_{k\in N}$ is also bounded below, and so, $\{H(\omega^k)\}_{k\in N}$ is convergent. The proof is completed.

The following result provides the boundedness of the sequence.

Theorem 1. *Suppose that the sequence $\{z^k = (x^k, y^k)\}_{k\in N}$ is generated by the Algorithm 3.1. Then $\{z^k\}_{k\in N}$ is bounded.*

Proof. From (5) and (7), we have $x^1 \in C$, $y^1 \in Q$, hence $\delta_C(x^1) = \delta_Q(y^1) = 0$. Follow (13), we have

$$H(\omega^1) \geq H(\omega^k) = F(x^k, y^k) + c_3\|x^k - \hat{x}^k\|^2 + c_4\|y^k - \hat{y}^k\|^2. \tag{14}$$

Therefore, the $\{H(\omega^k)\}_{k\in N}$ is bound.

As C and Q are compact, then $f = \delta_C(x)$ and $g = \delta_Q(y)$ are coercive.

We prove firstly by contradiction that the sequence $\{x^k\}_{k\in N}$ is bounded. Assuming that there exists a sequence $\{x^{k_j}\}_{k,j\in N} \subseteq \{x^k\}_{k\in N}$ satisfies $\lim_{j\to\infty} \|x^{k_j}\| = \infty$, then $\delta_C(x^{k_j}) \to \infty$, which yields a contradiction with the inequality (14). Hence $\{x^k\}_{k\in N}$ is bound. Similarly, $\{y^k\}_{k\in N}$ is bound. As c_3 and c_4 are nonnegative, follow from (14), we have $\{\hat{x}^k\}_{k\in N}$ and $\{\hat{y}^k\}_{k\in N}$ are both bound. Hence the sequence $\{z^k\}_{k\in N}$ and $\{\omega^k\}_{k\in N}$ are bound. The proof is completed.

The next result provides upper estimates for the limiting subgradients of the merit function H at ω^{k+1}.

Lemma 4. *For all $k > 0$, define*

$$\begin{cases} Q_x^{k+1} = A^T A(x^{k+1} - \hat{x}^k) + A^T(\hat{y}^k - y^{k+1}) + \dfrac{\lambda_k}{\alpha_k}(\nabla G_1(\hat{x}^k) - \nabla G_1(x^{k+1})), \\ Q_y^{k+1} = (y^{k+1} - \hat{y}^k) - A(x^{k+1} - \hat{x}^{k+1}) + \dfrac{\mu_k}{\beta_k}(\nabla G_2(\hat{y}^k) - \nabla G_2(y^{k+1})), \end{cases}$$

and

$$\begin{aligned} \xi_x^{k+1} &= Q_x^{k+1} + 2c_3(x^{k+1} - \hat{x}^{k+1}), \xi_y^{k+1} = Q_y^{k+1} + 2c_4(y^{k+1} - \hat{y}^{k+1}), \\ \xi_{\hat{x}}^{k+1} &= -2c_3(x^{k+1} - \hat{x}^{k+1}), \xi_{\hat{y}}^{k+1} = -2c_4(y^{k+1} - \hat{y}^{k+1}). \end{aligned}$$

Then, $\xi^{k+1} = (\xi_x^{k+1}, \xi_y^{k+1}, \xi_{\hat{x}}^{k+1}, \xi_{\hat{y}}^{k+1}) \in \partial H(\omega^{k+1})$, $(Q_x^{k+1}, Q_y^{k+1}) \in \partial F(x^{k+1}, y^{k+1})$, and exist $\rho_2 > 0$ such that $dist(0, \partial H(\omega^{k+1})) \leq \rho_2\|z^{k+1} - \hat{z}^{k+1}\|$.

Proof. The first order optimality condition of the (5) and (8) yield

$$0 \in \partial f(x^{k+1}) + \alpha_k A^T(A\hat{x}^k - \hat{y}^k) + \lambda_k(\nabla G_1(x^{k+1}) - \nabla G_1(\hat{x}^k)),$$

$$0 \in \partial g(y^{k+1}) + \beta_k(\hat{y}^k - A\hat{x}^{k+1}) + \mu_k(\nabla G_2(y^{k+1}) - \nabla G_2(\hat{y}^k)).$$

Equivalently, $-\alpha_k A^T(A\hat{x}^k - \hat{y}^k) - \lambda_k(\nabla G_1(x^{k+1}) - \nabla G_1(\hat{x}^k)) \in \partial f(x^{k+1})$. Multiplying both sides by $\frac{1}{\alpha_k}$ and from Remark 1, we get that

$$-A^T(A\hat{x}^k - \hat{y}^k) + \frac{\lambda_k}{\alpha_k}(\nabla G_1(\hat{x}^k) - \nabla G_1(x^{k+1})) \in \partial f(x^{k+1}).$$

Arguing similarly with y^{k+1}, we obtain

$$-(\hat{y}^k - A\hat{x}^{k+1}) + \frac{\mu_k}{\beta_k}(\nabla G_2(\hat{y}^k) - \nabla G_2(y^{k+1})) \in \partial g(y^{k+1}).$$

It is clear from the property of subgradient that

$$(Q_x^{k+1}, Q_y^{k+1}) \in \partial F(x^{k+1}, y^{k+1}) \quad \text{and} \quad \xi^{k+1} \in \partial H(\omega^{k+1}).$$

Then, follow from the proof strategy of the Lemma 3.3 in Chao [7], we have

$$Q_x^{k+1} \leq (\frac{\|B\|\mu l + L_1\|A\|^2\lambda_k}{\alpha\mu l})\|x^{k+1} - \hat{x}^{k+1}\| + \frac{\|B\|}{\beta}\|y^{k+1} - \hat{y}^{k+1}\|. \tag{15}$$

and

$$Q_y^{k+1} \leq (\frac{\|C\|\mu l + L_2\mu_k}{\beta\mu l})\|y^{k+1} - \hat{y}^{k+1}\| + \|C\|\|x^{k+1} - \hat{x}^{k+1}\|. \tag{16}$$

Hence, we obtain

$$\begin{aligned} dist(0, \partial H(\omega^{k+1})) &\leq \|\xi^{k+1}\| \\ &\leq \|\xi_x^{k+1}\| + \|\xi_y^{k+1}\| + \|\xi_{\hat{x}}^{k+1}\| + \|\xi_{\hat{y}}^{k+1}\| \\ &\leq (4c_3 + \frac{\|B\|\mu l + L_1\|A\|^2\lambda^+}{\alpha\mu l} + \|C\|)\|x^{k+1} - \hat{x}^{k+1}\| \\ &+ (4c_4 + \frac{\|C\|\mu l + L_2\mu^+}{\beta\mu l} + \frac{\|B\|}{\beta})\|y^{k+1} - \hat{y}^{k+1}\| \\ &\leq c(\|x^{k+1} - \hat{x}^{k+1}\| + \|y^{k+1} - \hat{y}^{k+1}\|), \end{aligned}$$

where $c = max\{4c_3 + \frac{\|B\|\mu l + L_1\|A\|^2\lambda^+}{\alpha\mu l} + \|C\|, 4c_4 + \frac{\|C\|\mu l + L_2\mu^+}{\beta\mu l} + \frac{\|B\|}{\beta}\}$.

Let $\rho_2 = \sqrt{2}c$, from the above inequality, we have

$$dist(0, \partial H(\omega^{k+1})) \leq c(\|x^{k+1} - \hat{x}^{k+1}\| + \|y^{k+1} - \hat{y}^{k+1}\|) \leq \rho_2\|z^{k+1} - \hat{z}^{k+1}\|. \tag{17}$$

The proof is completed.

Under the additional assumption that the sets C and Q are semi-algebraic sets, we can verify the convergence of the sequence $\{z^k\}_{k\in N}$.

Theorem 2. *Let C and Q be semi-algebraic sets, the sequence $\{z^k = (x^k, y^k)\}_{k\in N}$ generated by the Algorithm 3.1. Then $\sum_{k=1}^{\infty} \left\|z^{k+1} - z^k\right\| < \infty$, and as a sequence, we have $\{z^k\}_{k\in N}$ converges to a critical point of F.*

Proof. The proof is in the appendix.

The following theorem shows that under some mild conditions on the sets and matrix, the sequence $\{z^k\}_{k\in N}$ converge to the solution of the problem (1).

Theorem 3. *Suppose that the solution set of (1) is nonempty, the matrix A has full row-rank and the sets C and Q satisfy*

$$N_C(x^*) \cap N'(y^*) = \{0\}, \tag{18}$$

where $N^{'}(y^*) = \{-A^T v | v \in N_Q(y^*)\}$. Then, x^* is the solution of (1).

Proof. It follows from (5) that $x^k \in C$, for all $k = 0, 1, ...$ Moreover, since C is a closed set, we can obtain $x^* \in C$. Similarly, we have $y^* \in Q$. From Theorem 2, we see that $0 \in \partial F(z^*)$, i.e., $(0, 0) \in (\partial\delta_C(x^*) + A^T(Ax^* - y^*), \partial\delta_Q(y^*) + (y^* - Ax^*))$.

Equivalently, $-A^T(Ax^* - y^*) \in N_C(x^*), \quad Ax^* - y^* \in N_Q(y^*)$.

Hence, our assumption (18) implies that $-A^T(Ax^* - y^*) = 0$. According to A has full row-rank, we can obtain $Ax^* - y^* = 0$, i.e. $Ax^* = y^* \in Q$, thus x^* is the solution of (1). The proof is completed.

4 Conclusion

For non-convex split feasibility problem, an inertial Bregman proximal alternating linearized minimization method with line search is proposed in this paper. We proof the boundedness of the iterative sequence generated by the algorithm under a mild assumption. Additionally, under the assumption that the sets C and Q are semi-algebraic, we proof the global convergence of the sequence by using the Kurdyka-Łojasiewicz property of the auxiliary function. Finally, we provide the condition that ensure the sequence converges to the solution of NCSFP.

Appendix

A Proof of Lemma 3

Proof. According to (5), we have

$$\begin{aligned} & f(x^{k+1}) - \alpha_k \langle \nabla_x h(x^{k+1}, \hat{y}^k), x^k - x^{k+1} \rangle \\ & \leq f(x^k) + \frac{\mu}{2}\|\hat{x}^k - x^{k+1}\|^2 + \frac{\mu}{2}\|x^k - x^{k+1}\|^2 + \lambda_k [D_{G_1}(x^k, \hat{x}^k) - D_{G_1}(x^{k+1}, \hat{x}^k)], \end{aligned} \tag{19}$$

where, the inequality follows from (6).

Similarly, follow from (8) and (9), we obtain

$$\begin{aligned} & g(y^{k+1}) - \beta_k \langle \nabla_y h(\hat{x}^{k+1}, \hat{y}^{k+1})), y^k - y^{k+1} \rangle \\ \leq & g(y^k) + \frac{\mu}{2}\|\hat{y}^k - y^{k+1}\|^2 + \frac{\mu}{2}\|y^k - y^{k+1}\|^2 + \mu_k [D_{G_2}(y^k, \hat{y}^k) - D_{G_1}(y^{k+1}, \hat{y}^k)]. \end{aligned} \tag{20}$$

Follow from Lemma 3 and Remark 2, we can get

$$h(x^{k+1}, y^k) \leq h(x^k, y^k) + \langle \nabla_x h(x^k, y^k), x^{k+1} - x^k \rangle + \frac{\|A\|^2}{2}\|x^{k+1} - x^k\|^2, \tag{21}$$

$$h(x^{k+1}, y^{k+1}) \leq h(x^{k+1}, y^k) + \langle \nabla_y h(x^{k+1}, y^k), y^{k+1} - y^k \rangle + \frac{1}{2}\|y^{k+1} - y^k\|^2. \tag{22}$$

From (3), we obtain

$$\lambda_k[D_{G_1}(x^k,\hat{x}^k)-D_{G_1}(x^{k+1},\hat{x}^k)]\leq\lambda_k[\frac{L_1}{2}\|x^k-\hat{x}^k\|^2-\frac{\sigma_1}{2}\|x^{k+1}-\hat{x}^k\|^2], \tag{23}$$

$$\mu_k[D_{G_2}(y^k,\hat{y}^k)-D_{G_2}(y^{k+1},\hat{y}^k)]\leq\mu_k[\frac{L_2}{2}\|y^k-\hat{y}^k\|^2-\frac{\sigma_2}{2}\|y^{k+1}-\hat{y}^k\|^2]. \tag{24}$$

Adding (19)–(23) and $\alpha_k\leq 1$, $\beta_k\leq 1$ yields

$$\begin{aligned}&F(x^{k+1},y^{k+1})\leq F(x^k,y^k)+\langle x^{k+1}-x^k,\nabla_x h(x^k,y^k)-\nabla_x h(x^{k+1},\hat{y}^k)\rangle\\&+\langle y^{k+1}-y^k,\nabla_y h(x^{k+1},y^k)-\nabla_y h(\hat{x}^{k+1},\hat{y}^{k+1})\rangle+\lambda_k[\frac{L_1}{2}\|x^k-\hat{x}^k\|^2-\frac{\sigma_1}{2}\|x^{k+1}-\hat{x}^k\|^2]\\&+\mu_k[\frac{L_2}{2}\|y^k-\hat{y}^k\|^2-\frac{\sigma_2}{2}\|y^{k+1}-\hat{y}^k\|^2]+\frac{\|A\|^2}{2}\|x^{k+1}-x^k\|^2+\frac{1}{2}\|y^{k+1}-y^k\|^2\\&+\frac{\mu}{2}\|\hat{x}^k-x^{k+1}\|^2+\frac{\mu}{2}\|x^k-x^{k+1}\|^2+\frac{\mu}{2}\|\hat{y}^k-y^{k+1}\|^2+\frac{\mu}{2}\|y^k-y^{k+1}\|^2.\end{aligned} \tag{25}$$

From Lemma 1, we have

$$\begin{aligned}&\langle x^{k+1}-x^k,\nabla_x h(x^k,y^k)-\nabla_x h(x^{k+1},\hat{y}^k)\rangle\\&\leq\frac{1}{2}\|x^{k+1}-x^k\|^2+\frac{\|B\|^2}{2}\|x^k-x^{k+1}\|^2+\frac{\|B\|^2}{2}\|y^k-\hat{y}^k\|^2.\end{aligned} \tag{26}$$

Similarly,

$$\begin{aligned}&\langle y^{k+1}-y^k,\nabla_y h(x^{k+1},y^k)-\nabla_y h(\hat{x}^{k+1},\hat{y}^{k+1})\rangle\\&\leq\frac{1}{2}\|y^{k+1}-y^k\|^2+\frac{\|C\|^2}{2}\|x^{k+1}-\hat{x}^{k+1}\|^2+\frac{\|C\|^2}{2}\|y^k-\hat{y}^{k+1}\|^2.\end{aligned} \tag{27}$$

Substitute (26) and (27) into (25), we can get

$$\begin{aligned}F(x^{k+1},y^{k+1})\leq{}&F(x^k,y^k)+\frac{L_1\lambda_k}{2}\|x^k-\hat{x}^k\|^2-\frac{\|C\|^2}{2}\|x^{k+1}-\hat{x}^{k+1}\|^2\\&+(\frac{\mu^k L_2}{2}+\frac{\|B\|^2}{2})\|y^k-\hat{y}^k\|^2-\frac{\|C\|^2}{2}\|y^{k+1}-\hat{y}^{k+1}\|^2\\&+(\frac{1+\mu+\|A\|^2+\|B\|^2}{2})\|x^{k+1}-x^k\|^2+(\frac{\mu-\lambda_k\sigma_1}{2})\|x^{k+1}-\hat{x}^k\|^2\\&+(1+\frac{\mu+\|C\|^2}{2})\|y^{k+1}-y^k\|^2+(\frac{\mu-\mu_k\sigma_2}{2})\|y^{k+1}-\hat{y}^k\|^2.\end{aligned} \tag{28}$$

Using (7), we have

$$\begin{aligned}\|x^{k+1}-x^k\|^2&=\frac{1}{\alpha^2}\|(\hat{x}^{k+1}-x^{k+1})-\alpha(x^k-\hat{x}^k)\|^2\\&\leq\frac{1+\alpha}{\alpha}\|x^k-\hat{x}^k\|^2+\frac{1+\alpha}{\alpha^2}\|\hat{x}^{k+1}-x^{k+1}\|^2.\end{aligned} \tag{29}$$

Similarly,

$$\|y^{k+1}-y^k\|^2 \leq \frac{1+\beta}{\beta}\|y^k-\hat{y}^k\|^2+\frac{1+\beta}{\beta^2}\|\hat{y}^{k+1}-y^{k+1}\|^2. \tag{30}$$

Substitute (29)–(30) into (28), From the assumption, it holds that

$$\begin{aligned}
F(x^{k+1},y^{k+1}) \leq F(x^k,y^k) &+ [\frac{\lambda^+ L_1}{2}+\frac{(1+\alpha)(1+\mu+\|A\|^2+\|B\|^2)}{2\alpha}]\|x^k-\hat{x}^k\|^2 \\
&- [\frac{\lambda^-\sigma_1-\mu-(1+\alpha)(1+\mu+\|A\|^2+\|B\|^2)}{2\alpha^2}]\|x^{k+1}-\hat{x}^{k+1}\|^2 \\
&+ [\frac{\mu^+ L_2}{2}+\frac{(1+\beta)(3+\mu+\|C\|^2)+\|B\|^2\beta}{2\beta}]\|y^k-\hat{y}^k\|^2 \\
&- [\frac{\mu^-\sigma_2-\mu-(1+\beta)(2+\mu+\|C\|^2)}{2\beta^2}]\|y^{k+1}-\hat{y}^{k+1}\|^2.
\end{aligned}$$

This completes the proof.

B Proof of Theorem 2

Proof. According to Remark 1, $\delta_C(x)$ and $\delta_Q(y)$ are semi-algebraic functions. Combined with the fact that $\frac{1}{2}\|Ax-b\|^2$ is a semi-algebraic function, it follows that $F(x,y)$ is also a semi-algebraic function.

Since $\{\omega^k\}_{k\in N}$ is bounded, there exists a subsequence $\{\omega^{k_j}\}_{k\in N} \rightarrow \omega^*=(x^*,y^*,\hat{x}^*,\hat{y}^*)$. Then $\lim_{j\to\infty} H(\omega^{k_j})=H(\omega^*)$. From the convergence of $\{H(\omega^k)\}_{k\in N}$, we have $\lim_{k\to\infty} H(\omega^k)=H(\omega^*)$. Let's consider two scenarios.

(i) If there exist an integer $k_0\in N$ such that $H(\omega^{k_0})=H(\omega^*)$.

From (13) we have $x^{k+1}=\hat{x}^{k+1}, y^{k+1}=\hat{y}^{k+1}$. According to (29) and (30), we have $z^{k+1}=z^k, \forall k>k_0$. Thus the conclusion holds.

(ii)Assuming that $H(\omega^k)>H(\omega^*)$ for all $k\in N$.

We denote by Ω the set of all limit point of $\{\omega^k\}_{k\in N}$. Therefore, for any given $\varepsilon>0$, there exist $k_1\in N$ such that $dist(\omega^k,\Omega)<\varepsilon$ for all $k>k_1$.

Again, since $\lim_{k\to\infty} H(\omega^k)=H(\omega^*)$, for any given $\eta>0$, there exists $k_2\in N$ such that $H(\omega^*)<H(\omega^k)<H(\omega^*)+\eta$ for all $k>k_2$.

Consequently, for all $k>l=max\{k_1,k_2\}$, we have

$$\omega^k\in[H(\omega^*)<H(\omega^k)<H(\omega^*)+\eta]\cap\{\omega^k: dist(\omega^k,\Omega)<\varepsilon\}.$$

From (4) there exist a concativty function $\varphi(x)$ such that, for all $k>l$,

$$\varphi'(H(\omega^k)-H(\omega^*))\, dist(0,\partial H(\omega^k))\geq 1.$$

From the concavity of $\varphi(x)$, we have

$$\varphi(H(\omega^k)-H(\omega^*))-\varphi(H(\omega^{k+1})-H(\omega^*))\geq\varphi'(H(\omega^k)-H(\omega^*))(H(\omega^k)-H(\omega^{k+1})).$$

Involing $dist\left(0, \partial H\left(\omega^k\right)\right) \le \rho_2 \left\|z^k - \hat{z}^k\right\|$, we get that

$$\varphi^{'}\left(H\left(\omega^k\right) - H\left(\omega^*\right)\right) \ge \frac{1}{\rho_2 \left\|z^k - \hat{z}^k\right\|}. \tag{31}$$

Let $\Delta_{p,q} = \varphi\left(H\left(\omega^p\right) - H\left(\omega^*\right)\right) - \varphi\left(H\left(\omega^q\right) - H\left(\omega^*\right)\right)$, $p, q \in N$.

From (13) and (17), we have $\left\|z^{k+1} - \hat{z}^{k+1}\right\| \le \sqrt{s\Delta_{k,k+1}\left\|z^k - \hat{z}^k\right\|}$, where, $s = \frac{\rho_2}{\rho_1}$.

Summing up for $i = l+1, ..., k$, yields

$$\begin{aligned} 2\sum_{i=l+1}^{k} \|z^{i+1} - \hat{z}^{i+1}\| &\le \sum_{i=l+1}^{k} \|z^i - \hat{z}^i\| + s\sum_{i=l+1}^{k} \Delta_{i,i+1} \\ &= \sum_{i=l+1}^{k} \|z^{i+1} - \hat{z}^{i+1}\| + \|z^{l+1} - \hat{z}^{l+1}\| + s\Delta_{l+1,k+1}. \end{aligned}$$

Since $\varphi \ge 0$, we thus have

$$\begin{aligned} \sum_{i=l+1}^{k} \|z^{i+1} - \hat{z}^{i+1}\| &\le \|z^{l+1} - \hat{z}^{l+1}\| + s\Delta_{l+1,k+1} \\ &\le \|z^{l+1} - \hat{z}^{l+1}\| + s\varphi(F(z^{l+1}) - F(z^*)). \end{aligned}$$

Since the right hand-side of the inequality above does not depend on k at all, it easily shows that

$$\sum_{k=1}^{\infty} \|z^k - \hat{z}^k\| < \infty, \tag{32}$$

According to (29) and (30), $\sum_{k=1}^{\infty}\|x^{k+1} - x^k\| < \infty$,$\sum_{k=1}^{\infty}\|y^{k+1} - y^k\| < \infty$.

Hence, $\sum_{k=1}^{\infty}\|z^{k+1} - z^k\| \le \sum_{k=1}^{\infty}(\|x^{k+1} - x^k\| + \|y^{k+1} - y^k\|) < \infty$, It implies that $\left\{z^k\right\}$ is a Cauchy sequence and hence it is convergent.

We denote its limit point as z^*. It follows from Lemma 4 and (32) that $(Q_x^k, Q_y^k) \in \partial F(x^k, y^k)$ and $\lim\limits_{k\to\infty}(Q_x^k, Q_y^k) = (0, 0)$. According to the closeness of ∂F, we have that $0 \in \partial F(z^*)$. Hence, z^* is a critical point of F.

The proof is completed.

References

1. Attouch, H., Bolte, J., Redont, P., Soubeyran, A.: Proximal alternating minimization and projection methods for nonconvex problems: an approach based on the Kurdyka-Łojasiewicz inequality. Math. Oper. Res. **35**(2), 438–457 (2010)
2. Auslender, A.: Asymptotic properties of the Fenchel dual functional and applications to decomposition problems. J. Optim. Theory Appl. **73**(3), 427–449 (1992)
3. Bolte, J., Sabach, S., Teboulle, M.: Proximal alternating linearized minimization for nonconvex and nonsmooth problems. Math. Program. **146**(1–2), 459–494 (2014)

4. Bregman, L.: The relaxation method of finding the common points of convex sets and its application to the solution of problems in convex programming. USSR Comput. Math. Math. Phys. **7**(3), 200–217 (1967)
5. Byrne, C.: Iterative oblique projection onto convex sets and the split feasibility problem. Inverse. Probl. **18**(2), 441–453 (2002)
6. Censor, Y., Elfving, T.: A multiprojection algorithm using Bregman projections in a product space. Numer. Algor. **8**(2), 221–239 (1994)
7. Chao, M., Nong, F., Zhao, M.: An inertial alternating minimization with Bregman distance for a class of nonconvex and nonsmooth problems. J. Appl. Math. Comput. **69**(2), 1559–1581 (2023)
8. Chao, M., Zhang, Y., Jian, B.: An inertial proximal alternating direction method of multipliers for nonconvex optimization. Int. J. Comput. Math. **98**(6), 1199–1217 (2021)
9. Chen, C., Pong, T., Tan, L., Zeng, L.: A difference-of-convex approach for split feasibility with applications to matrix factorizations and outlier detection. J. Glob. Optim. **78**(1), 107–136 (2020)
10. Dang, Y., Sun, J., Xu, H.: Inertial accelerated algorithms for solving a split feasibility problem. J. Ind. Manag. Optim. **13**(3), 78–78 (2016)
11. Dong, Q., Yuan, H., Cho, Y., et al.: Modified inertial Mann algorithm and inertial CQ-algorithm for nonexpansive mappings. Optim. Lett. **12**(1), 87–102 (2018)
12. Gibali, A., Sabach, S., Voldman, S.: Non-convex split feasibility problems: models, algorithms and theory. Open. J. Math. Optim. **1**, 1–15 (2020)
13. Gao, X., Cai, X., Han, D.: A Gauss-Seidel type inertial proximal alternating linearized minimization for a class of nonconvex optimization problems. J. Glob. Optim. **76**(4), 863–887 (2020)
14. Guo, C., Zhao, J.: Two-step inertial Bregman proximal alternating linearized minimization algorithm for nonconvex and nonsmooth problems. arXiv:2306.07614 (2023)
15. Hanzely, F., Richtárik, P., Xiao, L.: Accelerated Bregman proximal gradient methods for relatively smooth convex optimization. Comput. Optim. Appl. **79**(2), 405–440 (2021)
16. Hao, Y., Zhao, J.: New Bregman projection algorithms for solving the split feasibility problem. J. Nonlinear. Variational. Anal. **7**(6), 971–984 (2023)
17. López, G., Martín-Márquez, V., Wang, F., Xu, H.: Solving the split feasibility problem without prior knowledge of matrix norms. Inverse. Probl. **28**(8), 374–389 (2012)
18. Liu, P., Jian, J., Shao, H., Wang, X., Xu, J., Wu, X.: A Bregman-style improved ADMM and its linearized version in the nonconvex setting: convergence and rate analyses. J. Oper. Res. Soc. China **12**(2), 298–340 (2024)
19. Mukkamala, C.M., Ochs, P., Pock, T., Sabach, S.: Convex-concave backtracking for inertial Bregman proximal gradient algorithms in nonconvex optimization. SIAM. J. Math. Data. Sci. **2**(3), 658–682 (2020)
20. Polyak, T.: Some methods of speeding up the convergence of iteration methods. Math. Phys. **4**(5), 1–17 (1964)
21. Qu, B., Xiu, N.: A note on the CQ algorithm for the split feasibility problem. Inverse. Probl. **21**(5), 1655–1665 (2005)
22. Shehu, Y., Gibali, A.: New inertial relaxed method for solving split feasibilities. Optim. Lett. **15**(6), 2109–2126 (2021)
23. Suantai, S., Kesornprom, S., Cholamjiak, P.: A new hybrid CQ algorithm for the split feasibility problem in Hilbert spaces and its applications to compressed sensing. Mathematics **7**(9), 789 (2019)

24. Xu, J., Chao, M.: An inertial Bregman generalized alternating direction method of multipliers for nonconvex optimization. J. Appl. Math. Comput. **68**(3), 1757–1783 (2022)
25. Zhao, J., Zhang, Y., Yang, Q.: Modified projection methods for the split feasibility problem and the multiple-sets split feasibility problem. Appl. Math. Comput. **219**(4), 1644–1653 (2012)

Convergence and Linear Convergence of Alternated Inertial Projection Algorithm for Split Feasibility Problems

Xiaojie Sun[1], Meixia Li[2(✉)], and Biao Qu[1]

[1] Institute of Operations Research, Qufu Normal University, Rizhao 276826, People's Republic of China

[2] School of Mathematics and Statistics, Weifang University, Weifang 261061, People's Republic of China

limeixia001@163.com

Abstract. In this article, an alternated inertial projection algorithm is introduced to solve the splitting feasibility problems, and the convergence of the iterative sequence generated by the proposed method is obtained under appropriate conditions. When the splitting feasibility problem satisfies some bounded linear regularity, it is proved that the generated sequences converge linearly. Finally, some numerical results demonstrate the feasibility and effectiveness of the proposed algorithm.

Keywords: Splitting feasible problems with multiple output sets · alternated internal projection algorithm · linear convergence

1 Introduction

Let H_1, H_2 be two real Hilbert spaces, $C_1 \subset H_1$ and $Q \subset H_2$ be two nonempty, closed and convex sets. The split feasibility problem (SFP) is formulated

$$\text{to find an element } \; x \in C_1 \; \text{ such that } \; Ax \in Q, \tag{1}$$

where $A : H_1 \to H_2$ is a bounded linear operator and $A^* : H_2 \to H_1$ be its adjoint. The SFP was first introduced by Censor and Elfving [3] for inverse problems which arised from phase retrievals and medical image reconstruction [1]. Recently, it was extended to systems biology [19] and electricity production [22,23]. In [1,2], to solve SFP, Byrne proposed a projection algorithm called CQ algorithm. It is defined by

$$x_{n+1} = P_{C_1}[x_n - \alpha A^T (I - P_Q) A x_n], \tag{2}$$

where P_{C_1} and P_Q are projections onto C_1 and Q respectively, and the stepsize α belongs to $(0, \frac{2}{\|A^T A\|})$. Subsequently, Yang [21] suggested a relaxed version by using the projections of half-spaces. But to obtain convergence, one must know the norm of the operator, which is usually not an easy job. To overcome

M. Li et al. (Eds.): TAMC 2025, LNCS 16084, pp. 396–407, 2026.
https://doi.org/10.1007/978-981-95-4839-2_31

this difficulty, Zhao et al. [24] introduced the two-step projection method for SFP by using the linesearch technique and proved its convergence under mild conditions. Subsequently, Gibali et al. [11] proposed a relaxation of CQ algorithm with the Armijo-linesearch in real Hilbert spaces. Later, Dang et al. [7] proposed the accelerated algorithms for solving SFP by using the inertial technique to speed up the convergence. And Gibali and Mai [10] introduced a modification of the relaxed CQ algorithm for solving SFP by the self-adaptive method and the inertial technique with a new stepsize and established the convergence theorem. In 2020, Shehu et al. [17] introduced algorithms with alternated inertial step to solve split feasibility problems, and gave global and linear convergence results. Suantai et al. [18] have suggested the inertial projection and contraction methods of split feasibility problem and have proved that the sequence in the proposed method converges weakly to a point of the SFP. Reich et al. [15] have presented and studied the following split feasibility problem with multiple output sets in Hilbert spaces. Chao et al. [5] proposed an improved random projection-type algorithm for solving multiple-sets split feasibility problems and proved the convergence and sublinear convergence rate of the algorithm.

Let H, H_i, $i = 1, 2, \cdots, N$, be real Hilbert spaces and let T_i: $H \to H_i, i = 1, 2, \cdots, N$, be bounded linear operators. Let C and Q_i be nonempty, closed and convex subsets of H and H_i, $i = 1, 2, \cdots, N$, respectively. Suppose that $S = C \cap (\cap_{i=1}^{N} T_i^{-1}(Q_i)) \neq \phi$. In this paper, we consider the following problem:

$$\text{to find an element } \ x \in C, \tag{3}$$

such that $T_i x \in Q_i$, for all $i = 1, 2, \cdots, N$.

As follows, we need to design an iterative algorithm to construct an element of S. In [15], Reich et al. defined the function $g : H \to R$, that is

$$g(x) = \max_{i=1,2,\cdots,N} f_i(x), \tag{4}$$

for all $x \in H$, where $f_i(x) = \frac{1}{2}\|(I - P_{Q_i})T_i x\|^2$. As you can see, a point $x^* \in H$ is a solution to (3), if and only if it is a solution to the problem

$$\min_{x \in C} g(x) \tag{5}$$

and the optimal value is zero, which is equivalent to

$$0 \in \nabla g(x^*) + N_C(x^*), \tag{6}$$

where $N_C(x)$ is the normal cone of C at the point x.

Inspired by these characteristics, Reich et al. [15] introduced the following iterative method

$$x_{n+1} = P_C[x_n - \gamma_n \sum_{i=1}^{N} T_i^*(I - P_{Q_i})T_i x_n], \tag{7}$$

and they established the weak convergence of the sequences generated by the iterative method. And they had to use the following assumption regarding the parameters γ_n:

$$0 < a \le \gamma_n \le b < \frac{2}{m \max\limits_{i=\overline{1,N}} \|T_i\|^2}, \tag{8}$$

where $i = \overline{1,N}$ denotes $i = 1, 2, \cdots, N$. The first step is to compute (or at least estimate) the operator norm of T_i.

In order to simplify the algorithm, and replace the sum of the index sets $\{1, 2, \cdots, N\}$ on the right-hand side of (7) with a suitable subset, Reich et al. [16] proposed a new sequence generated by

$$x_{n+1} = P_C[x_n - \gamma_n \sum_{i \in I(x_n)} \lambda_{i,n} T_i^*(I - P_{Q_i})T_i x_n], \tag{9}$$

where $I(x_n) = \{i : \|T_i x_n - P_{Q_i}(T_i x_n)\| = \max\limits_{i=1,2,\cdots,N} \|T_i x_n - P_{Q_i}(T_i x_n)\|\}$, $\lambda_{i,n} \ge 0$ for all $i \in I(x_n)$, $\sum\limits_{i \in I(x_n)} \lambda_{i,n} = 1$, and $\{\gamma_n\}$ is a sequence of nonnegative real numbers, which is defined by

$$\gamma_n = \begin{cases} \rho_n \dfrac{d_n^2}{\|\sum\limits_{i \in I(x_n)} \lambda_{i,n} T_i^*(I - P_{Q_i})T_i x_n\|^2}, \text{if} \|\sum\limits_{i \in I(x_n)} \lambda_{i,n} T_i^*(I - P_{Q_i})T_i x_n\| > 0, \\ 0, \quad \text{otherwise}. \end{cases} \tag{10}$$

where $d_n = \max\limits_{i=1,2,\cdots,N} \|T_i x_n - P_{Q_i}(T_i x_n)\|$, and $\{\rho_n\} \subset [a, b] \subset (0, 2)$.

Building on these developments, we propose a new alternated inertial projection algorithm for split feasible problems with multiple output sets. Our method combines the alternate inertial technique with an adaptive stepsize rule that does not require prior knowledge of operator norms. The paper is organized as follows. In Sect. 2, we review some concepts and lemmas that will be used to prove the main result. In Sect. 3, we present an iterative algorithm to solve the split feasibility problem with multiple output sets and establish the convergence and linear convergence of the proposed algorithms. In Sect. 4, the effectiveness of the algorithm is proved by numerical experiments.

2 Preliminaries

This section mainly expounds some basic theoretical knowledge needed. We use $\to$ and $\rightharpoonup$ to denote the strong convergence and weak convergence, respectively, and we use $\omega_\omega(x_i^k) = \{x : \exists x_i^{k_j} \rightharpoonup x\}$ to denote the weak ω-limit set of $\{x_i^k\}$.

For any $x \in H$, there exists a unique nearest point in C, denoted by $P_C x$, such that

$$\|x - P_C x\| \le \|x - y\|, \quad \forall y \in C. \tag{11}$$

It is well known that P_C is nonexpansive and firmly nonexpansive. And P_C has the some well-known properties (details please see [4]).

Definition 1. *[9] A mapping $T : C \to C$ is said to be demiclosed at 0 if, for any sequence $\{x_n\} \subset C$ which converges weakly to x and with $\|x_n - T(x_n)\| \to 0$, $T(x) = x$.*

Recall that a mapping $U : C \to C$ is said to be nonexpansive if $\|Ux - Uy\| \leq \|x - y\|$ for all $x, y \in C$. We denote the set of fixed points of U by $\text{Fix}(U)$, that is, $\text{Fix}(U) = \{x \in C | Ux = x\}$.

Lemma 1. *[9] Let T be a nonexpansive self-mapping of a closed and convex subset C of a Hilbert space H. Then the mapping $I - T$ is demiclosed, that is, whenever $\{x_n\}$ is a sequence in C which weakly converges to some $x \in C$ and the sequence $\{(I - T)x_n\}$ strongly converges to some y, it follows that $(I - T)x = y$.*

Definition 2. *The SFP (3) is said to satisfy the bounded linear regularity property if, for any $r > 0$ such that $S \cap \overline{B(0, r)} \neq \emptyset$ and there exists $\gamma_r > 0$ such that*

$$\gamma_r d_S(x) \leq d_{Q_i}(T_i x), i = 1, 2, \cdots, m, \text{ for any } x \in S \cap \overline{B(0, r)},$$

where $d_S(x) = \inf_{y \in S} \|x - y\|$, and $d_{Q_i}(T_i x) = \inf_{\overline{y} \in Q_i} \|T_i x - \overline{y}\|$.

3 Algorithm and Convergence Theorem

In this section, we give a new algorithm for solving (3) and prove the convergence of the algorithm. According to the definition of $f_i(x)$, we set $f_i(x_n) = \frac{1}{2}\|(I - P_{Q_i})T_i x_n\|^2$, and $\nabla f_i(x_n) = T_i^*(I - P_{Q_i})T_i x_n$.

Algorithm 1 *Alternated Inertial Projection Algorithm (abbr. AIPA)*

1. Choose $x_0, x_1 \in C$ as given starting points. $\{\theta_n\}$ and $\{\rho_n\}$are positive real sequences. Set $n = 1$.

2. Compute

$$w_n = \begin{cases} x_n, & n = even, \\ x_n + \theta_n(x_n - x_{n-1}), & n = odd. \end{cases} \tag{12}$$

3. Compute

$$x_{n+1} = P_C[w_n - \gamma_n \sum_{i \in I(w_n)} \lambda_{i,n} T_i^*(I - P_{Q_i})T_i w_n], \tag{13}$$

where $I(w_n) = \{i : \|T_i w_n - P_{Q_i}(T_i w_n)\| = \max_{i=1,2,\cdots,N} \|T_i w_n - P_{Q_i}(T_i w_n)\|\}, \lambda_{i,n} \geq 0$, for all $i \in I(w_n)$, $\sum_{i \in I(w_n)} \lambda_{i,n} = 1$. And $\{\gamma_n\}$ is defined by (10).

4. Set $n \leftarrow n + 1$, and goto 2.

Theorem 1. *Let H be a real Hilbert space. For $i = 1, 2, \cdots, N$, assume that C and Q_i are nonempty closed convex subsets of real Hilbert spaces H and H_i, $T_i : H \to H_i$ are bounded linear operators with their adjoint operators T_i^*, respectively. Suppose the solution set S of problem (3) is nonempty, $\{\rho_n\} \subset [a, b] \subset (0, 2)$ is non-decreasing and sequence $\{\theta_n\} \subset [0, \infty)$ such that $0 < \alpha \leq \rho_n(1 + \theta_n) \leq \beta < 2$, where a, b, α, β are real numbers. Then the sequence $\{x_n\}$ generated by Algorithm 1 converges to a point in S.*

Proof. Take n as even and pick a point z in S.
Using the fact that $I - P_C$ is firmly nonexpansive, we get

$$\begin{aligned}&\langle w_{n+1} - z, \sum_{i \in I(w_{n+1})} \lambda_{i,n+1} T_i^*(I - P_{Q_i}) T_i w_{n+1} \rangle \\ \geq &\sum_{i \in I(w_{n+1})} \lambda_{i,n+1} \|(I - P_{Q_i}) T_i w_{n+1}\|^2 \\ =&d_{n+1}^2.\end{aligned} \tag{14}$$

Then, from (13) and (14),

$$\begin{aligned}&\|x_{n+2} - z\|^2 \\ \leq&\|w_{n+1} - \gamma_{n+1} \sum_{i \in I(w_{n+1})} \lambda_{i,n+1} T_i^*(I - P_{Q_i}) T_i w_{n+1} - z\|^2 \\ =&\|w_{n+1} - z\|^2 + \gamma_{n+1}^2 \| \sum_{i \in I(w_{n+1})} \lambda_{i,n+1} T_i^*(I - P_{Q_i}) T_i w_{n+1}\|^2 \\ &- 2\gamma_{n+1} \langle w_{n+1} - z, \sum_{i \in I(w_{n+1})} \lambda_{i,n+1} T_i^*(I - P_{Q_i}) T_i w_{n+1} \rangle \\ \leq&\|w_{n+1} - z\|^2 - \gamma_{n+1}(2d_{n+1}^2 - \gamma_{n+1} \| \sum_{i \in I(w_{n+1})} \lambda_{i,n+1} T_i^*(I - P_{Q_i}) T_i w_{n+1}\|^2) \\ =&\|w_{n+1} - z\|^2 - 2 \frac{\rho_{n+1} d_{n+1}^4}{\| \sum_{i \in I(w_{n+1})} \lambda_{i,n+1} T_i^*(I - P_{Q_i}) T_i w_{n+1}\|^2} \\ &+ \frac{\rho_{n+1}^2 d_{n+1}^4}{\| \sum_{i \in I(w_{n+1})} \lambda_{i,n+1} T_i^*(I - P_{Q_i}) T_i w_{n+1}\|^2} \\ =&\|w_{n+1} - z\|^2 - \rho_{n+1}(2 - \rho_{n+1}) \frac{d_{n+1}^4}{\| \sum_{i \in I(w_{n+1})} \lambda_{i,n+1} T_i^*(I - P_{Q_i}) T_i w_{n+1}\|^2}.\end{aligned} \tag{15}$$

Now,

$$\begin{aligned}&\|w_{n+1} - z\|^2 \\ =&\|x_{n+1} + \theta_{n+1}(x_{n+1} - x_n) - z\|^2 \\ =&(1 + \theta_{n+1})\|x_{n+1} - z\|^2 - \theta_{n+1}\|x_n - z\|^2 + \theta_{n+1}(1 + \theta_{n+1})\|x_{n+1} - x_n\|^2.\end{aligned} \tag{16}$$

Using similar arguments in obtaining (15), we have

$$\begin{aligned}\|x_{n+1} - z\|^2 \leq&\|w_n - z\|^2 - \rho_n(2 - \rho_n) \frac{d_n^4}{\| \sum_{i \in I(w_n)} \lambda_{i,n} T_i^*(I - P_{Q_i}) T_i w_n\|^2} \\ =&\|x_n - z\|^2 - \rho_n(2 - \rho_n) \frac{d_n^4}{\| \sum_{i \in I(w_n)} \lambda_{i,n} T_i^*(I - P_{Q_i}) T_i w_n\|^2}.\end{aligned} \tag{17}$$

Put (16) and (17) into (15), we obtain

$$\begin{aligned}
&\|x_{n+2}-z\|^2\\
\leq&(1+\theta_{n+1})[\|x_n-z\|^2-\rho_n(2-\rho_n)\frac{d_n^4}{\|\sum\limits_{i\in I(w_n)}\lambda_{i,n}T_i^*(I-P_{Q_i})T_iw_n\|^2}]\\
&-\theta_{n+1}\|x_n-z\|^2+\theta_{n+1}(1+\theta_{n+1})\|x_{n+1}-x_n\|^2\\
&-\rho_{n+1}(2-\rho_{n+1})\frac{d_{n+1}^4}{\|\sum\limits_{i\in I(w_{n+1})}\lambda_{i,n+1}T_i^*(I-P_{Q_i})T_iw_{n+1}\|^2}\\
=&\|x_n-z\|^2-(1+\theta_{n+1})\rho_n(2-\rho_n)\frac{d_n^4}{\|\sum\limits_{i\in I(w_n)}\lambda_{i,n}T_i^*(I-P_{Q_i})T_iw_n\|^2}\\
&+\theta_{n+1}(1+\theta_{n+1})\|x_{n+1}-x_n\|^2\\
&-\rho_{n+1}(2-\rho_{n+1})\frac{d_{n+1}^4}{\|\sum\limits_{i\in I(w_{n+1})}\lambda_{i,n+1}T_i^*(I-P_{Q_i})T_iw_{n+1}\|^2}.
\end{aligned} \tag{18}$$

Then, we have

$$\begin{aligned}
\|x_{n+1}-x_n\|^2=&\|P_C(w_n-\gamma_n\sum_{i\in I(w_n)}\lambda_{i,n}T_i^*(I-P_{Q_i})T_iw_n)-P_C(x_n)\|^2\\
\leq&\|w_n-\gamma_n\sum_{i\in I(w_n)}\lambda_{i,n}T_i^*(I-P_{Q_i})T_iw_n-w_n\|^2\\
=&\frac{\rho_n^2d_n^4}{\|\sum\limits_{i\in I(w_n)}\lambda_{i,n}T_i^*(I-P_{Q_i})T_iw_n\|^2}.
\end{aligned} \tag{19}$$

Since $\{\rho_n\}$ is non-decreasing, put (19) into (18)

$$\begin{aligned}
&\|x_{n+2}-z\|^2\\
\leq&\|x_n-z\|^2-(1+\theta_{n+1})\rho_n(2-\rho_n)\frac{d_n^4}{\|\sum\limits_{i\in I(w_n)}\lambda_{i,n}T_i^*(I-P_{Q_i})T_iw_n\|^2}\\
&+\theta_{n+1}(1+\theta_{n+1})\frac{\rho_n^2d_n^4}{\|\sum\limits_{i\in I(w_n)}\lambda_{i,n}T_i^*(I-P_{Q_i})T_iw_n\|^2}\\
&-\rho_{n+1}(2-\rho_{n+1})\frac{d_{n+1}^4}{\|\sum\limits_{i\in I(w_{n+1})}\lambda_{i,n+1}T_i^*(I-P_{Q_i})T_iw_{n+1}\|^2}\\
\leq&\|x_n-z\|^2-\rho_n(1+\theta_{n+1})(2-\rho_{n+1}-\theta_{n+1}\rho_{n+1})\frac{d_n^4}{\|\sum\limits_{i\in I(w_n)}\lambda_{i,n}T_i^*(I-P_{Q_i})T_iw_n\|^2}
\end{aligned}$$

$$
\begin{aligned}
&- \rho_{n+1}(2-\rho_{n+1})\frac{d_{n+1}^4}{\|\sum_{i\in I(w_{n+1})}\lambda_{i,n+1}T_i^*(I-P_{Q_i})T_iw_{n+1}\|^2} \\
\leq &\|x_n - z\|^2.
\end{aligned} \tag{20}
$$

Therefore

$$
\|x_{n+2}-z\| \leq \|x_n - z\|. \tag{21}
$$

Hence, $\lim_{n\to\infty}\|x_{2n}-z\|$ exists and this implies that $\{x_{2n}\}$ is bounded. Consequently, from (20), we have

$$
\frac{d_n^2}{\|\sum_{i\in I(w_n)}\lambda_{i,n}T_i^*(I-P_{Q_i})T_iw_n\|} \to 0. \tag{22}
$$

For each $i \in I(w_n)$, we have

$$
\lambda_{i,n}d_n\|T_i^*(I-P_{Q_i})T_iw_n\| \leq \lambda_{i,n}d_n^2\|T_i\|, \tag{23}
$$

which implies that

$$
\begin{aligned}
d_n\sum_{i\in I(w_n)}\lambda_{i,n}\|T_i^*(I-P_{Q_i})T_iw_n\| \leq &d_n^2\sum_{i\in I(w_n)}\lambda_{i,n}\|T_i\| \\
\leq &d_n^2\max_{i\in I(w_n)}\{\|T_i\|\}.
\end{aligned} \tag{24}
$$

Combining this with the inequality

$$
\|\sum_{i\in I(w_n)}\lambda_{i,n}T_i^*(I-P_{Q_i})T_iw_n\| \leq \sum_{i\in I(w_n)}\lambda_{i,n}\|T_i^*(I-P_{Q_i})T_iw_n\|, \tag{25}
$$

we obtain

$$
d_n\|\sum_{i\in I(w_n)}\lambda_{i,n}T_i^*(I-P_{Q_i})T_iw_n\| \leq d_n^2\max_{i\in I(w_n)}\{\|T_i\|\}. \tag{26}
$$

Thus, we have

$$
d_n \leq \max_{i\in I(w_n)}\{\|T_i\|\}\frac{d_n^2}{\|\sum_{i\in I(w_n)}\lambda_{i,n}T_i^*(I-P_{Q_i})T_iw_n\|}, \tag{27}
$$

which, combined with (22), implies that $d_n \to 0$, that is, for each $i \in I(w_n)$,

$$
\lim_{n\to\infty}\|T_iw_n - P_{Q_i}T_iw_n\| = 0. \tag{28}
$$

Using the definition of $I(w_n)$, we obtain that, for all $i = 1, 2, \cdots, N$,

$$
\lim_{n\to\infty}\|T_iw_n - P_{Q_i}T_iw_n\| = 0. \tag{29}
$$

Since the sequence $\{x_{2n}\}$ is bounded, there exists a subsequence $\{x_{2n_j}\}$ of $\{x_{2n}\}$, such that $x_{2n_j} \rightharpoonup \overline{x}$ as $j \to \infty$. Since C is a closed and convex subset of H, from $\{x_{2n}\} \subset C$, we deduce that $\overline{x} \in C$. Furthermore, because T_i is a bounded linear operator, we have $T_i x_{2n_j} \rightharpoonup T_i\overline{x}$, for all $i = 1, 2, \cdots, N$. Thus according to Lemma 1 and (29), we get that $T_i\overline{x} \in \mathrm{Fix}(P_{Q_i})$, that is, $T_i\overline{x} \in Q_i$, for all $i = 1, 2, \cdots, N$. So $\overline{x} \in S$.

We next show that the sequence of odd terms $\{x_{2n+1}\}$ converges to $\overline{x}$. By similar arguments in obtaining (15), we obtain

$$\begin{aligned} \|x_{2n_j+1} - \overline{x}\| \leq& \|w_{2n_j} - \overline{x}\| \\ =& \|x_{2n_j} - \overline{x}\|. \end{aligned} \tag{30}$$

Thus,

$$x_{2n_j+1} \rightharpoonup \overline{x}. \tag{31}$$

Therefore, $x_n \rightharpoonup \overline{x}$, which completes the proof.

According to the bounded linear regularity property for (3) (see Definition 2) and the relevant reference, we can obtain the linear convergence of the sequences generated by Algorithm 1. Due to the space limitations, we only give the theorem and omit the proof.

Theorem 2. *Assume that the conditions of Theorem 1 hold, for simplicity, we assume that the inertial factor $\theta_n \equiv \theta$, (3) satisfies the bounded linear regularity property, and $\{x_n\}$ is generated by Algorithm 1. Then $\{x_n\}$ converges linearly to a point $p \in S$.*

4 Numerical Examples

In this section, the whole programs are written in Matlab 2016a. All the numerical results are carried out on a personal Lenovo computer with Intel(R) Core(TM) AMD Ryzen 5 4600U with Radeon Graphics CPU 2.10 GHZ RAM 8.00 GB.

We apply a real-world example which is based on Nash equilibrium concept and Nikaido-Isoda function. The function of Nikaido-Isoda has been used to models in electricity markets and others [6,20,23]. The Nikaido-Isoda is defined as

$$f(x, y) = \sum_{i=1}^{k} (f_i(x) - f_i(x[y_i])), \tag{32}$$

where $x[y_i]$ is a vector obtained by replacing component x_i with y_i.

There is a power production model. In the model, it is assumed that a company has k plants, and each plant can generate a type of power, such as nuclear, solar, wind, hydropower and thermoelectric power. The company obtains the maximum profit by seeking the corresponding production level of each factory. Set C_i as the policy set of factory i. That is, the production level x_i must be selected in C_i. The production price of factory i is an affine function given by $p_i(x_1, \cdots, x_k) = \alpha - \sum_{j=1}^{k} \tau_{ij} x_j$, for each i, α is a large positive number and $\tau_{ij} > 0$. This price function is displayed in [12]. The profit of factory i takes the form

$$f_i(x) = p_i(x_1, \cdots, x_k)x_i - c_i(x_i), \tag{33}$$

where $c_i(x_i)$ represents the production cost of the factory i.

In fact, the production level of each factory should meet a certain proportion. For example, the total output of a certain power type x_i and all other power types $\sum\limits_{j \neq i} x_j$ should be limited to a given percentage. It can be expressed as $l \leq \frac{x_i}{\sum\limits_{j \neq i} x_j} \leq u$, where l and u are given constants. Define the cost function as

$$c_i(x_i) = \frac{1}{2} r_i x_i^2 + q_i x_i, \qquad r_i > 0 \tag{34}$$

and the price function as

$$p_i(x_1, \cdots, x_k) = 30 - \sum_{j=1}^{k} \tau_{ij} x_j. \tag{35}$$

Note that r_j, q_j and each τ_{ij} are randomly generated in [0,20], [0,3] and $[0, \frac{1}{k}]$ [13]. For each i, we take the policy set $C_i = [0, 6]$.

We assume that the ratio of each type of electricity to all electricity production is less than thirty percent, which is represented by the constraints $0 \leq A_i(x) \leq 0.3$ for every $i = 1, 2, \cdots, m$, and

$$A_i(x) = \frac{\sum\limits_{l \in I_i} x_l}{\sum_{j=1}^{k} x_j}, \qquad i = 1, 2, \cdots, m, \tag{36}$$

where I_i is the set of factories that produce type i of electricity.

Let $p_i(x) = \frac{1}{2}\|A_i x - P_{Q_i}(A_i x)\|^2, i = 1, 2, \cdots, m$, and $p(x) = \max\limits_{i=1,2,\cdots,m} p_i(x)$.

By applying Algorithm 3.1 in [14] to the split feasibility problems with jointly constrained Nash equilibrium models, Dong et al. [8] proposed an algorithm (which we shall denote by Alg 3.1 in [8] here). Please see [8] for definitions of some notations used. The initial point $x_0 = x_1$ was randomly chosen in C. Let $m = 5, k = 50$ in this example. And choose $\lambda = 1, \rho_k = 2$ for Algorithm 3.1 in [8]. Now we compare the effect of Algorithm 1 and the Algorithm 3.1 in [8]. The choose of the part of parameters in Algorithm 1 is as follows.

Case (a) $\rho_k = 0.9, \theta_k = 0.4$;

Case (b) $\rho_k = 1.2, \theta_k = 0.5$.

The following figures display the results. From Figs. 1 and 2, we can see that the value of the function $p(x^k)$ in Algorithm 1 is always smaller than the value of the function $p(x^k)$ in Algorithm 3.1 of [8]. This indicates that Algorithm 1 outperforms Algorithm 3.1 of [8]. Further it demonstrates the feasibility and advantages of the Algorithm 1.

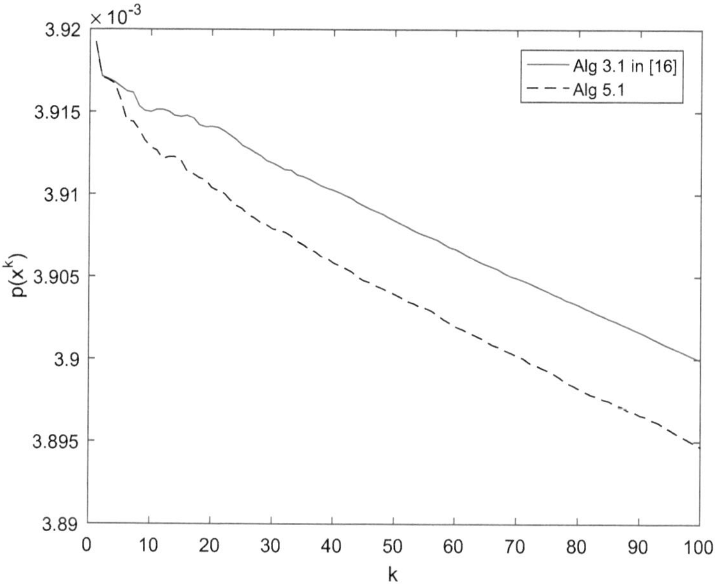

Fig. 1. Comparison of the electricity production market in case (a).

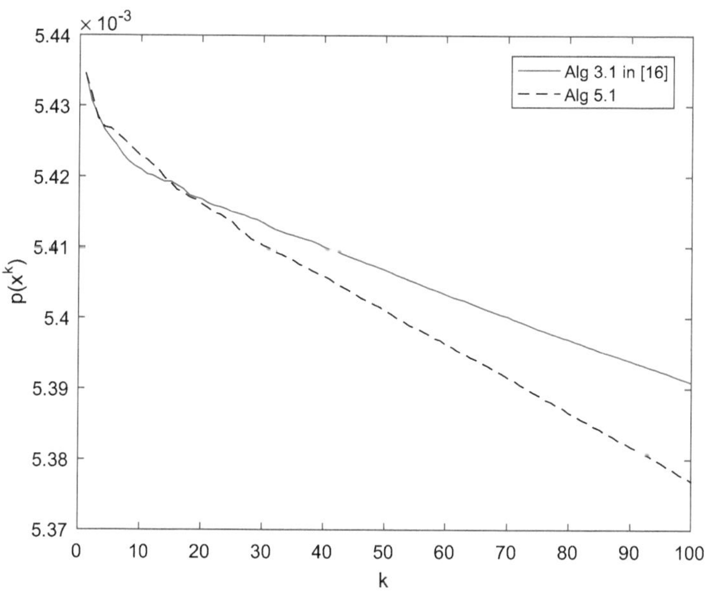

Fig. 2. Comparison of the electricity production market in case (b).

Acknowledgements. This project is supported by Natural Science Foundation of China (Grant No. 12301417, 11571120, and 12271309) and Shandong Provincial Natural Science Foundation (Grant No. ZR2020MA027, ZR2022MA034, ZR2024MA013, ZR2024MA004 and ZR2022MA038).

References

1. Byrne, C.: Iterative oblique projection onto convex sets and the split feasibility problem. Inverse Prob. **18**, 441–453 (2002)
2. Byrne, C.: A unified treatment of some iterative algorithms in signal processing and image reconstruction. Inverse Prob. **20**, 103–120 (2004)
3. Censor, Y., Elfving, T.: A multiprojection algorithm using Bregman projections in a product space. Numer. Algorithms **8**, 221–239 (1994)
4. Chang, S.: Some problems and results in the study of nonlinear analysis. Nonlinear Anal. Theory Methods Appl. **30**, 4197–4208 (1997)
5. Chao, W.J., Xia, F.-Q., Tu, K., Lu, J.: An improved random projection-type algorithm for solving multiple-sets split feasibility problems with application to medical radiotherapy. J. Comput. Appl. Math. **463**, 1–15 (2025)
6. Contreras, J., Klusch, M., Krawczyk, J.B.: Numerical solution to Nash Cournot equilibria in coupled constraint electricity markets. IEEE Trans. Power Syst. **19**, 195–206 (2004)
7. Dang, Y., Sun, J., Xu, H.: Inertial accelerated algorithms for solving a split feasibility problem. J. Ind. Manag. Optim. **13**, 1383–1394 (2017)
8. Dong, Q.L., Li, X., Rassias, T.M.: Two projection algorithms for a class of split feasibility problems with jointly constrained Nash equilibrium models. Optimization **70**, 871–897 (2021)
9. Geobel, K., Kirk, W.: Topics in Metric Fixed Point Theory. Cambridge Studies in Advanced Mathematics, vol. 28. Cambridge University Press, Cambridge (1990)
10. Gibali, A., Mai, D.T.: A new relaxed CQ algorithm for solving split feasibility problems in Hilbert spaces and its applications. J. Ind. Manag. Optim. **15**, 963–984 (2018)
11. Gibali, A., Liu, L.W., Tang, Y.C.: Note on the modified relaxation CQ algorithm for the split feasibility problem. Optim. Lett. **12**, 817–830 (2018)
12. Gorbachuk, V.M.: Cournot-Nash and Bertrand-Nash equilibria for a heterogeneous duopoly of differentiated products. Cybern. Syst. Anal. **46**, 25 (2010)
13. Iusem, A.N.: On some properties of paramonotone operator. J. Convex Anal. **5**, 269–278 (1998)
14. López, G., Martin-Márquez, V., Wang, F., Xu, H.K.: Solving the split feasibility problem without prior knowledge of matrix norms. Inverse Probl. **28**, 085004 (2012)
15. Reich, S., Truong, M.T., Mai, T.N.H.: The split feasibility problem with multiple output sets in Hilbert spaces. Optim. Lett. **14**(8), 2335–2353 (2020).
16. Reich, S., Tuyen, T.M., Ha, M.T.N.: An optimization approach to solving the split feasibility problem in Hilbert spaces. J. Global Optim. **79**, 837–852 (2021)
17. Shehu, Y., Dong, Q.L., Liu, L.L.: Global and linear convergence of alternated inertial methods for split feasibility problems. Revista de la Real Academia de Ciencias Exactas, Fsicas y Naturales. Serie A. Matemticas **115**, 1–26 (2021)
18. Suantai, S., Panyanak, B., Kesornprom, S., Cholamjiak, P.: Inertial projection and contraction methods for split feasibility problem applied to compressed sensing and image restoration. Optim. Lett. **16**, 1725–1744 (2022)

19. Wang, J.H., Hu, Y.H., Li, C., Yao, J.C.: Linear convergence of CQ algorithms and applications in gene regulatory network inference. Inverse Prob. **33**, 055017 (2017)
20. Wei, J.Y., Yves, S.: Spatial oligopolistic electricity models with Cournot generators and regulated transmission prices. Oper. Res. **47**, 102–112 (1999)
21. Yang, Q.Z.: The relaxed CQ algorithm for solving the split feasibility problem. Inverse Prob. **20**, 1261–1266 (2004)
22. Yen, L.H., Muu, L.D., Huyen, N.T.T.: An algorithm for a class of split feasibility problems: application to a model in electricity production. Math. Methods Oper. Res. **84**(3), 549–565 (2016).
23. Yen, L.H., Huyen, N.T.T., Muu, L.D.: A subgradient algorithm for a class of non-linear split feasibility problems: application to jointly constrained Nash equilibrium models. J. Global Optim. **73**(4), 849–868 (2019).
24. Zhao, J., Zhang, Y., Yang, Q.: Modified projection methods for the split feasibility problem and the multiple-sets split feasibility problem. Appl. Math. Comput. **219**, 1644–1653 (2012)

Maximizing the Difference of DR-Submodular Function on the Integer Lattice

Jingjing Tan[1], Cuiping Ge[1(✉)], Meixia Li[1], Meng Sun[1], and Ruiqi Yang[2]

[1] School of Mathematics and Statistics, Weifang University, Weifang 261061, People's Republic of China
wfgecuiping@163.com

[2] Beijing Institute for Scientific and Engineering Computing, Beijing University of Technology, Beijing 100124, People's Republic of China
yangruiqi@emails.bjut.edu.cn

Abstract. In this paper, we design two algorithms for the problem of maximizing the difference of two DR-submodular function $g-h$ (DDRS) on the integer lattice, where $g(\boldsymbol{x})$ is monotone and non-negative DR-submodular, and $h(\boldsymbol{x})$ is monotone and non-negative submodular with curvature c_h. Combining the lattice binary search with the threshold method, we present a one pass streaming algorithm for the unconstrained problems, and analyze the performance of the algorithm and obtain a provable approximation guarantees.

Keywords: Integer lattice · DR-submodular · Curvature · Streaming algorithm

1 Introduction

Submodular optimization problem is widely used in operational research, computer science, economics and other fields, and it is a hot topic of combinatorial optimization research. A lot of research results have been obtained on the maximization of set submodular functions [1–4,6]. Nemhauser and Wolsey [5] showed that for monotonic submodular functions with cardinality constraints, the $1-1/e$ -approximation can be achieved by using the greedy algorithm with a time complexity of $O(nk)$. Because of the need of practical problems such as sensor placement [7] and optimal budget allocation [8], the definition of submodular function is extended from set to integer lattice [9–12] domain. However, some subset selection problems in data science are not purely submodularity optimization problems, non-submodular functions are a natural extension of submodular functions [13–20,22]. Wang et al. [21] considered maximizing the difference of a non-negative normalized monotone submodular function and a non-negative linear function, they designed a bicriteria streaming algorithm.

M. Li et al. (Eds.): TAMC 2025, LNCS 16084, pp. 408–414, 2026.
https://doi.org/10.1007/978-981-95-4839-2_32

In this paper, we study the problem of maximizing the difference between two DR-submodular functions (DDRS) on the integer lattice. Let $E = \{e_1, e_2, \cdots, e_n\}$ be a ground set and $\boldsymbol{x}$ be an n-dimensional vector in $\mathbb{N}^E$. Denote $\mathcal{B} = \{\boldsymbol{x} \in \mathbb{N}^E : \boldsymbol{x} \leq \boldsymbol{B}\}$, $\mathcal{G}_{\boldsymbol{B}}$ is a set of monotone and non-negative DR-submodular functions and $\mathcal{H}_{\boldsymbol{B}}$ is a set of monotone and non-negative DR-submodular functions with curvature c_h. This problem can be characterized as follows:

$$\max_{\boldsymbol{x} \in \mathbb{N}^E} g(\boldsymbol{x}) - h(\boldsymbol{x}),$$

where the function $g(\boldsymbol{x}) \in \mathcal{G}_{\boldsymbol{B}}$ and $h(\boldsymbol{x}) \in \mathcal{H}_{\boldsymbol{B}}$. We first design a one pass streaming algorithm for the unconstrained model.

2 Preliminaries

In this section, we elucidate numerous symbols, conceptual definitions.

We use $[k]$ to denote all positive integers between 1 and k. The segment of coordinate $e_i \in E$ for $\boldsymbol{x}$ is represented as $\boldsymbol{x}(e_i)$. We employ $\boldsymbol{0}$ to depict the null vector and χ_{e_i} as the standard base vector. Thus, all components are at zero, except for the i-th component that stands at unity. For any subset $X \subseteq E$, denote χ_X be the characteristic vector of X, $\boldsymbol{x}(X)$ be the amounts to the summation of $\boldsymbol{x}(e_i)$ over all elements e_i belonging to $\boldsymbol{x}$. We define a multi-set $\{\boldsymbol{x}\}$, where e appears $\boldsymbol{x}(e)$ times, and use $|\{\boldsymbol{x}\}| := \boldsymbol{x}(E)$ to express the total amount of $\boldsymbol{x}(E)$. $\boldsymbol{x} \vee \boldsymbol{y}$ and $\boldsymbol{x} \wedge \boldsymbol{y}$ stand for the coordinate-wise maximum and minimum of vectors $\boldsymbol{x}$ and $\boldsymbol{y}$ respectively. Denote the difference of the multi-sets $\boldsymbol{x}$ and $\boldsymbol{y}$ as $\{\boldsymbol{x}\} \setminus \{\boldsymbol{y}\} := \{(\boldsymbol{x} \setminus \boldsymbol{y}) \vee \boldsymbol{0}\}$

A function $g : \mathbb{N}^E \to \mathbb{R}_+$ is known as monotone non-decreasing if and only if $g(\boldsymbol{x}) \leq g(\boldsymbol{y})$ can be satisfied for any $\boldsymbol{x} \leq \boldsymbol{y}$. A function g is deemed nonnegative and normalized if it invariably fulfills the conditions: $g(\boldsymbol{x}) \geq 0$ (for all $\boldsymbol{x} \in \mathbb{N}^E$) and $g(\boldsymbol{0}) = 0$.

Definition 1. A function $g : \mathbb{N}^E \to \mathbb{R}_+$ is thought of as diminishing return submodular (DR-submodular) if it qualifies the following inequality for any $e \in E$, and any vector $\boldsymbol{x}$ and $\boldsymbol{y}$ in $\mathbb{N}^E$ with $\boldsymbol{x} \leq \boldsymbol{y}$:

$$g(\boldsymbol{y} + \chi_e) - g(\boldsymbol{y}) \leq g(\boldsymbol{x} + \chi_e) - g(\boldsymbol{x}).$$

Definition 2. The curvature of a monotone non-decreasing DR-submondular function h on the inter lattice is defined as

$$c_h = 1 - \min_{e \in \boldsymbol{x}} \frac{h(\boldsymbol{B}) - h(\boldsymbol{B} - \chi_e)}{h(\chi_e)}$$

The value denoted by $g(\boldsymbol{x}|\boldsymbol{y})$ presents the marginal increase of a vector $\boldsymbol{x}$ concerning $\boldsymbol{y}$ expressed as:

$$g(\boldsymbol{x}|\boldsymbol{y}) = g(\boldsymbol{x} + \boldsymbol{y}) - g(\boldsymbol{y}).$$

3 Algorithms

We will design two combinatorial algorithms and analyze the performance of the two respectively in this section.

By designing a binary research algorithm, for each arriving element, we not only decide whether add it to the output solution, but also calculate the number added to the output solutions. The algorithm is designed as following.

Algorithm 1 Greedy algorithm for unconstrained DDRS on the integer lattice

Input: Data stream E, f, g, h, curvature c_h of h, $\lambda \geq 0$.
Output: a vector $\boldsymbol{x} \in \mathbb{N}^E$.

$\boldsymbol{x} \leftarrow \mathbf{0}$;
for $e \in E$ **do**
  find the level l_e with **BS Algorithm**$\Big(e, f, g, h, c_h, \boldsymbol{x}, \boldsymbol{B}, \lambda\Big)$;
  update $\boldsymbol{x} \leftarrow \boldsymbol{x} + l_e \chi_e$;
end for
return $\boldsymbol{x}$

Algorithm 2 BS Algorithm $(e, g, f, c_h, \boldsymbol{x}, \boldsymbol{B}, \lambda)$

Input: Data stream E, $e \in E$, $g \in \mathcal{G}_{\boldsymbol{B}}$, $h \in \mathcal{H}_{\boldsymbol{B}}$, $\boldsymbol{x} \in \mathbb{N}^E$, and $\lambda \geq 1$.
Output: l

$l_m \leftarrow 1$;
$l_n \leftarrow \boldsymbol{B}(e) - \boldsymbol{x}(e)$;
if $f(\chi_e|\boldsymbol{x}) < 0$ **then**
  return 0.
end if
if $f(l_n\chi_e|\boldsymbol{x}) \geq 0$ **then**
  return l_n
end if
while $l_m < l_n + 1$, **do**
  $a = \lfloor \frac{l_n + l_m}{2} \rfloor$
  if $\frac{f(a\chi_e|\boldsymbol{x})}{a} \geq 0$ **then**
    $l_m = a$,
  else
    $l_n = a$,
  end if
end while
return l_m

Lemma 1. *Suppose* $h \in \mathcal{H}_{\mathcal{B}}$ *with curvature* $c_h \in [0, 1)$*, for each* $\boldsymbol{x} \in \mathcal{B}$*, we have*

$$h(\boldsymbol{x}) - h(\mathbf{0}) \leq \sum_{e \in \{x\}} h(\chi_e|\mathbf{0}) \leq \frac{1}{1 - c_h}[h(\boldsymbol{x}) - h(\mathbf{0})].$$

Proof. Let $\boldsymbol{x} = \sum_{e \in \{x\}} \chi_e$. From the submodularity of function h, we have

$$h(\boldsymbol{x}) - h(\mathbf{0}) \leq \sum_{e \in \{x\}} h(\chi_e|\mathbf{0}). \tag{1}$$

According to the definition of curvature c_h, we can conclude that

$$h(\chi_e|\mathbf{0}) \leq \frac{1}{1 - c_h} h(\chi_e|\boldsymbol{B} - \chi_e). \tag{2}$$

The submodularity of function h indicate that

$$\sum_{e \in \{x\}} h(\chi_e|\boldsymbol{B} - \chi_e) \leq h(\boldsymbol{x}) - h(\mathbf{0}). \tag{3}$$

So, combining with the inequality (1)–(3), we have

$$\sum_{e \in \{x\}} h(\chi_e|\mathbf{0}\ ) \leq \frac{1}{1 - c_h}[h(\boldsymbol{x}) - h(\mathbf{0})],$$

which completes the proof.

We define function $f(\boldsymbol{x}) = g(\boldsymbol{x}) - \lambda \sum_{e \in \{x\}} h(\chi_e|\mathbf{0})$ for every $\boldsymbol{x} \in \mathcal{B}$. We can get the following conclusion.

Lemma 2. *Suppose* $f(\boldsymbol{x}) = g(\boldsymbol{x}) - \lambda \sum_{e \in \{x\}} h(\chi_e|\mathbf{0})$ *for every* $\boldsymbol{x} \in \mathcal{B}$*. Then* f *is DR-submodular.*

Proof. Let $\boldsymbol{x}$ and $\boldsymbol{y}$ are any two elements in $\mathcal{B}$ with $\boldsymbol{x} \leq \boldsymbol{y}$, for every $e \in E$, we have

$$\begin{aligned}
& f(\chi_e|\boldsymbol{x}) \\
&= f(\boldsymbol{x} + \chi_e) - f(\boldsymbol{x}) \\
&= g(\boldsymbol{x} + \chi_e) - \lambda \sum_{e \in \{x + \chi_e\}} h(\chi_e|\mathbf{0}) - g(\boldsymbol{x}) + \lambda \sum_{e \in \{x\}} h(\chi_e|\mathbf{0}) \\
&\geq g(\chi_e|\boldsymbol{y}) - \lambda h(\chi_e|\mathbf{0}) \\
&= f(\chi_e|\boldsymbol{y}),
\end{aligned}$$

which completes the proof.

From the above lemmas, we can obtain the following theorem.

Theorem 1. *Suppose $\boldsymbol{x}$ is an output of Algorithm 1, for any $\boldsymbol{y} \in \mathcal{B}$, we have*

$$g(\boldsymbol{x}) - h(\boldsymbol{x}) \geq \frac{\lambda - 1}{\lambda} g(\boldsymbol{y}) - \frac{\lambda - 1}{c_h} h(\boldsymbol{y}|\mathbf{0}) - h(\mathbf{0}).$$

Proof. Denote $f(\boldsymbol{x}) = g(\boldsymbol{x}) - \lambda \sum_{e \in \{x\}} h(\chi_e|\mathbf{0})$, for every $\boldsymbol{x} \in \mathcal{B}$. From Lemma 2, we know f is a submodular function. By the idea of algorithm design, if $f(\chi_e|\boldsymbol{x}) < 0$, we will be discard this element. Otherwise, we add χ_e with the amount of l_e which returned from Algorithm 2 to the current solution. From Algorithm 2, we can obtain that $f(l_e\chi_e|\boldsymbol{x}) \geq 0$ and $f((l_e+1)\chi_e|\boldsymbol{x}) < 0$. Let e be an arbitrary element of $\{\boldsymbol{y}\} \setminus \{\boldsymbol{x}\}$, we can conclude that $f(\chi_e|\boldsymbol{x}_e) < 0$ as $e \notin \{\boldsymbol{x}\}$, where $\boldsymbol{x}_e$ is the output of Algorithm 1 when element e coming. By the submodularity of f, we get

$$f(\chi_e|\boldsymbol{x}) \leq f(\chi_e|\boldsymbol{x}_e) < 0.$$

So,

$$\begin{aligned} 0 &\geq \sum_{e \in \{\boldsymbol{y}\} \setminus \{x\}} f(\chi_e|\boldsymbol{x}) \\ &\geq f(\boldsymbol{x} \vee \boldsymbol{y}) - f(\boldsymbol{x}) \\ &= g(\boldsymbol{x} \vee \boldsymbol{y}) - g(\boldsymbol{x}) - \lambda \sum_{e \in \{\boldsymbol{y}\} \setminus \{x\}} h(\chi_e|\mathbf{0}) \\ &\geq g(\boldsymbol{y}) - g(\boldsymbol{x}) - \lambda \sum_{e \in \{\boldsymbol{y}\}} h(\chi_e|\mathbf{0}) \end{aligned}$$

where the last inequality can be obtain from the monotonicity of g and h. From Lemma 1, we have

$$g(\boldsymbol{x}) \geq g(\boldsymbol{y}) - \lambda \sum_{e \in \{y\}} h(\chi_e|\mathbf{0}) \geq g(\boldsymbol{y}) - \frac{\lambda}{1 - c_h} h(y|\mathbf{0}). \tag{4}$$

Suppose $\boldsymbol{x}(E) = r$ and $\{\boldsymbol{x}\} = \{e_1, e_2, \cdots, e_{\widehat{r}}\}$. From the design of Algorithm 2, the amount l_e of χ_e which can be add to $\boldsymbol{x}$ is satisfying $f(l_e\chi_e|\boldsymbol{x}) \geq 0$ and $f((l_e+1)\chi_e|\boldsymbol{x}) < 0$, for each $i \in [\widehat{r}]$. Therefore, we receive

$$g(\boldsymbol{x}) - \lambda \sum_{e \in \{x\}} h(\chi_e|\mathbf{0}\ ) = f(\boldsymbol{x}\ ) = f(\mathbf{0}) + \sum_{i=1}^{\widehat{r}} h(l_{e_i}\chi_{e_i}|\{\boldsymbol{x}_{i-1}\}\ ) \geq 0.$$

Combining with the conclusion of Lemma 1, we have

$$g(\boldsymbol{x}) - \lambda(h(\boldsymbol{x}) - h(\mathbf{0})\ ) \geq g(\boldsymbol{x}) - \lambda \sum_{e \in \{x\}} h(\chi_e|\mathbf{0}) \geq 0,$$

that is

$$g(\boldsymbol{x}) - \lambda h(\boldsymbol{x}) \geq -\lambda h(\mathbf{0}) \tag{5}$$

Since $\lambda \geq 1$, calculation $(\lambda - 1)*$ (4) + (5), we obtain

$$g(\boldsymbol{x}) - h(\boldsymbol{x}) \geq \frac{\lambda - 1}{\lambda} g(\boldsymbol{y}\ ) - \frac{\lambda - 1}{c_h} h(\boldsymbol{y}|\mathbf{0}) - h(\mathbf{0}).$$

which complete the proof of Theorem 1.

4 Conclusions

In this paper, we present the development of combinatorial approximation algorithms designed to address the maximization problem of monotone non-negative DR-submodular function minus a monotone non-negative DR-submodular function with curvature in integer lattice. Simultaneously, we design greedy algorithm and threshold streaming algorithm for thc unconstraincd model, respectively.

Acknowledgements. The first author is supported by Natural Science Foundation of Shandong Province (No. ZR2022MA034) and National Natural Science Foundation of China (No. 12301417). The third author is supported by Natural Science Foundation of China (Nos. 11401438, 11571120). The fifth author is supported by National Natural Science Foundation of China (No. 12101587).

References

1. Chekuri, C., Quanrud, K.: Submodular function maximization in parallel via the multilinear relaxation. In: Proceedings of SODA, pp. 303–322 (2019)
2. Das, A., Kempe, D.: Algorithms for subset selection in linear regression. In: Proceedings of STOC, pp. 45–54 (2008)
3. Das, A., Kempe, D.: Submodular meets spectral: greedy algorithms for subset selection, sparse approximation and dictionary selection. In: Proceedings of ICML, pp. 1057–1064 (2011)
4. Ene, A., Nguyen, H.L.: Submodular maximization with nearly-optimal approximation and adaptivity in nearly-linear time. In: Proceedings of SODA, pp. 274–282 (2019)
5. Nemhauser, G.L., Wolsey, L.A., Fisher, M.L.: An analysis of approximations for maximizing submodular set functions. Math. Program. **14**, 265–294 (1978)
6. Feige, U.: A threshold of $\ln n$ for approximation set cover. J. ACM **32**, 65–82 (1998)
7. Krause, A., Singh, A., Guestrin, C.: Near-optimal sensor placements in gaussian processes: theory, efficient algorithms and empirical studies. J. Mach. Learn. Res. **9**, 235–284 (2008)
8. He, S., Zhang, J., Zhang, S.: Polymatroid optimization, submodularity, and joint replenishment games. Oper. Res. **60**, 128–137 (2012)
9. Yang, R.Q., Xu, D.C., Li, M., Xu, Y.C.: Thresholding methods for streaming submodular maximization with a cardinality constraint and its variants. Nonlinear Comb. Optim. **2019**, 123–140 (2019)
10. Yang, R.Q., Xu, D.C., Du, D.L., Xu, Y.C., Yan, X.H.: Maximization of constrained non-submodular functions. In: Proceedings of COCOON, pp. 615–626 (2019)

11. Zhang, Z.Z., Du, D.L., Jiang, Y.J., Wu, C.C.: Maximizing DR-submodular+supermodular functions on the integer lattice subject to a cardinality constraint. J. Global Optim. **80**, 595–616 (2021)
12. Soma, T., Yoshida, Y.: Maximization monotone submodular functions over the integer lattice. Math. Program. **172**, 539–563 (2018)
13. Kapralov, M., Post, I., Vondrák, J.: Online submodular welfare maximization: greedy is optimal. In: Proceedings of SODA, pp. 1216–1225 (2012)
14. Kempe, D., Kleinberg, J., Tardos, E.: Maximizing the spread of influence through a social network. In: Proceedings of KDD, pp. 137–146 (2003)
15. Khanna, R., Elenberg, E.R., Dimakis, A.G., Negahban S., Ghosh J.: Scalable greedy feature selection via weak submodularity. In: Proceedings of ICAIS, pp. 1560–1568 (2017)
16. Kuhnle, A., Smith, J.D., Crawford, V.G., Thai, M.T.: Fast maximization of non-submodular, monotonic functions on the integer lattice. In: Proceedings of ICML, pp. 2791–2800 (2018)
17. Tan, J., Xu, Y., Zhang, D., Zhang, X.: On streaming algorithms for maximizing a supermodular function plus a MDR-submodular function on the integer lattice. J. Comb. Optim. **45**(2), 1–19 (2023)
18. Sviridenko, M., Vondrák, J., Ward, J.: Optimal approximation for submodular and supermodular optimization with bounded curvature. In: Proceedings of SODA, pp. 1134–1148 (2015)
19. Iyer, R., Bilmes, J.: Algorithms for approximate minimization of the difference between submodular functions, with applications. In: Proceedings of the Twenty-Eighth Conference on Uncertainty in Artificial Intelligence, pp. 407–417 (2012)
20. Narasimhan, M., Bilmes, J.A.: A submodular-supermodular procedure with applications to discriminative structure learning. arXiv preprint arXiv:1207.1404 (2012)
21. Wang, Y.J., Xu, D.C., Du, D.L., Jiang, Y.J.: Bicriteria streaming algorithms to balance gain and cost with cardinality constraint. J. Comb. Optim. **44**, 2946–2962 (2022)
22. Lu, C., Yang, W.G., Gao, S.X.: Streaming algorithms for maximizing the? Difference of submodular functions and the sum of submodular and supermodular functions. Optim. Lett. **17**, 1643–1667 (2023)

Author Index

M. Li et al. (Eds.): TAMC 2025, LNCS 16084, pp. 415–416, 2026.
https://doi.org/10.1007/978-981-95-4839-2

MIX
Papier aus verantwortungsvollen Quellen
Paper from responsible sources
FSC® C105338

If you have any concerns about our products,
you can contact us on
ProductSafety@springernature.com

In case Publisher is established outside the EU,
the EU authorized representative is:
Springer Nature Customer Service Center GmbH
Europaplatz 3, 69115 Heidelberg, Germany

Printed by Libri Plureos GmbH
in Hamburg, Germany